Lecture Notes in Computer Science 4859

Commenced Publication in 1973
Founding and Former Series Editors:
Gerhard Goos, Juris Hartmanis, and Jan van Leeuwen

T0223181

K. Srinathan   C. Pandu Rangan
Moti Yung (Eds.)

Progress in Cryptology – INDOCRYPT 2007

8th International Conference on Cryptology in India
Chennai, India, December 9-13, 2007
Proceedings

 Springer

Volume Editors

K. Srinathan
International Institute of Information Technology
Center for Security, Theory and Algorithmic Research (C-STAR)
Gachibowli, Hyderabad, 500032, India
E-mail: srinathan@iiit.ac.in

C. Pandu Rangan
Indian Institute of Technology Madras
Department of Computer Science and Engineering
Chennai, 600036, India
E-mail: prangan55@yahoo.com

Moti Yung
Columbia University
Computer Science Department
New York, NY 10027, USA
E-mail: moti@cs.columbia.edu

Library of Congress Control Number: 2007939973

CR Subject Classification (1998): E.3, G.2.1, D.4.6, K.6.5, K.4, F.2.1-2, C.2

LNCS Sublibrary: SL 4 – Security and Cryptology

ISSN   0302-9743
ISBN-10   3-540-77025-9 Springer Berlin Heidelberg New York
ISBN-13   978-3-540-77025-1 Springer Berlin Heidelberg New York

Springer is a part of Springer Science+Business Media

springer.com

© Springer-Verlag Berlin Heidelberg 2007
Printed in Germany

Typesetting: Camera-ready by author, data conversion by Scientific Publishing Services, Chennai, India
Printed on acid-free paper   SPIN: 12196937   06/3180   5 4 3 2 1 0

Preface

INDOCRYPT 2007, the Eighth Annual International Conference on Cryptology in India, was organized by Cryptology Research Society India (CRSI) in cooperation with the Indian Institute of Technology, Madras. The conference was held in IIT Madras, December 10–12, 2007. INDOCRYPT 2007 was chaired by Vijayaraghavan, Executive Director, Society for Electronic Transactions and Security (SETS) and we had the privilege of serving as PC Co-chairs.

The conference received 104 submissions. Each paper was assigned at least three reviewers. After detailed discussions and deliberations, the PC identified 22 submissions for regular presentation and 11 submissions for short presentation.

The conference featured three invited lectures. S.V. Raghavan, IIT Madras, India, spoke on "Security of Critical Information Infrastructure." Venkatesan Ramaratnam, Microsoft Redmond, USA, delivered a lecture on "Cryptographic Applications of Rapid Mixing." Jonathan Katz, University of Maryland, presented a talk on the "Recent Research Trends in Signature Schemes." Besides the events of the conference, several pre-and post-conference tutorials were organized for the benefit of researchers in India and attendees of the conference. In the pre-conference tutorial Venkatesan Ramaratnam, Microsoft Redmond, USA, presented a tutorial on "Recent Snapshots of Cryptanalysis" and Ingrid Verbauwhede, Katholieke Universiteit Leuven, Belgium, conducted a tutorial on "Side Channel Attacks – An Overview." In the post-conference tutorial, Manoj M. Prabhakaran, University of Illinois, Urbana-Champaign, USA, presented a tutorial on "Theoretical Foundations of Public-Key Encryption" and Krzysztof Pietrzak, Centrum Voor Wisukunde en Informatica (CWI), The Netherlands conducted a tutorial on "Robust Combiners".

The success of INDOCRYPT 2007 is due to the contributions and support we received from several quarters. Our sincere thanks to many authors from around the world for submitting their papers. We are deeply grateful to the PC for their hard work, enthusiasm and continuous involvement in making a thorough and fair review of so many submissions. We received comments and clarifications instantly on several occasions, although the PC members and sub-reviewers were spread across of the globe with over 20 hours of difference in their time zones! Many thanks once again to the members for their great work. Thanks also to all external reviewers, listed on the following pages, for sparing their valuable time; We have already received several notes of appreciation from various authors for the detailed and thorough job they carried out while refereeing the submissions.

We would like to thank SETS for mobilizing funds from various government sectors and generously contributing to the smooth organization of the event. I thank DRDO, Bangalore and the Head of the Department of Computer Science, IIT Madras for supporting the travel and local hospitality for invited speakers and tutorial speakers under MOC grants the department received from DRDO.

Microsoft Research, India (MSRI) was a constant source of support and encouragement for various activities in cryptology in India, and for INDOCRYPT 2007 they volunteered to be a Platinum Sponsor of the event. The generous support from Hexaware and Prince Shri Venkateshwara Padmavathy Engineering College is warmly acknowledged. Specifically, we wish to thank Dr. Vijayaraghavan, Prof. S.V. Raghavan, Prof. Timothy A. Gonsalves, Mr. Eugene Xavier, Dr. K. Vasudevan for the financial support we received in a timely manner.

We wish to extend our sincere thanks to Mr. V. Veeraraghavan, Mr. E. Boopal and Mr. V.S. Balasundaram, who worked tiredlessly for three months attending to numerous details related to the events of the conference. Hi Tours did a very efficient job in providing tourism / stay-related logistics. It has been a pleasure working with Anna Kramer and Alfred Hoffman, our key contacts at Springer.

Last but not the least, our sincere thanks to Easychair.org. It is indeed a wonderful conference management system!

December 2007   K. Srinathan
                                                                C. Pandu Rangan
                                                                Moti Yung

INDOCRYPT 2007

December 10–12, 2007, IIT Madras, Chennai, India

Organized by
Cryptology Research Society of India (CRSI)

in cooperation with
Department of Computer Science and Engineering, IIT Madras, India.

General Chair

M.S. Vijayaraghavan, SETS, India

Program Co-chairs

K. Srinathan, IIIT, Hyderabad, India
C. Pandu Rangan, IIT Madras, Chennai, India
Moti Yung, Columbia University, New York, USA

Program Committee

Amit Sahai	University of California at Los Angles, USA
Anish Mathuria	Dhirubhai Ambani Ins. of Inf. & Comm. Tech., India
Bao Feng	Institute for Infocomm Research, Singapore
Bimal Roy	Indian Statistical Institute, Kolkotta, India
Debdeep Mukhopadhyay	Indian Institute of Technology Madras, India
Duncan S. Wong	City University, Hong Kong
Ed Dawson	Information Security Institute, QUT, Australia
K. Gopalakrishnan	East Carolina University, USA
Huaxiong Wang	Nanyang Technological University, Singapore
Kaoru Kurosawa	Ibaraki University, Japan
Krzysztof Pietrzak	Centrum Voor Wisukunde en Informatica, The Netherlands
Michel Abdalla	Ecole Normale Superieure, France
Moti Yung	Columbia University, USA
Pandu Rangan C.	Indian Institute of Technology Madras, India
Rei Safavi-Naini	University of Wollongong, Australia
Sanjit Chatterjee	Indian Statistical Institute, Kolkotta, India
Shailesh Vaya	Indian Institute of Technology Madras, India
Srinathan K.	IIIT, Hyderabad, India
Tatsuaki Okamoto	NTT Labs, Japan
Tsuyoshi Takagi	Future University Hakodate, Japan
Vassil Dimitrov	The University of Calgary, Canada

Organizing Committee

Boopal E.   Chennai, India
Veeraraghavan V.   Chennai, India

Reviewers

Michel Abdalla   Tetsuya Izu   Rei Safavi-Naini
Karl Abrahamson   Gonzalez Nieto Juan   Amit Sahai
Kazumaro Aoki   Shinsaku Kiyomoto   Somitra Sanadhya
Rana Barua   Kaoru Kurosawa   Sumanta Sarkar
Gary Carter   Hidenori Kuwakado   Taizo Shirai
Chris Charnes   Gatan Leurent   Franscesco Sica
Sanjit Chatterjee   Fagen Li   Michal Sramka
Hung-Yu Chien   Yi Lu   Kannan Srinathan
Joo Yeon Cho   Steve Lu   Makoto Sugita
Ashish Choudhary   Anish Mathuria   Tsuyoshi Takagi
Abhijit Das   Krystian Matusiewicz   Yasuo Takahashi
Ed Dawson   Rob McEvoy   Masahiko Takenaka
Vassil Dimitrov   Pradeep Kumar Mishra   Hidema Tanaka
Ratna Dutta   Shiho Moriai   Qiang Tang
Bao Feng   Debdeep Mukhopadhyay   Christophe Tartary
Matthieu Finiasz   Mathew Musson   Stefano Tessaro
Rosario Gennaro   Wakaha Ogata   Shigenori Uchiyama
K. Gopalakrishnan   Tatsuaki Okamoto   Shailesh Vaya
M. Choudary Gorantla   C. Pandu Rangan   Camille Vuillaume
Vipul Goyal   Krzysztof Pietrzak   Huaxiong Wang
Kishan Chand Gupta   David Pointcheval   Dai Watanabe
Yasuo Hatano   Havard Raddum   Duncan S. Wong
Keisuke Hakuta   Mohammad Reza   Mu-En Wu
Matt Henricksen   Reyhanitabar   Yongdong Wu
Kota Ideguchi   Bimal Roy   Moti Yung
Tetsu Iwata   Minoru Saeki   Sbastien Zimmer

Sponsors

SETS, India
Microsoft Research, India
Prince Shri Venkateshwara Padmavathy Engineering College, India

Table of Contents

X Short Presentation

Linearization Attacks Against Syndrome Based Hashes

Markku-Juhani O. Saarinen

Information Security Group
Royal Holloway, University of London
Egham, Surrey TW20 0EX, UK
m.saarinen@rhul.ac.uk

Abstract. In MyCrypt 2005, Augot, Finiasz, and Sendrier proposed FSB, a family of cryptographic hash functions. The security claim of the FSB hashes is based on a coding theory problem with hard average-case complexity. In the ECRYPT 2007 Hash Function Workshop, new versions with essentially the same compression function but radically different security parameters and an additional final transformation were presented. We show that hardness of average-case complexity of the underlying problem is irrelevant in collision search by presenting a linearization method that can be used to produce collisions in a matter of seconds on a desktop PC for the variant of FSB with claimed 2^{128} security.

Keywords: FSB, Syndrome Based Hashes, Provably Secure Hashes, Hash Function Cryptanalysis, Linearization Attack.

1 Introduction

A number of hash functions have been proposed that are based on "hard problems" from various branches of computer science. Recent proposals in this genre of hash function design include VSH (factoring) [3], LASH (lattice problems) [2], and the topic of this paper, Fast Syndrome Based Hash (FSB), which is based on decoding problems in the theory of error-correcting codes [1,6].

In comparison to dedicated hash functions designed using symmetric cryptanalysis techniques, "provably secure" hash functions tend to be relatively slow and do not always meet all of criteria traditionally expected of cryptographic hashes. An example of this is VSH, where only collision resistance is claimed, leaving the hash open to various other attacks [8].

Another feature of "provably secure" hash functions is that the proof is often a reduction to a problem with asymptotically hard worst-case or average-case complexity. Worst-case complexity measures the difficulty of solving pathological cases rather than typical cases of the underlying problem. Even a reduction to a problem with hard average complexity, as is the case with FSB, offers only limited security assurance as there still can be an algorithm that easily solves the problem for a subset of the problem space.

This common pitfall of provably secure cryptographic primitives is clearly demonstrated in this paper for FSB – it is shown that the hash function offers minimal preimage or collision resistance when the message space is chosen in a specific way.

K. Srinathan, C. Pandu Rangan, M. Yung (Eds.): Indocrypt 2007, LNCS 4859, pp. 1–9, 2007.

The remainder of this paper is structured as follows. Section 2 describes the FSB compression function. Section 3 gives the basic linearization method for finding pre-images and extends it to "alphabets". This is followed by an improved collision attack in Section 4 and discussion of attacks based on larger alphabets in Section 5.

Appendix A gives a concrete example of pre-image and collision attacks on a proposed variant of FSB with claimed 128-bit security.

2   The FSB Compression Function

The FSB compression function can be described as follows [1,6].

Definition 1. *Let $\mathcal{H}$ be an $r \times n$ binary matrix. The FSB compression function is a mapping from message vector* **s** *that contains w characters, each satisfying $0 \leq s_i < \frac{n}{w}$, to an r bit result as follows:*

$$\mathrm{FSB}(\mathbf{s}) = \bigoplus_{i=1}^{w} \mathcal{H}_{(i-1)\frac{n}{w}+s_i+1} \, ,$$

where $\mathcal{H}_i$ denotes column i of the matrix.

The FSB compression function is operated in Merkle–Damgård mode to process a large message [7,5]. The exact details of padding and chaining of internal state across compression function iterations are not specified.[1]

With most proposed variants of FSB, the character size $\frac{n}{w}$ is chosen to be 2^8, so that **s** can be treated as an array of bytes for practical implementation purposes. See Appendix A for an implementation example.

For the purposes of this paper, we shall concentrate on finding collisions and pre-images in the compression function. These techniques can be easily applied for finding full collisions of the hash function. The choice of $\mathcal{H}$ is taken to be a random binary matrix in this paper, although quasi-cyclic matrices are considered in [6] to reduce memory usage.

The final transformation proposed in [6] does not affect the complexity of finding collisions or second pre-images, although it makes first pre-image search difficult (equal to inverting Whirlpool [9]). Second pre-images can be easily found despite a strong final transform.

The security parameter selection in the current versions of FSB is based primarily on Wagner's generalized birthday attack [10,4]. The security claims are summarized in Table 1.

3   Linearization Attack

To illustrate our main attack technique, we shall first consider hashes of messages with binary values in each character: $s_i \in \{0, 1\}$ for $1 \leq i \leq w$. This message space is a small subset of all possible message blocks.

---

[1] Ambiguous definitions of algorithms makes experimental cryptanalytic work depend on guess-work on algorithm details. However, the attacks outlined in this paper should work, regardless of the particular details of chaining and padding.

Table 1. Parameterizations of FSB, as given in [6]. Line 6 (in bold) with claimed 2^{128} security was proposed for practical use. Pre-images and collisions can be found for this variant in a matter of seconds on a desktop PC.

Security	r	w	n	n/w
64-bit	512	512	131072	256
	512	450	230400	512
	1024	2^{17}	2^{25}	256
80-bit	512	170	43520	256
	512	144	73728	512
128-bit	**1024**	**1024**	**262144**	**256**
	1024	904	462848	512
	1024	816	835584	1024

We define a constant vector $\mathbf{c}$,

$$\mathbf{c} = \bigoplus_{i=1}^{w} \mathcal{H}_{(i-1)\frac{n}{w}+1},$$

and an auxiliary $r \times w$ binary matrix $\mathbf{A}$, whose columns $\mathbf{A}_i$, $1 \leq i \leq w$ are given by

$$\mathbf{A}_i = \mathcal{H}_{(i-1)\frac{n}{w}+1} \oplus \mathcal{H}_{(i-1)\frac{n}{w}+2}.$$

By considering how the XOR operations cancel each other out, it is easy to see that for messages of this particular type the FSB compression function is entirely linear:

$$\mathrm{FSB}(\mathbf{s}) = \mathbf{A} \cdot \mathbf{s} \oplus \mathbf{c}.$$

Note that in this paper $\mathbf{s}$ and $\mathbf{c}$ and other vectors are column vectors unless otherwise stated.

Furthermore, let us consider the case where $r = w$, and therefore $\mathbf{A}$ is a square matrix. If $\det \mathbf{A} \neq 0$ the inverse exists and we are able to find a pre-image $\mathbf{s}$ from the Hash $\mathbf{h} = \mathrm{FSB}(\mathbf{s})$ simply as

$$\mathbf{s} = \mathbf{A}^{-1} \cdot (\mathbf{h} \oplus \mathbf{c}).$$

If r is greater than w, the technique can still be applied to force given w bits of the final hash to some predefined value. Since the order of the rows is not relevant, we can simply construct a matrix that contains only the given w rows (i.e.. bits of the hash function result) of $\mathbf{A}$ that we are are interested in.

3.1 The Selection of Alphabet in a Preimage Attack

We note that the selection of $\{0, 1\}$ as the set of allowable message characters ("the alphabet") is arbitrary. We can simply choose any pair of values for each i so that $\mathbf{s}_i \in \{x_i, y_i\}$ and map each $x_i \mapsto 0$ and $y_i \mapsto 1$, thus creating a binary vector for the attack. The constant is then given by

$$\mathbf{c} = \bigoplus_{i=1}^{w} \mathcal{H}_{(i-1)\frac{n}{w}+x_i},$$

and columns of the $\mathbf{A}$ matrix are given by

$$\mathbf{A}_i = \mathcal{H}_{(i-1)\frac{n}{w}+x_i+1} \oplus \mathcal{H}_{(i-1)\frac{n}{w}+y_i+1}.$$

To invert a hash $\mathbf{h}$ we first compute

$$\mathbf{b} = \mathbf{A}^{-1}(\mathbf{h} \oplus \mathbf{c})$$

and then apply the mapping $s_i = x_i + b_i(y_i - x_i)$ on the binary result $\mathbf{b}$ to obtain a message $\mathbf{s}$ that satisfies $\mathrm{FSB}(\mathbf{s}) = \mathbf{h}$.

3.2 Invertibility of Random Binary Matrices

The binary matrices are essentially random for each arbitrarily chosen alphabet. Since the success of a pre-image attack depends upon the invertibility of the binary matrix $\mathbf{A}$, we note (without a proof) that the probability that an $n \times n$ random binary matrix has non-zero determinant and is therefore invertible in $\mathrm{GF}(2)$ is given by

$$p = \prod_{i=1}^{n}(1 - 2^{-i}) \approx 0.28879 \approx 2^{-1.792}$$

when n is even moderately large.

Two trials with two distinct alphabets are on the average enough to find an invertible matrix (total probability for 2 trials is $1 - (1-p)^2 \approx 0.49418$).

4   Finding Collisions When $r = 2w$

We shall expand our approach for producing collisions in $2w$ bits of the hash function result by controlling w message characters. This is twice the number compared to pre-image attack of Section 3.1. The complexity of the attack remains negligible – few simple matrix operations.

Assume that by selection of two distinct alphabets, $\{x_i, y_i\}$ and $\{x'_i, y'_i\}$, there are two distinct linear presentations for FSB, one containing the matrix $\mathbf{A}$ and constant $\mathbf{c}$ and the other one $\mathbf{A}'$ and $\mathbf{c}'$ correspondingly. To find a pair of messages $\mathbf{s}, \mathbf{s}'$ that produces a collision we must find a solution for $\mathbf{b}$ and $\mathbf{b}'$ in the equation

$$\mathbf{A} \cdot \mathbf{b} \oplus \mathbf{c} = \mathbf{A}' \cdot \mathbf{b}' \oplus \mathbf{c}'.$$

This basic collision equation can be manipulated to the form

$$(\mathbf{A} \mid \mathbf{A}') \cdot \begin{pmatrix} \mathbf{b} \\ \mathbf{b}' \end{pmatrix} = \begin{pmatrix} \mathbf{c} \\ \mathbf{c}' \end{pmatrix}.$$

The solution of the inverse $(\mathbf{A} \mid \mathbf{A}')^{-1}$ will allow us to compute the message pair $(\mathbf{b} \mid \mathbf{b}')^T$ that yields the same hash in $2w$ different message bits (since $r = 2w$ yields a square matrix in this case).

$$(\mathbf{A} \mid \mathbf{A}')^{-1} \cdot \begin{pmatrix} \mathbf{c} \\ \mathbf{c}' \end{pmatrix} = \begin{pmatrix} \mathbf{b} \\ \mathbf{b}' \end{pmatrix}.$$

The binary vector $(\mathbf{b} \mid \mathbf{b}')^T$ can then be split into two messages $\mathbf{s}$ and $\mathbf{s}'$ that produce the collision. For $1 \leq i \leq w$ we apply the alphabet mapping as follows:

$$\mathbf{s}_i = x_i + \mathbf{b}_i(y_i - x_i),$$
$$\mathbf{s}'_i = x'_i + \mathbf{b}'_i(y'_i - x'_i).$$

Here x_i, y_i and x'_i, y'_i represent the alphabets for $\mathbf{s}_i$ and $\mathbf{s}'_i$, respectively.

5 Larger Alphabets

Consider an alphabet of cardinality three, $\{x_i, y_i, z_i\}$. We can construct a linear equation in $GF(2)$ that computes the FSB compression function in this message space by using two columns for each message character $\mathbf{s}_i$. The linear matrix therefore has size $r \times 2w$. The constant $\mathbf{c}$ is computed as before as:

$$\mathbf{c} = \bigoplus_{i=1}^{w} \mathcal{H}_{(i-1)\frac{n}{w}+x_i},$$

and the odd and even columns are given by

$$A_{2i-1} = \mathcal{H}_{(i-1)\frac{n}{w}+x_i+1} \oplus \mathcal{H}_{(i-1)\frac{n}{w}+y_i+1},$$
$$A_{2i} = \mathcal{H}_{(i-1)\frac{n}{w}+x_i+1} \oplus \mathcal{H}_{(i-1)\frac{n}{w}+z_i+1}.$$

The message $\mathbf{s}$ must also be transformed into a binary vector $\mathbf{b}$ of length $2w$ via a selection function v:

$\mathbf{s}_i$	$v(\mathbf{s}_i)$
x_i	$(0,0)$
y_i	$(1,0)$
z_i	$(0,1)$

The binary vector $\mathbf{b}$ is constructed by concatenating the selection function outputs:

$$\mathbf{b} = (v(\mathbf{s}_1) \parallel v(\mathbf{s}_2) \parallel \cdots \parallel v(\mathbf{s}_w))^T.$$

We again arrive at a simple linear equation for the FSB compression function:

$$\text{FSB}(\mathbf{s}) = \mathbf{A} \cdot \mathbf{b} \oplus \mathbf{c}.$$

The main difference is that the message space is much larger, $3^w \approx 2^{1.585w}$. This construction is easy to generalize for alphabets of any size: $r \times (k-1)w$ size linear matrix is required for an alphabet of size k. However, we have not found cryptanalytic advantages in mapping hashes back to message spaces with alphabets larger than three.

5.1 Pre-image Search

It is easy to see that even if $\mathbf{A}$ is invertible, not all hash results are, since the solution of $\mathbf{b}$ may contain $v(\mathbf{s}_i) = (1,1)$ pairs. These do not map back to the message space in the selection function.

Given a random binary $\mathbf{b}$, the fraction of valid messages in the message space (alphabet of size 3) is given by $(3/4)^w = 2^{-0.415w}$. Despite this disadvantage, larger alphabets can be useful in attacks. We will illustrate this with an example.

Example. FSB parameters with $w = 64, n = 256 \times 64 = 16384$ and $r = 128$ is being used; 64 input bytes are processed into a 128 bit result. What is the complexity of a pre-image attack?[2]

Solution. We will use an alphabet of size 3. Considering both matrix invertibility (Section 3.2) and the alphabet mapping, the probability of successfully mapping the hash back to the alphabet is $0.28879 \times (3/4)^{64} = 2^{-28.4}$. We can precompute 2^{27} inverses $\mathbf{A}^{-1}$ for various message spaces offline, hence speeding up the time required to find an individual pre-image. There are also early-abort strategies that can be used to speed up the search.

Using these techniques, the pre-image search requires roughly 2^{28} steps in this case, compared to the theoretical 2^{128}.

5.2 Collision Search

Three-character alphabets can be used in conjunction with the collision attack outlined in Section 4. It is easy to see that it is possible to mix 3-character alphabets with binary alphabets. Each character position s_i that is mapped to a 3-character alphabet requires two columns in the linear matrix, whereas those mapped to a 2-character alphabet require only one column.

Generally speaking, the probability for finding two valid messages in each trial is $(3/4)^{2k} = 2^{-0.830k}$ when k characters in s and s' are mapped to 3-character alphabets.

Example. FSB parameters with $w = 170, n = 256 \times 170 = 43520$ and $r = 512$ is being used; 170 input bytes are processed into a 512-bit result. What is the complexity of collision search?[3]

Solution. We use a mixed alphabet; $k = 86$ characters are mapped to a 3-character alphabet and the remaining 84 characters are mapped to a binary alphabet. The linear matrix $\mathbf{A}$ therefore has $2 \times 86 + 84 = 256$ columns, and the combined matrix $(\mathbf{A} \mid \mathbf{A}')$ in the collision attack (similarly to 4) has size 512×512. Success of matrix inversion is $2^{-1.792}$. The probability of success in each trial is $2^{-0.830k-1.792} = 2^{-73.2}$, collision search has complexity of roughly 2^{73}.

6 Conclusions

We have shown that Fast Syndrome Based Hashes (FSB) are not secure against pre-image or collision attacks under the proposed security parameters. The attacks have been implemented and collisions for a variant with claimed 128-bit security can be found in less than a second on a low-end PC.

We feel that the claim of "provable security" is hollow in the case of FSB, where the security proof is based on a problem with hard average-case complexity, but which is almost trivially solvable for special classes of messages.

---

[2] The complexity of a collision attack in this case is negligible, as $r = 2w$ and the technique from Section 4 can be used.

[3] These security parameters are proposed for 80-bit security in [6] and reproduced in Table 1.

Acknowledgements

The author is thankful to N. Sendrier and A. Canteaut for hosting him during his visit to INRIA-Rocquencourt in June 2006, where this work was originated (although it took a year to mature to publishable form). The author would also like to thank Keith Martin and the INDOCRYPT Program Committee members for helpful comments. Financial support for this work was provided by INRIA, Nixu Ltd., and Academy of Finland.

References

1. Augot, D., Finiasz, M., Sendrier, N.: A family of fast syndrome based cryptographic hash functions. In: Dawson, E., Vaudenay, S. (eds.) DILS 2005. LNCS (LNBI), vol. 3615, pp. 64–83. Springer, Heidelberg (2005)
2. Bentahar, K., Page, D., Saarinen, M.-J.O., Silverman, J.H., Smart, N.: LASH. In: Proc. 2nd NIST Cryptographic Hash Workshop (2006)
3. Contini, S., Lenstra, A.K., Steinfeld, R.: VSH, an efficient and provably collision-resistant hash function. In: Vaudenay, S. (ed.) EUROCRYPT 2006. LNCS, vol. 4004, pp. 165–182. Springer, Heidelberg (2006)
4. Coron, J.-S., Joux, A.: Cryptanalysis of a provably secure cryptographic hash function. IACR ePrint 2004 / 013 (2004), Available at http://www.iacr.org/eprint
5. Damgård, I.B.: A design principle for hash functions. In: Brassard, G. (ed.) CRYPTO 1989. LNCS, vol. 435, pp. 416–427. Springer, Heidelberg (1990)
6. Finiasz, M., Gaborit, P., Sendrier, N.: Improved fast syndrome based cryptographic hash functions. In: ECRYPT Hash Function Workshop 2007 (2007)
7. Merkle, R.C.: A fast software one-way hash function. Journal of Cryptology 3, 43–58 (1990)
8. Saarinen, M.-J.O.: Security of VSH in the real world. In: Barua, R., Lange, T. (eds.) IN-DOCRYPT 2006. LNCS, vol. 4329, pp. 95–103. Springer, Heidelberg (2006)
9. Rijmen, V., Barreto, P.: "Whirlpool". Seventh hash function of ISO/IEC 10118-3:2004 (2004)
10. Wagner, D.: A generalized birthday problem. In: Yung, M. (ed.) CRYPTO 2002. LNCS, vol. 2442, pp. 288–304. Springer, Heidelberg (2002)

A   Appendix: A Collision and Pre-image Example

For parameter selection $r = 1024, w = 1024, n = 262144, s = 8192, n/w = 256$, the FSB compression function can be implemented in C as follows.

```
typedef unsigned char u8;        // u8 = single byte
typedef unsigned long long u64;  // u64 = 64-bit word

void fsb(u64 h[0x40000][0x10],   // "random" matrix
         u8 s[0x400],            // 1k message block
         u64 r[0x10])            // result
{
    int i, j, idx;

    for (i = 0; i < 0x10; i++)   // zeroise result
        r[i] = 0;
```

```
for (i = 0; i < 0x400; i++)        // process a block
  {
    idx = (i << 8) + s[i];         // index in H
    for (j = 0; j < 0x10; j++)
      r[j] ^= h[idx][j];           // xor over result
  }
}
```

Since the FSB specification does not offer any standard way of defining the "random" matrix $\mathcal{H}$ (or h[][] above), we will do so here using the Data Encryption Standard. Each 64-bit word h[i][j] is created by encrypting the 64-bit input value (i << 4) ^ j under an all-zero 56-bit key (00 00 00 00 00 00 00 00). The input and output values are handled in big-endian fashion. Some of the values are:[4]

```
Input to DES          Table Index        Value
0x0000000000000000    h[0x00000][0x0] = 0x8CA64DE9C1B123A7
0x0000000000000001    h[0x00000][0x1] = 0x166B40B44ABA4BD6
0x0000000000000002    h[0x00000][0x2] = 0x06E7EA22CE92708F
                         ....
0x0000000000000010    h[0x00001][0x0] = 0x5B711BC4CEEBF2EE
0x0000000000000011    h[0x00001][0x1] = 0x799A09FB40DF6019
0x0000000000000012    h[0x00001][0x2] = 0xAFFA05C77CBE3C45
                         ....
0x00000000003FFFFD    h[0x3FFFF][0xD] = 0x313C4BDBE2F7156A
0x00000000003FFFFE    h[0x3FFFF][0xE] = 0x19F32D6B2D9B57F5
0x00000000003FFFFF    h[0x3FFFF][0xF] = 0x804DB568319F4F8B
```

We shall define two 1024-byte message blocks that produce the same 1024-bit chosen output value in the FSB compression function, hence demonstrating the ease of preimage and collision search on a variant with claimed 2^{128} security. They were found in less than a second on an iBook G4 laptop.

The first message block uses the ASCII alphabet {A, C} or {0x41, 0x42}:

```
CAACACACCACAACACACACCACAACCCCCCACCAACACCAAACAAACACCAACACCACACCAA
ACACACCCCCAACCCAAAAAACCCACCACCCACCAAACACACCCCCCAACCACACCCAACACCA
AACCCACCCCCAACCCAAACAAAAACCCACAAAACACACCACCACCCCCACAACCCCACACAAA
AACCCCACCCCAACAACAAAAACAAAACCACACACACACCCCCAAACCCCCAAAAACCCACAAC
CAAACAACCCAAACACCAACCCCACACCCCAAAACCCAAAAAACACAAACCCCAACAAAACCAA
ACACCCCCCCCCAACAAAAACACCCACCCAACAAAAAAACACACCCCCCCAACCCACCCCAACA
AAAACCAACAACACCACCCCACCCCCACCACAAACACCCACCACCCAACCCCACCCAACAAAAC
ACCACCCCAACCCACAACCACCCAACACCAACACCAAAACACACCAAAACACCCAACACACCCC
CAAACACACACCACCACCACCACCCAAAAAAACCACACACCCAAAAAAACCCAAACCACCACCCA
CACAAACCCCAACCCAACCCAACCAACCACCAAAACCCAACCCCCAAAAAAACAACCAAACCCCA
AACACCCACAAACACCACCACAACAAAAACCAAACCCAAAAACCCACCACACACCCACACACAAA
CCACCCCAACCCCCAACAACCCCACACAACACAAACCACCCAACCCCAACCACAAAAACCCACC
ACAACCCAAACACACCCCAACAAACCAAACCCCACACCCAAAACCCCACACCACACACAAACAC
```

```
CACCCAAAAAACAACAACCACACACAACAAACCAAACAAAAAAAAAACCAAAAAACCCCCAACC
CACCCACCCACAAACAAAACCAAAAAAAACCCAAAAAAACCCAAAACCACAACCACCCCAACCA
CCCACCAAACAACAACCACACAAAAACACCCCACACCCCCCACCAACACAAAACCAAAAACCA
```

The second message block uses ASCII alphabet {A, H} or {0x41, 0x48}:

```
AHHHHAAAAAHAAAHAHAAAHAHHAHHAAHAAHHAHHHAAAAAHHAAHHHAHAHAAHAAAHHAA
AAAAHAHHAAAHAHHAHAAAHAAHAHAAAAHHHHHHHAAHAHAHAAAAAHAHHHHHAAHHHHAHAH
AAHAAAHAHAHHHHHAHHAHAHAAAHAHAAHAHHAAAAHAAHAAAHAAHHHHHAHAAHHAAHAH
HHAHAAHHHAAHAAAHHHHAHHHHAAHAAHAAAAAHAAHHAAAHAAHHHAAHAHAHHHAHAAHA
AHHAAAHHAAAAAHHAHAAAAAHAHAHHAHHAHAAHHAHAHAAHHHHAAHAHHHAAHHAHAAHH
AAHAHAAAHAHAAAHHAAAHAHHAHAHHAAAAAHHHHAAHAHAHHAHHHHHAAHHAAHHHHAHH
HHHAAAAAAAHHHAHAAAAHAAAHAAAAAAAHAAHHAHHAHHAHHAHHAAAAAAAHAHAAAHH
HAHHHHHHAHAAAHHAHAAHHHHAAHHAHHAHHAAHHHAHHAHHAAHHAAAHHAHAAHAHHHA
AAHAHAAAHAAHAAAAHHHHHAHHHHHAAHHHAAHHHAHHAAAHHHAHHAHAHAHHHHAAHAHHAHH
AAHAHAAHHAHHAAAAHHAHAHHHHHAAHHHAAHAAAHAAAHAAHHAHHAHHHHAHHHHHAHHHA
AHAHAAAAHHAAAAHHAAHHHHHHAAHAAHAAHHAAAHAHHAAAAAHHAAHAHHAHHAAHHHHAA
HHHAHHAAHAAHAAHAAHHHHHAAHAHHAHHAAHAAAAAHHAHHHHHAHAHHHHHHAHHHHHAAAA
HHHHHAAAAHHHAHHHHAHAAAAHHAHAAAHHAAAAHAHAHAAAAHHHHHHHAHAAHAAHAAAAHAA
HAAAHAHAHHHAHHAHHAHAHAAHAHHAAAAHAAAAHHAAHHHHAHHAAHHHAHAAAHAAAHHHAA
HAAHAAHAAAHAHHHAAHAHAHAAHAAAHAHHAHAAHHHAAHAAAAAHHAAAAHHHAHAHAAAAAH
AAAHAHAHHAAAAHHHHAAHHHAHAAHHHHHAHAAHHAHHHAAHAHHAHHHAAAAHHHAAHAAAAHH
```

The 1024-bit / 128-byte result of compressing either one of these blocks is:

```
Index       Hex                                     ASCII
00000000    5468697320697320    6120636f6c6c6973    |This is a collis|
00000010    696f6e20616e6420    7072652d696d6167    |ion and pre-imag|
00000020    6520666f72204661    73742053796e6472    |e for Fast Syndr|
00000030    6f6d652042617365    6420486173682e20    |ome Based Hash. |
00000040    4172626974726172    79207072652d696d    |Arbitrary pre-im|
00000050    6167657320636f6e    20626520666f756e    |ages can be foun|
00000060    6420696e2061206620    72616374696f6e20   |d in a fraction |
00000070    6f66206120736563    6f6e642120202020    |of a second!    |
```

A Meet-in-the-Middle Collision Attack
Against the New FORK-256

Markku-Juhani O. Saarinen

Information Security Group
Royal Holloway, University of London
Egham, Surrey TW20 0EX, UK
m.saarinen@rhul.ac.uk

Abstract. We show that a $2^{112.9}$ collision attack exists against the FORK-256 Hash Function. The attack is surprisingly simple compared to existing published FORK-256 cryptanalysis work, yet is the best known result against the new, tweaked version of the hash. The attack is based on "splitting" the message schedule and compression function into two halves in a meet-in-the-middle attack. This in turn reduces the space of possible hash function results, which leads to significantly faster collision search. The attack strategy is also applicable to the original version of FORK-256 published in FSE 2006.

Keywords: FORK-256, Hash Function Cryptanalysis, Meet-in-the-middle Attack.

1   Introduction

FORK-256 is a dedicated hash function that produces a 256-bit hash from a message of arbitrary size. The original version of FORK-256 was presented in the first NIST hash workshop and at FSE 2006 [1]. Several attacks have been outlined against this original version, namely:

- Matusiewicz, Contini, and Pieprzyk attacked FORK-256 by using the fact that the functions f and g in the step function were not bijective in the original version. They used microcollisions to find collisions of 2-branch FORK-256 and collisions of full FORK-256 with complexity of $2^{126.6}$ in [3].
- Independently, Mendel, Lano, and Preneel published the collision-finding attack on 2-branch FORK-256 using microcollisions and raised possibility of its expansion [5].
- At FSE 2007 [4], Matusiewicz et al. published the result of [3] and another attack which finds a collision with complexity of 2^{108} and memory of 2^{64}.

   In response to these attacks the authors of FORK-256 have recently proposed a new, tweaked version of FORK-256 [2], which is supposedly resistant to all before-mentioned attacks. We will present a simple attack which is the best currently known against the new version of FORK-256, and also applicable to the previous version.

K. Srinathan, C. Pandu Rangan, M. Yung (Eds.): Indocrypt 2007, LNCS 4859, pp. 10–17, 2007.

2   Description of New FORK-256

New FORK-256 (hereafter FORK-256) is a Merkle-Damgård hash with a 256-bit (8-word) internal state and a 512-bit (16-word) message block. Padding and chaining details are similar to those of the SHA and the MD families of hash functions.

FORK-256 is entirely built on shift, exlusive-or, and addition operations on 32-bit words. In this paper we use the following notation for these operations:

$x \oplus y$	Bitwise exclusive-or between x and y.
$x \boxplus y$	Equal to $(x + y) \mod 2^{32}$.
$x \boxminus y$	Equal to $(x - y) \mod 2^{32}$.
$x \lll y$	Circular left shift of 32-bit word x by y bits.

The compression function of FORK-256 consists of four independent "branches". Each one these branches takes in the 256-bit (8-word) chaining value and a 512-bit (16-word) message block to produce a 256-bit result. These four branch results are combined with the chaining value to produce the final compression function result. Figure 1 illustrates the branch structure.

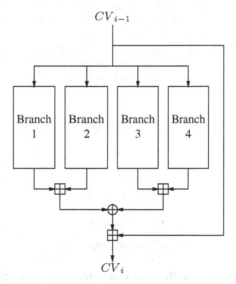

Fig. 1. Overall structure of four branches of FORK-256. Note that the lines are 256 bits wide; the addition symbols represent eight 32-bit modular additions in parallel.

The four branches are structurally equivalent, but differ in scheduling of the message words and round constants. Each branch is computed in eight steps, $0 \leq s \leq 7$. Each step utilizes two message words and two round constants.

The scheduling of the message block words $M[0 \dots 15]$ in each branch is given in Table 1. Round constants $\delta[0 \dots 15]$ are given in Table 2 and their schedule in Table 3. The original description uses auxiliary tables σ and ρ; for convenience we

Table 1. Message word schedule for FORK-256. It is easy to observe that in branch 2 and branch 3, $M[1]$ only affects the result in the last step. $M[14]$ is used in the last and next-to-last steps in branches 1 and 4, correspondingly. These observations are used in the attack.

Step	Branch 1		Branch 2		Branch 3		Branch 4	
s	$a_1^{(s)}$	$b_1^{(s)}$	$a_2^{(s)}$	$b_2^{(s)}$	$a_3^{(s)}$	$b_3^{(s)}$	$a_4^{(s)}$	$b_4^{(s)}$
0	$M[0]$	$M[1]$	$M[14]$	$M[15]$	$M[7]$	$M[6]$	$M[5]$	$M[12]$
1	$M[2]$	$M[3]$	$M[11]$	$M[9]$	$M[10]$	$M[14]$	$M[1]$	$M[8]$
2	$M[4]$	$M[5]$	$M[8]$	$M[10]$	$M[13]$	$M[2]$	$M[15]$	$M[0]$
3	$M[6]$	$M[7]$	$M[3]$	$M[4]$	$M[9]$	$M[12]$	$M[13]$	$M[11]$
4	$M[8]$	$M[9]$	$M[2]$	$M[13]$	$M[11]$	$M[4]$	$M[3]$	$M[10]$
5	$M[10]$	$M[11]$	$M[0]$	$M[5]$	$M[15]$	$M[8]$	$M[9]$	$M[2]$
6	$M[12]$	$M[13]$	$M[6]$	$M[7]$	$M[5]$	$M[0]$	$M[7]$	$M[14]$
7	$M[14]$	$M[15]$	$M[12]$	$M[1]$	$M[1]$	$M[3]$	$M[4]$	$M[6]$

Table 2. Round constants

$\delta[0] = \mathtt{0x428a2f98}$	$\delta[1] = \mathtt{0x71374491}$
$\delta[2] = \mathtt{0xb5c0fbcf}$	$\delta[3] = \mathtt{0xe9b5dba5}$
$\delta[4] = \mathtt{0x3956c25b}$	$\delta[5] = \mathtt{0x59f111f1}$
$\delta[6] = \mathtt{0x923f82a4}$	$\delta[7] = \mathtt{0xab1c5ed5}$
$\delta[8] = \mathtt{0xd807aa98}$	$\delta[9] = \mathtt{0x12835b01}$
$\delta[10] = \mathtt{0x243185be}$	$\delta[11] = \mathtt{0x550c7dc3}$
$\delta[12] = \mathtt{0x72be5d74}$	$\delta[13] = \mathtt{0x80deb1fe}$
$\delta[14] = \mathtt{0x9bdc06a7}$	$\delta[15] = \mathtt{0xc19bf174}$

use a ("left word"), b ("right word"), α ("left constant"), and β ("right constant") in this description as follows:

$$a_j^{(s)} = M[\sigma_j(2s)]$$
$$b_j^{(s)} = M[\sigma_j(2s+1)]$$
$$\alpha_j^{(s)} = \delta[\rho_j(2s)]$$
$$\beta_j^{(s)} = \delta[\rho_j(2s+1)]$$

FORK-256 uses two 32-bit Boolean functions f and g, which were redefined for the New FORK-256 to avoid microcollisions.

$$f(x) = x \oplus (x \lll 15) \oplus (x \lll 27)$$
$$g(x) = x \oplus ((x \lll 7) \boxplus (x \lll 25)).$$

Following the convention of the FORK-256 specification, let $CV_i[0..7]$ be the result of the compression function iteration i and $CV_0[0..7]$ the Initialization Vector, given in Table 4.

Table 3. Round constant schedule

Step	Branch 1		Branch 2		Branch 3		Branch 4	
s	$\alpha_1^{(s)}$	$\beta_1^{(s)}$	$\alpha_2^{(s)}$	$\beta_2^{(s)}$	$\alpha_3^{(s)}$	$\beta_3^{(s)}$	$\alpha_4^{(s)}$	$\beta_4^{(s)}$
0	$\delta[0]$	$\delta[1]$	$\delta[15]$	$\delta[14]$	$\delta[1]$	$\delta[0]$	$\delta[14]$	$\delta[15]$
1	$\delta[2]$	$\delta[3]$	$\delta[13]$	$\delta[12]$	$\delta[3]$	$\delta[2]$	$\delta[12]$	$\delta[13]$
2	$\delta[4]$	$\delta[5]$	$\delta[11]$	$\delta[10]$	$\delta[5]$	$\delta[4]$	$\delta[10]$	$\delta[11]$
3	$\delta[6]$	$\delta[7]$	$\delta[9]$	$\delta[8]$	$\delta[7]$	$\delta[6]$	$\delta[8]$	$\delta[9]$
4	$\delta[8]$	$\delta[9]$	$\delta[7]$	$\delta[6]$	$\delta[9]$	$\delta[8]$	$\delta[6]$	$\delta[7]$
5	$\delta[10]$	$\delta[11]$	$\delta[5]$	$\delta[4]$	$\delta[11]$	$\delta[10]$	$\delta[4]$	$\delta[5]$
6	$\delta[12]$	$\delta[13]$	$\delta[3]$	$\delta[2]$	$\delta[13]$	$\delta[12]$	$\delta[2]$	$\delta[3]$
7	$\delta[14]$	$\delta[15]$	$\delta[1]$	$\delta[0]$	$\delta[15]$	$\delta[14]$	$\delta[0]$	$\delta[1]$

Table 4. Initialization Vector

$$CV_0[0] = \texttt{0x6a09e667} \quad CV_0[1] = \texttt{0xbb67ae85}$$
$$CV_0[2] = \texttt{0x3c6ef372} \quad CV_0[3] = \texttt{0xa54ff53a}$$
$$CV_0[4] = \texttt{0x510e527f} \quad CV_0[5] = \texttt{0x9b05688c}$$
$$CV_0[6] = \texttt{0x1f83d9ab} \quad CV_0[7] = \texttt{0x5be0cd19}$$

Each branch j processes eight input words $R_j^{(0)}[t] = CV_i[t]$ to eight output words $R_j^{(8)}[t]$, $0 \le t \le 7$. Figure 2 illustrates the step function. For $0 \le s \le 7$:

$$t_1 = f(R_j^{(s)}[0] \boxplus a_j^{(s)})$$
$$t_2 = g(R_j^{(s)}[0] \boxplus a_j^{(s)} \boxplus \alpha_j^{(s)})$$
$$t_3 = g(R_j^{(s)}[4] \boxplus b_j^{(s)})$$
$$t_4 = f(R_j^{(s)}[4] \boxplus b_j^{(s)} \boxplus \beta_j^{(s)})$$
$$R_j^{(s+1)}[0] = R_j^{(s)}[7] \oplus (t_4 \lll 8)$$
$$R_j^{(s+1)}[1] = R_j^{(s)}[0] \boxplus a_j^{(s)} \boxplus \alpha_j^{(s)}$$
$$R_j^{(s+1)}[2] = R_j^{(s)}[1] \boxplus t_1$$
$$R_j^{(s+1)}[3] = (R_j^{(s)}[2] \boxplus (t_1 \lll 13)) \oplus t_2$$
$$R_j^{(s+1)}[4] = R_j^{(s)}[3] \oplus (t_2 \lll 17)$$
$$R_j^{(s+1)}[5] = R_j^{(s)}[4] \boxplus b_j^{(s)} \boxplus \beta_j^{(s)}$$
$$R_j^{(s+1)}[6] = R_j^{(s)}[5] \boxplus t_3$$
$$R_j^{(s+1)}[7] = (R_j^{(s)}[6] \boxplus (t_3 \lll 3)) \oplus t_4$$

The final result of the compression function for each word $0 \le t \le 7$ is

$$CV_{i+1}[t] = CV_i[t] \boxplus ((R_1^{(8)}t \boxplus R_2^{(8)}[t]) \oplus (R_3^{(8)}[t] \boxplus R_4^{(8)}[t])).$$

If i is the final iteration, CV_{i+1} is the final hash value.

3   Observations

Each branch of the compression function uses each message word $M[0\ldots15]$ exactly once. Due to diffusion properties of the step function, message words that are scheduled for the last steps do not affect all output words.

Consider the sixth output word of each branch, $R_j^{(8)}[5]$. The last step is defined as:

$$R_j^{(8)}[5] = R_j^{(7)}[4] \boxplus b_j^{(7)} \boxplus \beta_j^{(7)}.$$

Furthermore we "open up" $R_j^{(7)}[4]$ in the previous step:

$$R_j^{(7)}[4] = R_j^{(6)}[3] \oplus (g(R_j^{(6)}[0] \boxplus a_j^{(6)} \boxplus \beta_j^{(6)}) \lll 17).$$

Ignoring the round constants $\alpha_j^{(s)}$ and $\beta_j^{(s)}$, we can observe that the only message words in steps 6 and 7 affecting $R_j^{(8)}[5]$ are $a_j^{(6)}$ and $b_j^{(7)}$, the latter having a linear effect. Constants $b_j^{(6)}$ and $a_j^{(7)}$ have no effect in the computation of this word.

By thus inspecting the step function and the message word schedule in Table 1, it is easy to verify that $R_j[5]$ satisfies the following properties:

Branch 1: $R_1^{(8)}[5]$ is independent of $M[14] = a_1^{(7)}$.
Branch 2: $R_2^{(8)}[5]$ is linearly dependent on $M[1] = b_2^{(7)}$.
Branch 3: $R_3^{(8)}[5]$ is independent of $M[1] = a_3^{(7)}$.
Branch 4: $R_4^{(8)}[5]$ is independent of $M[14] = b_4^{(6)}$.

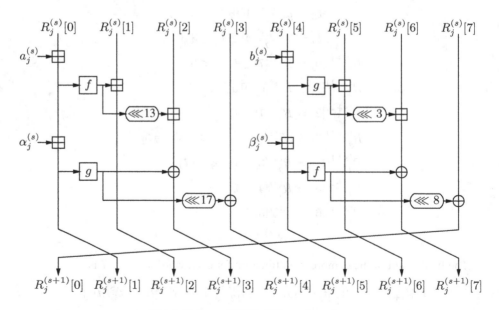

Fig. 2. The new FORK-256 step iteration

We shall use these simple observations to construct an attack against FORK-256. We note that due to the fact the message word schedule is shared between the old and new versions of FORK-256, the same four observations – and the same general attack – apply to both versions, although there are important technical differences between the old and the new version. The complexity of the attack is the same for both.

4   A Collision Attack

The main strategy of the attack is to use a fast method for finding messages that hash into a significantly smaller subset of possible hash values. We do this by forcing the sixth word of the compression function to remain constant over the hash function iteration, $CV_1[5] = CV_0[5]$, thereby generating hashes in a subset of size 2^{224}. Assuming uniform distribution, a full collision can be expected after $\sqrt{\frac{\pi}{2}} \times 2^{\frac{224}{2}} \approx 2^{112.3}$ hashes in the small subset have been found.

The value of $CV_1[5]$ is combined from the four branches and the initialization vector as follows:

$$CV_1[5] = CV_0[5] \boxplus ((R_1^{(8)}[5] \boxplus R_2^{(8)}[5]) \oplus (R_3^{(8)}[5] \boxplus R_4^{(8)}[5])).$$

By substituting $CV_1[5] = CV_0[5]$ and regrouping branches 2 and 3 on the left side and branches 1 and 4 on the right side, we obtain the following necessary and sufficient condition for $CV_1[5] = CV_0[5]$:

$$R_2^{(8)}[5] \boxminus R_3^{(8)}[5] = R_1^{(8)}[5] \boxminus R_4^{(8)}[5].$$

Our attack is based on choosing two message words $M[1]$ and $M[14]$ in a specific way to satisfy $CV_1[5] = CV_0[5]$, which is possible due to the observations given in the previous section. The values of the fourteen other message words are arbitrary and can be chosen at random (as long as they remain constant through the two phases of the attack). The two phases can be repeated any number of times to produce sufficient amount of hashes in the subset.

4.1   First Phase

Set $M[1] = 0$ and loop over $M[14] = 0, 1, 2, \cdots, 2^{32} - 1$. Compute branches 2 and 3 for each $M[14]$ to obtain $x = R_2^{(8)}[5] \boxminus R_3^{(8)}[5]$. Place x and $M[14]$ into a look-up table so that the value of $M[14]$ can be immediately retrieved based on the corresponding x value (i.e. $M[14]$ is indexed by x).

Note that since the mapping from $M[14]$ to x is not surjective, about $1/e \approx 36.8\%$ of the values of x will never occur (when the mapping is modeled as random). On the other hand, many x can be obtained with more than one value of $M[14]$. Using a straightforward lookup cannot handle the latter situation, but simple data structures with negligible expansion exist that can be used for these cases. The table does not need to be larger than 16 gigabytes (32 bits $\times 2^{32}$ entries).

4.2 Second Phase

Loop over the 2^{32} values of $M[1]$. Compute branches 1 and 4 for each $M[1]$ to obtain $y = R_1^{(8)}[5] \boxminus R_4^{(8)}[5] \boxplus M[1]$. The $M[1]$ term is included due to the linear dependence of $R_2^{(8)}[5]$ on it (this is also why $M[1]$ is set to zero in the first phase).

In each step, perform a look-up. If a match or matches $x = y$ are found, the necessary and sufficient condition is satisfied and we have found a message (or rather, a pair of $M[1]$ and $M[14]$ values) that produces one or more hashes that satisfy $CV_1[5] = CV_0[5]$.

4.3 Runtime Analysis

Each loop step in the second phase produces one match in the lookup table on average. This is due to the fact that even though the mapping is not surjective, there is a total of 2^{32} $M[14]$ entries in the table. Hence approximately 2^{32} hashes with the property are produced in the second phase.

Since computation of only two branches out of four are needed, the computational effort in the first and second phases is roughly equivalent to 2^{31} full hash computations each, or 2^{32} total. If the full 8 words in phase 1 are not stored, branches 2 and 3 need to be computed again to reproduce a full hash, bringing the total number to $3 * 2^{31}$. The average cost of producing a hash in the 2^{224} subset therefore is $\frac{3}{2}$ hash function invocations.

Unfortunately we have been unable to come up with a method of utilizing "memoryless" random-walk collision search methods such as those discussed in [6]. This is due to the fact that the algorithm outlined above only works in "batches" of 2^{32} to obtain a favorable average cost for each hash with the desired property $CV_1[5] = CV_0[5]$. The memory requirement is therefore equivalent to running time requirement, $\frac{3}{2}\sqrt{\frac{\pi}{2}} \times 2^{\frac{224}{2}} = 2^{112.9}$.

5 Further Work

The same observations about the effects of $M[1]$ and $M[14]$ on the final hash can be easily be adopted into a pre-image attack that recovers the values of these two message words with 2^{32} effort, rather than 2^{64} as expected in a brute-force search.

It may be possible to "fix" more than 32 bits by using additional words of keying material besides $M[1]$ and $M[14]$ in the attack. This would naturally lead to a more effective overall collision attack. Terms $M[0]$ and $M[5]$ appear to be good candidates as they are only used in steps 5 and 6 of branches 2 and 3, respectively, and are therefore not fully diffused at the end of step 7.

6 Conclusion

We have presented a $2^{112.9}$ collision attack against the new, improved version of the hash function FORK-256. This represents a speed improvement of factor $2^{15.4}$ over a straightforward collision search. The attack strategy is surprisingly simple, and can also be applied against the original version of FORK-256 in slightly modified form.

Acknowledgements

The author would thank Keith Martin and the INDOCRYPT Program Committee members for essential quality control and helpful comments. Financial support for this work was provided by Nixu Ltd. and Academy of Finland.

References

1. Hong, D., Chang, D., Sung, J., Lee, S., Hong, S., Lee, J., Moon, D., Chee, S.: A New Dedicated 256-Bit Hash Function: FORK-256. In: Robshaw, M. (ed.) FSE 2006. LNCS, vol. 4047, pp. 195–209. Springer, Heidelberg (2006)
2. Hong, D., Chang, D., Sung, J., Lee, S., Hong, S., Lee, J., Moon, D., Chee, S.: New FORK-256. Cryptology ePrint Archive 2007/185 (July 2007)
3. Matusiewicz, K., Contini, S., Pieprzyk, J.: Weaknesses of the FORK-256 Compression Function. Cryptology ePrint Archive 2006/317 (Second version) (November 2006)
4. Matusiewicz, K., Peyrin, T., Billet, O., Contini, S., Pieprzyk, J.: Cryptanalysis of FORK-256. In: Preproceeding of FSE 2007 (2007)
5. Mendel, F., Lano, J., Preneel, B.: Cryptanalysis of Reduced Variants of the FORK-256 Hash Function. In: Abe, M. (ed.) CT-RSA 2007. LNCS, vol. 4377, pp. 85–100. Springer, Heidelberg (2006)
6. van Oorschot, P., Wiener, M.: Parallel collision search with cryptanalytic applications. Journal of Cryptology 12, 1–28 (1999)

Multilane HMAC—
Security beyond the Birthday Limit

Kan Yasuda

NTT Information Sharing Platform Laboratories, NTT Corporation
3-9-11 Midoricho Musashino-shi, Tokyo 180-8585 Japan
yasuda.kan@lab.ntt.co.jp

Abstract. HMAC is a popular MAC (Message Authentication Code) that is based on a cryptographic hash function. HMAC is provided with a formal proof of security, in which it is proven to be a PRF (Pseudo-Random Function) under the condition that its underlying compression function is a PRF. Nonetheless, the security of HMAC is limited by a birthday attack, that is, HMAC using a compression function with n-bit output gets forged after about $2^{n/2}$ queries. In this paper we resolve this problem by introducing novel construction we call L-Lane HMAC. Our construction is provided with concrete-security reduction accomplishing a security guarantee well beyond the birthday limit. L-Lane HMAC requires more invocations to the compression function than the conventional HMAC, but the performance decline is smaller than those of previous constructs. In addition, L-Lane HMAC inherits the design principles of the original HMAC, such as single-key usage and off-the-shelf hash-function calls.

Keywords: message authentication code, hash function, birthday attack, multilane, NMAC, HMAC, failure-friendly.

1 Introduction

Birthday Attack on HMAC. The birthday paradox is a powerful principle that influences many sorts of cryptographic algorithms. Among those algorithms that are affected, MACs (Message Authentication Codes) of iterated structure are known to be vulnerable to birthday attacks [15]. The birthday attack on a MAC scheme detects an internal collision of the MAC algorithm and utilizes an "extension trick" to produce a forgery. For instance, HMAC [3] using a compression function $f : \{0,1\}^{n+m} \rightarrow \{0,1\}^n$ can be broken (forged) after $\mathcal{O}(2^{n/2})$ queries to its generation oracle. This is known as the *birthday limit*, for the quantity n corresponds to security parameters as explained below.

The value n determines the (maximum) size of the final output of HMAC algorithm. The final output for a given message M is called the *tag*. Note that a pair (M, tag) with a random value $\mathsf{tag} \in \{0,1\}^n$ gets verified as a valid one with a probability 2^{-n}. This means that an adversary who outputs such a pair would succeed in forgery with a probability 2^{-n}. This also implies that this type

K. Srinathan, C. Pandu Rangan, M. Yung (Eds.): Indocrypt 2007, LNCS 4859, pp. 18–32, 2007.

of adversary needs $\mathcal{O}(2^n)$ trials to the verification oracle in order to succeed in forgery with a good probability.

Moreover, the parameter n also specifies a key space. The keyed compression function $f_k : \{0,1\}^m \rightarrow \{0,1\}^n$, with $k \in \{0,1\}^n$, appears in the known security reduction [2] of HMAC, regardless of an actual size of HMAC key. The security proof states that HMAC relies on the pseudo-randomness of f_k. The pseudo-randomness is measured by the prf-advantage function $\mathrm{Adv}_f^{\mathrm{prf}}(t,q)$, where t and q are upper bounds of time resource and query numbers, respectively. If f_k is a "good" PRF (Pseudo-Random Function), then we expect $\mathrm{Adv}_f^{\mathrm{prf}}(t,2) \approx 2^{-n}$ for small t; the key exhaustive search would require $\mathcal{O}(2^n)$ computations of f.

Limited Security of NMAC/HMAC. HMAC is a derivative of more general construction called NMAC [3]. The security of NMAC is based on the cAU-PRF paradigm,[1] whose proof can be found in [2]. There, in an upper bound of the NMAC security, a term of order

$$\mathcal{O}\left(\frac{\ell q^2}{2^n}\right)$$

appears,[2] where ℓ is the maximum length of queries (the length is in blocks, and a block is m bits.)

So, for example, consider the case of $n - 128$, $m = 512$ and authenticating Blu-ray Discs (50GB $\approx 2^{38.6}$ bits). In such a case we have $\ell \approx 2^{29.6}$, and consequently the security guarantee of NMAC becomes vacuous when $q \geq 2^{49.2}$. While this number may be large enough in practice, it is far from being 128-bit security.

Bad news is that the security of NMAC is explicitly degraded when its underlying compression function f_k fails to be collision-resistant (thus increasing the quantity $\mathrm{Adv}_f^{\mathrm{prf}}(t,2)$). In such a case the birthday attacks become collision attacks, as recently reported for NMAC/HMAC with "weak" compression functions [9,6]. These shortcomings of the NMAC/HMAC designs are one of our motives for developing new construction that is birthday-resistant and hence failure-friendly.

We could, of course, start with a "wide-pipe" [10] compression function $f : \{0,1\}^{2n+m} \rightarrow \{0,1\}^{2n}$, use HMAC construction with an n-bit key and truncate its tag size from $2n$ to n bits. This would indeed yield a birthday-resistant MAC with full n-bit security, but it would be wasting f's capacity for resisting attacks. Such a function f deserves full $2n$-bit security, not n-bit.

Our Contributions: L-Lane NMAC/HMAC. We are interested in constructing a MAC scheme from a compression function $f : \{0,1\}^{n+m} \rightarrow \{0,1\}^n$ via a mode of operation that accomplishes (close to) full n-bit security. This goal is reached by our L-lane NMAC/HMAC construction. The basic idea of our

---

[1] The cAU-PRF construction can be viewed as a computational version of Carter-Wegman paradigm [17], as already pointed out by [2].

[2] Here, we are assuming that $\mathrm{Adv}_f^{\mathrm{prf}}(t,2) \approx 2^{-n}$.

construction is to utilize multiple internal "lanes" of Merkle-Damgård iteration, where each lane is kept of short length for maintaining reasonable performance. More precisely, the scheme works in the following ways:

1. **Security beyond the Birthday Limit.** L-Lane NMAC is provided with a proof of security, with a bound well beyond the birthday limit. That is, the problematic term changes to being of order[3]

$$O\left(\left(\frac{\ell q}{2^n}\right)^2\right).$$

   So in the above Blu-ray Disc example the vacuous limit changes to $q \geq 2^{98.4}$, much closer to 128-bit security. Also, the better bound gives us more "room" for q even if the compression function f_k fails to be a secure PRF, ensuring more resistance to collision attacks. In this sense, our construction is provided with failure-friendliness [10].

2. **Stateless and Deterministic Algorithms.** L-lane NMAC preserves an important notion of a conventional MAC: It avoids the use of counters or coins. Introducing a nonce element or randomness in a MAC scheme would also introduce inconvenience and troublesomeness to its system in practice.

3. **"More-than-Half" Efficiency.** L-lane NMAC requires more numbers of invocations to the compression function f than the original NMAC. This increase, however, is kept small. Roughly speaking, we construct a scheme that utilizes L-many lanes, where the length of each lane is about $1/(L-1)$-times the message length $|M|$. Therefore, L-Lane NMAC requires about

$$\left(1 + \frac{1}{L-1}\right)$$

   -times as many invocations to f as the original NMAC. For example, it becomes twice with $L = 2$, 1.5-times with $L = 3$ and 1.33-times with $L = 4$.

4. **"Independent" Parallelism.** L-Lane NMAC contains L-many lanes in its design. These lanes are "independently" parallelizable, that is, the output of the i-th block of a lane does not depend on the output of the $(i-1)$-th block of any other lane but does depend only on the message input to the i-th block itself and the output of the $(i-1)$-th block of the same lane. This "independence" gives us freedom of choosing an order of computation. For example, consider the case $L = 3$. We can:
   - compute the three lanes in parallel at once,
   - compute two of the three lanes in parallel first, and then compute the remaining one lane later, or
   - compute the three lanes one-by-one without any parallelism.

We also provide L-Lane HMAC construction. This is a derivative of L-Lane NMAC, with the following additional features:

---

[3] Again, we assume that $\mathrm{Adv}_f^{\mathrm{prf}}(t, 2) \approx 2^{-n}$.

5. Single-Key Usage. L-Lane NMAC requires $(L+3)$-many keys, an undesirable property in practice. In L-Lane HMAC construction, these keys are derived from a single key $k \in \{0,1\}^n$ using the pseudo-randomness of f.
6. Fixed IV and Merkle-Damgård Strengthening. L-Lane HMAC can call off-the-shelf hash functions that are implemented in the keyless Merkle-Damgård style, along with a fixed IV (Initial Vector) and a designated padding method (called the *Merkle-Damgård strengthening.*)

The only limitation by L-lane NMAC/HMAC, which is not present in the original NMAC/HMAC, is on the choice of a compression function $f : \{0,1\}^{n+m} \rightarrow \{0,1\}^n$ with the condition $m \geq 2n$. This restriction, however, is not a critical drawback in practice. In fact, most of existing compression functions, including md5, sha-1 and sha-256, clear this condition.

Previous Constructs. There are several known constructs of birthday-resistant MACs. These include XOR-MAC [5] based on a finite PRF, MACRX [4] based on universal hashing (with a finite PRF), and RMAC [8] based on a block cipher. These algorithms, however, are either *nonce-based* or *randomized.*

There are several approaches without counters nor coins. One is to construct a PRF $f'_{k'} : \{0,1\}^{2n} \rightarrow \{0,1\}^{2n}$ from a PRF $f_k : \{0,1\}^n \rightarrow \{0,1\}^n$ in a birthday-resistant way. Such construction includes Benes [1], Ω_t [13] and Feistel-6 [14]. These however require too many (4 or more) invocations to f, and consequently HMAC based on such an f' would be *inefficient.*

A more efficient approach relies on usage of two or more "streams" of data processing. This idea dates back to the design of the compression function RIPEMD and its application to Two-Track MAC [7]. A similar approach appears in the context of keyless hash functions as "Double-Pipe" hash [10] and "ℓ-Pipe" hash [16]. These constructs above come close to our L-Lane approach but differ in two points. One is that their performance *degrades by a factor of two or more*, as compared to their "single" versions. The other is that their two (or more) tracks/pipes are *dependant* each other, disabling out-of-order execution.

We also note that there are differences in reduction methodology. For example, the proofs in [10] are based on the assumption that the underlying compression functions are random functions. We take advantage of the presence of "secret" keys and avoid such usage of *random oracles*; our proofs are based on the sole assumption that the underlying compression function f is a PRF.

Outline of This Paper. In the following section we give notation, terminology and definitions that are necessary in this paper. We then give a general framework based on Doubly Injective Lengthening (DIL) in Sect. 3. This general framework, DIL-cAU-PRF construction, is applied to its simple instantiation we call "Two-Lane NMAC" in Sect. 4. We proceed to describing L-Lane NMAC for $L \geq 3$ in Sect. 5 and L-Lane HMAC in Sect. 6. Section 7 is devoted to discussions on the performance of our construction. Section 8 concludes this paper.

2 Preliminaries

Notation. Let m be a positive integer. The notation $\{0,1\}^m$ represents the set of m-bit strings, and $\{0,1\}^{m*}$ denotes the set of finite bit strings whose lengths in bits are multiples of m. Accordingly we write $\{0,1\}^*$ for the set of all finite bit strings. Note that the set $\{0,1\}^{m*}$ includes the empty string ε.

Given a finite bit string $x \in \{0,1\}^*$, $|x|$ stands for the length in bits of x. Given $x, y \in \{0,1\}^*$, $x\|y$ denotes the concatenation of x and y. Sometimes $x\|y$ is written simply xy. For $x, y \in \{0,1\}^*$ with $|x| = |y|$, we define $x \oplus y$ to be the bitwise exclusive-or of x and y. Given $x_1, \ldots, x_L \in \{0,1\}^*$ with $|x_1| = \cdots = |x_L|$, we write $\bigoplus_{i=1}^L x_i$ for $x_1 \oplus \cdots \oplus x_L$. When $L = 1$, we let $\bigoplus_{i=1}^L x_i \overset{\text{def}}{=} x_1$.

If x and y are two variables, the notation $x \leftarrow y$ indicates the operation of assigning the value of y to variable x. For a set X, $x \overset{\$}{\leftarrow} X$ indicates the operation of selecting an element uniformly at random from set X and assigning its value to variable x. We write $x_1, x_2, \ldots \overset{\$}{\leftarrow} X$ to mean the sequence of operations $x_1 \overset{\$}{\leftarrow} X, x_2 \overset{\$}{\leftarrow} X, \ldots$.

Given non-negative integers i and m, the notation $\langle i \rangle_m$ denotes the canonical m-bit encoding of integer i. The size m is assumed to be large enough in order for this symbol to make sense. Also, often m is omitted from the notation, and it is simply written as $\langle i \rangle$.

An adversary A is a probabilistic algorithm that may have access to one or more oracles. We use the symbol $A \Rightarrow x$ to indicate the event that adversary A outputs a value x. The notation $A^{\mathcal{O}} \Rightarrow x$ means that adversary A, having access to an oracle $\mathcal{O}$, outputs value x.

Security Definitions. We use the notions of PRF and cAU function. Recall that a secure PRF is a secure MAC.

1. Pseudo-Random Function (PRF). Let $f_k : X \to Y$ be a keyed function with $k \in K$. A prf-adversary A outputs either 0 or 1, having access to either the "real" oracle f or the "random" oracle \$. The real oracle chooses a key $k \overset{\$}{\leftarrow} K$ at the beginning of the overlying experiment and returns $f_k(x)$ upon a query x. The random oracle chooses a random function $\varphi : X \to Y$ at the beginning and upon a query x returns $\varphi(x)$. We then define

$$\mathrm{Adv}_f^{\mathrm{prf}}(A) \overset{\text{def}}{=} \Pr\left[A^f \Rightarrow 1\right] - \Pr\left[A^{\$} \Rightarrow 1\right].$$

Also, define

$$\mathrm{Adv}_f^{\mathrm{prf}}(t, q, \mu) \overset{\text{def}}{=} \max_A \mathrm{Adv}_f^{\mathrm{prf}}(A),$$

where max runs over all adversaries A having time complexity at most t, asking at most q queries to the oracle, each query being at most μ bits. By convention the time complexity includes the total execution time of the overlying experiment plus the code size of A. Also, the resource parameter μ is omitted when the domain X consists of fixed-length elements.

2. Computationally Almost Universal (cAU) Function. Again, let $f_k : X \to Y$ be a keyed function with $k \in K$. An au-adversary A simply outputs a pair of messages (x, x'). Define

$$\mathrm{Adv}_f^{\mathrm{au}}(A) \stackrel{\mathrm{def}}{=} \Pr\Big[f_k(x) = f_k(x') \wedge x \neq x' \,\Big|\, A \Rightarrow (x, x'), k \stackrel{\$}{\leftarrow} K\Big].$$

Also, define

$$\mathrm{Adv}_f^{\mathrm{au}}(\mu) \stackrel{\mathrm{def}}{=} \max_A \mathrm{Adv}_f^{\mathrm{au}}(A),$$

where max runs over all adversaries A that outputs a pair of messages, each message being at most μ bits. Note that time complexity t is omitted from the notation, for it is irrelevant in this context [2].

Lastly, we introduce the notation $\mathsf{time}(f)$. This represents the time resource needed to perform one computation of function f. We remark that in order for these definitions to make sense, we need to fix a model of computation, which is assumed to be the case throughout the paper.

3   General Framework

We introduce the notion of Doubly Injective Lengthening (DIL), which plays an important role in this work. It gives us a mechanism for resisting birthday attacks while maintaining reasonable performance.

Doubly Injective Lengthening (DIL). Fix an integer $L \geq 2$. Let

$$\Phi = \{\varphi_i\}_{1 \leq i \leq L}$$

be a family of functions $\varphi_i : \mathcal{M} \to X_i$. We say that Φ is *doubly injective* if $M \neq M'$ $(M, M' \in \mathcal{M})$ implies "$\varphi_i(M) \neq \varphi_i(M')$ and $\varphi_j(M) \neq \varphi_j(M')$ for some $1 \leq i < j \leq L$." Equivalently, Φ is doubly injective if "$\varphi_i(M) = \varphi_i(M')$ for at least $(L-1)$-many values of $i \in \{1, \ldots, L\}$" implies $M = M'$.

Given a DIL $\Phi = \{\varphi_i\}_{1 \leq i \leq L}$, define a function

$$\rho_\Phi(\mu) \stackrel{\mathrm{def}}{=} \max_{M, i} |\varphi_i(M)|,$$

where max is taken over $M \in \mathcal{M}$ with $|M| = \mu$ and over $i \in \{1, \ldots, L\}$. Note that the domain $\mathcal{M}$ and ranges X_i are assumed to be subsets of $\{0, 1\}^*$ so that $|M|$ and $|\varphi_i(M)|$ are well-defined.

With abuse of notation, we sometimes identify the above $\Phi = \{\varphi_i\}_{1 \leq i \leq L}$ with the function $\Phi : \mathcal{M} \to X_1 \times \cdots \times X_L$ in the obvious way. We treat Φ as a function family or a function interchangeably.

DIL-cAU-PRF Construction. Given a DIL $\Phi = \{\varphi_i\}_{1 \leq i \leq L}$ with $\varphi_i : \mathcal{M} \to X$, a cAU function $\boldsymbol{H} : K \times X \to Y$ and a PRF $\boldsymbol{G} : K' \times Y^L \to T$, we construct their compositions $\boldsymbol{H}^L \circ \Phi : K^L \times \mathcal{M} \to Y^L$ and $\boldsymbol{G} \circ \boldsymbol{H}^L \circ \Phi : K^L \times K' \times \mathcal{M} \to T$ as follows:

Function $\left(\boldsymbol{H}^L \circ \varPhi\right)_{k_1,\dots,k_L}(M)$

  $x_i \leftarrow \varphi_i(M)$ for $i = 1, \dots, L$

  $y_i \leftarrow \boldsymbol{H}_{k_i}(x_i)$ for $i = 1, \dots, L$

  Output $(y_1, \dots, y_L)$

Function $\left(\boldsymbol{G} \circ \boldsymbol{H}^L \circ \varPhi\right)_{k_1,\dots,k_L,k'}(M)$

  $(y_1, \dots, y_L) \leftarrow \left(\boldsymbol{H}^L \circ \varPhi\right)_{k_1,\dots,k_L}(M)$

  tag $\leftarrow \boldsymbol{G}_{k'}(y_1, \dots, y_L)$

  Output tag

See also Fig. 1 for a pictorial description of the latter composition. We show that this composition $\boldsymbol{G} \circ \boldsymbol{H}^L \circ \varPhi$ is a secure PRF with such reduction as avoids birthday attacks:

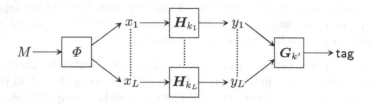

Fig. 1. DIL-cAU-PRF construction

Theorem 1. *We have*

$$\mathrm{Adv}^{\mathrm{prf}}_{\boldsymbol{G} \circ \boldsymbol{H}^L \circ \varPhi}(t, q, \mu) \leq \mathrm{Adv}^{\mathrm{prf}}_{\boldsymbol{G}}(t, q) + \binom{q}{2} \cdot \left(\mathrm{Adv}^{\mathrm{au}}_{\boldsymbol{H}}(\mu')\right)^2,$$

where $q \geq 2$ and $\mu' = \rho_\varPhi(\mu)$.

Proof. This theorem immediately follows from the two lemmas below.   □

Lemma 1. *We have*

$$\mathrm{Adv}^{\mathrm{prf}}_{\boldsymbol{G} \circ \boldsymbol{H}^L \circ \varPhi}(t, q, \mu) \leq \mathrm{Adv}^{\mathrm{prf}}_{\boldsymbol{G}}(t, q) + \binom{q}{2} \cdot \mathrm{Adv}^{\mathrm{au}}_{\boldsymbol{H}^L \circ \varPhi}(\mu),$$

where $q \geq 2$.

Proof. This is just a direct application of the cAU-PRF construction, whose proof can be found in [2].   □

Lemma 2. *We have*

$$\mathrm{Adv}^{\mathrm{au}}_{\boldsymbol{H}^L \circ \varPhi}(\mu) \leq \left(\mathrm{Adv}^{\mathrm{au}}_{\boldsymbol{H}}(\mu')\right)^2,$$

where $\mu' = \rho_\varPhi(\mu)$.

Proof. Let A be an au-adversary attacking $\boldsymbol{H}^L \circ \varPhi$, that outputs a pair of messages, each message being at most μ bits. Without loss of generality we assume that A always outputs a fixed pair (M, M'), with $M, M' \in \mathcal{M}$ and $M \neq M'$. The condition $M \neq M'$ implies that there exist $1 \leq \alpha < \beta \leq L$ such that $\varphi_\alpha(M) \neq \varphi_\alpha(M')$ and $\varphi_\beta(M) \neq \varphi_\beta(M')$, because $\varPhi$ is doubly injective. Write

$x_i \leftarrow \varphi_i(M)$, $x'_i \leftarrow \varphi_i(M')$, $y_i \leftarrow \boldsymbol{H}_{k_i}(x_i)$ and $y'_i \leftarrow \boldsymbol{H}_{k_i}(x'_i)$. Then the au-advantage of A attacking $\boldsymbol{H}^L \circ \varPhi$ can be bounded as:

$$\mathrm{Adv}^{\mathrm{au}}_{\boldsymbol{H}^L \circ \varPhi}(A) = \Pr\left[y_i = y'_i \text{ for } i = 1, \ldots, L \,\middle|\, k_1, \ldots, k_L \xleftarrow{\$} K\right]$$

$$\leq \Pr\left[y_\alpha = y'_\alpha \wedge y_\beta = y'_\beta \,\middle|\, k_1, \ldots, k_L \xleftarrow{\$} K\right]$$

$$= \Pr\left[\boldsymbol{H}_{k_\alpha}(x_\alpha) = \boldsymbol{H}_{k_\alpha}(x'_\alpha) \wedge \boldsymbol{H}_{k_\beta}(x_\beta) = \boldsymbol{H}_{k_\beta}(x'_\beta) \,\middle|\, k_\alpha, k_\beta \xleftarrow{\$} K\right].$$

The last line above clarifies the fact that the two events "$y_\alpha = y'_\alpha$" and "$y_\beta = y'_\beta$" are independent:

$$\Pr\left[y_\alpha = y'_\alpha \wedge y_\beta = y'_\beta \,\middle|\, k_\alpha, k_\beta \xleftarrow{\$} K\right] = \Pr\left[y_\alpha = y'_\alpha \,\middle|\, k_\alpha \xleftarrow{\$} K\right]$$

$$\times \Pr\left[y_\beta = y'_\beta \,\middle|\, k_\beta \xleftarrow{\$} K\right].$$

Here the first probability in the product can be bounded as

$$\Pr\left[y_\alpha = y'_\alpha \,\middle|\, k_\alpha \xleftarrow{\$} K\right] = \Pr\left[\boldsymbol{H}_{k_\alpha}(x_\alpha) = \boldsymbol{H}_{k_\alpha}(x'_\alpha) \,\middle|\, k_\alpha \xleftarrow{\$} K\right] \leq \mathrm{Adv}^{\mathrm{au}}_{\boldsymbol{H}}(\mu'),$$

owing to the conditions $x_\alpha \neq x'_\alpha$ and $|x_\alpha|, |x'_\alpha| \leq \mu' \stackrel{\mathrm{def}}{=} \rho_\varPhi(\mu)$. Similarly for the second term we obtain $\Pr\left[y_\beta = y'_\beta \,\middle|\, k_\beta \xleftarrow{\$} K\right] \leq \mathrm{Adv}^{\mathrm{au}}_{\boldsymbol{H}}(\mu')$, which together gives us the desired bound $\mathrm{Adv}^{\mathrm{au}}_{\boldsymbol{H}^L \circ \varPhi}(A) \leq \left(\mathrm{Adv}^{\mathrm{au}}_{\boldsymbol{H}}(\mu')\right)^2$.   □

4   Two-Lane NMAC

Two-Lane NMAC is one of the simplest instantiations of the DIL-cAU-PRF paradigm, using a compression function $f : \{0,1\}^{n+m} \rightarrow \{0,1\}^n$ with $m \geq 2n$. Two-Lane NMAC is also used as a building component of L-Lane NMAC ($L \geq 3$) in Sect. 5.

Description of Two-Lane NMAC. See Fig. 2 for an illustration. In the following we describe how each component $\varPhi$, $\boldsymbol{H}$ or $\boldsymbol{G}$ is instantiated in Two-Lane NMAC.

1. Trivial Instantiation of DIL $\varPhi$. The diagonal map $\varPhi : M \mapsto (M, M)$ is used in Two-Lane NMAC. This corresponds to a trivial instantiation of DIL with $L = 2$ and $\varphi_1, \varphi_2 : \mathcal{M} \rightarrow \mathcal{M}$ being the identity map. In such a case the condition $M \neq M'$ immediately yields $\varphi_1(M) = \varphi_2(M) = M \neq M' = \varphi_1(M') = \varphi_2(M')$.
2. Merkle-Damgård Iteration for cAU $\boldsymbol{H}$. Given a compression function $f : \{0,1\}^{n+m} \rightarrow \{0,1\}^n$, we obtain a keyed function $f_k : \{0,1\}^m \rightarrow \{0,1\}^n$ with $k \in \{0,1\}^n$ by defining $f_k(x) \stackrel{\mathrm{def}}{=} f(k\|x)$ for $x \in \{0,1\}^m$. We then iterate f in the Merkle-Damgård style to obtain the component $\boldsymbol{H}_k : \{0,1\}^{m*} \rightarrow \{0,1\}^n$, as follows:

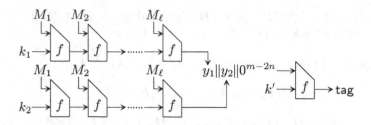

Fig. 2. Two-Lane NMAC

Function $H_k(M)$
    Divide $M = M_1 \| M_2 \| \cdots \| M_\ell$ with $M_i \in \{0,1\}^m$
    $v_1 \leftarrow f_k(M_1)$
    $v_i \leftarrow f(v_{i-1} \| M_i)$ for $i = 2, \ldots, \ell$
    Output v_ℓ.

We set $H_k(\varepsilon) \stackrel{\text{def}}{=} k$. The only problem of this construction is that its domain is restricted to $\{0,1\}^{m*}$. We can make it accept an arbitrary-length message $M \in \{0,1\}^*$ by an appropriate padding. Any one-to-one padding works here, because the composition of a cAU function and a one-to-one padding is again a cAU function. For example, the canonical 10^* works, and the popular Merkle-Damgård strengthening $10^* \| \langle |\mu| \rangle$ works as well.

3. **Compression Function f in Place of PRF G.** For the component $G_{k'}$, we simply use the compression function $f_{k'} : \{0,1\}^m \to \{0,1\}^n$ with $k' \in \{0,1\}^n$ as is. This is feasible owing to the assumption that f is a PRF with $m \geq 2n$. More precisely, we define

$$G_{k'}(y_1, y_2) \stackrel{\text{def}}{=} f_{k'}(y_1 \| y_2 \| 0^{m-2n})$$

for $y_1, y_2 \in \{0,1\}^n$, so that $\mathrm{Adv}_G^{\mathrm{prf}}(t,q) = \mathrm{Adv}_f^{\mathrm{prf}}(t,q)$.

Security of Two-Lane NMAC. We first point out a well-known result that the function H constructed from f via Merkle-Damgård is indeed a cAU function:

Lemma 3. *If f is a PRF, then H, constructed from f via the Merkle-Damgård iteration as above, is cAU. More concretely, we have*

$$\mathrm{Adv}_H^{\mathrm{au}}(\mu) \leq (2\ell - 1) \cdot \mathrm{Adv}_f^{\mathrm{prf}}(t,2) + \frac{1}{2^n},$$

where $\ell = \lceil \mu/m \rceil$ and $t = 4\ell \cdot \mathrm{time}(f)$.

Proof. This result is given in [2]. □

We are now ready to show our security result of Two-Lane NMAC:

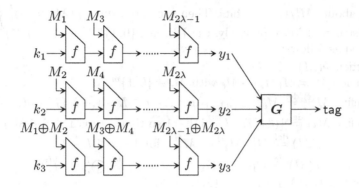

Fig. 3. L-Lane NMAC, case $L = 3$

Theorem 2. *The upper-bound security of Two-Lane NMAC is given by*

$$\mathrm{Adv}^{\mathrm{prf}}_{\text{Two-Lane NMAC}}(t,q,\mu) \leq \mathrm{Adv}^{\mathrm{prf}}_{f}(t,q) + 2q^2\left(\ell \cdot \mathrm{Adv}^{\mathrm{prf}}_{f}(t',2) + 2^{-n-1}\right)^2,$$

where $\ell = \lceil \mu/m \rceil$ and $t' = 4\ell \cdot \mathrm{time}(f)$.

Proof. By Theorem 1 and the above lemma, we get

$$\mathrm{Adv}^{\mathrm{prf}}_{\text{Two-Lane NMAC}}(t,q,\mu) \leq \mathrm{Adv}^{\mathrm{prf}}_{f}(t,q) + \binom{q}{2} \cdot \left((2\ell-1) \cdot \mathrm{Adv}^{\mathrm{prf}}_{f}(t',2) + \frac{1}{2^n}\right)^2$$

$$\leq \mathrm{Adv}^{\mathrm{prf}}_{f}(t,q) + \frac{q^2}{2} \cdot \left(2\ell \cdot \mathrm{Adv}^{\mathrm{prf}}_{f}(t',2) + 2^{-n}\right)^2$$

$$= \mathrm{Adv}^{\mathrm{prf}}_{f}(t,q) + 2 \cdot q^2\left(\ell \cdot \mathrm{Adv}^{\mathrm{prf}}_{f}(t',2) + 2^{-n-1}\right)^2,$$

where $\ell = \lceil \mu/m \rceil$ and $t' = 4\ell \cdot \mathrm{time}(f)$. $\qquad\square$

5 L-Lane NMAC ($L \geq 3$)

A disadvantage of Two-Lane NMAC is its performance. It requires about twice as many invocations to f as the conventional NMAC does. This becomes problematic especially with long messages. We resolve this problem by increasing the number of lanes from 2 to $L \geq 3$ but making each lane $1/(L-1)$-times shorter at the same time. See Fig. 3 for the case $L = 3$.

Description of L-Lane NMAC. In L-Lane NMAC, each component Φ, H or G is instantiated in the following ways:

1. Instantiation of DIL via Parity Code. Intuitively, this DIL works as follows: a message M is divided into $(L-1)$-many pieces $\varphi_1(M)$, ..., $\varphi_{L-1}(M)$, each

being about $|M|/(L-1)$ bits. Then the checksum $\varphi_L(M) = \bigoplus_{i=1}^{L-1} \varphi_i(M)$ is constructed. More precisely, the DIL $\Phi : \{0,1\}^{m*} \rightarrow (\{0,1\}^{m*})^L$ is constructed as follows:

Function $\Phi(M)$

> Divide $M = M_1 M_2 \cdots M_\ell$ with $M_i \in \{0,1\}^m$
>
> Define $M_j^{(i)} \stackrel{\text{def}}{=} M_{i+(j-1)(L-1)}$ for $1 \le i \le L-1$, $i + (j-1)(L-1) \le \ell$
>
> Define $\lambda(i) \stackrel{\text{def}}{=} \lfloor (\ell - i)/(L-1) \rfloor + 1$ so that j runs $1 \le j \le \lambda(i)$
>
> Define $\varphi_i(M) \stackrel{\text{def}}{=} M_1^{(i)} M_2^{(i)} \cdots M_{\lambda(i)}^{(i)}$ for $1 \le i \le L-1$
>
> Define $\varphi_L(M) \stackrel{\text{def}}{=} C_1 C_2 \cdots C_{\lambda(1)}$ where $C_j \stackrel{\text{def}}{=} \bigoplus_{i=1}^{L-1} M_j^{(i)}$
>
> Output $(\varphi_1(M), \ldots, \varphi_{L-1}(M), \varphi_L(M))$.

A problem of the DIL Φ above is that its domain is restricted to $\{0,1\}^{m*}$. It can be resolved by applying the canonical padding 10^* to message M before inputting it to Φ. Observe that the composition of a DIL and a one-to-one padding is again a DIL.

2. cAU $\boldsymbol{H}$ via Merkle-Damgård. There is nothing new in the construction of $\boldsymbol{H}$; the cAU function $\boldsymbol{H}$ is constructed from the compression function f, via the Merkle-Damgård iteration, as in Two-Lane NMAC. Also, note that we do not have to deal with padding here, for it is taken care of by the DIL Φ.

3. PRF $\boldsymbol{G}$ from f or Two-Lane NMAC. The construction varies depending on two cases:

   - If $m \ge Ln$, then we can use f_k in place of $\boldsymbol{G}_{k'}$, just as in Two-Lane NMAC.

   - If $m < Ln$, then we cannot use f_k as is anymore. Making the condition $m \ge Ln$ an assumption (as we do for $m \ge 2n$) is not realistic. This obstacle is solved by building $\boldsymbol{G}_{k'}$ via Two-Lane NMAC from f. This method is feasible, because the resulting scheme is still birthday-resistant, and the performance decline is not sharp, for the Two-Lane NMAC $\boldsymbol{G}$ processes only Ln-bit data. Note, though, that now $\boldsymbol{G}$ requires three keys internally (not just the single key k'.)

Security of L-Lane NMAC. Our upper-bound security result of L-Lane NMAC is essentially the same as that of Two-Lane NMAC. We just need to verify that the function family Φ via parity code is indeed doubly injective:

Lemma 4. *The function family $\Phi = \{\varphi_i\}_{1 \le i \le L}$ constructed above via parity code is doubly injective.*

Proof. Let $M, M' \in \{0,1\}^{m*}$. Suppose we know that the condition $\varphi_i(M) = \varphi_i(M')$ holds for all $i = 1, \ldots, L$ except for some $\alpha \in [1, L]$ (i.e., we do not know if the condition $\varphi_\alpha(M) = \varphi_\alpha(M')$ holds). We show that $\varphi_\alpha(M) = \varphi_\alpha(M')$ indeed holds, which implies $M = M'$. For, write $\varphi_i(M) = M_1^{(i)} \cdots M_{\lambda(i)}^{(i)}$ and $\varphi_i(M') = M_1^{'(i)} \cdots M_{\lambda'(i)}^{'(i)}$. It can be directly verified that $\lambda(\alpha) = \lambda'(\alpha)$. Also, we have $M_j^{(\alpha)} = \bigoplus_{i \ne \alpha} M_j^{(i)} = \bigoplus_{i \ne \alpha} M_j^{'(i)} = M_j^{'(\alpha)}$ for all $j = 1, \ldots, \lambda(\alpha)$. Therefore we obtain $\varphi_\alpha(M) = \varphi_\alpha(M')$, as desired. $\square$

Therefore, just as in Two-Lane NMAC, we obtain

$$\mathrm{Adv}^{\mathrm{prf}}_{L\text{-Lane NMAC}}(t, q, \mu) \leq \mathrm{Adv}^{\mathrm{prf}}_{G}(t, q) + 2q^2 \left(\lambda \cdot \mathrm{Adv}^{\mathrm{prf}}_{f}(t', 2) + 2^{-n-1} \right)^2 ,$$

where $\lambda = \lceil \mu/(m(L-1)) \rceil$ and $t' = 4\lambda \cdot \mathrm{time}(f)$. If $m \geq Ln$, then we use f in place of G, so in the above formula we also replace "G" with "f." On the other hand, if $m < Ln$, then we use Two-Lane NMAC in place of G, which takes Ln-bit data as its input. Consequently, in such a case we have

$$\mathrm{Adv}^{\mathrm{prf}}_{G}(t, q) = \mathrm{Adv}^{\mathrm{prf}}_{\text{Two-Lane NMAC}}(t, q, Ln)$$

$$\leq \mathrm{Adv}^{\mathrm{prf}}_{f}(t, q) + 2q^2 \left(\lambda' \cdot \mathrm{Adv}^{\mathrm{prf}}_{f}(t', 2) + 2^{-n-1} \right)^2 ,$$

where $\lambda' = \lceil Ln/m \rceil$ and t' the same as before. Overall, the upper bound security of L-Lane NMAC $(m < Ln)$ is given by

$$\mathrm{Adv}^{\mathrm{prf}}_{L\text{-Lane NMAC}}(t, q, \mu) \leq \mathrm{Adv}^{\mathrm{prf}}_{f}(t, q) + 4q^2 \left(\lambda \cdot \mathrm{Adv}^{\mathrm{prf}}_{f}(t', 2) + 2^{-n-1} \right)^2 ,$$

where $\lambda = \max \left\{ \lceil \mu/(m(L-1)) \rceil, \lceil Ln/m \rceil \right\}$ and t' the same as before. So in all the cases the construction is birthday-resistant.

6 L-Lane HMAC

L-Lane NMAC significantly improves performance over Two-Lane NMAC but still has two problematic features. One is that it requires too many keys (either $L + 1$ or $L + 3$). To manage so many keys (even when $L = 2, 3$) is usually troublesome and unwelcome in practical cryptographic applications. The other problem is that hash functions are most often available in software libraries as a form of keyless Merkle-Damgård style with a fixed IV, disabling us from keying directly the compression function via its chaining variable.

In this section we introduce slightly modified version we call L-Lane HMAC, in which the above two problems are resolved. Also, L-Lane HMAC allows hash functions to be implemented with so called the Merkle-Damgård strengthening, and we analyze how this padding method does (not) affect the security.

Single Key. A non-desirable feature of L-Lane NMAC is that it utilizes $L + 1$ (or $L + 3$) keys. These keys can be derived from a single key $k \in \{0, 1\}^n$, as

$$k_i \stackrel{\mathrm{def}}{=} f(\mathsf{IV}\|k\| \langle i \rangle_{m-n})$$

for $i = 1, \ldots, L + 1$ (or $L + 3$). Here $\mathsf{IV} \in \{0, 1\}^n$ denotes the initial vector defined by an implementation of hash function, and we are assuming the pseudo-randomness of the function $f^*_k : \{0, 1\}^{m-n} \to \{0, 1\}^n$ defined by $f^*_k(x) \stackrel{\mathrm{def}}{=} f(\mathsf{IV}\|k\|x)$. This would add a term $\mathrm{Adv}^{\mathrm{prf}}_{f^*}(t, L + 1)$ (or $\mathrm{Adv}^{\mathrm{prf}}_{f^*}(t, L + 3)$) to the upper bound formula. Clearly this does not affect the birthday-attack resistance.

Merkle-Damgård Implementation. Hash functions are usually implemented with a padding function of the form $10^* \| \langle|\mu|\rangle$, called the Merkle-Damgård strengthening. This would introduce L-many extra blocks of computation (one in each lane). The reduction proof still works through, because the Merkle-Damgård strengthening can be incorporated into cAU function H (to create another cAU function.) It would only affect the coefficient λ (increase by 1.)

Also, the very last invocation of f (that outputs the tag) may have to be iterated via the Merkle-Damgård iteration due to this padding, if there is not enough room for padding in $\{0,1\}^{m-2n}$. However, since the input to the last block is just (part of) padding, it would still be birthday-resistant. See [18] for an analysis of this type of construction and its security.

Security of L-Lane HMAC. The above modifications do not lose a formal proof of security. For example, consider the case G is just f and there is enough room in $\{0,1\}^{m-2n}$ for the Merkle-Damgård strengthening. Then, an upper bound of such L-Lane HMAC is given by

$$\mathrm{Adv}^{\mathrm{prf}}_{L\text{-Lane HMAC}}(t,q,\mu) \leq \mathrm{Adv}^{\mathrm{prf}}_f(t,q) + 2q^2 \left(\lambda \cdot \mathrm{Adv}^{\mathrm{prf}}_f(t',2) + 2^{-n-1}\right)^2$$
$$+ \mathrm{Adv}^{\mathrm{prf}}_{f*}(t, L+1),$$

where $\lambda = \lceil \mu/m \rceil + 1$ and t' the same as before. This still guarantees that such L-Lane HMAC is birthday-resistant. Also, upper bounds for other cases can be obtained similarly.

7   Performance Issues

The problem of multilane construction is its performance. In this section we briefly discuss issues upon implementing L-Lane NMAC/HMAC.

Choosing Optimal Value of L. Theoretically, increasing the value of L would decrease the number of invocations to f, especially with long messages. However, increasing value L could cause various drawbacks.

One is inefficiency with short messages. In L-Lane HMAC, L-many lanes are invoked regardless of the size of a message. Also, the number of invocations to f in Two-Lane NMAC G would become consuming (Note that with such a large L the component G is most likely built via Two-Lane NMAC.)

Another is that increasing L would require more registers in its implementation. Since current CPUs are equipped with a limited number of registers, value L should be chosen accordingly.

The value $L = 3$ (150%) seems to be adequate for many purposes. The value $L = 4$ (133%) may fit in some scenarios, but increasing the value more than $L \geq 5$ would probably cause more troubles than benefit. Plus, the (theoretical) increase in performance becomes relatively small with $L \geq 5$ (125%, 120%, ...)

Table 1. Parallel implementation of hash functions [12,11]

	par.	cycles/byte	rel.		par.	cycles/byte	rel.
	1	5.53	100%		1	23.73	100%
MD5	2	4.31	78%	SHA-256	2	20.59	87%
	3	3.66	66%		3	22.14	93%
	1	9.73	100%		1	36.5	100%
SHA-1	2	8.30	85%	SHA-512	2	22.1	61%
	3	8.73	90%				

Parallelism. Table 1 lists figures from previous results of implementing hash functions, with Pentium III (for MD5, SHA-1 and SHA-256) [12] and with Pentium 4 (for SHA-512) [11], showing the performance gain by parallel implementation. From these numbers we expect that it should be feasible to suppress the increase of computational cost in L-Lane NMAC/HMAC, especially when $L = 2, 3$.

8   Concluding Comments

In this paper we introduce a new paradigm we call DIL-cAU-PRF construction that is quite effective to accomplish resistance to birthday attacks. By its structure and security reduction, it also makes collision attacks difficult, thus providing a failure-friendly design. Extra cost to pay is the decrease in performance, but it is significantly small as compared to previous constructs.

Acknowledgments. The author would like to express his gratitude to the INDOCRYPT 2007 referees for their valuable comments. The author is also grateful to Kazumaro Aoki for having discussions on material for writing Sect. 7.

References

1. Aiello, W., Venkatesan, R.: Foiling birthday attacks in length-doubling transformations — Benes: A non-reversible alternative to Feistel. In: Maurer, U.M. (ed.) EUROCRYPT 1996. LNCS, vol. 1070, pp. 307–320. Springer, Heidelberg (1996)
2. Bellare, M.: New proofs for NMAC and HMAC: Security without collision-resistance. In: Dwork, C. (ed.) CRYPTO 2006. LNCS, vol. 4117, pp. 602–619. Springer, Heidelberg (2006)
3. Bellare, M., Canetti, R., Krawczyk, H.: Keying hash functions for message authentication. In: Koblitz, N. (ed.) CRYPTO 1996. LNCS, vol. 1109, pp. 1–15. Springer, Heidelberg (1996)
4. Bellare, M., Goldreich, O., Krawczyk, H.: Stateless evaluation of pseudorandom functions: Security beyond the birthday barrier. In: Wiener, M.J. (ed.) CRYPTO 1999. LNCS, vol. 1666, pp. 270–287. Springer, Heidelberg (1999)

5. Bellare, M., Guérin, R., Rogaway, P.: XOR MACs: New methods for message authentication using finite pseudorandom functions. In: Coppersmith, D. (ed.) CRYPTO 1995. LNCS, vol. 963, pp. 15–28. Springer, Heidelberg (1995)
6. Contini, S., Yin, Y.L.: Forgery and partial key-recovery attacks on HMAC and NMAC using hash collisions. In: Lai, X., Chen, K. (eds.) ASIACRYPT 2006. LNCS, vol. 4284, pp. 37–53. Springer, Heidelberg (2006)
7. den Boer, B., Rompay, B.V., Preneel, B., Vandewalle, J.: New (two-track-)MAC based on the two trails of RIPEMD. In: Vaudenay, S., Youssef, A.M. (eds.) SAC 2001. LNCS, vol. 2259, pp. 314–324. Springer, Heidelberg (2001)
8. Jaulmes, É., Joux, A., Valette, F.: On the security of randomized CBC-MAC beyond the birthday paradox limit: A new construction. In: Daemen, J., Rijmen, V. (eds.) FSE 2002. LNCS, vol. 2365, pp. 237–251. Springer, Heidelberg (2002)
9. Kim, J., Biryukov, A., Preneel, B., Hong, S.: On the security of HMAC and NMAC based on HAVAL, MD4, MD5, SHA-0 and SHA-1. In: De Prisco, R., Yung, M. (eds.) SCN 2006. LNCS, vol. 4116, pp. 242–256. Springer, Heidelberg (2006)
10. Lucks, S.: A failure-friendly design principle for hash functions. In: Roy, B. (ed.) ASIACRYPT 2005. LNCS, vol. 3788, pp. 474–494. Springer, Heidelberg (2005)
11. Matsui, M., Fukuda, S.: How to maximize software performance of symmetric primitives on Pentium III and 4 processors. In: Gilbert, H., Handschuh, H. (eds.) FSE 2005. LNCS, vol. 3557, pp. 398–412. Springer, Heidelberg (2005)
12. Nakajima, J., Matsui, M.: Performance analysis and parallel implementation of dedicated hash functions. In: Knudsen, L.R. (ed.) EUROCRYPT 2002. LNCS, vol. 2332, pp. 165–180. Springer, Heidelberg (2002)
13. Patarin, J.: Improved security bounds for pseudorandom permutations. In: ACM Conference on Computer and Communications Security, pp. 142–150 (1997)
14. Patarin, J.: About Feistel schemes with six (or more) rounds. In: Vaudenay, S. (ed.) FSE 1998. LNCS, vol. 1372, pp. 103–121. Springer, Heidelberg (1998)
15. Preneel, B., van Oorschot, P.C.: On the security of iterated message authentication codes. IEEE Transactions on Information Theory 45(1), 188–199 (1999)
16. Speirs, W.R., Molloy, I.: Making large hash functions from small compression functions. Cryptology ePrint Archive, 2007/239 (2007)
17. Wegman, M.N., Carter, L.: New hash functions and their use in authentication and set equality. J. Comput. Syst. Sci. 22(3), 265–279 (1981)
18. Yasuda, K.: "Sandwich" is indeed secure: How to authenticate a message with just one hashing. In: Pieprzyk, J., Ghodosi, H., Dawson, E. (eds.) ACISP 2007. LNCS, vol. 4586, pp. 355–369. Springer, Heidelberg (2007)

On the Bits of Elliptic Curve
Diffie-Hellman Keys

David Jao[1], Dimitar Jetchev[2], and Ramarathnam Venkatesan[3,4]

[1] University of Waterloo, Waterloo ON N2L3G1, Canada
djao@math.uwaterloo.ca
[2] Dept. of Mathematics, University of California at Berkeley, Berkeley, CA 94720
jetchev@math.berkeley.edu
[3] Microsoft Research India Private Limited, "Scientia", No:196/36,
2nd Main Road, Sadashivnagar, Bangalore – 560080, India
[4] Microsoft Research, 1 Microsoft Way, Redmond WA 98052
venkie@microsoft.com

Abstract. We study the security of elliptic curve Diffie-Hellman secret keys in the presence of oracles that provide partial information on the value of the key. Unlike the corresponding problem for finite fields, little is known about this problem, and in the case of elliptic curves the difficulty of representing large point multiplications in an algebraic manner leads to new obstacles that are not present in the case of finite fields. To circumvent this obstruction, we introduce a *small* multiplier version of the hidden number problem, and we use its properties to analyze the security of certain Diffie-Hellman bits. We suggest new character sum conjectures that guarantee the uniqueness of solutions to the hidden number problem, and provide some evidence in support of the conjectures by showing that they hold on average in certain cases. We also present a Gröbner basis algorithm for solving the hidden number problem and recovering the Diffie-Hellman secret key when the elliptic curve is defined over a constant degree extension field and the oracle is a coordinate function in the polynomial basis.

1   Introduction

The Diffie-Hellman scheme is a fundamental protocol for public key exchange between two parties. Its original definition over finite fields is based on the hardness of computing the map $g, g^a, g^b \mapsto g^{ab}$ for $g \in \mathbb{F}_p^*$, while its elliptic curve analogue depends on the difficulty of computing $P, aP, bP \mapsto abP$ for points P on an elliptic curve.

A natural question in this context is whether an adversary can compute some *partial information* about g^{ab} (resp. abP) for the finite field (resp. the elliptic curve) case. In studying this problem for the finite field case, Boneh and Venkatesan [4] formulated the *hidden number problem* (HNP) and showed that a solution to the HNP allows one to reduce the question of computing partial information to the question of computing the key itself (see also [24,15]). For example, using these techniques one can show that computing $\mathrm{MSB}_k(g^{ab})$ is tantamount to

K. Srinathan, C. Pandu Rangan, M. Yung (Eds.): Indocrypt 2007, LNCS 4859, pp. 33–47, 2007.

computing g^{ab} itself for $k \geq 5\sqrt{\log p}$. In addition, the hidden number problem has turned out to be of cryptanalytic interest in its own right. For attacks on cryptosystems using partial information, see [20,23,21,16,24,17,22]. Thus an important motivation for the problem we consider is to find elliptic curve analogues of these attacks.

It is natural to ask the analogous question for elliptic curve Diffie-Hellman bits, namely, can we prove that partial information about elliptic curve Diffie-Hellman keys over a fixed curve E is unpredictable if we assume that the Diffie-Hellman problem for E is hard? Unfortunately, very little is known about this question. If one is allowed to look for a related curve with a hard Diffie-Hellman problem, then Boneh and Shparlinksi [3] provide an affirmative answer. While having formal proofs of the security of Diffie-Hellman bits is the most important application, it is also desirable from a cryptanalytic point of view to have practical algorithms for solving the corresponding hidden number problem (defined in Section 2.1). However, there are two fundamental obstructions which render the question much more difficult in the case of elliptic curves.

In the finite field case, one views elements of $\mathbb{F}_p$ as integers, embeds them in lattices equipped with the Euclidean metric and applies lattice reduction algorithms. In the elliptic curve case, no useful metrics are available; this represents the first fundamental obstruction. Furthermore, point multiplication on elliptic curves transforms the coordinates of a point via rational polynomials whose degrees grow exponentially in the size of the multiplier. This means that in general one can only write down explicit algebraic expressions in the case of small multipliers. This complexity constraint introduces the second fundamental obstruction—it is not even clear if the hidden number problem has a unique solution at all when the random multipliers are constrained to lie within small intervals. (By contrast, if one is allowed to use arbitrary multipliers, it is very easy to establish uniqueness in both the finite field and elliptic curve cases.) To deal with this obstruction, we introduce new character sums, conjecture some non-trivial estimates which are sufficient to prove uniqueness, and prove that our conjecture holds on average in the case of quadratic residuosity of the x coordinate. We also prove an upper bound on the number of solutions for any uniformly distributed output function, under the assumption of the Generalized Riemann Hypothesis. Although this approach falls short of the goal of actually recovering the value of abP via partial information, we feel that it remains a valuable first step given the lack of other results in this area.

We present a complete recovery algorithm for the hidden number problem in the case of curves over constant degree extensions, using Gröbner bases and elimination ideals. At present we are only able to implement our solution using oracles that provide outputs of length approximately $1/3$ that of the (compressed) input point itself, e.g. 50 bits of output in the case of a 160-bit base field. Given recent progress and widespread interest in Gröbner bases algorithms, we may in the near future be able to recover Diffie-Hellman keys using less information (e.g. 32 bits of output for a 160-bit base field). However, obtaining results comparable to the

the finite field case (where we output $O(\sqrt{\log p})$ bits) seems to be fundamentally out of reach., and in certain cases is even presumed to be infeasible (see [2]).

2   Preliminaries

Let $q = p^k$ where p is a prime. We view $\mathbb{F}_q$ as a vector space over $\mathbb{F}_p$ and identify $\mathbb{F}_q$ with $\mathbb{F}_p^k$ using a polynomial basis. For a point P on an elliptic curve E over $\mathbb{F}_q$, let $x(P)$ and $y(P)$ be the x and y-coordinates of P, respectively, and let $x_0(P)$ denote the first coordinate in the vector representation of $x(P)$.

2.1   Partial Diffie-Hellman Bits

To extract partial information about points on elliptic curves, we consider a map $\mathbf{Bits}_\ell : E(\mathbb{F}_q) \rightarrow \{0,1\}^\ell$ which will assume one of the following three types:

1. Algebraic: $\mathbf{Bits}_\ell(P) = x_0(P)$;
2. Analytic: $\mathbf{Bits}_\ell(P) = \chi(x(P))$ for a suitable character $\chi : \mathbb{F}_q^\times \rightarrow \mathbb{C}^\times$;
3. MSB: $\mathbf{Bits}_\ell(P) = \mathrm{MSB}_\ell(x(P))$, which is the ℓ most significant bits of $x(P)$ expressed in binary.

Given a point $P \in E(\mathbb{F}_q)$ and two multiples aP and bP, let

$$\mathrm{PDH}_E(P, aP, bP) = \mathbf{Bits}_\ell(abP).$$

To study the security of the function PDH_E, we assume that there is a hidden point Q on E and an oracle $\mathcal{A}$ to compute the function $r \mapsto \mathbf{Bits}_\ell(rQ)$. We refer to r as the *multiplier*. One can then state the general *Multiplier Elliptic Curve Hidden Number Problem* (M-EC-HNP).

Multiplier-EC-Hidden-Number-Problem: Given an oracle $\mathcal{A}$ to compute the map $r \mapsto \mathbf{Bits}_\ell(rQ)$, recover the point Q.

Here the value of r may be chosen either by an adversary or randomly. In our setting, one queries the oracle many times so that one gets a total of $ck \log p$ output bits, for some $c > 1$. The hidden number problem is related to the problem of showing that PDH_E is secure, because a solution to the hidden number problem allows an adversary to determine abP given an oracle for PDH_E.

To solve the M-EC-HNP problem, one needs to address the following two questions:

Uniqueness: Is the underlying solution unique? If not, can the solutions at least be narrowed down to a small list?

Reconstruction: Is there an efficient algorithm to solve M-EC-HNP?

In most cases, one can easily show uniqueness if the queries are allowed to use large multipliers r. Unfortunately, these multipliers lead to division polynomials of exponentially large degree applied to Q, which cannot be handled using the techniques of Section 3. For this reason, any reconstruction algorithm based on these methods will be limited to small multipliers r with $r < O((\log p)^d)$. By contrast, large multipliers for the analogous HNP over $\mathbb{F}_p$ pose no critical problems, and this difference represents a fundamental new restriction in the elliptic curve context. To analyze the statistical behavior of the output values for general oracles, we can apply the techniques of [18] which make use of the Generalized Riemann Hypothesis (see Section 5). However, these methods turn out to be insufficient for establishing uniqueness. To show uniqueness for the analytic case (only), we present a new character sum conjecture (and supporting evidence). Note that our algebraic map is significantly different from the finite field trace map used for bit extraction (see [13]) because the multipliers act via rational polynomial functions on the hidden point.

In the finite field case, the statistical properties and the pseudorandom number generators can be studied via estimates of character sums over *large* intervals (see [9,10,6,7,8,12]).

Remark 2.1. In the finite field case, one can use metrics and the LLL lattice reduction algorithm (see [19]) to reconstruct the secret efficiently (see [4] and [5]). However, without such metrics, there are no analogous reconstruction algorithms in the elliptic curve case. Nonetheless, we give a reconstruction algorithm using Gröbner bases algorithms in the algebraic case when k is small (see Section 3). In the analytic case, we use our character sum conjectures mentioned above to link the problem of solving M-EC-HNP to that of decoding certain error-correcting codes (see Section 4.2).

Remark 2.2. A detailed account on the general hidden number problem is given in [25]. The slightly more general hidden number problem for elliptic curves (as discussed in [2]) is the following:

EC-HNP: Let E be an elliptic curve over a finite field $\mathbb{F}_q$. Recover a point $P \in E(\mathbb{F}_q)$ given k pairs $(Q_i, \mathrm{MSB}_\ell(x(P + Q_i)))$ for some $\ell > 0$ and for k points $Q_1, \ldots, Q_k \in E(\mathbb{F}_q)$ chosen independently and at random.

3   Algebraic Case with Low Degree Extensions

In this section we outline an efficient reconstruction algorithm for the elliptic curve hidden number problem in the case of $\mathbf{Bits}_\ell(P) = x_0(P)$ over field extensions of constant degree. Since the technique is more transparent in the case of low degree extensions, we first illustrate the algorithm for degree 2 and degree 3 extensions before addressing the general case. Our method makes use of small multipliers and for this reason is limited to constant degree extensions.

3.1   Elliptic Curves over Finite Field Extensions of Degree 2

Suppose E is an elliptic curve over $\mathbb{F}_{p^2}$ given by a Weierstrass equation $y^2 = x^3 + \alpha x + \beta$ with $\alpha, \beta \in \mathbb{F}_{p^2}$. We will solve the M-EC-HNP in the algebraic case where $\mathbf{Bits}_\ell(P) = x_0(P)$ and $\ell = \lfloor \log_2 p \rfloor$.

Proposition 3.1. *Let $\ell = \lfloor \log_2 p \rfloor$ and $\mathbf{Bits}_\ell(P) = x_0(P)$. There exists an efficient algorithm (polynomial in $\log p$) for solving the M-EC-HNP.*

Proof. Let w be a generator for $\mathbb{F}_{p^2}/\mathbb{F}_p$, where $w^2 = u$ for some non-square element $u \in \mathbb{F}_p^\times$. Let $Q \in E(\mathbb{F}_{p^2})$ be the point which we are about to recover. It suffices to recover $\underline{x} = (x_0, x_1)$. The key ingredient for the proof is the observation that the coordinate $x(2Q)$ is expressible as a rational function purely of the coordinate x. More precisely, we have the point doubling formula [26, III.2.3]

$$x(2Q) = \frac{x^4 - 2\alpha x^3 - 8\beta x - \alpha^2}{4(x^3 + \alpha x + \beta)}.$$

We substitute $x = x_0 + wx_1$ into the right hand side and use $w^2 = u$ to write down

$$x(2Q) = \frac{P_0(x_0, x_1) + wP_1(x_0, x_1)}{Q_0(x_0, x_1) + wQ_1(x_0, x_1)},$$

where $P_0(x_0, x_1)$ and $P_1(x_0, x_1)$ are polynomials defined over $\mathbb{F}_p$ of degrees at most 4 and $Q_0(x_0, x_1)$ and $Q_1(x_0, x_1)$ are rational polynomials of degrees 3. Next, we rationalize the denominators to obtain

$$x(2Q) = \frac{P_0 Q_0 - uP_1 Q_1}{Q_0^2 - uQ_1^2} + w\frac{P_1 Q_0 - P_0 Q_1}{Q_0^2 - uQ_1^2}.$$

If $x(2Q) = (x_0', x_1')$ for some $x_0' \in \mathbb{F}_p$ and $x_1' \in \mathbb{F}_p$ then

$$x_0' = \frac{P_0(x_0, x_1)Q_0(x_0, x_1) - uP_1(x_0, x_1)Q_1(x_0, x_1)}{Q_0^2(x_0, x_1) - uQ_1^2(x_0, x_1)}.$$

This formula provides a way of patching together the partial data. Indeed, x_0 is recovered directly as $x_0 = \text{MSB}_{\lfloor \log_2 p \rfloor}(x(Q))$. One also knows the value of $x_0' = \text{MSB}_{\lfloor \log_2 p \rfloor}(x(2Q))$, so in order to recover x_1 one needs to find a zero over $\mathbb{F}_p$ of the polynomial

$$F(X) =$$
$$P_0(x_0, X)Q_0(x_0, X) - uP_1(x_0, X)Q_1(x_0, X) - x_0'Q_0(x_0, X)^2 - ux_0'Q_1(x_0, X)^2.$$

The explicit formula for P_0, P_1, Q_0, Q_1 show that F has constant degree (independent of E and p) and non-zero leading coefficient. Since we know that the hidden point Q exists, the polynomial must have a solution over $\mathbb{F}_p$. Computing the $\mathbb{F}_p$-roots can be solved in polynomial time using standard algorithms. This solves the M-EC-HNP in this particular case.

Remark 3.1. The solution of the M-EC-HNP in this case implies the security of the Diffie-Hellman bits for algebraic output functions on degree 2 field extensions. Indeed, if $\mathcal{A}$ is an oracle which computes $x_0(abP)$ from an input (P, aP, bP) then solving M-EC-HNP means that one could reconstruct the secret abP.

3.2 Elliptic Curves over Extensions of Degree 3

Let E be an elliptic curve over $\mathbb{F}_{p^3}$ given by a Weierstrass equation

$$E \; : \; y^2 = x^3 + \alpha x + \beta, \; \alpha, \beta \in \mathbb{F}_{p^3}.$$

We will show how to solve efficiently the M-EC-HNP in the algebraic case. The proof will be similar to the previous case of extensions of degree two, except that it will involve more technicalities. In what follows, $x_0(P)$ may be naturally extended by considering $\text{trace}(x(P))$.

Proposition 3.2. *Let* $\ell = \lfloor \log_2 p \rfloor$ *and* $\textbf{Bits}_\ell(P) = x_0(P)$. *There exists an efficient algorithm (polynomial in* $\log p$*) for solving the M-EC-HNP.*

We first fix some choice for representing elements of the finite field. Let w be a generator for the field extension $\mathbb{F}_{p^3}/\mathbb{F}_p$. Without loss of generality (and to avoid some technical difficulties), choose w so that it is a root of an irreducible polynomial (over $\mathbb{F}_p$) whose quadratic term is zero, i.e., $w^3 - uw - v = 0$.

Proof. Let Q be the hidden point which we wish to recover. We write $x(Q) = (x_0, x_1, x_2)$ and $y(Q) = (y_0, y_1, y_2)$. Let $\mathcal{A}$ be an oracle which computes the function $r \mapsto x_0(rQ)$ for any $P \in E(\mathbb{F}_{p^3})$. We make three queries to $\mathcal{A}$ with $P = Q, 2Q$ and $3Q$, respectively. We use the fact that $x(2Q)$ and $x(3Q)$ are both rational functions of $x = x(Q)$. Let

$$x_0(Q) = s_1, \quad x_0(2Q) = s_2, \quad x_0(3Q) = s_3.$$

We will show how to put this information together, so that we can recover a finite (constant in p) list of candidates for the point Q.

The query $x_0(3Q)$. According to [26, Ex.3.7], the multiplication-by-3 map on E is given (as a rational function on the coordinates of Q) by

$$x(3Q) = \frac{\phi_3(x, y)}{\psi_3^2(x, y)},$$

where

$$\psi_3(x, y) = 3x^4 + 6\alpha x^2 + 12\beta x - \alpha^2$$

and

$$\phi_3(x, y) = 8y^2(x^6 + 5\alpha x^4 + 20\beta x^3 - 5\alpha^2 x^2 - 4\alpha\beta x - 8\beta^2 - \alpha^3) =$$
$$= 8(x^3 + \alpha x + \beta)(x^6 + 5\alpha x^4 + 20\beta x^3 - 5\alpha^2 x^2 - 4\alpha\beta x - 8\beta^2 - \alpha^3).$$

Writing $\alpha = \alpha_0 + w\alpha_1 + w^2\alpha_2$, $\beta = \beta_0 + w\beta_1 + w^2\beta_2$ and $x = x_0 + wx_1 + w^2 x_2$ we can express

$$\frac{\phi_3}{\psi_3^2} = \frac{P_0(x_0, x_1, x_2) + wP_1(x_0, x_1, x_2) + w^2 P_2(x_0, x_1, x_2)}{Q_0(x_0, x_1, x_2) + wQ_1(x_0, x_1, x_2) + w^2 Q_2(x_0, x_1, x_2)}, \tag{3.1}$$

where the P_i's and Q_i's are polynomials with coefficients in $\mathbb{F}_p$. The next step is to write the above rational function as

$$\frac{\phi_3}{\psi_3^2} = r_0(x_0, x_1, x_2) + wr_1(x_0, x_1, x_2) + w^2 r_2(x_0, x_1, x_2),$$

where r_i's are rational functions over $\mathbb{F}_p$ which are explicitly computable in terms of α, β and w. To do this, we need to multiply the numerator and denominator of (3.1) by a suitable factor so that the denominator becomes a polynomial in x_0, x_1 and x_2 with coefficients in $\mathbb{F}_p$. Since w is a root of the polynomial $z^3 - uz - v = 0$ defined over $\mathbb{F}_p$ the rationalizing factor will be

$$
\begin{aligned}
F &= (Q_0 + w_1 Q_1 + w_1^2 Q_2)(Q_0 + w_2 Q_1 + w_2^2 Q_2) \\
&= Q_0^2 + Q_0 Q_1 (w_1 + w_2) + Q_0 Q_2 (w_1^2 + w_2^2) + Q_1^2 w_1 w_2 \\
&\quad + Q_1 Q_2 w_1 w_2 (w_1 + w_2) + Q_2^2 w_1^2 w_2^2 \\
&= (Q_0^2 + 2u Q_0 Q_2 + u Q_1^2 + 2v Q_1 Q_2 + u^2 Q_2^2) \\
&\quad + w(-Q_0 Q_1 + 2u Q_1 Q_2 + v Q_2^2) + w^2(-Q_0 Q_2 + Q_1^2 - u Q_1^2),
\end{aligned}
$$

where w_1 and w_2 are the other two roots of the above polynomial of degree 3 and for obtaining the last equality we have used $w + w_1 + w_2 = 0$, $w_1 w_2 w_3 = v$ and $w^3 - uw - v = 0$. Notice that $F\psi_3^2$ is a polynomial in x_0, x_1, x_2 of degree 24 defined over $\mathbb{F}_p$, and $F\phi_3$ (defined over $\mathbb{F}_{p^3}$) has degree 25. Thus, if we write $F\phi_3 = p_0 + w p_1 + w^2 p_2$ where p_i's are polynomials in x_0, x_1, x_2 defined over $\mathbb{F}_p$ then we have $r_i = s_i/(F\psi_3^2)$ and the degree of the denominator of r_0 is at most 25, whereas the degree of its numerator is 24. The query $x_0(3Q) = s_3$ gives us the value of the function $r_0(x_0, x_1, x_2)$ at the triple $(x_0, x_1, x_2) \in \mathbb{F}_p^3$ which we are looking for.

The query $x_0(2Q)$. The point doubling formula reads as

$$
x(2Q) = \frac{x^4 - 2\alpha x^3 - 8\beta x - \alpha^2}{4(x^3 + \alpha x + \beta)}.
$$

Since $\alpha = \alpha_0 + w\alpha_1 + w^2 \alpha_2$, $\beta = \beta_0 + w\beta_1 + w^2 \beta_2$ and $x = x_0 + wx_1 + w^2 x_2$, we can express

$$
x(2Q) = \frac{R_0(x_0, x_1, x_2) + w R_1(x_0, x_1, x_2) + w^2 R_2(x_0, x_1, x_2)}{T_0(x_0, x_1, x_2) + w T_1(x_0, x_1, x_2) + w^2 T_2(x_0, x_1, x_2)}.
$$

As in the case of multiplication by 3, we rationalize the above function by multiplying the numerator and denominator by

$$
\begin{aligned}
F &= (T_0^2 + 2u T_0 T_2 + u T_1^2 + 2v T_1 T_2 + u^2 T_2^2) + \\
&\quad + w(-T_0 T_1 + 2u T_1 T_2 + v T_2^2) + w^2(-T_0 Q_2 + T_1^2 - u T_2^2)
\end{aligned}
$$

and write it in the form

$$
q_0(x_0, x_1, x_2) + w q_1(x_0, x_1, x_2) + w^2 q_2(x_0, x_1, x_2),
$$

where the q_i are $\mathbb{F}_p$-rational functions whose denominators have degree 9 and whose numerators have degree at most 10. As in the previous case, the query $x_0(2Q) = s_2$ gives us the value of the function $q_0(x_0, x_1, x_2)$ at the triple $(x_0, x_1, x_2) \in \mathbb{F}_p^3$.

Recovering $x(Q)$. We recover $x_0 = s_1$ from the query $x_0(Q) = s_1$. The query $x_0(2Q) = s_2$ gives us a polynomial relation $G(x_1, x_2) = 0$ over $\mathbb{F}_p$ between the (yet) unknown x_1 and x_2 coming from

$$q_0(s_1, x_1, x_2) = s_2.$$

Note that the degree of G is at most 10. Similarly, the query $x_0(3Q)$ gives us a polynomial relation $H(x_1, x_2) = 0$ over $\mathbb{F}_p$ coming from

$$r_0(s_1, x_1, x_2) = s_3$$

of degree at most 25. Thus, (x_1, x_2) is a simultaneous solution over $\mathbb{F}_p$ of G and H. We determine the solutions by taking the resultant $\mathrm{Res}(G, H)$, which is a polynomial in a single variable of degree at most 250. Since we are only interested in the $\mathbb{F}_p$-solutions, it suffices to factor the resultant over $\mathbb{F}_p$ and look up the linear factors. Thus, we obtain a finite (constant in p) set of possible solutions (x_1, x_2), which proves the proposition.

3.3   Elliptic Curves over $\mathbb{F}_q$

Let $q = p^k$. Let E be an elliptic curve over $\mathbb{F}_q$ given by a Weierstrass equation

$$E \; : \; y^2 = x^3 + \alpha x + \beta, \; \alpha, \beta \in \mathbb{F}_q.$$

We will describe an algorithm to solve the M-EC-HNP for $\ell = \lfloor \log_2 p \rfloor$ and $\mathbf{Bits}_\ell(P) = x_0(P)$ which will generalize the previous cases of extensions of degrees two and three.

Let w be a generator for the field extension $\mathbb{F}_q / \mathbb{F}_p$ and let

$$f(z) = z^k - u_1 z^{k-1} - \cdots - u_k$$

be the minimal polynomial for w over $\mathbb{F}_p$. As before, suppose that we have an oracle $\mathcal{A}$ which computes $x_0(P')$ given P', aP', bP'. Our goal is to recover $x(Q)$ given

$$\langle x_0(mQ) \; : \; m = 1, 2, \ldots, k \rangle.$$

Let $x = x(Q) = x_0 + wx_1 + \cdots + w^{k-1}x_{k-1}$. The main idea is to interpret the above data as a system of polynomial equations with coefficients in $\mathbb{F}_p$ and degrees bounded independently of $\log p$, and to use a Gröbner basis algorithm to solve the system and thereby compute $x_0, x_1, \ldots, x_{k-1}$. To compute the equations in the system, we use the division polynomials from [26, §III, Ex.3.7] to find $x(mQ) = \varphi_m(Q) / \psi_m(Q)^2$. Next, we observe that φ_m / ψ_m^2 is a rational function on x defined over $\mathbb{F}_q$, and we write it (after rationalizing the denominators) as

$$r_0^{(m)}(x_0, x_1, \ldots, x_{k-1}) + wr_1^{(m)}(x_0, x_1, \ldots, x_{k-1}) + \cdots$$
$$+ w^{k-1} r_{k-1}^{(m)}(x_0, x_1, \ldots, x_{k-1}),$$

where $r_i^{(m)}(x_0, x_1, \ldots, x_{k-1})$ are rational functions defined over $\mathbb{F}_p$ whose denominators have degrees $k \deg(\psi_m^2) = 2k(m^2 - 1)$ and whose numerators have degrees $m^2 + (k-1)(m^2 - 1) = 2k(m^2 - 1) + 1$ (here, we are using that the single-variate polynomial ψ_m has degree $m^2 - 1$ and ϕ_m has degree m^2). Next, if $x_0(mQ) = s_m$ then we obtain the equation

$$r_0^{(m)}(x_0, x_1, \ldots, x_{k-1}) = s_m, \ \forall m = 1, \ldots, k,$$

which gives us a polynomial equation over $\mathbb{F}_p$

$$g_m(x_0, x_1, \ldots, x_{k-1}) = 0,$$

whose degree is bounded by $2k(m^2 - 1) + 1$. We then use a Gröbner basis algorithm to try to compute a Gröbner basis for the ideal

$$I = \langle g_1, \ldots, g_k \rangle \subset \mathbb{F}_p[x_0, \ldots, x_{k-1}],$$

which will allow us to solve for $x_0, x_1, \ldots, x_{k-1}$.

Wide practical interest in Gröbner bases and continued improvements in algorithmic implementations have pushed the limits of what can be solved by these systems. For a 160-bit EC system it may soon be possible to solve the problem using outputs of $\ell \in [16, 32]$ bits per iteration (for example, by using a degree ten extension, and outputting one co-ordinate of $x(P)$).

4   Analytic Case

We now consider the case of an output function which is equal to a group character (such as the quadratic residuosity character). Let $\chi \colon \mathbb{F}_q^\times \to \mathbb{C}^\times$ be a nontrivial character of the multiplicative group $\mathbb{F}_q^\times$. Given a point P, we will look at the values $\chi(x(P))$ defined by the character χ. For completeness, if $x(P) = 0$ we set $\chi(0) = 0$.

4.1   Our Conjecture on Character Sums

Our conjecture is the following:

Conjecture 4.1. Let $P, P' \in E(\mathbb{F}_q)$ be two points, such that $x(P) \neq x(P')$. There exists $\varepsilon > 0$, such that for every $B = \Omega((\log q)^2)$,

$$\left| \sum_{r \leq B} \chi(x(rP))\chi(x(rP')) \right| = O(B^{1-\varepsilon}).$$

Although this bound suffices for our needs, the actual bound may be closer to $B^{0.5}$. Note that classical character sums over finite fields traditionally take the form $\sum_{x \in \mathbb{F}_q} \chi(f(x))$ for some polynomial $f(x)$. General character sums over $\mathbb{F}_q$ have been considered by Deligne in [11], but very little is known over short intervals. Viewed as a sum over some function field, the above sum does not include all polynomials of small degree, but only those that correspond to the map $P \mapsto rP$.

We obtain the following immediate corollary of the above conjecture.

Corollary 4.2. *Assuming Conjecture 4.1, let P and P' be two points in $E(\mathbb{F}_q)$ with $x(P) \neq x(P')$ and let χ be the quadratic character. Let $B = \Omega((\log q)^2)$ For randomly a chosen $r \in \{1, \ldots, B\}$*

$$\left| \mathrm{Prob}_r \left[\chi(x(rP)) \neq \chi(x(rP')) \right] - \frac{1}{2} \right| = O(B^{-\varepsilon}).$$

Remark 4.1. The purpose of this conjecture is to show that knowing enough of the values of partial bits (namely, character values) of $x(r_iP)$ for small multipliers suffices to uniquely identify the point. If we assume Conjecture 4.1, and choose $P, P' \in E(\mathbb{F}_q)$ such that $x(P) \neq x(P')$, then for B and χ as above and random integers $r_1, \ldots, r_t \in \{1, B\}$, the values $\{\chi(x(r_iP))\}_{i=1}^t$ and $\{\chi(x(r_iP'))\}_{i=1}^t$ will be distinct with high probability.

4.2   Relationship to an Error-Correcting Code

Proofs of many hard-core bit theorems involve problems related to error correcting codes, and the hard core property of the bit is equivalent to finding efficient decoding algorithms for certain codes (see [1]). In our case this connection exists as well, but it is unclear if the code admits an efficient decoding algorithm. However, if decoding should turn out to be intractable then the code may be of independent interest in cryptography. Therefore, it seems worthwhile to mention the resulting code here.

We define a binary code that uses small multiples of points on elliptic curves for encoding. We fix a finite field $\mathbb{F}_q$ and a bound $B \leq O(\log q)^2$. Our choice of code corresponds to a selection of a random sequence $\mathbf{c} = c_1, \ldots, c_t$ with $1 \leq c_i \leq B$ and a character $\chi \colon \mathbb{F}_q^\times \to \{\pm 1\}$. The parameter t will be the length of the code words, and $\mathbb{F}_q$ will roughly correspond to the message space in the following manner. Given an easily invertible map $m \mapsto P$ to map messages into points on elliptic curves, our algorithm to encode m works as follows: let P be its image on the curve and let $P, P_1, \ldots, P_t$ be the sequence of nodes visited by the walk specified by $\mathbf{c}$. Our encoding of m is the sequence $\chi(x(P_1)), \ldots, \chi(x(P_t))$.

Our character sum assumptions imply that the minimum distance of the code is $(\frac{1}{2} - \varepsilon)t$. It is clear that any decoding algorithm that maps an uncorrupted codeword into the point P can be used to solve the hidden number problem using the analytic bit extractor, while correcting a corrupted codeword will yield a proof for the pseudo-randomness of Diffie-Hellman bits.

4.3   Proof of Our Conjecture on Average

We provide some evidence in support of Conjecture 4.1 by showing that the conjecture holds on average for the quadratic character of Corollary 4.2, in the sense that

$$\left| \sum_{P \in E} \sum_{r \leq B} \chi(x(rP))\chi(x(rP')) \right| \leq (\#E) \cdot O(B^{1/2})$$

for a fixed $P' \in E$ when $\chi \colon \mathbb{F}_q^\times \to \{\pm 1\}$ is the quadratic character. We start with the identity

$$\left| \sum_{P \in E} \sum_{r \leq B} \chi(x(rP))\chi(x(rP')) \right| = \left| \sum_{r \leq B} \chi(x(rP')) \right| \cdot \left| \sum_{P \in E} \chi(x(P)) \right|,$$

and we will prove that

$$\left| \sum_{P \in E} \chi(x(P)) \right| = O(\sqrt{\#E}).$$

Clearly this bound is sufficient to finish the proof. To prove it, observe that

$$\left| \sum_{P \in E} \chi(x(P)) \right| = \left| \sum_{x \in \mathbb{F}_q} (1 + \chi(x^3 + \alpha x + \beta))\chi(x) \right| = \left| \sum_{x \in \mathbb{F}_q} \chi(x^4 + \alpha x^2 + \beta x) \right|,$$

where $y^2 = x^3 + \alpha x + \beta$ is the equation for E. Let C denote the curve $y^2 = x^4 + \alpha x^2 + \beta x$. Then C is singular if and only if $\beta = 0$ or $4\alpha^3 + 27\beta^2 = 0$. The latter possibility may be excluded since E is nonsingular. Hence we are left with two cases to consider. If C is nonsingular, then the claim follows from the work of Deligne [11]. On the other hand, if $\beta = 0$, then $\alpha \neq 0$, and

$$\left| \sum_{x \in \mathbb{F}_q} \chi(x^4 + \alpha x^2) \right| = \left| \sum_{x \in \mathbb{F}_q} \chi(x^2 + \alpha) \right|.$$

Now the curve C' given by $y^2 = x^2 + \alpha$ is again nonsingular, so [11] again gives the desired bound.

5   Expander Graphs and Character Sums

In this section we formalize the small multiplier hidden number problem in terms of graph theory by defining a graph whose edges correspond to pairs of points related by small multipliers. This graph is isomorphic (as a graph) to a certain Cayley graph for $(\mathbb{Z}/N\mathbb{Z})^\times$. Under the assumption of the Generalized Riemann Hypothesis, we establish its eigenvalue separation. Our graph is directed, but standard techniques allow us to infer the rapid mixing of directed graphs by analyzing its undirected version. The rapid mixing of the graph implies, on average and with high probability, an upper bound on the number of solutions to the small multiplier hidden number problem, for *any* function **Bits**$_\ell$ whose output values are uniformly distributed over the points of the elliptic curve.

Let $q = p^k$ and assume that the number of points $N = \#E(\mathbb{F}_q)$ is prime (this is a reasonable cryptographic assumption).

5.1   Constructing the Graph G_E and the Subgraph G'

Let $m = O((\log q)^d)$ for $d > 2$ and some sufficiently large (but absolute) implied constant, and S_m be the set of all prime numbers less than or equal to m. Define a directed graph G_E with nodes consisting of the points $Q \in E(\mathbb{F}_q)$ and edges of the form $\{Q, rQ\}$ for every prime $r \in S_m$.

Consider the subgraph G' with vertices consisting of all points $P \neq O_E$. Since N is prime, this graph is isomorphic (only as a graph) to the Cayley graph of $(\mathbb{Z}/N\mathbb{Z})^\times$ with respect to S_m. To establish such an isomorphism, choose a generator Q of $E(\mathbb{F}_q)$ and a primitive element $g \in (\mathbb{Z}/N\mathbb{Z})^\times$ and map each vertex via $sQ \mapsto g^s$ and each edge via $\{Q, sQ\} \mapsto \{g, g^s\}$. Using the arguments of [18], one shows under GRH that the graph G' is $\#S_m$-regular and connected. Specifically, the eigenvalues of the adjacency matrix are the character sums $\lambda_\chi = \sum_{p \in S_m} \chi(\bar{p}) = \sum_{p \leq m} \chi(\bar{p})$, where $\chi \colon (\mathbb{Z}/N\mathbb{Z})^\times \to \mathbb{C}^\times$ varies over the characters of $(\mathbb{Z}/N\mathbb{Z})^\times$ and $\bar{p}$ denotes the image of the prime p in $(\mathbb{Z}/N\mathbb{Z})^\times$. The eigenvector corresponding to the eigenvalue λ_χ is $e_\chi = (\chi(x))_{x \in (\mathbb{Z}/N\mathbb{Z})^\times}$. The largest eigenvalue, corresponding to the trivial character, is $\lambda_{\text{triv}} = \pi(m)$. Hence, to show that G' has good expansion properties, we need an estimate on λ_χ. Such an estimate can be obtained using the methods of [18]. More precisely, under the Generalized Riemann Hypothesis, one can show that $\lambda_\chi < C(N)\sqrt{\pi(m)}$ for some constant $C(N)$ (depending only on N) with $\lim_{N \to \infty} C(N) = 0$, whenever χ is a non-trivial character.

5.2   Distributional Properties

Consider a pseudorandom number generator that initializes P_0 to some random point on E and then performs the following steps:

1. Choose $r_i \in [1, B]$ at random (where $B = O((\log p)^2)$).
2. Set $P_{i+1} = r_i P_i$.
3. Output $\mathbf{Bits}_\ell(x(P_{i+1}))$.

Given a sequence of ℓ-bit strings $h_1, h_2, \ldots, h_L$, what is the probability that the generator will output this sequence? Using the methods of [14], suitably adapted to our situation, we prove the following proposition, which provides a satisfactory bound as long as the second eigenvalue of the normalized adjacency matrix of the graph G' is small (which is the case for large enough N). This bound implies a corresponding upper bound on the number of solutions to the related hidden number problem under the GRH assumption (although it would take some effort to work out what exactly the corresponding bound is).

Proposition 5.1. *Let $h_1, h_2, \ldots, h_L$ be a sequence of values of the function*

$$\mathbf{Bits}_\ell \colon E(\mathbb{F}_q) \setminus \{O_E\} \to \{0,1\}^\ell,$$

such that the sets $F_i := \mathbf{Bits}_\ell^{-1}(h_i)$ have size $\mu_i v$, where $v = \#V(G_E)$. Let A be the normalized adjacency matrix of G', with second largest eigenvalue λ_2. The number of random walks on G' of length L, such that the i-th node in the walk is equal to h_i is bounded by $\prod_{i=1}^{L} M_i$, where $M_i = \sqrt{\mu_i^2 + \lambda_2^2 + 2\mu_i\sqrt{1 - \mu_i\lambda_2}}$.

Proof. Let $\lambda_1 \geq \lambda_2 \geq \cdots \geq \lambda_v$ be the eigenvalues of A. We denote by $e_1, \ldots, e_v$ the corresponding eigenvectors. The eigenvalue $\lambda_1 = 1$ is the trivial eigenvalue and its eigenvector is $e_1 = (1, 1, \ldots, 1)$. Let $V_1 \subset \mathbb{R}^v$ be the subspace spanned by e_1 and $V_2 \subset \mathbb{R}^v$ be the subspace spanned by $e_2, \ldots, e_v$. The spaces V_1 and V_2 are orthogonal to each other and are both preserved by A.

One can then write a given vector $X \in \mathbb{R}^v$ as $X = X_1 + X_2$, where $X_1 \in V_1$ and $X_2 \in V_2$. For each $i = 1, \ldots, L$, denote by P_i the projection operator to the set F_i. In other words, $P_i X$ is the vector $Y \in \mathbb{R}^v$, whose coordinates Y_j for each $1 \leq j \leq v$ are given by $Y_j = X_j$ if the j-th node of G' is a point in F_i and $Y_j = 0$ otherwise.

The proof is based on the observation that if $X = e_1/v$, then the j-th component of the vector $Y = \prod_{i=1}^{L}(PA_i)X$ is exactly the probability that a random walk of length L ends in the j-th node of G' in such a way that for each $i = 1, \ldots, L$ the walk has passed through F_i at the i-th step. Therefore, the probability that a random walk lands in the set F_i at the i-th step is given by

$$P(\text{walk passes through } F_1, \ldots, F_L) = \sum_{j=1}^{v} |Y_j| \leq \sqrt{v} \|Y\| = \sqrt{v} \left\| \prod_{i=1}^{L}(PA_i)X \right\|,$$

where $\|Y\|$ is the L^2-norm of Y, i.e. $\|Y\| = \sqrt{\sum_{i=1}^{v} |Y_i|^2}$.

We will be done if we find an upper bound for $\|P_i AU\|/\|U\|$ for arbitrary vectors $U \in \mathbb{R}^v$ and projection operators P_i. Let $U = U_1 + U_2$ for $U_1 \in V_1$ and $U_2 \in V_2$. Since $AU_1 = U_1$ and $P_i^2 = P_i$, we obtain

$$\|P_i AU\| = \|P_i(P_i U_1 + AU_2)\| \leq \|P_i U_1 + AU_2\|.$$

Our goal is to give an upper bound of $\|P_i U_1 + AU_2\|$ in terms of $\|U\| = \|U_1 + U_2\|$. Since $P_i U_1$ is no longer a vector in $V_1 = (V_2)^{\perp}$, we need to estimate the cosine of the angle between $P_i U_1$ and AU_2 and then use the law of cosines to express the sum in terms of this estimate. Let θ_i be the angle between U_1 and $P_i U_1$. Then

$$\cos \theta_i = \frac{U_1 \cdot P_i U_1}{\|U_1\| \|P_i U_1\|} = \frac{|F_i|}{\sqrt{|F_i|}\sqrt{|G|}} = \sqrt{\frac{|F_i|}{|G|}} = \sqrt{\mu_i}.$$

In particular, we have $0 \leq \theta_i \leq \frac{\pi}{2}$. Let ϕ_i be the angle between $P_i U_1$ and AU_2. Since $\phi_i \leq \frac{\pi}{2} + \theta_i \leq \pi$, it follows that $-\cos \phi_i \leq -\cos\left(\frac{\pi}{2} + \theta_i\right)$ and therefore

$$\|P_i U_1 + AU_2\|^2 = \|P_i U_1\|^2 + \|AU_2\|^2 - 2\|P_i U_1\| \|AU_2\| \cos \phi_i$$
$$\leq \|P_i U_1\|^2 + \|AU_2\|^2 - 2\|P_i U_1\| \|AU_2\| \cos\left(\frac{\pi}{2} + \theta_i\right).$$

But

$$-\cos\left(\frac{\pi}{2} + \theta_i\right) = \sin \theta_i = \sqrt{1 - \cos^2 \theta_i} = \sqrt{1 - \mu_i}.$$

Moreover, $\|P_i U_1\| \leq \mu_i |G|$ and $\|AU_2\| \leq \lambda_2 |G|$, so

$$\|P_i U_1 + AU_2\|^2 \leq (\mu_i^2 + \lambda_2^2 + 2\mu_i \sqrt{1 - \mu_i}\lambda_2)|G|.$$

Thus,

$$P(\text{walk passes through } F_1, \ldots, F_L) \leq \prod_{i=1}^{L} \sqrt{\mu_i^2 + \lambda_2^2 + 2\mu_i\sqrt{1 - \mu_i}\lambda_2}.$$

References

1. Akavia, A., Goldwasser, S., Safra, S.: Proving hard-core predicates using list decoding. In: FOCS 2003. Proceedings of the 44th Annual IEEE Symposium on Foundations of Computer Science, p. 146. IEEE Computer Society, Washington, DC (2003)
2. Boneh, D., Halevi, S., Howgrave-Graham, N.: The modular inversion hidden number problem. In: Boyd, C. (ed.) ASIACRYPT 2001. LNCS, vol. 2248, pp. 36–51. Springer, Heidelberg (2001)
3. Boneh, D., Shparlinski, I.: On the unpredictability of bits of the elliptic curve Diffie-Hellman scheme. In: Kilian, J. (ed.) CRYPTO 2001. LNCS, vol. 2139, pp. 201–212. Springer, Heidelberg (2001)
4. Boneh, D., Venkatesan, R.: Hardness of computing the most significant bits of secret keys in Diffie-Hellman and related schemes. In: Koblitz, N. (ed.) CRYPTO 1996. LNCS, vol. 1109, pp. 129–142. Springer, Heidelberg (1996)
5. Boneh, D., Venkatesan, R.: Rounding in lattices and its cryptographic applications. In: Proceedings of the Eighth Annual ACM-SIAM Symposium on Discrete Algorithms, pp. 675–681. ACM, New York (1997)
6. Bourgain, J.: New bounds on exponential sums related to the Diffie-Hellman distributions. C.R. Math. Acad. Sci. Paris 338(11), 825–830 (2004)
7. Bourgain, J.: Estimates on exponential sums related to the Diffie-Hellman distributions. Geom. Funct. Anal. 15(1), 1–34 (2005)
8. Bourgain, J.: On an exponential sum related to the Diffie-Hellman cryptosystem. Int. Math. Res. Not., pages Art. ID 61271, 15 (2006)
9. Canetti, R., Friedlander, J., Konyagin, S., Larsen, M., Lieman, D., Shparlinski, I.: On the statistical properties of Diffie-Hellman distributions. Israel J. Math. 120, 23–46 (2000)
10. Canetti, R., Friedlander, J., Shparlinski, I.: On certain exponential sums and the distribution of Diffie-Hellman triples. J. London Math. Soc (2), 59(3), 799–812 (1999)
11. Deligne, P.: Cohomologie étale. In: de Boutot, J.F., Grothendieck, A., Illusie, L., Verdier, J.L. (eds.) Séminaire de Géométrie Algébrique du Bois-Marie SGA $4\frac{1}{2}$, Avec la collaboration. Lecture Notes in Mathematics, vol. 569, Springer, Berlin (1977)
12. Friedlander, J., Shparlinski, I.: On the distribution of the power generator. Math. Comp (electronic) 70(236), 1575–1589 (2001)
13. Galbraith, S., Hopkins, H., Shparlinski, I.: Secure bilinear Diffie-Hellman bits. In: Wang, H., Pieprzyk, J., Varadharajan, V. (eds.) ACISP 2004. LNCS, vol. 3108, pp. 370–378. Springer, Heidelberg (2004)
14. Goldreich, O., Impagliazzo, R., Levin, L., Venkatesan, R., Zuckerman, D.: Security preserving amplification of hardness. In: 31st Annual Symposium on Foundations of Computer Science, vol. I, II, pp. 318–326. IEEE Comput. Soc. Press, Los Alamitos, CA (1990)
15. González Vasco, M.I., Shparlinski, I.: On the security of Diffie-Hellman bits. In: Cryptography and computational number theory, Progr. Comput. Sci. Appl. Logic, vol. 20, pp. 257–268. Birkhäuser, Basel (2001)

16. González Vasco, M.I., Shparlinski, I.: Security of the most significant bits of the Shamir message passing scheme. Math. Comp (electronic) 71(237), 333–342 (2002)
17. Howgrave-Graham, N., Nguyen, P., Shparlinski, I.: Hidden number problem with hidden multipliers, timed-release crypto, and noisy exponentiation. Math. Comp (electronic) 72(243), 1473–1485 (2003)
18. Jao, D., Miller, S.D., Venkatesan, R.: Do all elliptic curves of the same order have the same difficulty of discrete log? In: Roy, B. (ed.) ASIACRYPT 2005. LNCS, vol. 3788, pp. 21–40. Springer, Heidelberg (2005)
19. Lenstra, A.K., Lenstra Jr., H.W., Lovász, L.: Factoring polynomials with rational coefficients. Math. Ann. 261(4), 515–534 (1982)
20. Nguyen, P.: The dark side of the hidden number problem: lattice attacks on DSA. In: Cryptography and computational number theory, Progr. Comput. Sci. Appl. Logic, Birkhäuser, Basel, vol. 20, pp. 321–330 (2001)
21. Nguyen, P., Shparlinski, I.: The insecurity of the digital signature algorithm with partially known nonces. J. Cryptology 15(3), 151–176 (2002)
22. Nguyen, P., Shparlinski, I.: The insecurity of the elliptic curve digital signature algorithm with partially known nonces. Des. Codes Cryptogr. 30(2), 201–217 (2003)
23. Shparlinski, I.: On the generalised hidden number problem and bit security of XTR. In: Bozta, S., Sphparlinski, I. (eds.) Applied Algebra, Algebraic Algorithms and Error-Correcting Codes. LNCS, vol. 2227, pp. 268–277. Springer, Heidelberg (2001)
24. Shparlinski, I.: Cryptographic applications of analytic number theory. In: Progress in Computer Science and Applied Logic, Complexity lower bounds and pseudorandomness, vol. 22, Birkhäuser Verlag, Basel (2003)
25. Shparlinski, I.: Playing 'hide-and-seek' with numbers: the hidden number problem, lattices and exponential sums. In: Public-key cryptography, Proc. Sympos. Appl. Math., vol. 62, pp. 153–177. Amer. Math. Soc., Providence, RI (2005)
26. Silverman, J.: The arithmetic of elliptic curves. In: Graduate Texts in Mathematics, vol. 106, Springer, New York (1992) Corrected reprint of the 1986 original

A Result on the Distribution of Quadratic Residues with Applications to Elliptic Curve Cryptography

Muralidhara V.N. and Sandeep Sen

Department of Computer Science and Engineering,
Indian Institute of Technology, Delhi
Hauz Khas, New Delhi 110 016, India
{murali,ssen}@cse.iitd.ernet.in

Abstract. In this paper, we prove that for any polynomial function f of fixed degree without multiple roots, the probability that all the $(f(x+1), f(x+2), ..., f(x+\kappa))$ are quadratic non-residue is $\approx \frac{1}{2^\kappa}$. In particular for $f(x) = x^3 + ax + b$ corresponding to the elliptic curve $y^2 = x^3 + ax + b$, it implies that the quadratic residues $(f(x+1), f(x+2), ...$ in a finite field are sufficiently randomly distributed. Using this result we describe an efficient implementation of El-Gamal Cryptosystem. that requires efficient computation of a mapping between plain-texts and the points on the elliptic curve.

1 Introduction

The distribution of quadratic residues is an interesting problem in Number theory and has many practical applications including Cryptography and Random number generation. In particular it is conjectured to be random and there are many constructions based on this conjecture [1,2]. Peralta [2] proves that for any randomly chosen $x \in F_q$, the probability of $(x+1, x+2, ..., x+\kappa)$ matching any particular quadratic sequence of length κ is in the range $\frac{1}{2^\kappa} \pm \kappa \frac{3+\sqrt{q}}{q}$. In this paper we prove a similar result for the sequence $(f(x+1), f(x+2), ..., f(x+\kappa))$, for any polynomial function f of fixed degree without multiple roots. In particular for $f(x) = x^3 + ax + b$ corresponding to the elliptic curve $y^2 = x^3 + ax + b$, it implies that the quadratic residues $(f(x+1), f(x+2), ...$ in a finite field are sufficiently randomly distributed.

The main motivation for this work is Elliptic Curve El-Gamal Cryptosystem and Koblitz's mapping from the message units to points on an elliptic curve. In the following sections we briefly describe these two methods.

1.1 El-Gamal Cryptosystem

We start with a fixed publicly known finite field K, an elliptic curve E/K defined over it and a base point $B \in E/K$ (we refer to [5] for basic definitions and

K. Srinathan, C. Pandu Rangan, M. Yung (Eds.): Indocrypt 2007, LNCS 4859, pp. 48–57, 2007.
© Springer-Verlag Berlin Heidelberg 2007

notations). Each user chooses a random integer b, which is kept secret, and computes the point $x = bB$ which is the public key. To send a message P to Bob, Alice chooses a random integer k and sends the pair of points $(kB, P + k(bB))$ (where bB is Bob's public key) to Bob. To read the message, Bob multiplies the first point in the pair by his secret key b and subtracts the result from the second point: $P + k(bB) - b(kB)$ that yields P.

One of the commonly used ECC, El-Gamal Cryptosystem requires a mapping from the message units to points on an elliptic curve, i.e., we need an efficient algorithm which computes a mapping between the points on an elliptic curve and a plain-text which forms the basis of encryption and decryption routines.

To date no polynomial time deterministic algorithm is known for this problem. However we do have polynomial time randomized algorithms. We sketch such an algorithm due to Koblitz [5] that makes the following assumptions:

- F_q is a field with p^n elements ($p > 3$, prime).
- κ is a large enough integer so that we are satisfied with the failure probability $\frac{1}{2^\kappa}$ when we attempt to embed a plain text message m.
- Message units are integers between 0 and $M - 1$.
- The finite field is chosen in such a way that $q > \kappa \cdot M$.
- An integer $m = \sum_{i=0}^{n-1} a_i p^i$ is mapped to $(a_0, a_1, ..., a_{n-1}) \in F_q$.

KOBLITZ'S ALGORITHM

1. Given m, find an element $x \in F_q$ corresponding to $m\kappa + 1$ and compute $f(x) = x^3 + ax + b$ and check whether $f(x)$ is a quadratic residue.
   (This can be easily done because an element $\alpha \in F_q$ is a quadratic residue if and only if $\alpha^{(q-1)/2} = 1$).
2. If $f(x)$ is a quadratic residue then we can find a y such that $y^2 = x^3 + ax + b$ and we map m to $P_m = (x, y)$.
   (There are polynomial time probabilistic algorithms to find the square roots in finite fields of odd order [5]).
3. If $f(x)$ is not a quadratic residue then we try points corresponding to $m\kappa + j$, $1 < j \le \kappa$ till we find an x such that $f(x)$ is a quadratic residue.

Suppose x_1 is the integer corresponding to the point $x \in F_q$. We can recover m from the point $P_m = (x, y)$ by dividing $x_1 - 1$ by κ ($x_1 - 1 = m\kappa + j$, $0 \le j < \kappa$).

It has been conjectured that the probability that the above algorithm would fail to find an embedding of a given plain-text message is $\approx \frac{1}{2^\kappa}$, where κ is the number of repetitions of step 3. If quadratic residues in a finite field of odd order are randomly distributed then in fact the κ events (in the above algorithm) are independent and hence the probability that the algorithm would fail is exactly $\frac{1}{2^\kappa}$. In section 4 we show the existence of such finite fields; however we do not know of any efficient construction of finite fields in which quadratic residues are randomly distributed. A naive modification would be to map a message to a point on the curve by choosing a random field element; by Hasse's theorem [5,6], we will succeed with probability about 1/2. But the drawback of such an approach would be that we have to send the random element with each message for decryption, thereby increasing the message expansion factor considerably.

1.2   Previous Results

There is an extensive literature on the distribution of quadratic residues and non residues over finite Fields [1,2]. In particular, Peralta [2] proves that for any randomly chosen $x \in F_q$, the probability of $(x + 1, x + 2, ..., x + \kappa)$ matching any particular binary sequence of length κ is in the range $\frac{1}{2^\kappa} \pm \kappa \frac{3+\sqrt{q}}{q}$. We note that we are interested in the sequence $(f(x + 1), f(x + 2), ..., f(x + \kappa))$ where $f(x) = x^3 + ax + b$ for an elliptic curve $y^2 = x^3 + ax + b$. In the rest of the paper $\chi(x)$ denote the characteristic function defined as

$$\chi(x) = \begin{cases} -1 & \text{if x is a quadratic non-residue} \\ 0 & \text{if x is zero} \\ 1 & \text{if x is quadratic residue} \end{cases}$$

To prove our main theorem we prove the following lemma,

Lemma 1. *Let $g(x)$ be any polynomial of degree d which don't have multiple roots and k be a positive integer such that $dk < p$. If $i_1, i_2, ..., i_m$ ($m \le k$), be any m distinct integers between 1 and k, then*

$$\mid \sum_{x \in F_q} \chi(g'(x)) \mid \le d' \sqrt{q} \tag{1}$$

where $g'(x) = \prod_{j=1}^{m} g(x + i_j))$ and $d' = dm - 1$, the degree of g'.

We note that C. Mauduit and A. Sárközy [1] prove results on the pseudorandom properties of distribution of quadratic residues of arithmetic progression (not exactly for the sequence that we are looking at). In the process, they prove that for any $g(x) \in F_q[X]$ polynomial of degree d that does not have multiple roots, then

$$\mid \sum_{x \in F_q} \chi(g(x)) \mid \le 9d\sqrt{q} \log q \tag{2}$$

So with an additional constraint $dk < p$, our bound is better by a factor $O(\log q)$.

1.3   Main Result

In this paper we address the following problem:

Let $S = \{(a_0, a_1, ..., a_{n-1}) : a_i \in \mathbb{Z}_p, 0 \le i < n, p \ prime > 3\}$. Order the elements of S^* in a reverse lexicographic order, that is,

$$x_1 \quad = (1, 0, 0, \ldots, 0).$$
$$x_2 \quad = (2, 0, 0, \ldots, 0).$$
$$\vdots$$
$$x_p \quad = (0, 1, 0, \ldots, 0).$$
$$x_{p+1} = (1, 1, 0, \ldots, 0).$$
$$\vdots$$
$$x_{2p+1} = (1, 2, 0, \ldots, 0).$$
$$\vdots$$
$$x_{p^n - 1} = (p-1, p-1, p-1, \ldots, p-1).$$

Let $a, b \in S$ be two fixed elements. Given any $\kappa \in N$ can we bound the number of κ-sub-sequences

$$\langle x_{l+1}, x_{l+2}, \ldots, x_{l+\kappa} \rangle, \qquad 0 \le l < p^n - \kappa - 1$$

such that all of $x_{l+i}^3 + a x_{l+i} + b$, $1 \le i \le \kappa$ are quadratic non-residues by $\approx \frac{p^n - 1}{2^\kappa}$?

If the answer to this question is yes, then in Koblitz's algorithm, we can begin with a random element x_l, $0 \le l \le p^n - \kappa$. The probability that all of $x_{l+i}^3 + a x_{l+i} + b$, $1 \le i \le \kappa$ are quadratic non-residues is $\approx \frac{1}{2^\kappa}$ which will mean that the conjecture is correct.

In the following sections we prove a somewhat weaker version of this. We prove that if we choose a random element $x \in F_{p^n}$ then the probability that all of $(x+i)^3 + a(x+i) + b$, $1 \le i \le \kappa$ are quadratic non-residues is $\approx \frac{1}{2^\kappa}$. Note that if $x = x_r$ and if $p|(r+1)$ then $x+1 = x_{r-p+1}$ else $x+1 = x_{r+1}$. Hence by randomly picking an element in the above sequence and adding 1 to it repeatedly we may be able to get at most p consecutive elements in the sequence.

By exploiting this result, we propose a provably efficient alternative to Koblitz scheme in section 3 that requires similar computations as Koblitz's original method and the (expected) message expansion factor is also identical.

We formally prove the following theorem in next section,

Theorem 1. *Let $g(x)$ be any polynomial of degree d which don't have multiple roots. If κ is any positive integer $< \frac{p}{d}$ then*

$$| \{ x \in F_q \mid g(x+1), g(x+2), \ldots, g(x+\kappa) \text{ are quadratic non-residues} \} |$$

is between $\frac{q - \mu_\kappa(a,b)}{2^\kappa} - (d\kappa - 1)\sqrt{q}$ and $\frac{q - \mu_\kappa(a,b)}{2^\kappa} + (d\kappa - 1)\sqrt{q}$, where $q = p^n$ and $0 \le \mu_\kappa(a,b) \le 3$.

2   Proof of the Main Result

Let F_{p^n} be a field with $q = p^n$ (p is a prime > 3) elements. We define a relation $\sim$ on F_{p^n} as $x \sim y$ iff there is a non-negative integer k such that $x - y = 1 + 1 + \ldots + 1$, where 1 is added k times. This is an equivalence relation on F_{p^n}. Each equivalence class will have p (characteristic of F_{p^n}) elements.

Let α, β, γ be three distinct elements in the same equivalence class. We may assume that if μ_1 and μ_2 are least positive integers such that $\alpha = \beta + \mu_1, \alpha = \gamma + \mu_2$ then $\mu_1 < \mu_2$. Let $k_\alpha, k_\beta, k_\gamma$ be least positive integers such that $\alpha = \beta + k_\beta$, $\beta = \gamma + k_\gamma, \gamma = \alpha + k_\alpha$. Now $k_\alpha, k_\beta, k_\gamma$ are such that $\alpha = k_\beta + k_\gamma + k_\alpha + \alpha$ and $k_\alpha + k_\beta + k_\gamma = p$ (in general it can be any multiple of p, but the assumption $\mu_1 < \mu_2$ makes it p). Hence one of $k_\alpha, k_\beta, k_\gamma$ is $> p/3$.

So we may assume that $k_\alpha > p/3$. Let k be any positive integer $< p/3$. Now for any m integers, $i_1, i_2, \ldots, i_m$ such that $1 \le i_1 < i_2 < \ldots < i_m \le k$ the following should hold.

$$\alpha - i_1 \notin \{\beta - i_j \mid 1 \le j \le m\} \cup \{\gamma - i_j \mid 1 \le j \le m\}$$

Hence if α, β, γ are three distinct elements in the same equivalence class and $i_1, i_2, \ldots, i_m$ are any m integers such that $1 \le i_1 < i_2 < \ldots < i_m \le k < p/3$ then one of the following holds.

$$\gamma - i_1 \notin \{\alpha - i_j \mid 1 \le j \le m\} \cup \{\beta - i_j \mid 1 \le j \le m\}$$
$$\alpha - i_1 \notin \{\beta - i_j \mid 1 \le j \le m\} \cup \{\gamma - i_j \mid 1 \le j \le m\}$$
$$\beta - i_1 \notin \{\gamma - i_j \mid 1 \le j \le m\} \cup \{\alpha - i_j \mid 1 \le j \le m\}$$

This observation will be used to prove the following Lemma.

Lemma 2. *Let $g(x)$ be any polynomial of degree d which don't have multiple roots and k be a positive integer such that $dk < p$. If $i_1, i_2, \ldots, i_m$ $(m \le k)$, be any m distinct integers between 1 and k, then $\prod_{j=1}^{m} g(x + i_j)$ cannot be written as $h(x)^2$ for some $h(x) \in F_{p^n}[X]$.*

Proof. Let $\alpha_1, \alpha_2, \ldots \alpha_d$ be the roots of g in the splitting field. From the definition these must be distinct. Also note that $\alpha_1 - i, \alpha_2 - i, \alpha_d - i$ are the roots of $g(x + i) \; \forall i, \; 1 \le i \le k$.

If there exists a polynomial $h(x) \in F_{p^n}[X]$ such that

$$\prod_{j=1}^{m} g(x + i_j) = h(x)^2 \tag{3}$$

then m has to be even (as degree of $\prod_{j=1}^{m} g(x + i_j)$ is dm and degree of $h(x)^2$ is even). So the multiplicity of any root of $\prod_{j=1}^{m} g(x + i_j)$ is even. As i_j's are distinct $\alpha_i - i_a \ne \alpha_i - i_b$ for $\forall a, b \le m$, it follows that multiplicity of any root of $\prod_{j=1}^{m} g(x + i_j)$ is $\le d$, hence should be 2.

From the arguments given before this lemma, at least one of $\alpha_1 - i_1, \alpha_2 - i_1, \alpha_d - i_1$ can not be of multiplicity 2 (if α_i's are not in the same equivalence class then this is trivially true), which is a contradiction.

The proof of Lemma 1, follows from Lemma 2 and Weil's theorem on finite fields[6]. Now we are ready to prove Theorem 1.

Proof of Theorem 1. Let $A(x) = \prod_{i=1}^{\kappa}(1 - \chi(g(x + i)).$[1]
$S = \{x \in F_q \mid g(x + 1), g(x + 2), \ldots, g(x + \kappa)$ are quadratic non-residues$\}$

---

[1] Similar idea was used in [3] and was suggested to us by Radhakrishnan[9].

$S' = \{x \in F_q \mid \text{at least one of } g(x+1)., g(x+\kappa) \text{ is a quadratic residues}\}$
$S'' = \{x \in F_q \mid g(x+1) = g(x+2) =, ..., = g(x+\kappa) = 0\}.$

Clearly, $F_q = S \cup S' \cup S''$ and

$$A(x) = \begin{cases} 2^\kappa & \text{if } x \in S, \\ 0 & \text{if } x \in S' \\ 1 & \text{if } x \in S'' \end{cases}$$

Let $\mid S \mid = N$ and denote $\mu_\kappa(a, b) = \mid S'' \mid$. Note that $\alpha \in S'' \implies g(\alpha + 1) = g(\alpha + 2) =, ..., = g(\alpha + \kappa) = 0 \implies \alpha - 1, \alpha - 2, ..., \alpha - \kappa$ are roots of $g(x)$ and hence $0 \leq \mu_\kappa(a, b) \leq d$ and $\mu_\kappa(a, b) = 0$ if $\kappa > d$.

Now

$$\sum_{x \in F_q} A(x) = 2^\kappa N + \mu_\kappa(a, b) \tag{4}$$

Notice that

$$A(x) = 1 + \sum_{m=1}^{\kappa} (-1)^m \sum_{1 \leq i_1 < i_2 < ... < i_n \leq \kappa} \chi(g(x+i_1)g(x+i_2)...g(x+i_n))$$

Hence

$$N2^\kappa + \mu_\kappa(a, b) - q = \sum_{x \in F_q} \sum_{m=1}^{\kappa} (-1)^m \sum_{1 \leq i_1 < i_2 < ... < i_n \leq \kappa} \chi(g(x+i_1)g(x+i_2)...g(x+i_n))$$

$$= \sum_{m=1}^{\kappa} (-1)^m \sum_{1 \leq i_1 < i_2 < ... < i_n \leq \kappa} \sum_{x \in F_q} \chi(g(x+i_1)g(x+i_2)...g(x+i_n))$$

By taking modulus,

$$\mid N2^\kappa + \mu_\kappa(a, b) - q \mid \leq \mid \sum_{m=1}^{\kappa} (-1)^m \sum_{1 \leq i_1 < i_2 < ... < i_n \leq \kappa} \sum_{x \in F_q} \chi(g(x+i_1)g(x+i_2)...g(x+i_n)) \mid$$

By triangular inequality,

$$\mid N2^\kappa + \mu_\kappa(a, b) - q \mid \leq \sum_{m=1}^{\kappa} \sum_{1 \leq i_1 < i_2 < ... < i_n \leq \kappa} \mid \sum_{x \in F_q} \chi(g(x+i_1)g(x+i_2)...g(x+i_n)) \mid$$

Applying Lemma 1

$$\leq \sum_{m=1}^{\kappa} {}^\kappa C_m (dm - 1)\sqrt{q}$$

$$< (d\kappa - 1)2^\kappa \sqrt{q}$$

Hence $\frac{q - \mu_\kappa(a,b)}{2^k} - (d\kappa - 1)\sqrt{q} < N < \frac{q - \mu_\kappa(a,b)}{2^k} + (d\kappa - 1)\sqrt{q}$

Corollary 1. *Let $y^2 = x^3 + ax + b$ be an elliptic curve over F_{p^n} (p is a prime > 3). Let $g(x) = x^3 + ax + b$. If κ is any positive integer $< \frac{p}{3}$ then for a randomly chosen $x \in F_q$ the probability that all of $g(x+1), g(x+2), ..., g(x+\kappa)$ are quadratic non-residues is $\approx \frac{1}{2^\kappa}$*

For $g(x) = x^3 + ax + b$, the result follows from Theorem 1.

3   A Modified ECC

Here we propose a modification for El-Gamal Cryptosystem with Koblitz's method for Embedding plain-texts on to the points of elliptic curve which exploits the result of the previous section. More specifically, given a plain-text message, we try to map it to a random point on the Elliptic Curve by choosing an initial random shift in Koblitz's algorithm.

3.1   Key Generation

We suppose that all parties have agreed upon an elliptic curve $E/K : y^2 = x^3 + ax + b$ over a finite field $K = F_{p^n}$ and $p > 3$, a point P of high order on it and a failure factor $\kappa(< p/3)$. Let $r_1, r_2, \ldots, r_t$ be t randomly chosen integers between 1 and p^n and they are made public.

Each party A does the following:

> – Choose a random integer a.
> – a is A's **Secret Key**.
> – aP is A's **Public Key**.

3.2   Encryption

To send a message m to Alice, Bob does the following:

> – Choose a random integer μ and s, $1 \leq s \leq t$.
> – Obtain Alice's public key aP and Compute μaP a point on the elliptic curve.
> – Find $x \in F_q$ corresponding to $m\kappa + r_s + 1$.
>   If $x^3 + ax + b$ is a quadratic residue (or zero) then find a y such that $y^2 = x^3 + ax + b$ and take $P(m, r_s) = (x, y)$ else try with points corresponding to $m\kappa + r_s + j, 1 < j \leq \kappa$.
> – Send $(\mu P, P(m, r_s) + \mu aP)$ and s.

The probability that Bob fails to find $P(m, r_s)$ with the shift corresponding to the random number r_s is $\frac{1}{2^\kappa}$ by Corollary 1. If he fails, then he tries with some other random number s for $1 \leq j \leq t$. If he fails with all the $r_1, r_2, \ldots, r_t$ then he would try with some random r's until he succeeds and he would send this r along with the message (This will happen with negligible probability $(1/2)^{t\kappa}$).

3.3   Decryption

To recover the message m, Alice does the following:

- Multiply the first point in the above pair by her secret key a and subtracts the results from the second point to get the point
$$P(m,r) \ = \ (P(m,r) + \mu aP) - a\mu P.$$
(Here r is one of the public r_i's or is sent with the message).
Let $P(m,r) = (x,y)$.
- Find x_1, the integer corresponding to x.
- m is obtained by dividing $x_1 - 1 - r$ by κ.
$$\text{i.e. } x_1 - 1 - r \ = \ m\kappa + j, \ 0 \le j < \kappa.$$

Our modified encryption and decryption schemes require similar computations as Koblitz's original method and the (expected) message expansion factor is also identical. Moreover, our method has following advantages over the original method:

1. The probability that Bob fails to encrypt a message with the shift corresponding to a random number μ is provably $\frac{1}{2^\kappa}$.
2. The *failure factor* κ can be small, because even if we fail with one *randomshift* we can try with another *random shift*. Small *failure factor* κ implies that the message units can be large, as $M\kappa < q$ where is message units m are such that $m < M$.
3. Random Embedding: The point $P(m,r)$ not only depends on m, it also depends on r, so even for a fixed message the point corresponding to the message will be different on different occasions. This prevents an eavesdropper from guessing the message. The usual procedure is to pad random bits, but strictly speaking it does not really make the message random.

4   Randomizing the Distribution of Quadratic Residues in a Finite Field

In this section we would like to address the question: *Can we Randomize the distribution of quadratic residues in a finite field?* The following theorem says that the answer is yes.

Theorem 2. *Let S be a set with p^n elements, p an odd prime, n any natural number. Given $x_1, x_2, \ldots, x_{\frac{p^n-1}{2}}$ in S, there exists two binary operations $\oplus$ and $\star$ such that $(S, \oplus, \star)$ is a field and the quadratic residues are precisely these x_i's.*

Proof. Let $(F_q, +, *)$ be a field with q elements where $q = p^n$ and β be a fixed non-residue in F_q. Let $a_i, i = 1, 2, \ldots, (p^n - 1)/2$ be the nonzero elements of F_q, written as n-tuples of elements of F_q, whose first nonzero coordinate lies in $\{1, 2, \ldots, (p-1)/2\}$, listed reverse lexicographically. Let $S = \{x_1, x_2, \ldots, x_{\frac{p^n-1}{2}}, y_1, y_2, \ldots, y_{\frac{p^n-1}{2}}, O\}$.

We define a bijection ϕ from S to F_q as[2],

$$O \mapsto 0$$
$$x_i \mapsto a_i^2$$
$$y_i \mapsto \beta a_i^2, \qquad 1 \le i \le \frac{p^n - 1}{2}.$$

With this we define two binary operations $\oplus$ and $\star$ on S as

$$a \oplus b = \phi^{-1}\{\phi(a) + \phi(b)\}$$
$$a \star b = \phi^{-1}\{\phi(a) * \phi(b)\}, \quad \forall a, b \in S$$

It can be easily verified that $(S, \oplus, \star)$ is a field and the quadratic residues in $S(+, \star)$ are x_i's.

Both ϕ and ϕ^{-1} can be found in polynomial time, however finding ϕ^{-1} involves finding square roots, which is very costly (compared to addition , multiplication, inversion) as it takes $O(\log p^n)$ operations. We note that each elliptic curve operation involves 6 additions, 3 multiplications and 1 inversion (field operations). Since implementation of El-Gamal Cryptosystem involves computing scalar multiplication, kP which would take $2 \log k$ elliptic curve operations, this method is not practical.

5   Weil's Theorem

Theorem 3. *(Weil's Theorem) Let $f(x) \in F_q[X]$ be any polynomial of positive degree that is not a square of any of polynomial.($f(x) \ne h^2(x)$ for all $h(x) \in F_q[X]$). Let d be the number of distinct roots of $f(x)$ in splitting field over F_q, then we have*

$$|\sum_{x \in F_q} \chi(f(x))| \le (d-1)\sqrt{q} \tag{5}$$

where

$$\chi(x) = \begin{cases} -1 & \text{if } x \text{ is a quadratic non-residue} \\ 0 & \text{if } x \text{ is zero} \\ 1 & \text{if } x \text{ is quadratic residue} \end{cases}$$

For proof, the reader is referred to [6]

References

1. Mauduit, C., Sárközy, A.: On finite pseudorandom binary sequences 1: Measure of pseudorandomness, the Legendre symbol. Acta Arith. 82, 365–377 (1997)
2. Peralta, R.: On the distribution of quadratic residues and nonresidues modulo a prime number. Mathematics of Computation 58(197), 433–440 (1992)

---

[2] This construction was pointed out by an anonymous reviewer for an earlier version of the paper.

3. Babai, L., G'al, A., Koll'ar, J., R'onyai, L., Szab'o, T., Wigderson, A.: Extremal Bipartite Graphs and Superpolynomial Lowerbounds for Monotone Span Programs. In: Proc. ACM STOC 1996, pp. 603–611 (1996)
4. Gallant, R., Lambert, R., Vanstone, S.: Improving the parallelized Pollard lambda search on binary anomalous curves. Mathematics of Computation 69, 1699–1705 (2000)
5. Koblitz, N.: A Course in Number theory and Cryptography. Springer, New York (1994)
6. Lidl, R., Niederreiter, H., Cohn, P.M.: Encyclopedia of Mathematics and its Applications20-Finite Fields. Cambridge University Press, Cambridge (1997)
7. Menezes, A.: Elliptic Curve Public Key Cryptosystems. Kluwer Academic Publishers, Dordrecht (1996)
8. Pollard, J.: Monte Carlo methods for index computation mod p. Mathematics of computation 32, 918–924 (1978)
9. Radhakrishnan, J.: Private Communication
10. Van Oorschot, P., Wiener, M.: Parallel collision search with cryptanalytic applications. Journal of Cryptology 12, 1–28 (1999)
11. Wiener, M., Zuccherato, R.: Faster attacks on elliptic curve cryptosystems. In: Tavares, S., Meijer, H. (eds.) SAC 1998. LNCS, vol. 1556, pp. 190–200. Springer, Heidelberg (1999)

Related-Key Attacks on the Py-Family of Ciphers and an Approach to Repair the Weaknesses*

Gautham Sekar, Souradyuti Paul, and Bart Preneel

Katholieke Universiteit Leuven, Dept. ESAT/COSIC,
Kasteelpark Arenberg 10,
B–3001, Leuven-Heverlee, Belgium
{gautham.sekar,souradyuti.paul,bart.preneel}@esat.kuleuven.be

Abstract. The stream cipher TPypy has been designed by Biham and Seberry in January 2007 as the strongest member of the Py-family ciphers, after weaknesses in the other members Py, Pypy, Py6 were discovered. One main contribution of the paper is the detection of related-key weaknesses in the Py-family of ciphers including the strongest member TPypy. Under related keys, we show a distinguishing attack on TPypy with data complexity $2^{192.3}$ which is lower than the previous best known attack on the cipher by a factor of 2^{88}. It is shown that the above attack also works on the other members TPy, Pypy and Py. A second contribution of the paper is design and analysis of two fast ciphers RCR-64 and RCR-32 which are derived from the TPy and the TPypy respectively. The performances of the RCR-64 and the RCR-32 are 2.7 cycles/byte and 4.45 cycles/byte on Pentium III (note that the speeds of the ciphers Py, Pypy and RC4 are 2.8, 4.58 and 7.3 cycles/byte). Based on our security analysis, we conjecture that no attacks lower than brute force are possible on the RCR ciphers.

1   Introduction

Timeline – The Py-Family of Ciphers

- **April 2005, Design.** The ciphers Py and Py6, designed by Biham and Seberry, were submitted to the ECRYPT project for analysis and evaluation in the category of software based stream ciphers [4]. The impressive speed

---

* This work was supported in part by the Concerted Research Action (GOA) Ambiorics 2005/11 of the Flemish Government, by the IAP Programme P6/26 BCRYPT of the Belgian State (Belgian Science Policy), and in part by the European Commission through the IST Programme under Contract IST-2002-507932 ECRYPT. The first author is supported by an IWT SoBeNeT project. The second author is supported by an IBBT (Interdisciplinary Institute for Broadband Technology) project. The information in this document reflects only the authors' views, is provided as is and no guarantee or warranty is given that the information is fit for any particular purpose. The user thereof uses the information at its sole risk and liability.

K. Srinathan, C. Pandu Rangan, M. Yung (Eds.): Indocrypt 2007, LNCS 4859, pp. 58–72, 2007.
© Springer-Verlag Berlin Heidelberg 2007

of the cipher Py in software (about 2.5 times faster than the RC4) made it one of the fastest and most attractive contestants.

– **March 2006, Attack (at FSE 2006).** Paul, Preneel and Sekar reported distinguishing attacks with $2^{89.2}$ data and comparable time against the cipher Py [18]. Crowley [7] later reduced the complexity to 2^{72} by employing a Hidden Markov Model.

– **March 2006, Design (at the Rump session of FSE 2006).** A new cipher, namely Pypy, was proposed by the designers to rule out the aforementioned distinguishing attacks on Py [5].

– **May 2006, Attack (presented at Asiacrypt 2006).** Distinguishing attacks were reported against Py6 with 2^{68} data and comparable time by Paul and Preneel [19].

– **October 2006, Attack (presented at Eurocrypt 2007).** Wu and Preneel showed key recovery attacks against the ciphers Py, Pypy, Py6 with chosen IVs. This attack was subsequently improved by Isobe *et al.* [11].

– **January 2007, Design.** Three new ciphers TPypy, TPy, TPy6 were proposed by the designers [3]; the ciphers can very well be viewed as the strengthened versions of the previous ciphers Py, Pypy and Py6 where the above attacks should not apply. So far there exist no published attacks on TPypy, TPy and TPy6.

February 2007, Attack. Sekar, Paul and Preneel published distinguishing attacks on Py, Pypy, TPy and TPypy with data complexities 2^{281} each [23].

– **June 2007, Attack (to be presented at ISC 2007).** Sekar, Paul and Preneel showed new weaknesses in the stream ciphers TPy and Py. Exploiting these weaknesses distinguishing attacks on the ciphers are constructed where the best distinguisher requires 2^{275} data and comparable time.

– **July 2007, Attack and Design (presented at WEWoRC 2007).** Sekar, Paul and Preneel mounted distinguishing attacks on TPy6 and Py6 with 2^{233} data and comparable time each [22]. Moreover, they have modified TPy6 to design two new ciphers TPy6–A and TPy6–B which were claimed to be free from all attacks excluding brute force ones.[1]

Contribution of the paper. The list that orders the Py-family of ciphers in terms of increasing security is: Py6→Py→ Pypy → TPy6 → TPy → TPypy (the strongest). The ciphers are normally used with 32-byte keys and 16-byte initial values (or IV). However, the key size may vary from 1 to 256 bytes and the IV from 1 to 64 bytes. The ciphers were claimed by the designers to be free from related-key and distinguishing attacks [3,4,5].

(i) *Related-key Weaknesses.* One major contribution of the paper is the discovery of related-key attacks due to weaknesses in the key scheduling algorithms of the Py-family of ciphers. The main idea behind a related-key attack is that, the attacker, who chooses a relation f between a pair of keys key_1 and key_2 (e.g., $key_1 = f(key_2)$) rather than the actual values of the keys, is able to extract

---

[1] It has been reported very recently that Tsunoo *et al.* showed a distinguishing attack on TPypy with a data complexity of 2^{199} [25].

secret information from a cryptosystem using the relation f [2,13]. Related-key weakness is a cause for concern in a protocol where key-integrity is not guaranteed or when the keys are generated manually rather than from a pseudorandom number generator [12]. Related-key weaknesses are not new in the literature. The usefulness of such type of attacks was first outlined by Knudsen in [14,15]; since then a good deal of research has been spent on related-key weaknesses on block ciphers [2,12,13,16]. The related-key weaknesses of a block cipher can be translated into attacking hash functions based on that particular block cipher and vice versa [9,10,17,20,26,27].

On the other hand, discovery of related-key weaknesses of stream ciphers is not very common in the literature, mainly due to the heavy operations executed in one-time key-scheduling algorithms compared to the operations performed in iterative block ciphers. However, there is an example where related-key weaknesses of the stream cipher RC4 are used to break the WEP protocol with practical complexity [8]. Furthermore, there is a growing tendency by the designers nowadays to build hash functions from stream ciphers [6] instead of building them from block ciphers. In such attempts, related-key weaknesses of stream ciphers need to be addressed carefully.

In the paper, we show that, when used with the identical IVs of 16 bytes each, if two long keys key_1 and key_2 of 256 bytes each, are related in the following manner,

1. $key_1[16] \oplus key_2[16] = 1$,
2. $key_1[17] \neq key_2[17]$ and
3. $key_1[i] = key_2[i] \ \forall i \notin \{16, 17\}$

then the above relation, exploiting the weaknesses of the key setup algorithms of Py-family of ciphers (i.e., TPypy, TPy, Pypy, Py), propagates through the IV setup algorithms and finally induces biases in the outputs at the 1st and the 3rd rounds. Such related key pairs are used to build a distinguisher for each of the aforementioned ciphers with $2^{193.7}$ output words and comparable time (note that, in total, there are 2^{2048} such pairs, while our distinguisher needs any $2^{193.7}$ randomly chosen pairs of keys). This result constitutes the best attack on the strongest member of the Py-family of ciphers TPypy; they are also shown to be effective on the other members TPy, Pypy and Py (see Table 1). These related-key attacks work with any IV-size ranging from 16 to 64 bytes. However, the attack complexities increase with shorter keys. Note that the usage of long keys in the Py-family of ciphers makes it very attractive to be used as fast hash functions (e.g., by replacing of the key with the message). In such cases, these related-key weaknesses can turn out to be serious impediments.

(ii) *The Ciphers RCR-32 and RCR-64.* Finally, we make simple modifications to the ciphers TPypy and TPy to build two new ciphers RCR-32 and RCR-64 respectively. In the modified designs, the key scheduling algorithms of RCR-32 and RCR-64 are identical with those of the TPypy and the TPy. The changes are made *only* to the round functions where *variable rotations* are replaced with *constant rotations*. Our extensive analyses show that the modifications not only

Table 1. Attacks on the Py-family of stream ciphers ('X' denotes that the attack does not work)

Attack	Py6	Py	Pypy	TPy6	TPy	TPypy
Crowley [7]	X	2^{72}	X	X	2^{72}	X
Isobe et al. [11]	X	2^{24}	2^{24}	X	X	X
Paul et al. [18]	X	2^{88}	X	X	2^{88}	X
Paul-Preneel [19]	2^{68}	X	X	2^{68}	X	X
Sekar et al. [21]	X	2^{275}	X	X	2^{275}	X
Sekar et al. [22]	2^{233}	X	X	2^{233}	X	X
Sekar et al.[23]	X	2^{281}	2^{281}	X	2^{281}	2^{281}
Wu-Preneel [29]	X	2^{24}	2^{24}	X	X	X
Related key (this paper)	X	$2^{193.7}$	$2^{193.7}$	X	$2^{193.7}$	$2^{193.7}$

free the Py-family ciphers from *all* the existing attacks, it also improves on the performance of the ciphers without exposing them to new weaknesses (see Sect. 5 for an elaborate security analysis). As a result, the cipher RCR-64 goes on to become one of the *the fastest* stream ciphers published in the literature (approximately 2.7 cycles per byte on Pentium III). The names are chosen to reflect the functionalities involved in the ciphers. For example, RCR-64 denotes *Rolling, Constant Rotation and 64 bits output/round.*

2   Description of the Stream Ciphers TPypy, TPy, Pypy and Py

Each of the Py-family of ciphers is composed of three parts: (1) a key setup algorithm, (2) an IV setup algorithm and (3) a round function or pseudorandom bit generation algorithm (PRBG). The first two parts are used for the initial one-time mixing of the secret key and the IV. These parts generate a pseudorandom internal state composed of (1) a permutation P of 256 elements, (2) a 32-bit array Y of 260 elements and (3) a 32-bit variable s. The key/IV setup uses two intermediate variables: (1) a fixed permutation of 256 elements denoted by *internal_permutation* and (2) a variable EIV whose size is equal to that of the IV. The round function, which is executed iteratively, is used to update the internal state (i.e., P, Y and s) and to generate pseudorandom output bits. The key setup algorithms of the TPypy, the TPy, the Pypy and the Py are identical. Notation for different parts of the four ciphers is provided in Table 2.

Due to space constraints, the KS, the IVS_1, the IVS_2, the RF_1 and the RF_2, as mentioned in Table 2, are described in the full version of the paper [24]. The details of the algorithms can also be found in [3,4,5].

Table 2. Description of the ciphers TPypy, TPy, Pypy and Py

	TPypy	TPy	Pypy	Py
Key Setup	KS	KS	KS	KS
IV Setup	IVS_1	IVS_1	IVS_2	IVS_2
Round Function	RF_1	RF_2	RF_1	RF_2

3   Notation and Convention

The notation and the convention followed in the paper are described below.

- The pseudorandom bit generation algorithm of a stream cipher is denoted by PRBG.
- The outputs generated when key_1 and key_2 are used are denoted by O and Z respectively.
- $O_{(b)}^a$ (or $Z_{(b)}^a$) denotes the bth bit ($b = 0$ is the least significant bit or lsb) of the second output word generated at round a when key_1 (or key_2) is used. We do not use the first output word anywhere in our analysis.
- P_1^a, Y_1^{a+1} and s_1^a are the inputs to the PRBG at round a when key_1 is used. It is easy to see that when this convention is followed the O^a takes a simple form: $O^a = (s \oplus Y^a[-1]) + Y^a[P^a[208]]$. The same applies to key_2.
- $Y_1^a[b]$, $P_1^a[b]$ denote the bth elements of array Y_1^a and P_1^a respectively, when key_1 is used.
- $Y_1^a[b]_i$, $P_1^a[b]_i$ denote the ith bit of $Y_1^a[b]$, $P_1^a[b]$ respectively.
- The operators '+' and '−' denote *addition modulo* 2^{32} and *subtraction modulo* 2^{32} respectively, except when used with expressions which relate two elements of array P. In this case they denote *addition and subtraction over* $\mathbb{Z}$.
- The symbol '⊕' denotes bitwise *exclusive-or*, ∩ denotes set intersection and ∪ denotes set union.

4   Related-Key Weaknesses in the Py-Family of Ciphers

We first choose two keys, key_1 and key_2 (each key is 256 bytes long), such that,

C1. $key_1[16] \oplus key_2[16] = 1$ (without loss of generality, assume lsb of $key_1[16]$ is 1),
C2. $key_1[17] \neq key_2[17]$ and **C3.** $key_1[i] = key_2[i]$ $\forall i \notin \{16, 17\}$.

Now we observe that the above relation between the keys can be traced through various parts of the Py-family of ciphers.

4.1   Propagation of the Weaknesses Through the Key Setup Algorithm

For key_1 and key_2, the values of the variable s through Algorithm A are tabulated in Table 3. The Algorithm A is a part of the key setup algorithm KS (described in the full version of the paper [24]).

```
Algorithm A
for(j=0; j<keysizeb; j++)
  {
    s = s + key[j];
    s0 = internal_permutation[s&0xFF];
    s = ROTL32(s, 8) ^ (u32)s0;
  }
```

Table 3. The variable s after rounds 15, 16 and 17 of Algorithm A

End of round	s (using key_1)	s (using key_2)
15	$s_{1,15}^A$	$s_{2,15}^A = s_{1,15}^A$
16	$s_{1,16}^A$	$s_{2,16}^A = s_{1,16}^A - \delta_1$ (say)
17	$s_{1,17}^A$	$s_{2,17}^A = s_{1,17}^A$ if $key_2[17] = key_1[17] + \delta_1$

If x is a 32-bit variable, let $B(x)$ denote the least significant byte of x. In Table 3,

$$\delta_1 = s_{1,16}^A - s_{2,16}^A \tag{1}$$
$$= ROTL32((s_{1,15}^A + key_1[16]), 8) \oplus ip[B(s_{1,15}^A + key_1[16])] \tag{2}$$
$$- ROTL32((s_{2,15}^A + key_2[16]), 8) \oplus ip[B(s_{2,15}^A + key_2[16])], \tag{3}$$

where ip denotes *internal_permutation*.

Now, if $key_2[17] = key_1[17] + \delta_1$ (call this the event D_1), it is observed from Algorithm A that the following equation is satisfied:

$$s_{1,17}^A = s_{2,17}^A.$$

For event D_1 to occur, δ_1 should be an 8-bit integer. Running simulation, it is determined that

$$Pr[|\delta_1| = 8] \approx \frac{1}{2}.$$

Hence,

$$Pr[D_1] \approx 2^{-9}. \tag{4}$$

If $s_{1,17}^A = s_{2,17}^A$, then in the subsequent rounds of Algorithm A, the s_1^A and s_2^A remain the same, that is, $s_{1,k}^A = s_{2,k}^A$, where $k = 18, 19, ..., 255$.

  Given that the D_1 occurs, that is, $s_1^A = s_2^A$ at the end of Algorithm A, or $s_{1,255}^A = s_{2,255}^A$, we now trace the values of s through Algorithm B which forms

```
Algorithm B
 for(j=0; j<keysizeb; j++)
   {
     s = s + key[j];
     s0 = internal_permutation[s&0xFF];
     s ^= ROTL32(s, 8) + (u32)s0;
   }
```

Table 4. s after rounds 15, 16 and 17 of Algorithm B given event D_1 occurs

End of round	s (using key_1)	s (using key_2)
15	$s_{1,15}^B$	$s_{2,15}^B = s_{1,15}^B$
16	$s_{1,16}^B$	$s_{2,16}^B = s_{1,16}^B - \delta_2$ (say)
17	$s_{1,17}^B$	$s_{2,17}^B = s_{1,17}^B$ if $key_2[17] = key_1[17] + \delta_2$

another part of the key setup. Table 4 compares the values of s after rounds 15, 16 and 17 of Algorithm B when key_1 and key_2 are used.

In Table 4,

$$\delta_2 = s_{1,16}^B - s_{2,16}^B$$
$$= ROTL32((s_{1,15}^B + key_1[16]), 8) \oplus ip[B(s_{1,15}^B + key_1[16])]$$
$$- ROTL32((s_{2,15}^B + key_2[16]), 8) \oplus ip[B(s_{2,15}^B + key_2[16])]. \quad (5)$$

Now, given event D_1 occurs, i.e., $s_1^A = s_2^A$ at the end of Algorithm A, if $\delta_2 = \delta_1$ (call this the event D_2), we will have $key_2[17] = key_1[17] + \delta_2$ and hence from Algorithm B, the following equation is satisfied:

$$s_{1,17}^B = s_{2,17}^B.$$

For event D_2 to occur, δ_2 should be an 8-bit integer. Running simulation, it is determined that

$$Pr[|\delta_2| = 8] \approx \frac{1}{2^{2.4}}.$$

Hence,

$$Pr[D_2|D_1] \approx 2^{-10.4} \Rightarrow Pr[D_2 \cap D_1] \approx Pr[D_1] \cdot 2^{-10.4} \approx 2^{-19.4}. \quad (6)$$

If $s_{1,17}^B = s_{2,17}^B$, then in the subsequent rounds of Algorithm B, the s_1^B and s_2^B remain the same, that is, $s_{1,k}^B = s_{2,k}^B$, where $k = 18, 19, ..., 255$.

Given that the $D_2 \cap D_1$ occurs, that is, $s_1^B = s_2^B$ at the end of Algorithm B , or $s_{1,255}^B = s_{2,255}^B$, the values of s and Y are traced through Algorithm C which forms the final part of the key setup. In the full version of the paper we compare the values of s and Y after rounds 15, 16 and 17 of Algorithm C when key_1 and key_2 are used [24]. Since Algorithm C and the corresponding table have striking similarities with Algorithm A and Table 3, they are described in the full version [24] and we provide only the results of our analysis. Now, given that the event $D_2 \cap D_1$ occurs, i.e., $s_1^B = s_2^B$ at the end of Algorithm B, if $\delta_3 = \delta_1$ (call this the event D_3), we will have $key_2[17] = key_1[17] + \delta_3$ and hence from Algorithm C, the following equation is satisfied:

$$s_{1,17}^C = s_{2,17}^C.$$

For event D_3 to occur, δ_2 should be an 8-bit integer. Running simulation, it is determined that

$$Pr[|\delta_3| = 8] \approx \frac{1}{2}.$$

Hence,

$$Pr[D_3|D_2 \cap D_1] \approx 2^{-9} \Rightarrow Pr[D_3 \cap D_2 \cap D_1] \approx Pr[D_2 \cap D_1] \cdot 2^{-9} \approx 2^{-28.4}. \quad (7)$$

If $s_{1,17}^C = s_{2,17}^C$, then in the subsequent rounds of Algorithm C, the s_1^C and s_2^C remain the same, that is, $s_{1,k}^C = s_{2,k}^C$, where $k = 18, 19, ..., 255$ and $Y_1[j] = Y_2[j]$, where $j \neq 13$.

4.2   Propagation of the Weaknesses Through the IV Setup

Given that the $D_3 \cap D_2 \cap D_1$ occurs, i.e., $s_1^C = s_2^C$ at the end of Algorithm C, or $s_{1,255}^C = s_{2,255}^C$, and $Y_1[i] = Y_2[i]$ ($i \neq 13$), we now trace the variables s, Y, P and EIV through the first part of the IV setup. We now consider Algorithm D which is a part of the IV setup. It is to be noted that s, Y (obtained after the key setup) and the iv are the basic elements used in the IV setup to define the P and the EIV and to update the s and the Y. We now model our attack in such a way that the same IV is used with both the keys. Prior to the execution of Algorithm D, the only elements of array Y which are used in the first part of the IV setup are $Y[0]$, $Y[1]$, $Y[YMININD]$ and $Y[YMAXIND]$. Since $Y[13]$ is not used, it follows that P_1 (that is, P when key_1 is used) and P_2 (that is, P when key_2 is used) are identical.

```
Algorithm D
 for(i=0; i<ivsizeb; i++)
  {
    s = s + iv[i] + Y(YMININD+i);
    u8 s0 = P(s&0xFF);
    EIV(i) = s0;
    s = ROTL32(s, 8) ^ (u32)s0;
  }
```

In Algorithm D as well, $Y[13]$ is not used to update the s or define the EIV when the IV is of the recommended size of 16 bytes. For longer IVs, we can induce the first difference in the keys (that is, where the least significant bits alone differ) according to the size of the IV. An example is provided in the full version [24]. It is to be noted that, if the IV-size is N bytes, the first difference in the keys should be induced nowhere: neither (1) in the first $N-1$ bytes (i.e., key bytes 0 to $N-1$), nor (2) in the last $N-3$ bytes (i.e., key bytes $260-N$ to 256). Otherwise, it is immaterial as to where the first difference is set (i.e., anywhere

```
Algorithm E
 for(i=0; i<ivsizeb; i++)
  {
    s = s + iv[i] + Y(YMAXIND-i);
    /*s = s + EIV((i+ivsizeb-1)mod ivsizeb) + Y(YMAXIND-i); for IVS1.*/
    u8 s0 = P(s&0xFF);
    EIV(i) += s0;
    s = ROTL32(s, 8) ^ (u32)s0;
  }
```

from byte N to $259 - N$) – in all the cases, bias induced will be approximately identical (this is established from a large number of experiments).

We now consider Algorithm E. Again, $Y[13]$ is not used to update the s or the EIV (for both IVS_1 and IVS_2). Hence, at the end of Algorithm E, we have $s_1 = s_2$, $EIV_1 = EIV_2$, $P_1 = P_2$ and $Y_1[i] = Y_2[i]$ (where $i \neq 13$). With this result, we now proceed to the second part of the IV setup.

In the second part of the IV setup (that is, for IVS_2), when $i = 16$ ($i = 17$ for IVS_1), the s generated using key_1 and key_2 are different due to the difference in $Y[13]$. This causes the EIVs to be different in the following round and hence $P_1 \neq P_2$. In the subsequent rounds, the mixing becomes more random with the result that at the end of 260 rounds, we have $Y_1[j] = Y_2[j]$ where $j \in \{-3, ..., 12\}$.

```
IV setup part-2
 for(i=0; i<260; i++)
  {
    u32 x0 = EIV(0) = EIV(0) ^ (s&0xFF);
    rotate(EIV);
    swap(P(0), P(x0));
    rotate(P);
    Y(YMININD)=s=(s ^ Y(YMININD))+Y(x0);
    /*s=ROTL32(s,8)+Y(YMAXIND);
    Y(YMININD)+=s^Y(x0); for IVS1.*/
    rotate(Y);
  }
```

This result holds only if $x0 \neq 13$ when $i = 0, ..., 15$. The probability that this occurs is $(\frac{255}{256})^{j+4} \approx 1$ when $j \in \{-3, ..., 12\}$. With this result, we now analyze the keystream generation algorithm.

4.3   Propagation of the Weaknesses Through the Round Function

Here, we consider only the round function RF_1 (see the full version [24]). The formulas for the lsb of the outputs generated at rounds 1 and 3 when key_1 (the output words are denoted by O) and key_2 (the output words are denoted by Z) are used are given below.

$$O_{(0)}^1 = s_{1(0)}^1 \oplus Y_1^1[-1]_0 \oplus Y_1^1[P_1^1[208]]_0, \tag{8}$$

$$O_{(0)}^3 = s_{1(0)}^3 \oplus Y_1^3[-1]_0 \oplus Y_1^3[P_1^3[208]]_0, \tag{9}$$

$$Z_{(0)}^1 = s_{2(0)}^1 \oplus Y_2^1[-1]_0 \oplus Y_2^1[P_2^1[208]]_0, \tag{10}$$

$$Z_{(0)}^3 = s_{2(0)}^3 \oplus Y_2^3[-1]_0 \oplus Y_2^3[P_2^3[208]]_0. \tag{11}$$

Let C_1, C_2, C_3 and C_4 denote $Y_1^1[P_1^1[208]]_0$, $Y_1^3[P_1^3[208]]_0$, $Y_2^1[P_2^1[208]]_0$ and $Y_2^3[P_2^3[208]]_0$ respectively. Each row in Table 5 gives the conditions on the elements of P_1 and P_2 which when simultaneously satisfied gives $C_1 \oplus C_2 \oplus C_3 \oplus C_4 = 0$. The corresponding probabilities are also given. From Table 5, it follows that events G_2, G_3 and G_4 can be ignored when compared to G_1. We now state the following theorem.

Theorem 1. $s_1^1 = s_1^3$ when the following conditions are simultaneously satisfied.

1. $P_1^2[116] \equiv -18 \bmod 32$ (event E_1),
2. $P_1^3[116] \equiv -18 \bmod 32$ (event E_2),
3. $P_1^2[72] = P_1^3[239] + 1$ (event E_3),
4. $P_1^2[239] = P_1^3[72] + 1$ (event E_4).

Proof. The formulas for s_1^2 and s_1^3 are given below:

$$s_1^2 = ROTL32(s_1^1 + Y_1^2[P_1^2[72]] - Y_1^2[P_1^2[239]], P_1^2[116] + 18 \bmod 32), \tag{12}$$

$$s_1^3 = ROTL32(s_1^2 + Y_1^3[P_1^3[72]] - Y_1^3[P_1^3[239]], P_1^3[116] + 18 \bmod 32). \tag{13}$$

Condition 1 (i.e., $P_1^2[116] \equiv -18 \bmod 32$) reduces (12) to

$$s_1^2 = s_1^1 + Y_1^2[P_1^2[72]] - Y_1^2[P_1^2[239]].$$

Therefore, (13) becomes

$$s_1^3 = ROTL32(s_1^1 + \sum_{i=2}^3 (Y_1^i[P_1^i[72]] - Y_1^i[P_1^i[239]]), P_1^3[116] + 18 \bmod 32). \tag{14}$$

Now, condition 3 (i.e., $P_1^2[72] = P_1^3[239] + 1$) and condition 4 ($P_1^2[239] = P_1^3[72] + 1$) together imply $\sum_{i=2}^3 (Y_1^i[P_1^i[72]] - Y_1^i[P_1^i[239]]) = 0$ and hence reduce (14) to

$$s_1^3 = ROTL32(s_1^1, P_1^3[116] + 18 \bmod 32). \tag{15}$$

Table 5. When G_j $(1 \leq j \leq 4)$ occurs, $C_1 \oplus C_2 \oplus C_3 \oplus C_4 = 0$

Event	Conditions	Probability	Result
G_1	$P_1^1[208] = P_1^3[208] + 2$, $P_2^1[208] = P_2^3[208] + 2$	2^{-16}	$C_1 = C_2, C_3 = C_4$
G_2	$P_1^1[208] = P_2^1[208]$, $P_1^1[208], P_2^1[208] \leq 12$, $P_1^3[208] = P_2^3[208]$, $P_1^3[208], P_2^3[208] \leq 12$	$2^{-24.6}$	$C_1 = C_3, C_2 = C_4$
G_3	$P_1^1[208] = P_1^3[208] + 2$, $2 \leq P_1^1[208] \leq 12$, $P_1^3[208] \leq 10$, $P_2^1[208] = P_2^3[208] + 2$, $2 \leq P_2^1[208] \leq 12$, $P_2^3[208] \leq 10$	$2^{-25.4}$	$C_1 = C_4, C_2 = C_3$
G_4	$G_2 \cap G_1$	Negligible $(\ll 2^{-25})$	$C_1 = C_2 = C_3 = C_4$

Now, when event E_2 (that is, $P_1^3[116] \equiv -18 \mod 32$) occurs, (15) becomes

$$s_1^3 = ROTL32(s_1^1, 0) = s_1^1. \tag{16}$$

This completes the proof. □

Now, $s_1^1 = s_1^3 \Rightarrow s_{1(0)}^1 = s_{1(0)}^3$ and $Pr[E_1] \approx Pr[E_2] \approx 2^{-5}$ and $Pr[E_3] \approx Pr[E_4] \approx 2^{-8}$. The four events E_1, E_2, E_3 and E_4 are assumed to be independent to facilitate calculation of bias. The actual value without independence assumption is in fact more, making the attack marginally stronger. Hence, $Pr[E_1 \cap E_2 \cap E_3 \cap E_4] = 2^{-26}$. Similarly, we have $s_2^1 = s_2^3$ when the following conditions are simultaneously satisfied.

1. $P_1^2[116] \equiv -18 \mod 32$ (event E_5), **2.** $P_2^3[116] \equiv -18 \mod 32$ (event E_6),
3. $P_2^2[72] = P_2^3[239] + 1$ (event E_7),   **4.** $P_2^2[239] = P_2^3[72] + 1$ (event E_8).

Again, $s_2^1 = s_2^3 \Rightarrow s_{2(0)}^1 = s_{2(0)}^3$ and

$$Pr[\cap_{i=1}^8 E_i] = \frac{1}{2^{52}}. \tag{17}$$

From the analysis in Sect. 4.1 and 4.2, when $D_3 \cap D_2 \cap D_1$ occurs, $Y_1^1[j] = Y_2^1[j]$ where $j \in \{-3, ..., 12\}$. $Y_1^1[i] = Y_2^1[i] \Rightarrow Y_1^1[-1]_0 = Y_2^1[-1]_0$ and $Y_1^1[-1]_0 = Y_1^1[1]_0 = Y_2^1[1]_0 = Y_2^3[-1]_0$. Therefore, from equations (8), (9), (10) and (11), we observe that

$$O_{(0)}^1 \oplus O_{(0)}^3 \oplus Z_{(0)}^1 \oplus Z_{(0)}^3 = 0 \tag{18}$$

holds when the following events simultaneously occur.

1. $D_3 \cap D_2 \cap D_1$, **2.** $\cap_{i=1}^8 E_i$ and **3.** G_1.

In the following section, we calculate the probability that (18) is satisfied.

4.4   The Distinguisher

Let L denote the event $(\cap_{i=1}^8 E_i) \cap (D_3 \cap D_2 \cap D_1) \cap (G_1)$. From (7), (17) and Table 5, we get: $Pr[L] = 2^{-52} \cdot 2^{-28.4} \cdot 2^{-16} = 2^{-96.4}$. Assuming randomness of the outputs when event L does not occur (concluded from a large number of experiments), we have:

$$Pr[O_{(0)}^1 \oplus O_{(0)}^3 \oplus Z_{(0)}^1 \oplus Z_{(0)}^3 = 0] = \frac{1}{2}(1 + \frac{1}{2^{96.4}}). \tag{19}$$

To compute the number of samples required to establish an optimal distinguisher with advantage greater than 0.5, we use the following equation:

$$n = 0.4624 \cdot \frac{1}{p^2} \tag{20}$$

from [1,18]. Here, $p = 2^{-97.4}$. Therefore, the number of samples is $2^{193.7}$.

4.5   Attacks with Shorter Keys

The related-key attacks described in the previous sections can be applied with shorter keys also. However, the data complexity of the distinguisher increases exponentially as key size decreases. For example, when the key size is 128 bytes, the distinguisher works with $2^{229.7}$ data and comparable time. For 64-byte key size, the data complexity of the distinguisher is $2^{247.7}$.

5   New Stream Ciphers – RCR-32 and RCR-64

As mentioned in Sect. 1, in the last couple of years, the Py-family of ciphers have come under several cryptanalytic attacks. In spite of the weaknesses, the ciphers retain some attractive features such as modification of the internal states with clever use of rolling arrays and fast mixing of several arithmetic operations. This motivates us to explore the possibility of designing new ciphers that retain all the good properties of the Py-family and yet are secure against all the existing and new attacks.

In this section, we propose two new ciphers, RCR-32 (*R*olling, *C*onstant *R*otation, *32*-bit output per round) and RCR-64 derived from TPypy and Tpy, which are shown to be secure against all the existing attacks on the TPypy and TPy. The speeds of execution of the RCR-64 and the RCR-32 in software are 2.7 cycles and 4.45 cycles per byte which are better than the performances of the TPy (2.8 cycles/byte) and the TPypy (4.58 cycles/byte) respectively.

The key/IV setup algorithms of the RCR-64 and the RCR-32 are identical with those of the TPy and the TPypy. The PRBGs of the RCR-64 and the RCR-32 are also very similar to those of the TPy and the TPypy. The only changes in the PRBGs are that: the *variable rotation* of the quantity s is replaced by a *constant rotation* of 19. Single round of RCR-32 and RCR-64 are shown in Algorithm 1.

5.1   Security Analysis

Due to restrictions on the page limit, the security analysis has been provided in the full version of the paper [24].

Algorithm 1. Round functions of RCR-32 and RCR-64

Require: $Y[-3, ..., 256]$, $P[0, ..., 255]$, a 32-bit variable s
Ensure: 64-bit random output (for RCR-64) or 32-bit random output (for RCR-32)
    /*Update and rotate P*/
1: swap $(P[0], P[Y[185]\&255])$;
2: rotate (P);
    /* Update s*/
3: $s+ = Y[P[72]] - Y[P[239]]$;
4: $s = ROTL32(s, 19)$; /***Tweak** - the variable s undergoes a *constant*, non-zero rotation.*/
    /* Output 4 or 8 bytes (the least significant byte first)*/
5: output $((ROTL32(s, 25) \oplus Y[256]) + Y[P[26]])$;/* This step is skipped for RCR-32.*/
6: output $((\quad s \quad \oplus Y[-1]) + Y[P[208]])$;
    /* Update and rotate Y*/
7: $Y[-3] = (ROTL32(s, 14) \oplus Y[-3]) + Y[P[153]]$;
8: rotate(Y);

---

6 Future Work and Conclusion

In this paper, for the first time, we detect weaknesses in the key scheduling algorithms of several members of the Py-family. Precisely, we build distinguishing attacks with data complexities 2^{193} each. Furthermore, we modify the ciphers TPypy and TPy to generate two fast ciphers, namely RCR-32 and RCR-64, in an attempt to rule out all the attacks against the Py-family of ciphers. We conjecture that attacks lower than brute force are not possible on RCR ciphers.

Our present work leaves room for interesting future work. The usage of long keys and IVs (e.g., possibility of 256-byte keys and 64-byte IVs) in RCR ciphers makes them good candidates to be used as hash functions. One can also try to combine a MAC and an encryption algorithm in a single primitive using RCR ciphers. It seems worthwhile to address these issues in future.

References

1. Baignères, T., Junod, P., Vaudenay, S.: How Far Can We Go Beyond Linear Cryptanalysis? In: Lee, P.J. (ed.) ASIACRYPT 2004. LNCS, vol. 3329, pp. 432–450. Springer, Heidelberg (2004)
2. Biham, E.: New Types of Cryptanalytic Attacks Using Related Keys. J. Cryptology 7(4), 229–246 (1994)
3. Biham, E., Seberry, J.: Tweaking the IV Setup of the Py Family of Ciphers – The Ciphers Tpy, TPypy, and TPy6 (January 25, 2007), Published on the author's webpage at http://www.cs.technion.ac.il/~biham/
4. Biham, E., Seberry, J.: Py (Roo): A Fast and Secure Stream Cipher using Rolling Arrays. ecrypt submission (2005)
5. Biham, E., Seberry, J.: Pypy (Roopy): Another Version of Py. ecrypt submission (2006)

6. Chang, D., Gupta, K., Nandi, M.: RC4-Hash: A New Hash Function based on RC4 (Extended Abstract). In: Barua, R., Lange, T. (eds.) INDOCRYPT 2006. LNCS, vol. 4329, Springer, Heidelberg (2006)

7. Crowley, P.: Improved Cryptanalysis of Py. In: Workshop Record of SASC 2006 - Stream Ciphers Revisited, ECRYPT Network of Excellence in Cryptology, Leuven, Belgium, pp. 52–60 (February 2006)

8. Fluhrer, S., Mantin, I., Shamir, A.: Weaknesses in the Key Scheduling Algorithm of RC4. In: Vaudenay, S., Youssef, A.M. (eds.) SAC 2001. LNCS, vol. 2259, pp. 1–24. Springer, Heidelberg (2001)

9. Handschuh, H., Knudsen, L., Robshaw, M.: Analysis of SHA-1 in Encryption Mode. In: Naccache, D. (ed.) CT-RSA 2001. LNCS, vol. 2020, pp. 70–83. Springer, Heidelberg (2001)

10. Handschuh, H., Naccache, D.: SHACAL. In: First Nessie Workshop, Leuven (2000)

11. Isobe, T., Ohigashi, T., Kuwakado, H., Morii, M.: How to Break Py and Pypy by a Chosen-IV Attack. eSTREAM, ECRYPT Stream Cipher Project, Report 2006/060

12. Kelsey, J., Schneier, B., Wagner, D.: Related-key cryptanalysis of 3-WAY, Biham-DES, CAST, DES-X, NewDES, RC2, and TEA. In: Han, Y., Quing, S. (eds.) ICICS 1997. LNCS, vol. 1334, pp. 233–246. Springer, Heidelberg (1997)

13. Kelsey, J., Schneier, B., Wagner, D.: Key-Schedule Cryptoanalysis of IDEA, G-DES, GOST, SAFER, and Triple-DES. In: Koblitz, N. (ed.) CRYPTO 1996. LNCS, vol. 1109, pp. 237–251. Springer, Heidelberg (1996)

14. Knudsen, L.R.: Cryptanalysis of LOKI. In: Matsumoto, T., Imai, H., Rivest, R.L. (eds.) ASIACRYPT 1991. LNCS, vol. 739, pp. 22–35. Springer, Heidelberg (1993)

15. Knudsen, L.R.: Cryptanalysis of LOKI91. In: Zheng, Y., Seberry, J. (eds.) AUSCRYPT 1992. LNCS, vol. 718, pp. 196–208. Springer, Heidelberg (1993)

16. Knudsen, L.: A key-schedule weakness in SAFER K-64. In: Coppersmith, D. (ed.) CRYPTO 1995. LNCS, vol. 963, pp. 274–286. Springer, Heidelberg (1995)

17. Dunkelman, O., Biham, E., Kellar, N.: A Simple Related-Key Attack on the Full SHACAL-1. In: Abe, M. (ed.) CT-RSA 2007. LNCS, vol. 4377, Springer, Heidelberg (2006)

18. Paul, S., Preneel, B., Sekar, G.: Distinguishing Attacks on the Stream Cipher Py. In: Robshaw, M. (ed.) FSE 2006. LNCS, vol. 4047, pp. 405–421. Springer, Heidelberg (2006)

19. Paul, S., Preneel, B.: On the (In)security of Stream Ciphers Based on Arrays and Modular Addition. In: Robshaw, M. (ed.) FSE 2006. LNCS, vol. 4047, pp. 69–83. Springer, Heidelberg (2006)

20. Research and Development in Advanced Communication Technologies in Europe, RIPE Integrity Primitives: Final Report of RACE Integrity Primitives Evaluation (R1040), RACE (June 1992)

21. Sekar, G., Paul, S., Preneel, B.: New Weaknesses in the Keystream Generation Algorithms of the Stream Ciphers TPy and Py. In: Garay, J.A., Lenstra, A.K., Mambo, M., Peralta, R. (eds.) Information Security Conference 2007. LNCS, vol. 4779, pp. 249–262. Springer, Heidelberg (2007)

22. Sekar, G., Paul, S., Preneel, B.: Attacks on the Stream Ciphers TPy6 and Py6 and Design of New Ciphers TPy6-A and TPy6-B. In: WEWoRC-Western European Workshop on Research in Cryptology (2007)

23. Sekar, G., Paul, S., Preneel, B.: Weaknesses in the Pseudorandom Bit Generation Algorithms of the Stream Ciphers TPypy and TPy, available at http://eprint.iacr.org/2007/075.pdf

24. Sekar, G., Paul, S., Preneel, B.: Related-key Attacks on the Py-family of Ciphers and an Approach to Repair the Weaknesses, available at http://www.cosic.esat.kuleuven.be/publications/article-932.pdf
25. Tsunoo, Y., Saito, T., Kawabata, T., Nakashima, H.: Distinguishing Attack against TPypy. Selected Areas in Cryptography (to appear, 2007)
26. Wang, X., Yao, A., Yao, F.: Cryptanalysis on SHA-1. Cryptographic Hash Workshop, NIST, Gaithersburg (2005)
27. Wang, X., Yin, Y.L., Yu, H.: Finding Collisions in the Full SHA-1. In: Shoup, V. (ed.) CRYPTO 2005. LNCS, vol. 3621, pp. 17–36. Springer, Heidelberg (2005)
28. Wang, X., Yu, H.: How to Break MD5 and Other Hash Functions. In: Cramer, R.J.F. (ed.) EUROCRYPT 2005. LNCS, vol. 3494, pp. 19–35. Springer, Heidelberg (2005)
29. Wu, H., Preneel, B.: Differential Cryptanalysis of the Stream Ciphers Py, Py6 and Pypy. In: Naor, M. (ed.) Eurocrypt 2007. LNCS, vol. 4515, pp. 276–290. Springer, Heidelberg (2007)

Related-Key Differential-Linear Attacks on Reduced AES-192

Wentao Zhang[1], Lei Zhang[2], Wenling Wu[2], and Dengguo Feng[2]

[1] State Key Laboratory of Information Security,
Graduate University of Chinese Academy of Sciences, Beijing 100049, P.R. China
zhangwt06@yahoo.com
[2] State Key Laboratory of Information Security,
Institute of Software, Chinese Academy of Sciences, Beijing 100080, P.R. China
{zhanglei1015,wwl,feng}@is.iscas.ac.cn

Abstract. In this paper, we study the security of AES-192 against related-key differential-linear cryptanalysis, which is the first attempt using this technique. Among our results, we present two variant attacks on 7-round AES-192 and one attack on 8 rounds using a 5-round related-key differential-linear distinguisher. One key point of the construction of the distinguisher is the special property of MC operation of AES. Compared with the best known results of related-key impossible differential attacks and related-key rectangle attacks on AES-192, the results presented in this paper are not better than them, but the work is a new attempt, and we hope further work may be done to derive better results in the future.

Keywords: AES, cryptanalysis, related-key, differential-linear attack.

1 Introduction

AES [11] supports 128-bit block size with three different key lengths (128, 192 and 256 bits). Because of the importance of AES, it is very necessary to constantly reevaluate its security under various cryptanalytic techniques. In this paper, we study the security of 192-bit key version of AES (AES-192) against related-key differential-linear cryptanalysis.

Related-key attacks [1] allow an attacker to obtain plaintext-ciphertext pairs by using related (but unknown) keys. The attacker first searches for possible weaknesses of the encryption and key schedule algorithms, then choose an appropriate relation between keys and make two encryptions using the related keys expecting to derive the unknown key information. We can see that the attacker is authorized more power in a related-key environment than in a non-related-key environment. Therefore, for many block ciphers which have some weaknesses in its key schedule, including AES, IDEA and Shacal-1 etc, better cryptanalytic results have been derived in related-key environments compared with the known results in non-related-key environments.

If we view the expanded keys as a sequence of words, then the key schedule of AES-192 applies a non-linear transformation once every six words, whereas

K. Srinathan, C. Pandu Rangan, M. Yung (Eds.): Indocrypt 2007, LNCS 4859, pp. 73–85, 2007.
© Springer-Verlag Berlin Heidelberg 2007

the key schedules of AES-128 and AES-256 apply non-linear transformations once every four words. This property brings better and longer related-key differentials of AES-192, so directly makes AES-192 more susceptible to related-key differential-sort attacks than AES-128 and AES-256.

In the last few years, the security of AES-192 against related-key differential-sort attacks has drawn much attention from cryptology researchers [7,2,13,3,8]. In [7], Jakimoski et al. presented related-key differential attacks on 6-round AES-192 using a 4-round differential. In [7,2,13], impossible differential attacks on 7- and 8-round AES-192 were presented using almost the same impossible differential distinguisher, but the attack complexity is improved one by one. In [3,8], the security of AES-192 against the related-key boomerang attack were studied or improved. Hitherto, the best known related-key attack on AES-192 is due to Jongsung Kim et al.[8], which can attack up to 10-round AES-192.

Modern block ciphers try to avoid good long statistical properties in order to resist traditional attacks such as differential [4] and linear cryptanalysis [10], but usually good short properties still exist. Thus, assuming the attacker can construct two consecutive short distinguishers both with good statistical properties, and the two distinguishers can be combined together to form a long statistical property, then the attacker can attack more rounds using this combined long distinguisher. Up to now, there are mainly three combined attacks among cryptanalytic techniques of block ciphers (only considering non-related-key environment), they are impossible-differential attack [5], boomerang attack [12] and differential-linear attack [9]. Impossible differential attack and boomerang attack can both be regarded as combined differential-sort attacks. When mounting an impossible differential attack, the attacker needs to construct two segments of differentials both with probability 1, one in the forward direction, and the other in the reverse direction, where the intermediate differences contradict each other. While when mounting a boomerang attack, the attacker also needs to construct two segments of differentials with some probability (not always 1), then combines two suits of the two differentials (thus four short differentials) in a certain manner to derive a long distinguisher. Differential-linear attack is a combination of differential and linear attacks. When mounting a differential-linear attack, the attacker first constructs a differential with probability 1, then constructs a linear approximation, in which the differential segment creates the linear approximation also with probability 1.

As we mentioned above, there are several research results on the security of AES-192 against related-key impossible differential attack and related-key boomerang attack. But we have not see any research result on the security of AES-192 against differential-linear attack. Therefore, we study this problem in this paper.

Amongst our results, we can attack up to 8-round AES-192. We summarize our results along with the best known ones under related-key boomerang attack [8] and related-key impossible differential attack of AES-192 [13] in Table 1.

Here is the outline. In Section 2, we present a formalized description of differential-linear cryptanalysis. In Section 3, we give a brief description of AES and some notations. In Section 4, we construct a 5-round related-key differential-linear distinguisher. Using this distinguisher, Section 5 presents an attack on 7-round AES-192; Section 6 presents an attacks on 8 rounds; And Section 7 presents a second attack on 7 rounds. Finally, Section 8 summarizes this paper.

Table 1. Comparison of Some Previous Attacks with Our Attacks

Source	Number of Rounds	Data Complexity	Time Complexity	Number of Keys	Attack Type
Ref [8]	8	2^{94}RK-CP	2^{120}	2	
	9	2^{85}RK-CP	2^{182}	64	RK Boomerang
	10	2^{125}RK-CP	2^{182}	256	
	10	2^{124}RK-CP	2^{183}	64	
Ref [13]	7	2^{52}RK-CP	2^{80}	2	
	8	$2^{64.5}$RK-CP	2^{177}	2	RK Imp.Diff
	8	2^{88}RK-CP	2^{153}	2	
	8	2^{112}RK-CP	2^{136}	2	
This paper	7	2^{22}RK-CP	2^{187}	2	
	8	2^{118}RK-CP	2^{165}	2	RK Diff-Linear
	7	2^{70}RK-CP	2^{130}	2	

RK – Related-key, CP – Chosen plaintext,
Time complexity is measured in encryption units.

2 Differential-Linear Cryptanalysis

In this section, we give a formalized description of differential-linear cryptanalysis. As for related-key differential-linear cryptanalysis, we only need to modify the corresponding variables.

Differential-linear cryptanalysis can also be regarded as linear cryptanalysis whose scenario is modified from known plaintext attack into chosen plaintext attack. For a r-round block cipher, first applying a r_1-round differential characteristic from the first to the r_1-th round, then applying a r_2-round linear approximation from the $(r_1 + 1)$-th round to the $(r_1 + r_2)$-th round, as to 1R-attack, note that $r_1 + 1 + r_2 = r$.

In the following, we give a formalization of differential-linear cryptanalysis in the framework of linear cryptanalysis using the concepts above. The following r_2-round linear approximation with bias ε is applied from the $(r_1 + 1)$-th round to the $(r - 1)$-th round:

$$In_{r_1+1} \cdot \Gamma I_{r_1+1} \oplus Out_{r-1} \cdot \Gamma O_{r-1} \oplus K \cdot \Gamma K = 0 \qquad (1)$$

where In_i, Out_i denote the input and output of the i-th round, K is the cipher key, and $\Gamma I_i, \Gamma O_i, \Gamma K$ represent their masks respectively.

Assuming that the subkey of the last round k_r is known, then we can get Out_{r-1} from k_r and the ciphertext C. Let $Out_{r-1} = Dec(C, k_r)$, then we have:

$$In_{r_1+1} \cdot \Gamma I_{r_1+1} \oplus Dec(C, k_r) \cdot \Gamma O_{r-1} \oplus K \cdot \Gamma K = 0 \tag{2}$$

Let equation (2) and another (2) for two different plaintexts be XORed, then:

$$(In_{r_1+1} \oplus In^*_{r_1+1}) \cdot \Gamma I_{r_1+1} \oplus (Dec(C, k_r) \oplus Dec(C^*, k_r)) \cdot \Gamma O_{r-1} = 0 \tag{3}$$

If the two plaintexts are chosen appropriately and make

$$(In_{r_1+1} \oplus In^*_{r_1+1}) \cdot \Gamma I_{r_1+1} = 0 \tag{4}$$

Then we have the following linear approximation:

$$(Dec(C, k_r) \oplus Dec(C^*, k_r)) \cdot \Gamma O_{r-1} = 0 \tag{5}$$

Equation (5) has a bias of $2\varepsilon^2$ as calculated by Piling-up Lemma. Thus, from Ref.[10] , $8 \times (2\varepsilon^2)^{-2}$ chosen plaintext pairs are required for a high success rate attack.

Now, let us look at equation (4). In Langford and Hellman's original paper [9] and many other related papers, the difference in the bits (or bytes) of In_{r_1+1} masked by ΓI_{r_1+1} are chosen to be zero, thus obviously equation (4) holds. But notice that equation (4) only requires that the parity of the bits difference masked by ΓI_{r_1+1} equals to zero, rather than directly fixing the bits difference to zero. In [6], the authors also expressed this observation, but their work did not use such a point. In this paper, we use this point to construct a 4-round differential and a 1-round linear approximation which make equation (4) hold, otherwise only a 3-round differential can be reached because of the good diffusion of AES. Moreover, the key point of the construction of the 4-round differential is the use of one special property of MC operation of AES, which will be explained in detail later.

3 Description of AES

The AES algorithm encrypts or decrypts data blocks of 128 bits by using keys of 128, 192 or 256 bits. The 128-bit plaintexts and the intermediate state are treated as byte matrices of size 4×4, which is shown as follows:

0	4	8	12
1	5	9	13
2	6	10	14
3	7	11	15

Each round is composed of four operations:

- SubBytes(SB): applying the S-box on each byte.
- ShiftRows(SR): cyclically shifting each row (the i'th row is shifted by i bytes to the left, $i = 0, 1, 2, 3$).

- MixColumns(MC): multiplication of each column by a constant 4×4 matrix M over the field $GF(2^8)$, where M is

$$\begin{pmatrix} 02 \ 03 \ 01 \ 01 \\ 01 \ 02 \ 03 \ 01 \\ 01 \ 01 \ 02 \ 03 \\ 03 \ 01 \ 01 \ 02 \end{pmatrix}$$

- AddRoundKey(ARK): XORing the state and a 128-bit subkey.

The MixColumns operation is omitted in the last round, and an additional AddRoundKey operation is performed before the first round. We also assume that the MixColumns operation is omitted in the last round of the reduced-round variants.

The number of rounds is dependent on the key size, 10 rounds for 128-bit keys, 12 for 192-bit keys and 14 for 256-bit keys.

The key schedule of AES-192 takes the 192-bit secret key and expands it to thirteen 128-bit subkeys. The expanded key is a linear array of 4-byte words and is denoted by $G[4 \times 13]$. Firstly, the 192-bit secret key is divided into 6 words $G[0], G[1] \ldots G[5]$. Then, perform the following:

For $i = 6, \ldots 51$, do

If ($i \equiv 0 \mod 6$), then $G[i] = G[i-6] \oplus SB(G[i-1] \lll 8) \oplus RCON[i/6]$

Else $G[i] = G[i-6] \oplus G[i-1]$

where $RCON[\cdot]$ is an array of predetermined constants, $\lll$ denotes rotation of a word to the left by 8 bits.

3.1 Notations

In the rest of this paper, we will use the following notations: x_i^I denotes the input of the i'th round, while x_i^S, x_i^R, x_i^M and x_i^O respectively denote the intermediate values after the application of SubBytes, ShiftRows, MixColumns and AddRoundKey operations of the i'th round. Obviously, $x_{i-1}^O = x_i^I$ holds.

Let k_i denote the subkey in the i'th round, and the initial whitening subkey is k_0. In some cases, the order of the MixColumns and the AddRoundKey operation in the same round is changed, which is done by replacing the subkey k_i with an equivalent subkey w_i, where $w_i = MC^{-1}(k_i)$.

Let $(x_i)_{Col(l)}$ denote the l'th column of x_i, where $l = 0, 1, 2, 3$. And $(x_i)_j$ the j'th byte of $x_i (j = 0, 1, \ldots 15)$, here Column(0) includes bytes 0,1,2 and 3, Column(1) includes bytes 4,5,6 and 7, etc.

Let λ denote a 8-bit mask, which specifies the bits involved in a linear approximation. The symbol " $\cdot$ " denote a bitwise AND operation.

4   A 5-Round Related-Key Differential-Linear Distinguisher

In our following attacks, we use two kinds of differences between the two related keys. Choosing the first one, we can apply the 5-round related-key differential-linear distinguisher from the very beginning. While choosing the second one, we will apply the distinguisher from the second round.

In this section, we only consider the first case, and deal with the second one later.

We choose the difference between the two related keys as follows:
$((0,0,0,0),(0,0,0,0),(a,0,0,0),(a,0,0,0),(0,0,0,0),(0,0,0,0))$.

Hence, the subkey differences in the first 8 rounds are as presented in Table 2 according to the key schedule of AES-192.

Table 2. The first Subkey Differences

Round(i)	$\Delta k_{i,Col(0)}$	$\Delta k_{i,Col(1)}$	$\Delta k_{i,Col(2)}$	$\Delta k_{i,Col(3)}$
0	$(0,0,0,0)$	$(0,0,0,0)$	$(a,0,0,0)$	$(a,0,0,0)$
1	$(0,0,0,0)$	$(0,0,0,0)$	$(0,0,0,0)$	$(0,0,0,0)$
2	$(a,0,0,0)$	$(0,0,0,0)$	$(0,0,0,0)$	$(0,0,0,0)$
3	$(0,0,0,0)$	$(0,0,0,0)$	$(a,0,0,0)$	$(a,0,0,0)$
4	$(a,0,0,0)$	$(a,0,0,0)$	$(0,0,0,b)$	$(0,0,0,b)$
5	$(a,0,0,b)$	$(0,0,0,b)$	$(a,0,0,b)$	$(0,0,0,b)$
6	$(0,0,c,b)$	$(0,0,c,0)$	$(a,0,c,b)$	$(a,0,c,0)$
7	$(0,0,c,b)$	$(0,0,c,0)$	$(0,d,c,b)$	$(0,d,0,b)$
8	$(a,d,c,0)$	$(0,d,0,0)$	$(0,d,c,b)$	$(0,d,0,b)$

a, b, c and d are non-zero byte differences.

Now, assuming the subkey differences are as presented in Table 2, we will build a 5-round related-key differential-linear distinguisher in the following. Firstly, a 4-round related-key differential, then a 1-round related-key linear approximation.

Choosing plaintext pairs (P, P^*) with difference $(0,0,0,0),(0,0,0,0),(a,0,0,0),(a,0,0,0)$. Then, the input difference Δx_1^I is canceled by the whitening subkey difference. The zero difference Δx_1^I is preserved through all the operations until the AddRoundKey operation of the second round, as the subkey difference of the first round is zero. Thus, we can get $\Delta x_3^I = \Delta k_2 = ((a,0,0,0),(0,0,0,0),(0,0,0,0),(0,0,0,0))$, where only one byte is active. Then the next three operations in the third round will convert the active byte to a complete column of active bytes, and after the AddRoundKey operation with k_3, we will get $\Delta x_3^O = ((N,N,N,N),(0,0,0,0),(a,0,0,0),(a,0,0,0))$, where N denotes a non-zero byte (possibly distinct). Applying the SubBytes and ShiftRows operations of the 4'th round, Δx_4^O will evolve into $\Delta x_4^R = ((\delta,0,0,0),(0,0,0,N),(N,0,N,0),(N,N,0,0))$, where δ specially denotes the non-zero byte in position 0 of Δx_4^R, we can find that only one byte is active both in Column 0 and Column 1. Hence, after the following MC operation, we get $\Delta x_4^M = ((02\cdot\delta,\ \delta,\ \delta,\ 03\cdot\delta),(N,N,N,N),(?,?,?,?),(?,?,?,?))$. Finally, applying the key addition with k_4, we get $\Delta x_4^O = ((?,\delta,\delta,N),(?,N,N,N),(?,?,?,?),(?,?,?,?))$. Notice that the byte difference is equal in byte position 1 and 2 of Δx_4^O, ie., the following equation holds with probability 1:

$$\Delta(x_4^O)_1 = \Delta(x_4^O)_2 \neq 0 \tag{6}$$

In the following SB operation of the 5'th round, we use a linear approximation. δ is a non-zero byte, so there are $(2^{24} - 2^{16})$ possible values for the 4-byte element $((x_4^O)_1, (x_4^{*O})_1, (x_4^O)_2, (x_4^{*O})_2)$ which satisfies $(x_4^O)_1 \oplus (x_4^{*O})_1 = (x_4^O)_2 \oplus (x_4^{*O})_2 \neq 0$. Through calculation, we find that for every possible 8-bit linear mask $\lambda \in \{1, 2, \ldots, 255\}$, 8389632 elements satisfy the following relation out of all the $(2^{24} - 2^{16})$ 4-byte elements:

$$\lambda \cdot \{(x_5^S)_1 \oplus (x_5^{*S})_1 \oplus (x_5^S)_2 \oplus (x_5^{*S})_2\} = 0 \tag{7}$$

Thus, the bias of the above equation is $\frac{8389632}{2^{24}-2^{16}} - \frac{1}{2} \approx 2^{-9}$.

For convenience, we write equation (7) as follows:

$$\lambda \cdot \{\Delta(x_5^S)_1 \oplus \Delta(x_5^S)_2\} = 0 \tag{8}$$

Because the value of subkeys keep unchanged in the attacks, after the SR operation and AR operation (XOR with w_5) in the 5'th round, we finally get the following 5-round differential-linear distinguisher with a bias of about 2^{-9}:

$$\lambda \cdot \{\Delta(x_5^W)_{13} \oplus \Delta(x_5^W)_{10}\} = 0 \tag{9}$$

In the following attacks, we will fix $\lambda = 0x01$.

Using this 5-round related-key differential-linear distinguisher, we will present some attacks on 7- and 8-round AES-192 in the following sections.

5   Attacking 7-Round

At first, we assume the values of $a, b, c,$ and d are all known, ie., we have two related keys K_1 and K_2 with the required subkey differences listed in Table 2. We will deal with the conditions on the related keys to achieve these subkey differences at the end of this section.

In order to calculate the difference in the position of bytes 10 and 13 of x_5^W, we need to know 8 bytes in positions 2,3,5,6,8,9,12,15 of w_6 and all the 16 bytes of k_7. According to the key schedule of AES-192, we can calculate the first two columns of k_6 (thus the first two columns of w_6) if k_7 is known. So we only need to guess 4 bytes in positions 8,9,12,15 of w_6 and all the 16 bytes of k_7.

The attack procedure is as follows:

5.1   The Attack Procedure

1. Generate m plaintext pairs, for each plaintext pair (P, P^*), $P \oplus P^* = (0,0,0,0), (0,0,0,0), (a,0,0,0), (a,0,0,0)$.
2. Ask for the encryption of the plaintext pairs, one plaintext P under K_1, and the other plaintext P^* under K_2.
3. Initialize an array of 2^{160} counters to zeroes.
4. Guess the value of the subkey bytes in positions 8,9,12,15 of w_6 and all the 16 bytes of k_7, then perform the followings: For each ciphertext pair, decrypt to get the intermediate two bytes $(x_5^W)_{10}$ and $(x_5^W)_{13}$. Calculate $(0x01) \cdot \{\Delta(x_5^W)_{13} \oplus \Delta(x_5^W)_{10}\}$, if it equals to 0, increment the counter in the array which relates to the 20 guessing bytes of the subkeys.

5. Assume the highest entry is T_{max}, and the lowest entry T_{min}, if $|T_{max} - m/2| > |T_{min} - m/2|$, then adopt the key candidate corresponding to T_{max}. If $|T_{max} - m/2| < |T_{min} - m/2|$, then adopt the key candidate corresponding to T_{min}.

5.2 Analysis of the Attack Complexity

Based on Matsui's rule of thumb, approximately $m = 2^3 \times (2^{-9})^{-2} = 2^{21}$ plaintext pairs are needed. So the data complexity is 2^{22} chosen plaintexts. And the time complexity is about $2^{160} \times 2^{22} \times 2/7 \approx 2^{180}$ 7-round AES encryptions.

In the above attack, we assumed that the values of a, b, c and d are known. Here, the value a can be chosen by the attacker. The value c can be calculated from b and $(k_5)_{15} = (k_7)_3 \oplus (k_7)_7$. And the value d can be calculated from c and $(k_7)_6$. The value b is the result of application of SubBytes operation, so there are 127 possible values of b given the value of a. Hence, we only need to repeat the attack for all the possible values of b. Therefore, the total time complexity is multiplied by 2^7, ie. 2^{187}, and the data complexity remains unchanged.

6   Attacking 8-Round

To attack 8-round AES-192, we adopt another key difference between the two related keys as follows: $((a, 0, 0, 0), (0, 0, 0, 0), (a, 0, 0, 0), (0, 0, 0, 0), (0, 0, 0, 0), (0, 0, 0, 0))$. The corresponding subkey differences in the first 8 rounds are as presented in Table 3, which will be used in the attacks in this section.

Table 3. The Second Subkey Differences

Round(i)	$\Delta k_{i,Col(0)}$	$\Delta k_{i,Col(1)}$	$\Delta k_{i,Col(2)}$	$\Delta k_{i,Col(3)}$
0	$(a,0,0,0)$	$(0,0,0,0)$	$(a,0,0,0)$	$(0,0,0,0)$
1	$(0,0,0,0)$	$(0,0,0,0)$	$(a,0,0,0)$	$(a,0,0,0)$
2	$(0,0,0,0)$	$(0,0,0,0)$	$(0,0,0,0)$	$(0,0,0,0)$
3	$(a,0,0,0)$	$(0,0,0,0)$	$(0,0,0,0)$	$(0,0,0,0)$
4	$(0,0,0,0)$	$(0,0,0,0)$	$(a,0,0,0)$	$(a,0,0,0)$
5	$(a,0,0,0)$	$(a,0,0,0)$	$(a,0,0,0)$	$(a,0,0,0)$
6	$(a,0,0,b)$	$(0,0,0,b)$	$(a,0,0,b)$	$(0,0,0,b)$
7	$(a,0,0,b)$	$(0,0,0,b)$	$(a,0,c,b)$	$(a,0,c,0)$
8	$(0,0,c,b)$	$(0,0,c,0)$	$(a,0,c,b)$	$(a,0,c,0)$

a, b and c are non-zero byte differences.

As in section 3, we can construct a 5-round differential-linear distinguisher between rounds 2-6 using a very similar approach. Assume the input difference $\Delta x_1^M = (0,0,0,0), (0,0,0,0), (a,0,0,0), (a,0,0,0)$, then the following relation holds with a bias of about 2^{-9}:

$$\lambda \cdot \{\Delta(x_6^W)_{13} \oplus \Delta(x_6^W)_{10}\} = 0 \tag{10}$$

In the following attack, we need to guess the corresponding subkey bytes in round 1, 7 and 8. In order to decrease the amount of guessing subkey bytes, we make two restrictions on the data pairs which are finally used in counting. First, we restrict that the byte difference in positions 8,10,11,12,13,15 of x_6^O are all zero. Thus, to calculate $\Delta(x_6^W)_{13}$ and $\Delta(x_6^W)_{10}$, we only need to know the values of $(x_6^O)_9$ and $(x_6^O)_{14}$ because of the linearity of MC operation. Also if this first restriction holds, then the byte difference in positions 2,3,8,9,12,15 of x_6^W are all known, which is equal to the corresponding byte of Δw_7. Second, we restrict that the byte difference in positions 0,3,9,10,14,15 of x_7^O are all zero. Thus, to meet the condition on the 6 bytes of Δx_6^W, we only need to know the value in byte positions 1,2,8,11,12,13 of x_7^O. Also if this second restriction holds, then the byte difference in positions 0,2,3,5,6,7 of ciphertexts are all known. The attack procedure is also illustrated in Figure 1, where ? denotes an unknown byte, * denotes a byte we need to know.

$$
\begin{pmatrix} a & 0 & ? & ? \\ ? & 0 & 0 & ? \\ ? & ? & 0 & 0 \\ 0 & ? & ? & 0 \end{pmatrix}
\xrightarrow{AR}
\begin{pmatrix} 0 & 0 & ? & ? \\ ? & 0 & 0 & ? \\ ? & ? & 0 & 0 \\ 0 & ? & ? & 0 \end{pmatrix}
\rightarrow
$$

$$
\text{Round 1} : \xrightarrow{SB}
\begin{pmatrix} 0 & 0 & ? & ? \\ ? & 0 & 0 & ? \\ ? & ? & 0 & 0 \\ 0 & ? & ? & 0 \end{pmatrix}
\xrightarrow{SR}
\begin{pmatrix} 0 & 0 & ? & ? \\ 0 & 0 & ? & ? \\ 0 & 0 & ? & ? \\ 0 & 0 & ? & ? \end{pmatrix}
\xrightarrow[\textbf{Prob.}]{MC}
\begin{pmatrix} 0 & 0 & a & a \\ 0 & 0 & 0 & 0 \\ 0 & 0 & 0 & 0 \\ 0 & 0 & 0 & 0 \end{pmatrix}
\xrightarrow{AR}
\begin{pmatrix} 0 & 0 & 0 & 0 \\ 0 & 0 & 0 & 0 \\ 0 & 0 & 0 & 0 \\ 0 & 0 & 0 & 0 \end{pmatrix}
\rightarrow
$$

$$\longrightarrow \cdots\cdots\cdots\cdots \text{ the 5-round differential-linear distinguisher } \cdots\cdots\cdots\cdots \longleftarrow$$

$$
\text{Round 6} : \cdots\cdots\cdots\cdots\cdots\cdots\cdots\cdots\cdots\cdots\cdots\cdots\cdots\cdots\cdots
\longleftarrow
\begin{pmatrix} ? & ? & ? & ? \\ ? & ? & ? & * \\ ? & ? & * & ? \\ ? & ? & ? & ? \end{pmatrix}
\xleftarrow[\textbf{Prob.}]{MC^{-1}}
\begin{pmatrix} ? & ? & 0 & 0 \\ ? & ? & * & 0 \\ ? & ? & 0 & * \\ ? & ? & 0 & 0 \end{pmatrix}
\longleftarrow
$$

$$
\text{Round 7} : \xleftarrow{SB^{-1}}
\begin{pmatrix} ? & ? & 0 & 0 \\ ? & ? & * & 0 \\ ? & ? & 0 & * \\ ? & ? & 0 & 0 \end{pmatrix}
\xleftarrow{SR^{-1}}
\begin{pmatrix} ? & ? & 0 & 0 \\ ? & * & 0 & ? \\ 0 & * & ? & ? \\ 0 & ? & ? & 0 \end{pmatrix}
\xleftarrow{AR}
\begin{pmatrix} ? & ? & \Delta(w_7)_8 & \Delta(w_7)_{12} \\ ? & * & \Delta(w_7)_9 & ? \\ \Delta(w_7)_2 & * & ? & ? \\ \Delta(w_7)_3 & ? & ? & \Delta(w_7)_{15} \end{pmatrix}
\xleftarrow[\textbf{Prob.}]{MC^{-1}}
\begin{pmatrix} 0 & * & * & * \\ * & * & 0 & * \\ * & * & 0 & 0 \\ 0 & * & * & 0 \end{pmatrix}
\longleftarrow
$$

$$
\text{Round 8} : \xleftarrow{SB^{-1}}
\begin{pmatrix} 0 & * & * & * \\ * & * & 0 & * \\ * & * & 0 & 0 \\ 0 & * & * & 0 \end{pmatrix}
\xleftarrow{SR^{-1}}
\begin{pmatrix} 0 & * & * & * \\ * & 0 & * & * \\ 0 & 0 & * & * \\ 0 & 0 & * & * \end{pmatrix}
\xleftarrow{AR}
\begin{pmatrix} 0 & * & * & * \\ * & 0 & * & * \\ c & c & * & * \\ b & 0 & * & * \end{pmatrix}
$$

Fig. 1. Related-key Differential-linear Attack on 8-round AES-192

The attack procedure is as follows:

1. Choose a structure of 2^{64} plaintexts, which have certain fixed values in 8 byte positions 0,3,4,5,9,10,14,15 and take all the 2^{64} possible values in the other 8 byte positions. Generate n structures $S_1, S_2, \ldots, S_n$.

2. Compute n structures $S_1^*, S_2^*, \ldots, S_n^*$ by XORing the plaintexts in $S_1, S_2, \ldots, S_n$ with a 128-bit value $((a, 0, 0, 0), (0, 0, 0, 0), (0, 0, 0, 0), (0, 0, 0, 0))$.

3. Ask for the encryption of the pool S_i under K_1, and of the pool S_i^* under K_2 $(i = 1, 2, \ldots, n)$. Denote the ciphertexts of the pool S_i by T_i, and the ciphertexts of the pool S_i^* by T_i^*.

4. For all ciphertexts $C_i \in T_i$, compute $D_i = C_i \oplus ((0,0,c,b),(0,0,c,0),(0,0,0,0),(0,0,0,0))$. Select all the ciphertext pairs (D_i, C_i^*), where $C_i^* \in T_i^*$, satisfying that D_i and C_i^* have equal values in byte positions 0,2,3,5,6 and 7.

5. Guess the 8 byte values in positions 1,2,6,7,8,11,12 and 13 of k_0, and perform the followings:

   (a) For each pair remained after step 4, encrypt the corresponding plaintexts to get the difference of column 2 of x_1^M using the guess of bytes 2,7,8,13 of k_0. If the column difference is $(a, 0, 0, 0)$, then keep the pair remained. If not, discard the pair.

   (b) For each pair remained after step 5.a, calculate the difference of column 3 of x_1^M using the guess of bytes 1,6,11,12 of k_0. If the column difference is $(a, 0, 0, 0)$, then keep the pair remained. If not, discard the pair.

   (c) Initialize an array of 2^{144} counters to zeroes.

   (d) Guess the 10 byte values in positions 1,4,8,9,10,11,12,13,14 and 15 of k_8, and perform the followings:

      i. For each pair remained after step 5.b, decrypt the ciphertexts to get the difference in positions 2 and 3 of x_7^W using the guess of $(k_8)_{10}$ and $(k_8)_{13}$. And check if both $\Delta(x_7^W)_2 = \Delta(w_7)_2$ and $\Delta(x_7^W)_3 = \Delta(w_7)_3$ hold. If not, then discard the data pair.

      ii. For each pair remained after step 5.d.i, decrypt to get the difference in positions 8 and 9 of x_7^W using the guess of $(k_8)_8$ and $(k_8)_{15}$. And check if both $\Delta(x_7^W)_8 = \Delta(w_7)_8$ and $\Delta(x_7^W)_9 = \Delta(w_7)_9$ hold. If not, then discard the data pair.

      iii. For each pair remained after step 5.d.ii, decrypt to get the difference in positions 12 and 15 of x_7^W using the guess of $(k_8)_9$ and $(k_8)_{12}$. And check if both $\Delta(x_7^W)_{12} = \Delta(w_7)_{12}$ and $\Delta(x_7^W)_{15} = \Delta(w_7)_{15}$ hold. If not, then discard the data pair.

      iv. From $(k_8)_{col(2)}$ and $(k_8)_{col(3)}$, we can calculate $(k_7)_{col(1)}$, then we can know $(w_7)_5$ and $(w_7)_6$. Together with the guess of $(k_8)_1$, $(k_8)_4$, $(k_8)_{11}$ and $(k_8)_{14}$, we can do the followings: for each pair remained after step 5.d.iii, decrypt to get the two bytes in positions 9 and 14 of x_6^O. Because the byte difference in the other three bytes are all zero in the last two columns of x_6^O, we can calculate the difference in positions 10 and 13 of x_6^W, and calculate $(0x01) \cdot \{\Delta(x_6^W)_{13} \oplus \Delta(x_6^W)_{10}\}$, if it equals to 0, increment the counter in the array which relates to the 18 bytes of the subkeys.

      v. Assume the highest entry is T_{max}, and the lowest entry T_{min}, if $|T_{max} - m/2| > |T_{min} - m/2|$, then adopt the key candidate corresponding to T_{max}. If $|T_{max} - m/2| < |T_{min} - m/2|$, then adopt the key candidate corresponding to T_{min}.

6.1   Analysis of the Attack Complexity

For each plaintext pair $P \in S_i$ and $P^* \in S_i^*$, we have $P \oplus P^* = ((a, ?, ?, 0), (0, 0, ?, ?), (?, 0, 0, ?), (?, ?, 0, 0))$, where ? denotes any byte value. Thus, from the two pools of S_i and S_i^*, 2^{128} plaintext pairs can be derived, therefore we can derive $2^{128}m$ plaintext pairs in all. After the filtering in step 4, there remains about $2^{80}m$ pairs. Then, after the filtering in step 5.a and 5.b, about $2^{16}m$ pairs will remain for the 8 bytes guess of k_0, and all the remained pairs satisfy the required input difference of the 5-round differential-linear distinguisher. After step 5.d.i, about m pairs will remain for the 8 bytes guess of k_0 and 2 bytes guess of k_8. After step 5.d.ii, about $2^{-16}m$ pairs will remain for the 8 bytes guess of k_0 and 4 bytes guess of k_8. After step 5.d.iii, about $2^{-32}m$ pairs will remain for the 8 bytes guess of k_0 and 6 bytes guess of k_8. Now, the remaining pairs can be used to do counting, $2^{-32}m = 2^{21}$ pairs are needed, so $m = 2^{53}$. Therefore, the data complexity is about 2^{118} plaintexts.

The time complexity is dominated by step 5.a, step 5.b and step5.d.iv. Step 5.a requires about $2^{80+53+32+1} = 2^{166}$ one-round encryptions. Step 5.b requires about $2^{48+53+64+1} = 2^{166}$ one-round encryptions. Step 5.d.iv requires about $2^{-32+53+64+80+1} = 2^{166}$ two-round decryptions. Therefore, the whole time complexity is about 2^{165} 8-round AES encryptions.

In the above attack, we assumed that the values of a, b and c are known. Here, the value a can be chosen by the attacker. The value b can be calculated from a and $(k_5)_{12} = (k_7)_0 \oplus (k_7)_4 = (k_8)_4 \oplus (k_8)_{12}$. And the value c can be calculated from b and $(k_7)_7 = (k_8)_{11} \oplus (k_8)_{15}$. Hence, we need not to guess the values of b and c.

7   Another Attack on 7-Round AES-192

From the 8-round attack in section 5, we can naturally get a truncated attack on 7 rounds.

In this truncated 7-round attack, we restrict that the 6 byte difference in positions 2,3,8,9,12 and 15 take certain fixed values. And we need to guess two subkey bytes in positions 5 and 6 of k_7, and decrypt to get $(x_6^M)_{10}$ and $(x_6^M)_{13}$. We need to repeat this attack for all the possible values of b given a, and all the possible values of c given a and b. In all, the data complexity is about 2^{70} plaintexts, the time complexity is about 2^{130} 7-round AES encryptions.

8   Summary

We applied the related-key differential-linear cryptanalysis to AES-192. Amongst our results, we presented two variant attacks on 7-round AES-192, and an attack on 8 rounds. The comparison of our attack results and other related-key attack results on AES-192 are summarized in Table 1. The best known related-key impossible differential attacks also can reach up to 8 rounds, and the attack complexity is better than that in our paper. Moreover, the best known related-key boomerang attacks can reach up to 10 rounds! However, our attack in this paper is a new try, it is the first investigation on the strength of AES-192 against differential-linear cryptanalysis in related-key environments.

We stress that the key point of our 5-round distinguisher is the special property of MC operation of AES, ie., the coefficients in two bytes of one column are equal. The same property is used in the related-key differential attack on AES-192 [13], but the attack can only reach 7 rounds.

Finally, we point out that maybe longer effective related-key differential-linear distinguisher exists for AES, or using some auxiliary techniques (eg., enhanced differential-linear attack [6], multi-linear approximation), better attack results may be derived. And we expect further research on related-key differential-linear cryptanalysis of AES.

Acknowledgment

We would like to thank anonymous referees for their helpful comments and suggestions. The research presented in this paper is supported by the National Natural Science Foundation of China (No. 90604036, 60603018); the National Basic Research 973 Program of China (No. 2004CB318004); the National High Technology Research and Development 863 Program of China (No. 2007AA01Z470).

References

1. Biham, E.: New Types of Cryptanalytic Attacks Using Related Keys. Journal of Cryptology 7(4), 229–246 (1994)
2. Biham, E., Dunkelman, O., Keller, N.: Related-Key Impossible Differential Attacks on 8-Round AES-192. In: Pointcheval, D. (ed.) CT-RSA 2006. LNCS, vol. 3860, pp. 21–33. Springer, Heidelberg (2006)
3. Biham, E., Dunkelman, O., Keller, N.: Related-Key Boomerang and Rectangle Attacks. In: EUROCRYPT 2005. LNCS, vol. 3557, pp. 507–525. Springer, Heidelberg (2005)
4. Biham, E., Shamir, A.: Differential cryptanalysis of DES-like cryptosystems. Journal of Cryptology 4(1), 3–72 (1991)
5. Biham, E., Biryukov, A., Shamir, A.: Cryptanalysis of Skipjack Reduced to 31 Rounds. In: Stern, J. (ed.) EUROCRYPT 1999. LNCS, vol. 1592, pp. 12–23. Springer, Heidelberg (1999)
6. Biham, E., Dunkelman, O., Keller, N.: Enhancing Differential-Linear Cryptanalysis. In: Zheng, Y. (ed.) ASIACRYPT 2002. LNCS, vol. 2501, pp. 254–266. Springer, Heidelberg (2002)
7. Jakimoski, G., Desmedt, Y.: Related-Key Differential Cryptanalysis of 192-bit Key AES Variants. In: Matsui, M., Zuccherato, R.J. (eds.) SAC 2003. LNCS, vol. 3006, pp. 208–221. Springer, Heidelberg (2004)
8. Kim, J., Hong, S., Preneel, B.: Related-Key Rectangle Attacks on Reduced AES-192 and AES-256. In: Encryption 2007. LNCS, vol. 4593, pp. 225–241. Springer, Heidelberg (2007)
9. Langford, S.K., Hellman, M.E.: Differential-linear cryptanalysis. In: Desmedt, Y.G. (ed.) CRYPTO 1994. LNCS, vol. 839, pp. 17–25. Springer, Heidelberg (1994)

10. Matsui, M.: Linear Cryptanalysis Method for DES Cipher. In: Helleseth, T. (ed.) EUROCRYPT 1993. LNCS, vol. 765, pp. 386–397. Springer, Heidelberg (1994)
11. National Institute of Standards and Technology. Advanced Encryption Standard (AES), FIPS Publication 197 (November 26, 2001),
    Available at http://csrc.nist.gov/encryption/aes
12. Wagner, D.: The Boomerang Attack. In: Knudsen, L.R. (ed.) FSE 1999. LNCS, vol. 1636, pp. 156–170. Springer, Heidelberg (1999)
13. Zhang, W., Wu, W., Zhang, L., Feng, D.: Improved Related-Key Impossible Differential Attacks on Reduced-Round AES-192. In: Cryptography 2006. LNCS, vol. 4356, pp. 15–27 (2007)

Improved Meet-in-the-Middle Attacks on Reduced-Round DES*

Orr Dunkelman, Gautham Sekar, and Bart Preneel

[1] Katholieke Universiteit Leuven
Department of Electrical Engineering ESAT/SCD-COSIC
[2] Kasteelpark Arenberg 10, B-3001 Leuven-Heverlee, Belgium
{orr.dunkelman,gautham.sekar,bart.preneel}@esat.kuleuven.be

Abstract. The Data Encryption Standard (DES) is a 64-bit block cipher. Despite its short key size of 56 bits, DES continues to be used to protect financial transactions valued at billions of Euros. In this paper, we investigate the strength of DES against attacks that use a limited number of plaintexts and ciphertexts. By mounting meet-in-the-middle attacks on reduced-round DES, we find that up to 6-round DES is susceptible to this kind of attacks. The results of this paper lead to a better understanding on the way DES can be used.

1 Introduction

The Data Encryption Standard (DES) is a well known and widely deployed cipher since its standardization in 1977. Its wide deployment, even today, makes it a target for repeated analyses, as the security of many electronic transactions still relies on DES. The cipher is a Feistel block cipher with 16 rounds, 64-bit block and 56-bit key.

Due to its importance, DES has received a great deal of cryptanalytic attention. However, besides using the complementation property, there were no short-cut attacks against the cipher until differential cryptanalysis was applied to the full DES in 1991 [2].

In [3], Chaum and Evertse presented several meet in the middle attacks on reduced variants of DES. They showed that the first six round of DES are susceptible to meet-in-the-middle attacks, such as rounds 2–8. They also showed that their approach cannot be extended to more than seven rounds of DES.

In 1987 Davies described a known plaintext attack on DES [6]. The attack obtains 16 linear equations of the key bits given sufficiently many known plaintexts by examining the bits that are shared by neighboring S-boxes. Davies' on the full DES requires more plaintexts than the entire code book. For 8-round

* This work was supported in part by the Concerted Research Action (GOA) Ambiorics 2005/11 of the Flemish Government and by the IAP Programme P6/26 BCRYPT of the Belgian State (Belgian Science Policy). The second author is supported by an IWT SoBeNeT project.

K. Srinathan, C. Pandu Rangan, M. Yung (Eds.): Indocrypt 2007, LNCS 4859, pp. 86–100, 2007.
© Springer-Verlag Berlin Heidelberg 2007

DES, the attack requires about 2^{40} known plaintexts. In [7] these results were slightly improved but still could not attack the full DES faster than exhaustive key search. In 1994 Biham and Biryukov [1] improved the attack to be applicable to the full DES. Their variant of the attack requires 2^{50} known plaintexts and has a running time of 2^{49} encryptions on an average. A chosen ciphertext variant of the attack is presented in [13]; it has a data complexity of 2^{45} chosen plaintexts.

The first attack on DES that is faster than exhaustive key search was presented in [2]. The attack, differential cryptanalysis, requires 2^{47} chosen plaintexts. The attack examines pairs of plaintexts and ciphertexts, trying to find a pair that satisfies some differential (i.e., given some input difference between the two plaintexts, the output difference of the two ciphertexts is as predicted). Once such a pair is found, the key can be deduced.

In [14] another attack on DES is presented, linear cryptanalysis. The linear attack on DES uses 2^{43} known plaintexts and deduces the key by checking whether some linear relation between the plaintext and the ciphertext is satisfied. This attack was later improved by Shimoyama and Kaneko by exploiting nonlinear relations as well [19]. The improved attack has a data complexity of $2^{42.6}$ known plaintexts. Using chosen plaintexts, Knudsen and Mathiassen reduced the data complexity in [14] by a factor of 2.

Even after DES was theoretically broken, it was claimed that DES was still secure, as it was not possible to mount these attacks in practice. RSA Data Security Inc. has issued several "DES Challenges" during the mid '90s. In each such challenge, RSA published a plaintext and its ciphertext encrypted using DES under some unknown key, and offered a prize of several thousand US dollars for whoever finds the secret key [4]. The first exhaustive key search took about 75 days and the key was found using 14,000–80,000 computers over the Internet [20]. Ever since, the time required for each new DES challenge has been reduced. In 1997 the Electronic Frontier Foundation (EFF) built a special purpose machine that costs 250,000 US dollars which retrieved the key in 56 hours by means of exhaustive key search [9]. Today, using a COPACOBANA machine an exhaustive key search of DES can be performed in 17 days for the cost of less than 9000 Euros [12].

A new approach was presented by Raddum and Semaev for solving sparse systems of non-linear equations in [17] and used them to attack up to 4 rounds of DES. With 16 known plaintext-ciphertext pairs, their techniques produce an equation system with 1080 variables and 2048 non-linear equations from 4-round DES. While their methods work on 5 or more rounds of DES, they are too complex to consider in practice.

Recently, Courtois and Bard claimed that practical algebraic attacks are possible for reduced-round versions of DES [5]. Their attack represents DES as a system of multivariate equations with the key bits as unknowns and tries to solve the system using SAT solvers. Their technique can find the key with up to six rounds of DES faster than exhaustive key search.

Motivation behind our work: Despite the well known weaknesses of DES, the cipher is still widely deployed and used. Besides, DES-like ciphers are being suggested as a solution for encryption in RFID systems [16].

All the existing attacks on DES either use a long time (exhaustive key search) or use a very large number of plaintexts. This motivated us to investigate how many rounds of DES can be broken using the meet-in-the-middle technique, using one (or very few) plaintexts. We aimed at finding the best attacks on reduced-round DES. The results of this paper shed more light on the security of DES, leading to a better understanding on the way DES can be used.

Contribution of this paper: We improve the attacks due to Chaum and Evertse [3] by performing the meet-in-the-middle in a slightly different manner than done earlier. Rather than guessing all the key bits that are required to produce some value, our approach guesses actual intermediate encryption values, thus saving the need to guess many key bits to obtain the value of an intermediate encryption bit.

The new approach reduces the time complexity of the meet-in-the-middle attacks, as it allows for guessing significantly less number of key bits. Moreover, by obtaining several known plaintexts, one can increase the number of intermediate encryption bits that are guessed. This follows from the fact that even if with only one of the known plaintexts, a specific key guess has no possible intermediate encryption value which fits the meet-in-the-middle condition, then the key guess is necessarily wrong.

Another possible use of our approach is in the chosen text scenario, where by fixing some bits of the plaintext (or the ciphertext), it is possible to force the intermediate values of several plaintext/ciphertext pairs to a specific value. This leads to a reduction in the number of bits that the attacker needs to guess (across several plaintext/ciphertext pairs).

This approach may be used to improve other meet in the middle attacks. To the best of our knowledge this is the first case where the attacker guesses intermediate encryption values rather than keys in a meet-in-the-middle attack. In this paper, we also provide insights into how our attacks might be extended to attack DES with more than 6 consecutive rounds using a similar approach as described above.

We compare the results of our attack with other attacks in in Table 1. We note that for differential and linear cryptanalysis we used a lower bound based on a linear attack with one active S-box in the round before and a round after the approximation is used. We used a similar lower bound for a differential attack on DES (taking into consideration a 3R attack). We also note that these attacks have two properties which make them inferior to our results: first of all, these attacks are statistical, i.e., while our approach ensures finding the key, statistical attacks may fail. In the table we mentioned the complexities of these attacks with at least 90% success rate. The second property is that the mentioned time complexities for these attacks is the time complexity required to retrieve several key bits, while our complexities are mentioned for finding the entire key.

Table 1. Comparison of Attacks on Reduced-Round DES

Rounds	Attack	Data Complexity	Time Complexity
4	Differential	16 CP	Negligible[†]
	Linear	52 KP	$> 2^{13.7}$ [†]
	Algebraic ([5])	1 KP	2^{46}
	MitM ([3])	1 KP	2^{35} [†]
	MitM (Section 4.1)	1 KP	$2^{32.0}$
	MitM (Section 4.2)	15 KP	$2^{20.0}$
	MitM (Section 4.3)	6 CC	$2^{19.3}$
5	Differential	64 CP	$> 2^{11.7}$ [†]
	Linear	72 KP	$> 2^{13.8}$ [†]
	Algebraic ([5])	3 KP	$2^{54.3}$
	MitM ([3])	1 KP	$2^{45.5}$ [†]
	MitM (Section 5)	51 KP	$2^{35.5}$
	MitM (Section 5)	28 KP	$2^{37.9}$
	MitM (Section 5.1)	8 CP	2^{30}
6	Differential	256 CP	$2^{13.7}$
	Linear	> 104 KP	$2^{13.9}$ [†]
	Algebraic ([5])	N/A	$2^{50.1}$
	MitM ([3])	1 KP	$2^{52.9}$ [†]
	MitM (Section 6)	1 KP	$2^{51.8}$

KP — Known Plaintexts, CP — Chosen Plaintexts,
CC — Chosen Ciphertexts, N/A — Not Available
Time complexity is given in full encryption units.
[†] — The attack retrieves only parts of the key.

The paper is organized as follows: Section 2 describes DES. In Sect. 3 we give an alternative description of DES and give the notations used in this paper. Our attack on 4-round DES is described in Sect. 4. Our results on 5-round and 6-round DES are described in Sect. 5 and Sect. 6, respectively. Finally, we present our conclusions and a few open problems in Sect. 7.

2   Description of DES

The Data Encryption Standard (DES) was accepted as the American standard in 1977 and became a de-facto standard for most protocols around the world [15]. DES is a 16-round Feistel block cipher, which accepts a 64-bit block and encrypts it under a 56-bit key. The input is divided into two halves, left and right, each consisting of 32 bits. The round function is applied 16 times to the two halves. In each round, the right half enters the F-function of DES along with the round's subkey. The output of F is XORed to the left half. Then, the two halves are swapped. We give the outline of DES in Figure 1.

Let $IP(x)$ be the operation of permuting a vector $x \in \{0, 1\}^{64}$ according to the initial permutation, and let $FP(x)$ be the final permutation. These permutations satisfy $FP = IP^{-1}$. As both $IP(\cdot)$ and $FP(\cdot)$ have no cryptographic effect in block modes such as ECB or CBC, we disregard their existence. Let L_{in}, R_{in}

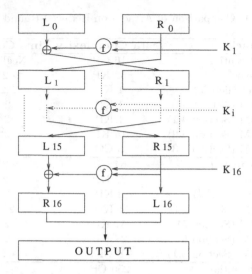

Fig. 1. General structure of the Data Encryption Standard

be the left and right halves, respectively, entering the round, and let L_{out}, R_{out} be the left and right halves that the round outputs. Then, the round function is denoted by $(L_{out}, R_{out}) = Round_{K_r}(L_{in}, R_{in})$ and $K_r \in \{0,1\}^{48}$ is the round subkey. Given this setting, one round of DES (without the swap of the Feistel construction) is represented by $R_{out} = R_{in}, L_{out} = L_{in} \oplus F(R_{in}, K_r)$.

Note that the Initial Permutation and the Final Permutation are omitted in Figure 1. The F-function of DES accepts an input of 32 bits along with a 48-bit subkey. The input is expanded into 48 bits (by duplicating 16 of the 32 input bits), and the expanded input is XORed with the subkey. The 48-bit outcome is divided into eight groups of six bits each. Each group enters a 6x4 S-box which is

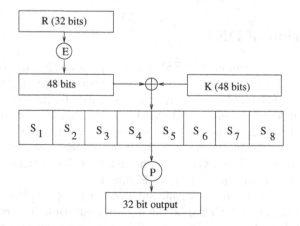

Fig. 2. F-function of DES

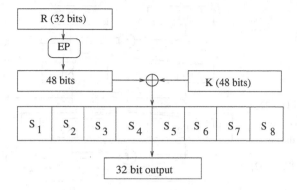

Fig. 3. An alternative description of DES F-function

a nonlinear look up table. The same eight S-boxes $S1, S2, \ldots, S8$ are applied in the same order in each round. The output of the S-boxes is permuted according to some permutation table P, and becomes the output of F. The outline of F is given in Fig. 2.

The key schedule algorithm of DES takes as an input the 56-bit user supplied key, K, and produces 16 subkeys, $K_1, \ldots, K_{16}$, where each subkey is 48 bits long. The algorithm uses two tables namely, Permuted Choice-1 (*PC-1*) and Permuted Choice-2 (*PC-2*). For most applications discussed in this paper, the details of how the subkeys are derived are not important, therefore, we omit its full description and refer the reader to [15].

3   An Alternative Description of DES and Notations Used

Since this paper is based on [3], we retain the same alternative description of DES used by Chaum and Evertse. In their alternative description of DES, IP, FP, *PC-1* are not used and E, P are combined into one table EP. This model makes the description of the results more clear, while not affecting the correctness of the result. The F-function of the alternative description is illustrated in Fig. 3.

Let K denote the full 56-bit user supplied key. Following [15], we use the big endian notations, i.e., 'bit 1' is the most significant bit of the key, and 'bit 56' is the least significant bit of the key. We denote the i-th subkey by K_i. Finally, let Y be some variable (an intermediate encryption value or a key). We use $Y[a–b]$ to denote bits $a, \ldots, b$ of Y.

4   Meet-in-the-Middle Attack on 4-Round DES

We now describe our attack on 4-round DES. First, we start with a short description of meet-in-the-middle attacks. Let $\mathcal{M}$ denote the message space and $\mathcal{K}$ denote the key space. Suppose that $G_K, H_K : \mathcal{M} \times \mathcal{K} \to \mathcal{M}$ are two block ciphers and let $F_K = H_K \circ G_K$. In a meet-in-the-middle attack, the attacker

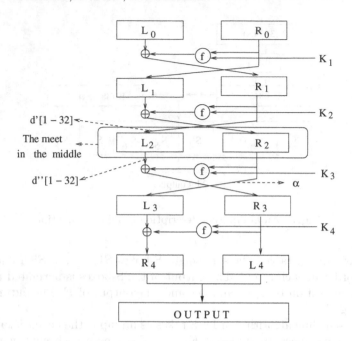

Fig. 4. 4-Round DES

tries to deduce K from a given plaintext ciphertext pair $c = F_K(p)$ by trying to solve

$$G_K(p) = H_K^{-1}(c). \qquad (1)$$

In some of the cases, the equation is not tested for all the bits of the intermediate encryption value, but rather to only some of them.

Let $d'[1\text{-m}] = G_K(p)$ and $d''[1\text{-m}] = H_K^{-1}(p)$. In our attack on 4-round DES, G_K consists of the first 2 rounds of DES and H_K contains of rounds 3 and 4. Let us consider $d'[9\text{-}12]$ and $d''[9\text{-}12]$ as illustrated in Fig. 4.

It was observed in [3] that in order to compute $d'[9\text{-}12]$ and $d''[9\text{-}12]$, it is sufficient to guess only 37 key bits. Thus, if for a key guess the computed values of $d'[9\text{-}12]$ and $d''[9\text{-}12]$ disagree, then the key guess cannot be correct (as it leads to contradiction) and can be discarded.

Our main observation is the fact that the values of $d'[9\text{-}12]$ and $d''[9\text{-}12]$ can be computed by guessing less key bits in exchange for guessing internal bits. Consider $d'[9\text{-}12]$, this value is equal to:

$$d'[9\text{-}12] = L_0[9\text{-}12] \oplus S_3[EP(R_0)[13\text{-}18] \oplus K_1[13\text{-}18]] \qquad (2)$$

and $d''[9\text{-}12]$ is equal to

$$d''[9\text{-}12] = L_4[9\text{-}12] \oplus S_3[EP(L_3)[13\text{-}18] \oplus K_3[13\text{-}18]]. \qquad (3)$$

Let $L_3 = [\alpha_1\text{-}\alpha_{32}]$, then

$$EP(L_3)[13\text{-}18] = [\alpha_{17}\alpha_1\alpha_{15}\alpha_{23}\alpha_{26}\alpha_5]. \qquad (4)$$

If we guess $K_1[13-18]$ and $K_3[13-18]$, the only remaining unknowns in the computation of $d''[9-12]$ are $[\alpha_{17}\alpha_1\alpha_{15}\alpha_{23}\alpha_{26}\alpha_5]$.

Consider α_{17}. In order to compute this bit we can either guess key bits $K_4[25-30]$ or guess α_{17} directly. Thus, a different attack algorithm for the meet-in-the-middle attack would be to guess all the 37 key bits suggested by Chaum and Evertse, besides the 6 bits which compose $K_4[25-30]$. For each guess of the 31 key bits, the attacker tries the two possibilities of α_{17}. If for both values the equality $d'[9-12] = d''[9-12]$ is not achieved, then the guess of the 31 bits is necessarily wrong. As for a specific (wrong) guess of the key and of α_{17} the probability of equality is $1/16$, the probability that a wrong 31-bit key guess has at least one α_{17} for which the equality is satisfied is $1 - (15/16)^2 \approx 1/8$. Hence, the attacker can guess 31 bits, and by trying the two possibilities of α_{17} reduce the number of remaining candidates to 2^{28}. From this point, the attacker can either repeat Chaum and Evertse's original attack or use a more advanced approach. In Table 2 we list the required key bits for determining $d'[9-12]$ and $d''[9-12]$, and note which of the key bits determine only one of them.

For example, the attacker can guess several α_i values simultaneously, thus reducing the number of possible keys (in exchange for increasing the probability that a wrong key remains). For example, if the attacker guesses two intermediate encryption bits, the probability that a key remains is $1 - (15/16)^4 \approx 2^{-2.1}$. For three and four intermediate bits the remaining probabilities are $2^{-1.3}$ and $2^{-0.6}$, respectively. This approach can lead, in the extreme, to the following meet-in-the-middle attack:

4.1 A Meet-in-the-Middle Attack with One Known Plaintext

We first define a procedure to analyze a meet-in-the-middle attack on a specific S-box. Attacking Sx in round 2 means that we guess the key which enters this S-box, as well as Sx in round 4 (in order to determine their outputs). We also need to know the 6 bits which enter this S-box, i.e., we need to know the output of 6 S-boxes in round 1. For example, performing a meet-in-the-middle on $S3$ of round 2 involves guessing $K_1[1-12]$, $K_1[19-24]$, $K_2[13-18]$, $K_4[13-18]$ (a total of 19 bits), and guessing 3 intermediate encryption values (α_{17}, α_{23}, α_{26}). Thus, it is expected that after such an analysis, of the 2^{19} possible values for the 19-bit key, only $2^{17.7}$ values remain. Similarly, one can define a meet-in-the-middle attack on Sx in round 3 (while guessing the key of Sx in round 1, and the output of 6 S-boxes in round 4).

To describe the attack algorithm, we give the sequence of attacked S-boxes. For each step, we give the number of additional key bits to be guessed, along with the number of intermediate bits that the attacker has to guess, and the number of remaining key guesses after the S-box is attacked. The attacker can retrieve the full key using about $2^{32.0}$ 4-round DES encryptions by attacking the following sequence of S-boxes:

| Round | S-box | Number of Guessed | | Number of Remaining |
		Key Bits	Intermediate Bits	Key Guess
2	$S3$	19	3	$2^{19} \cdot 2^{-1.3} = 2^{17.7}$
3	$S2$	+3	4	$2^{17.7} \cdot 2^3 \cdot 2^{-0.6} = 2^{20.1}$
2	$S1$	+2	4	$2^{20.1} \cdot 2^2 \cdot 2^{-0.6} = 2^{21.5}$
3	$S4$	+3	3	$2^{21.5} \cdot 2^3 \cdot 2^{-1.3} = 2^{23.2}$
$2^{\dagger}$	$S4$	+1	3	$2^{23.2} \cdot 2^1 \cdot 2^{-1.3} = 2^{22.9}$
3	$S3$	-	3	$2^{22.9} \cdot 2^{-1.3} = 2^{21.6}$
2	$S2$	-	4	$2^{21.6} \cdot 2^{-0.6} = 2^{21.0}$
3	$S1$	-	4	$2^{21.0} \cdot 2^{-0.6} = 2^{20.4}$
2	$S8$	+9	2 $(-2)^{\ddagger}$	$2^{20.4} \cdot 2^9 \cdot 2^{-4} = 2^{25.4}$
3	$S5$	+5	1 $(-5)^{\ddagger}$	$2^{25.4} \cdot 2^5 \cdot 2^{-8} = 2^{22.4}$
3	$S6$	+4	2 $(-5)^{\ddagger}$	$2^{22.4} \cdot 2^4 \cdot 2^{-7} = 2^{19.4}$
2	$S7$	+4	1 $(-4)^{\ddagger}$	$2^{19.4} \cdot 2^4 \cdot 2^{-7} = 2^{16.4}$
3	$S7$	+3	2 $(-5)^{\ddagger}$	$2^{16.4} \cdot 2^3 \cdot 2^{-7} = 2^{12.4}$
3	$S8$	+2	1 $(-9)^{\ddagger}$	$2^{12.4} \cdot 2^2 \cdot 2^{-12} = 2^{2.4}$
Exhaustively search the remaining $2^{3.4}$ keys.				
$\dagger$ — At this point the entire half of the key is known.				
$\ddagger$ — The $(-i)$ means that there i bits that were earlier guessed are now known (and can be used to discard wrong guesses).				

4.2 Using Multiple Known Plaintexts

If several plaintext/ciphertext pairs are at the disposal of the attacker, they can be used to deduce the value of the first 19 guessed bits in a more efficient way. The attacker uses the first plaintext/ciphertext pair to reduce the number of possible keys to $2^{17.7}$. Then, using the next plaintext/ciphertext pair, he repeats the analysis (with less candidates for the 19 bits of the key). As the probability that a key remains after each iteration of the analysis is $1 - (15/16)^8 \approx 0.4$, the number of trials t required for discarding all the wrong keys satisfies: $2^{19} \cdot 0.4^t < 1$. Thus, after 15 plaintext/ciphertext pairs, we expect to have only the right value for 19 key bits, which can then be used to retrieve the remaining key bits in a similar manner.

The time complexity of the attack in this case is about 2^{20} full 4-round DES encryptions (there are 2^{19} keys, and 2^3 intermediate values to check for each of them).

4.3 Using Chosen Ciphertexts

It is also possible to use chosen ciphertexts to improve the data complexity of the known plaintext attack. If we choose the ciphertexts in such a way that the intermediate encryption bits which are guessed are the same for all the ciphertexts, we actually improve the filtering each new plaintext/ciphertext pair offers. This follows the fact that in the known plaintext scenario, each plaintext/ciphertext pair may "allow" a key guess to pass due to a different value in the intermediate

Table 2. Key bits determining the 'middle' bits of 4-round DES

Round/S-box	Key bits	Bit determined	Bits appearing once [†]
1/3	5, 9, 13, 20, 24, 27		24
1/8	30, 33, 37, 43, 47, 51	α_{17}	30, 33, 37, 43, 47, 53
3/3	2, 8, 12, 16, 23, 27		
4/1	2, 7, 11, 17, 20, 23	α_1	7, 11, 17
4/2	6, 9, 12, 16, 21, 27	α_5	6, 21
4/4	5, 8, 13, 19, 22, 26	α_{15}	19, 22, 26
4/6	29, 36, 39, 46, 51, 54	α_{23}	29, 36, 39, 46, 51, 54
4/7	31, 34, 40, 45, 50, 55	α_{26}	31, 34, 40, 45, 50, 55
Bits of K not affecting (1)	1,3,4,10,14,15,18,25,28,32,38,41,42,44,48,49,52,56		

[†] — These bits appear only once in computing d' and d''.

encryption values. In the chosen ciphertext scenario, the attacker guesses the 19 key bits. A key which is not discarded, but has less than 8 possible intermediate encryption values (which is the case for most of the keys), is tested with the next plaintext/ciphertext pair only with the intermediate encryption values which satisfied the meet-in-the-middle condition earlier.

Thus, a given key has probability 0.6 to be discarded with the first plaintext/ciphertext pair, probability 0.32 to pass to the next pair with only one candidate value for the intermediate encryption bits, probability 0.074 to pass to the next pair with two possible values in the intermediate encryption values, and so forth. Thus, it is expected that the next pair discards 15 out of 16 remaining keys with one value, and about 14 out of 16 keys remaining with one value (while reducing the number of possible intermediate encryption values of most of them to 1). We conclude that 6 chosen ciphertexts are sufficient to find the first 19 key bits (from where by repeating the previous attacks we can find the rest of the key). The running time of the attack is $2^{19.3}$ encryptions.

5   Attack on 5-Round DES

In this case, G_K is a block cipher consisting of the first 2 rounds and H_K contains rounds 3, 4 and 5 of DES. Let us consider the intermediary bits $d'[41–44]$ and $d''[41–44]$.

We present the results of our analysis of 5-round DES in Table 3. It was observed in [3] that in order to compute $d'[41–44]$ and $d''[41–44]$, it is sufficient to guess only 47 key bits. Thus, if for a key guess the computed values of $d'[41–44]$ and $d''[41–44]$ disagree, then the key is necessarily wrong, and can be discarded.

Again, the values of $d'[41–44]$ and $d''[41–44]$ can be computed by guessing less key bits in exchange for guessing the values of intermediate bits. Consider $d'[41–44]$, this value is equal to:

$$d'[41–44] = R_0[9–12] \oplus S_3[EP(R_1)[13–18] \oplus K_2[13–18]] \tag{5}$$

Table 3. Key bits determining the 'middle' bits of 5-round DES

Round/S-box	Key bits	Bit determined	Bits appearing once [†]
1/1	2,6,12,15,18,25	β_1	2, 12
1/2	1,4,7,11,16,22	β_5	16
1/4	3,8,14,17,21,28	β_{15}	3, 17
1/5	32,38,42,48,53,56	β_{17}	
1/6	31,34,41,46,49,52	β_{23}	34, 46
1/7	29,35,40,45,50,54	β_{26}	40, 50, 54
2/3	6,10,14,21,25,28		
4/3	1,4,10,14,18,25		
5/1	4,9,13,19,22,25	γ_1	9, 13, 19
5/2	1,8,11,14,18,23	γ_5	23
5/4	7,10,15,21,24,28	γ_{15}	24
5/5	32,35,39,45,49,55	γ_{17}	39,55
5/6	31,38,41,48,53,56	γ_{23}	
5/7	29,33,36,42,47,52	γ_{26}	33, 36, 47
Bits of K not affecting (1)		5,20,26,27,30,37,43,44,51	

[†] — These bits appear only once in computing d' and d''.

and $d''[41\text{--}44]$ is equal to

$$d''[41\text{--}44] = L_5[9\text{--}12] \oplus S_3[EP(L_4)[13\text{--}18] \oplus K_4[13\text{--}18]]. \qquad (6)$$

Let $R_1 = [\beta_1\text{--}\beta_{32}]$, $L_4 = [\gamma_1\text{--}\gamma_{32}]$. Then,

$$EP(R_1)[13\text{--}18] = [\beta_{17}\beta_1\beta_{15}\beta_{23}\beta_{26}\beta_5], \qquad (7)$$

$$EP(L_4)[13\text{--}18] = [\gamma_{17}\gamma_1\gamma_{15}\gamma_{23}\gamma_{26}\gamma_5]. \qquad (8)$$

If we guess $K_2[13\text{--}18]$ and $K_4[13\text{--}18]$, the only unknowns left in the computations of $d'[41\text{--}44]$ and $d''[41\text{--}44]$ are $[\beta_{17}\beta_1\beta_{15}\beta_{23}\beta_{26}\beta_5]$ and $[\gamma_{17}\gamma_1\gamma_{15}\gamma_{23}\gamma_{26}\gamma_5]$.

Consider β_1. In order to compute this bit we can either guess 2 key bits (of $K_1[1\text{--}6]$) or guess β_1 directly. Thus, a different attack algorithm for the meet-in-the-middle attack would be to guess all the 47 key bits suggested by Chaum and Evertse, besides the 2 key bits. For each guess of the 45 key bits, the attacker tries the two possibilities of β_1. If for both values the equality $d'[41\text{--}44] = d''[41\text{--}44]$ is not achieved, then the guess of the 45 bits is necessarily wrong. As for a specific (wrong) guess of the key and of β_1 the probability of equality is 1/16, the probability that a wrong 45-bit key has at least one β_1 for which the equality is satisfied is $1 - (15/16)^2 \approx 1/8$. Hence, the attacker can guess the 45 bits, and by trying the two possibilities of β_1 reduce the number of remaining candidates to 2^{42}. From this point, the attacker can either repeat Chaum and Evertse's original attack or use a more advanced approach along similar lines as the method described in Sect. 4.2.

However, a more efficient attack exists. We note that there are many bits which are used twice in determining the values of $\beta_{17}, \beta_{23}, \beta_{26}$ and γ_{17}, γ_{23}, and γ_{26}. However, it is still more efficient to guess the value of $\beta_{23}, \beta_{26}, \gamma_{17}$, and γ_{26} than guessing these bits directly. More precisely, to determine β_{17} and γ_{23} it is sufficient to guess 8 key bits, that along with the 24 bits required for the S-boxes affected by bits 1–28 of the key, are sufficient to determine the values of $d'[41-44]$ and $d''[41-44]$ up to the value of the four intermediate bits.

Thus, we can optimize the known plaintext attack to guess 32 bits of the key, along with 4 intermediate encryption bits. The data complexity of the attack in that case is 51 known plaintexts, with time complexity of about $2^{35.5}$ 5-round DES encryptions. It is possible to guess 4 more key bits in order to determine the value of β_{23}, thus reducing the data complexity of the attack to 28 known plaintexts while the time complexity is increased to $2^{37.9}$.

5.1   Using Chosen Plaintexts

The attacker can choose the plaintexts such that the values of β_{17}, β_{23} and β_{26} are the same for all the plaintexts. Then, the attacker guesses these bits as part of the key, but can now correlate between the various plaintexts in a much stronger way.

By guessing for each possible guess of the 24 key bits affecting S-boxes S1, S2, S3, and S4, and of the three fixed bits β_{17}, β_{23} and β_{26}, the attacker tries the 8 possibilities for the unknown γ bits. The data complexity of this attack is 8 chosen plaintexts, and the time complexity is about 2^{30} 5-round DES encryptions.

6   Attack on 6-Round DES

In this case, G_K is a block cipher consisting of the first 3 rounds and H_K contains rounds 4, 5 and 6 of DES. Let us consider the intermediary bits $d'[5-8]$ and $d''[5-8]$.

The analysis of 6-round DES proceeds along the same lines as the analysis of 4-round DES presented in Sect. 4. We present the results of our analysis of 6-round DES in Table 4.

It was observed in [3] that in order to compute $d'[5-8]$ and $d''[5-8]$, it is sufficient to guess 54 key bits. Thus, if for a key guess the computed values of $d'[5-8]$ and $d''[5-8]$ disagree, then the key is necessarily wrong, and can be discarded.

Here again, the values of $d'[5-8]$ and $d''[5-8]$ can be computed by guessing less key bits in exchange for guessing internal bits. We have,

$$d'[5-8] = R_0[5-8] \oplus S_2[EP(R_1)[7-12] \oplus K_2[7-12]] \qquad (9)$$

and $d''[5-8]$ is equal to

$$d''[5-8] = R_6[5-8] \oplus S_2[EP(L_4)[7-12] \oplus K_4[7-12]]$$
$$\oplus S_2[EP(L_6)[7-12] \oplus K_6[7-12]]. \qquad (10)$$

Table 4. Key bits determining the 'middle' bits of 6-round DES

Round/S-box	Key bits	Bit determined	Bits appearing once [†]
1/1	2,6,12,15,18,25	β_1	
1/3	5,9,13,20,24,27	β_{12}	
1/5	32,38,42,48,53,56	β_{17}	
1/6	31,34,41,46,49,52	β_{21}	
1/7	29,35,40,45,50,54	β_{28}	
1/8	30,33,37,43,47,51	β_{29}	
2/2	2,5,8,12,17,23		
4/2	6,9,12,16,21,27		
5/1	4,9,13,19,22,25	γ_1	4, 19
5/3	3,6,12,16,20,27	γ_{12}	
5/5	32,35,39,45,49,55	γ_{17}	
5/6	31,38,41,48,53,56	γ_{21}	
5/7	29,33,36,42,47,52	γ_{28}	36
5/8	30,37,40,44,50,54	γ_{29}	
There are no key bits of round 6 that appear only once in computing d' and d''.			
Bits of K not affecting (1)	7,28		

[†] — These bits appear only once in computing d' and d''.

We have,

$$EP(R_1)[7\text{--}12] = [\beta_{21}\beta_{29}\beta_{12}\beta_{28}\beta_{17}\beta_1], \tag{11}$$

$$EP(L_4)[7\text{--}12] = [\gamma_{21}\gamma_{29}\gamma_{12}\gamma_{28}\gamma_{17}\gamma_1]. \tag{12}$$

If we guess $K_2[7\text{--}12]$, $K_4[7\text{--}12]$ and $K_6[7\text{--}12]$, the unknowns left in the computations of $d'[5\text{--}8]$ and $d''[5\text{--}8]$ are $[\beta_{21}\beta_{29}\beta_{12}\beta_{28}\beta_{17}\beta_1]$ and $[\gamma_{21}\gamma_{29}\gamma_{12}\gamma_{28}\gamma_{17}\gamma_1]$.

Consider γ_1. In order to compute this bit we can either guess 2 key bits (of $K_5[1\text{--}6]$) or guess γ_1 directly. Thus, a different attack algorithm for the meet-in-the-middle attack would be to guess all the 54 key bits suggested by Chaum and Evertse, besides the 2 key bits. For each guess of the 52 key bits, the attacker tries the two possibilities of γ_1. If for both values the equality $d'[5\text{--}8] = d''[5\text{--}8]$ is not achieved, then the guess of the 52 bits is necessarily wrong. As for a specific (wrong) guess of the key and of γ_1 the probability of equality is $1/16$, the probability that a wrong 52-bit key has at least one γ_1 for which the equality is satisfied is $1 - (15/16)^2 \approx 1/8$. Hence, the attacker can guess the 52 bits, and by trying the two possibilities of γ_1 reduce the number of remaining candidates to 2^{49}. Now, using similar techniques as in Sect. 4.1, the number of 6-round DES encryptions to retrieve the full key can be calculated to be $2^{51.8}$.

7   Conclusions and Open Problems

In this paper, we have found hitherto unknown weaknesses in block ciphers with up to 6 rounds of DES. We use the meet-in-the-middle technique and improve the time complexities (at the cost of few plaintexts) of similar attacks on DES by Chaum and Evertse [3]. We obtained that the time complexities for key search in the case of 4, 5 and 6-round DES are 2^{20}, $2^{35.5}$ and $2^{51.8}$ using 15, 51 and 1 known plaintexts respectively. With 6 chosen ciphertexts and 8 chosen plaintexts the time complexities in the case of 4-round and 5-round attacks are $2^{19.3}$ and 2^{30} respectively.

Our research leaves room for alluring open problems. It can be seen from Table 2, Table 3 and Table 4 that Chaum and Evertse have considered bits of the key K that do not appear in the first columns of these tables; we have considered bits of K that appear only once (and sometimes twice) in the first columns. Hence, a natural extension will be to experiment with bits which appear more times in the first columns of these tables. This technique could be tried on DES with higher number of rounds. Another extension of the attacks described in this paper follows a suggestion in [3] by which one may try to change the tables defining the S-boxes. By either of these methods, it could be possible to cryptanalyse DES variants consisting of 8 or more rounds.

Acknowledgments

The authors wish to thank the anonymous reviewers of Indocrypt-2007 for their constructive comments on our work.

References

1. Biham, E., Biryukov, A.: An Improvement of Davies' Attack on DES. Journal of Cryptology 10(3), 195–206 (1997)
2. Biham, E., Shamir, A.: Differential Cryptanalysis of the Data Encryption Standard. Springer, Heidelberg (1993)
3. Chaum, D., Evertse, J.-H.: Cryptanalysis of DES with a Reduced Number of Rounds: Sequences of Linear Factors in Block Ciphers. In: Williams, H.C. (ed.) CRYPTO 1985. LNCS, vol. 218, pp. 192–211. Springer, Heidelberg (1986)
4. CNET News.com, Users take crack at 56-bit crypto (1997), Available on-line at http://news.com.com/2100-1023-278658.html?legacy=cnet
5. Courtois, N.T., Bard, G.V.: Algebraic Cryptanalysis of the Data Encryption Standard (2006), Available on-line at http://eprint.iacr.org/2006/402.pdf
6. Davies, D.W.: Investigation of a Potential Weakness in the DES Algorithm, private communications (1987)
7. Davies, D.W., Murphy, S.: Pairs and Triplets of DES S-Boxes. Journal of Cryptology 8(1), 1–25 (1995)
8. Diffie, W., Hellman, M.E.: Exhaustive Cryptanalysis of the NBS Data Encryption Standard. Computer 10(6), 74–84 (1977)
9. Electronic Frontier Foundation, Cracking DES, Secrets of Encryption Research, Wiretap Politics & Chip Design, O'reilly (1998)
10. Hellman, M.E.: A Cryptanalytic Time-Memory Tradeoff. IEEE Transactions on Information Theory 26(4), 401–406 (1980)
11. Knudsen, L.R., Mathiassen, J.E.: A Chosen-Plaintext Linear Attack on DES. In: Schneier, B. (ed.) FSE 2000. LNCS, vol. 1978, pp. 262–272. Springer, Heidelberg (2001)

12. Kumar, S., Paar, C., Pelzl, J., Pfeiffer, G., Schimmler, M.: Breaking Ciphers with COPACOBANA - A Cost-Optimized Parallel Code Breaker. In: Goubin, L., Matsui, M. (eds.) CHES 2006. LNCS, vol. 4249, pp. 101–118. Springer, Heidelberg (2006)
13. Kunz-Jacques, S., Muller, F.: New Improvements of Davies-Murphy Cryptanalysis. In: Roy, B. (ed.) ASIACRYPT 2005. LNCS, vol. 3788, pp. 425–442. Springer, Heidelberg (2005)
14. Matsui, M.: Linear Cryptanalysis Method for DES Cipher. In: Helleseth, T. (ed.) EUROCRYPT 1993. LNCS, vol. 765, pp. 386–397. Springer, Heidelberg (1994)
15. National Bureau of Standards, Data Encryption Standard, Federal Information Processing Standards Publications No. 46 (1977)
16. Poschmann, A., Leander, G., Schramm, K., Paar, C.: New Light-Weight DES Variants Suited for RFID Applications. In: Proceedings of Fast Software Encryption 14. LNCS, Springer, Heidelberg (to appear, 2007)
17. Raddum, H., Semaev, I.: New Technique for Solving Sparse Equation Systems (2006), Available on-line at http://eprint.iacr.org/2006/475.pdf
18. Shamir, A.: On the Security of DES. In: Williams, H.C. (ed.) CRYPTO 1985. LNCS, vol. 218, pp. 280–281. Springer, Heidelberg (1986)
19. Shimoyama, T., Kaneko, T.: Quadratic Relation of S-box and Its Application to the Linear Attack of Full Round DES. In: Krawczyk, H. (ed.) CRYPTO 1998. LNCS, vol. 1462, pp. 200–211. Springer, Heidelberg (1998)
20. RSA Data Security, Team of Universities, Companies and Individual Computer Users Linked Over the Internet Crack RSA's 56-Bit DES Challenge (1997), Available on-line at http://www.rsasecurity.com/news/pr/970619-1.html
21. Wiener, M.J.: Efficient DES Key Search, Technical Report TR-244, Carleton University. In: Stinson, D.R. (ed.) CRYPTO 1993. LNCS, vol. 773, Springer, Heidelberg (1994)

Probabilistic Perfectly Reliable and Secure Message Transmission – Possibility, Feasibility and Optimality

Kannan Srinathan[2], Arpita Patra[1], Ashish Choudhary[1,*], and C. Pandu Rangan[1,**]

[1] Dept of Computer Science and Engineering
IIT Madras, Chennai India 600036
arpita@cse.iitm.ernet.in, ashishc@cse.iitm.ernet.in, rangan@iitm.ernet.in
[2] Center for Security, Theory and Algorithmic Research
International Institute of Information Technology
Hyderabad India 500032
shankar@research.iiit.ac.in, srinathan@iiit.ac.in

Abstract. We study the interplay of network connectivity and the issues related to feasibility and optimality for *probabilistic perfectly reliable message transmission* (PPRMT) and *probabilistic perfectly secure message transmission* (PPSMT) in a *synchronous* network under the influence of a *mixed* adversary who possesses *unbounded* computing power and can corrupt different set of nodes in Byzantine, omission, failstop and passive fashion simultaneously. Our results show that that *randomness helps in the possibility of multiphase PPSMT and significantly improves the lower bound on communication complexity for both PPRMT and PPSMT protocols!!*

Keywords: Probabilistic Reliability, Information Theoretic Security, Fault Tolerance.

1 Introduction

We study the fundamental problem of *probabilistic perfectly reliable message transmission* (PPRMT), where two non-faulty players, the sender **S** and the receiver **R** are part of a synchronous network modeled as a undirected graph, a part of which may be under the influence of a unbounded computational powerful *mixed* adversary which is denoted by three tuple (t_b, t_o, t_f, t_p) and can corrupt t_b, t_o, t_f and t_p nodes in Byzantine, omission, failstop and passive fashion respectively. **S** intends to transmit a message m chosen from a finite field $\mathbb{F}$ to **R** using

* Work Supported by Project No. CSE/05-06/076/DITX/CPAN on Protocols for Secure Communication and Computation Sponsored by Department of Information Technology, Government of India.
** Work Supported by Project No. CSE/05-06/076/DITX/CPAN on Protocols for Secure Communication and Computation Sponsored by Department of Information Technology, Government of India.

K. Srinathan, C. Pandu Rangan, M. Yung (Eds.): Indocrypt 2007, LNCS 4859, pp. 101–122, 2007.
© Springer-Verlag Berlin Heidelberg 2007

some protocol such that **R** should *correctly* obtain **S**'s message with probability at least $(1 - \delta)$ for arbitrarily small $0 < \delta < 1/2$. The problem of *probabilistic perfectly secure message transmission* (PPSMT) is same as PPRMT except that the adversary should not get any information about the message.

Intuitively, the allowance of a small probability of error in the transmission should result in improvements in both the fault tolerance as well as the efficiency aspects of reliable and secure protocols. What exactly is the improvement? — this is the central question addressed in this paper. More specifically, we address the following in the context of PPRMT and PPSMT: (i) When is a protocol possible in the given network (Possibility) (ii) Once the existence of a protocol is ensured, what is the minimum communication complexity required by any protocol to reliably/securely send a message (Optimality), (iii) Finally, how to design such protocol which satisfies the proven minimum communication complexity bound (Feasibility). Finally, we compare our results with the existing results for *perfectly reliable message transmission* (PRMT) and *perfectly secure message transmission* (PSMT) and show that randomness and probabilistic approaches lead to improved communication, phase[1] and computational complexities. Moreover results on mixed adversaries reveal *higher* level of fault tolerance in the underlying network.

The problem of PPRMT and PPSMT in the presence of static[2] threshold Byzantine adversary was first defined and solved by Franklin *et al* [4]. As one of the key results, they have proved, that over undirected graphs PPRMT (PPSMT) is possible if and only if PRMT (PSMT) is possible!!! Subsequent works on PPRMT and PPSMT include [14,5].

1.1   Our Contribution

Any reliable/secure protocol is analyzed by the following parameters: the connectivity requirement of the network, the number of phases required by the protocol, the total number of field elements communicated by **S** and **R** throughout the protocol and the computation done by **S** and **R**. There is a *trade-off* among these parameter which is well studied in the literature for PRMT and PSMT [9,13]. In this paper we try to understand this trade-off for PPRMT and PPSMT in the presence of a *mixed* adversary, which is done for the *first* time in the literature of PPRMT and PPSMT[3]. The contribution of our paper is four-fold and can be summarized as follows: **(a)** We characterize single phase PPRMT and multiphase PPSMT protocols in the presence of mixed adversary and show that in many practical scenarios, our characterization shows higher level of fault tolerance in the underlying network, while the extant results offer no such insight. **(b)** We prove the lower bound on the communication complexity of any single phase PPRMT and multiple phase PPSMT protocol tolerating mixed adversary. **(c)** We also design

---

[1] A phase is a send from **S** to **R** or vice-versa.

[2] By static adversary, we mean an adversary that decides on the set of players to corrupt before the start of the protocol.

[3] PRMT and PSMT in the presence of mixed adversary is studied in [7].

polynomial time *bit optimal* single phase PPRMT and four phase PPSMT protocols whose communication complexity satisfy our proven lower bounds. Our *single* phase PPRMT protocol has a *special* property that it achieves reliability with *constant* overhead when considered with *only* Byzantine adversary. Similarly our *four* phase PPSMT protocol has a *special* property that it achieves *secrecy* with *constant* overhead when considered with *only* Byzantine adversary. **(d)** Finally, we also compare our bit optimal PPRMT and PPSMT protocols with the existing bit optimal PRMT and PSMT protocols and cite many practical scenarios where no bit optimal PRMT or PSMT protocol exist but bit optimal PPRMT and PPSMT protocol do exists thus showing the power of allowing negligible error probability in the reliability of the protocols (without sacrificing secrecy).

1.2 Network Model

Following [3], we abstract away the network and concentrate on solving PPRMT and PPSMT problem for a single pair of processors, the *sender* **S** and the *receiver* **R**, connected by n parallel bi-directional channels $w_1, w_2, \ldots, w_n$ called wires such that an adversary having unbounded computing power can corrupt upto t_b, t_o, t_f and t_p wires in Byzantine, omission[4], failstop[5] and passive fashion respectively. Moreover, we assume that the wires that are under the control of the adversary in Byzantine, omission, failstop and passive fashion are mutually disjoint. Note that there is a difference between fail-stop and omission error[6]. If some value is sent over all the wires then it is said to be "broadcast"[7].

2 Probabilistic Perfectly Reliable Message Transmission

Here we completely characterize the set of tolerable adversaries, prove the lower bound for communication complexity of any single phase PPRMT protocol and present efficient/optimal protocol for single phase PPRMT.

---

[4] We say that a player P is under the control of an adversary in omission fashion, if the adversary can block the working of P at will at any time during the execution of the protocol. Also, as long as P is alive, it will follow the instructions of the protocol honestly. The adversary can eavesdrop the data/computation by P but cannot make P to deviate from the proper execution of the protocol. However, a blocked P can again become alive at some later stage of the protocol.

[5] We say that a player P is under the control of an adversary in a fail-stop manner, if the adversary can force P to *crash* at will at any time during the execution of the protocol. However, as long as P is alive, it will honestly follow the protocol. Also once P is crashed, it will not become alive again.

[6] The fail-stop error models a hardware failure caused by any natural calamity or manual shutdown. Also the nodes which are fail-stop corrupted cannot be passively listened by the adversary. On the other hand, nodes corrupted by omission adversary has listening capability. Thus omission adversary can be considered as a combination of fail-stop and passive adversary with the exception that unlike fail-stop error, a node which is crashed once by omission error may become alive during later stages of the protocol.

[7] Any information which is "broadcast" over at least $2t_b + t_o + t_f + 1$ wires will be recovered correctly at the receiving end (the receiver can output the majority).

2.1   Characterization for PPRMT

The existing characterization for PPRMT tolerating Byzantine adversary is:

Theorem 1 ([4]). *PPRMT between* **S** *and* **R** *against a* t_b *active Byzantine adversary is possible iff the network is* $(2t_b + 1)$-*(**S**,**R**)-connected.*

The characterization for PPRMT tolerating mixed adversary is as follows:

Theorem 2. *PPRMT between* **S** *and* **R** *against a mixed adversary* (t_b, t_o, t_p, t_f) *is possible iff the network is* $(2t_b + t_o + t_f + 1)$-*(**S**,**R**)-connected.*

Proof: If part: Consider a network which is $(2t_b + t_o + t_f + 1)$-(**S**,**R**)-connected. To send a message m, **S** simply *broadcasts* m to **R** over $2t_b + t_o + t_f + 1$ wires. It is easy to see that **R** will receive m with probability one by taking majority[8]. Only if part: Assume that a PPRMT protocol Π exists in a network $\mathcal{N}$ that is not $(2t_b + t_o + t_f + 1)$-(**S**,**R**)-connected. Consider the network $\mathcal{N}'$, induced by $\mathcal{N}$, on deleting $(t_o + t_f)$ vertices from a minimal vertex cutset of $\mathcal{N}$ (this can be viewed as an adversary blocking the communication over $t_o + t_f$ wires). It follows that $\mathcal{N}'$ is not $(2t_b + 1)$-(**S**,**R**)-connected. Evidently, if Π is a PPRMT protocol on $\mathcal{N}$, then Π' is a PPRMT protocol on $\mathcal{N}'$, where Π' is the protocol Π restricted to the players in $\mathcal{N}'$. However, from Theorem 1, Π' is non-existent. Thus Π is impossible too.   □

Significance of Theorem 2: *Theorem 2 strictly generalizes Theorem 1 because we obtain the latter by substituting* $t_o = t_f = 0$. *Now consider a network, which is* 4-*(**S**,**R**)-connected. From Theorem 1, on this network, any PPRMT protocol can tolerate one Byzantine fault. However, according to Theorem 2, it is possible to tolerate one additional faulty player, which can be either omission or fail-stop faulty. Thus our characterization shows more fault tolerance in comparison to the existing results.*
    In the sequel, we show that allowance of negligible error probability in transmission reduces the communication lower bound markedly in comparison to perfect transmission.

2.2   Lower Bound on Communication Complexity of Single Phase PPRMT Protocol

We now prove the lower bound on the communication complexity of any single phase PPRMT protocol tolerating mixed adversary.

Theorem 3. *Any single phase PPRMT protocol, from* **S** *to* **R** *over n wires, communicates* $\Omega(\frac{n\ell}{n-(t_b+t_o+t_f)})$ *field elements to reliably transmit (with high probability)* ℓ *field elements.*

Proof: In any single phase PPRMT protocol, the concatenation of the information sent over n wires can be viewed as a (probabilistic) error correcting code

---

[8] The protocol described here is a naive protocol which does not take the advantage of allowing small error probability in the reliability.

which can correct t_b Byzantine errors and $t_o + t_f$ erasures with an arbitrarily high probability. Without loss of generality, the domain of the set of possible values of the data sent along the wire can be assumed to be the same for all the wires. Let $\mathbb{S}$ be the set of possible values of the data sent along the wires. Thus, each codeword can be viewed as concatenation of n elements from $\mathbb{S}$ which can be represented by $n \log |\mathbb{S}|$ bits. Now, the removal of any $(t_b + t_o + t_f)$ elements from each of the codewords which corresponds to an adversary blocking $t_b + t_o + t_f$ wires (a Byzantine adversary can also block communication) should result in shortened codewords that are all distinct. For if any two were identical, the original codewords could have differed only in at most $(t_b + t_o + t_f)$ elements implying that there exist two codewords c_1 and c_2 and an adversarial strategy such that the receiver's view is the *same* on the receipt of c_1 and c_2. Specifically, without loss of generality assume that c_1 and c_2 differ only in their last $(t_b + t_o + t_f)$ elements. That is, $c_1 = \alpha \circ \beta$ and $c_2 = \alpha \circ \gamma$, where $\circ$ denotes concatenation and $|\beta| = |\gamma| = (t_b + t_o + t_f)$ elements. Now, consider the two cases: (a) c_1 is sent and the adversary corrupts it to $\alpha \circ \perp$ by completely blocking the last $(t_b + t_o + t_f)$ elements (wires) and (b) c_2 is sent and the adversary again corrupts it to $\alpha \circ \perp$. Thus, **R** can not distinguish between the receipt of c_1 and c_2 with probability greater than $\frac{1}{2}$, which violates the PPRMT communication property (in any PPRMT protocol, receiver should be able to receive the message with probability more than $\frac{1}{2}$). Therefore, all shortened codewords containing $n - (t_b + t_o + t_f)$ elements from $\mathbb{S}$ are distinct. This implies that there are same number of shortened and original codewords. But the number of shortened codewords can be at most $C = |\mathbb{S}|^{(n-(t_b+t_o+t_f))}$. Now each shortened codeword can be represented by $\log C = (n - (t_b + t_o + t_f)) \log |\mathbb{S}|$ bits. Since, for error-correcting we need to communicate the longer codeword containing $n \log |\mathbb{S}|$, reliable communication of shortened codeword of $k = \log C$ bits incurs a communication cost of at least $n \log |\mathbb{S}|$ bits. Hence communicating a single bit incurs communicating $\frac{n}{(n-(t_b+t_o+t_f))}$ bits. So to communicate ℓ elements from a field $\mathbb{F}$, represented by $\ell \log |\mathbb{F}|$ bits, $\Omega(\frac{n\ell}{(n-(t_b+t_o+t_f))} \log |\mathbb{F}|)$ bits need to be sent. Since $\log |\mathbb{F}|$ bits represents one field element from $\mathbb{F}$, communicating ℓ elements from $\mathbb{F}$ requires a communication complexity of $\Omega(\frac{n\ell}{(n-(t_b+t_o+t_f))})$ field elements.

Note: *In any PPRMT protocol designed in a field $\mathbb{F}$, the size of the field depends upon the error probability δ of the protocol (we show this in next section)[9].*

Single Phase PRMT vs Single Phase PPRMT: *While the lower bound on the communication complexity of any single phase PRMT tolerating mixed*

---

[9] From Theorem 3, any PPRMT protocol to send ℓ field elements from $\mathbb{F}$ need to communicate $\Omega(\frac{n\ell}{(n-(t_b+t_o+t_f))} log|\mathbb{F}|)$ bits. Thus the communication complexity of any single phase PPRMT protocol is a function of δ (since $|\mathbb{F}|$ is a function of δ), though it is not explicitly mentioned in the expression derived in Theorem 3. It should also be noted that communication complexity explicitly depends upon the message size ℓ.

adversary is $\Omega(\frac{n\ell}{(n-(2t_b+t_o+t_f)})$ *[11], the same for PPRMT is* $\Omega(\frac{n\ell}{(n-(t_b+t_o+t_f)})$ *(Theorem 3). This clearly brings forth the power of randomization.*

2.3   Single Phase Bit Optimal PPRMT Protocol

We now present an optimal single phase PPRMT protocol **PPRMT_Single_Phase**, which delivers $(t_b+1)n$ field elements by communicating $O(n^2)$ field elements in single phase with (arbitrarily) high probability where $n = 2t_b+t_o+t_f+1$. **PPRMT_Single_Phase** achieves reliability with *constant* overhead, when considered with only Byzantine adversary. The message block is represented by $\mathbf{M} = [m_1\,m_2\,\cdots\,m_n\,m_{n+1}\,m_{n+2}\cdots m_{2n}\cdots m_{t_b n+1}\,m_{t_b n+2}\cdots m_{t_b n+n}]$. Before the protocol, we describe a novel technique, called as **Extrapolation Technique** which we use in designing single phase PPRMT protocol **PPRMT_Single_Phase**.

Extrapolation Technique: We visually represent $\mathbf{M}$ as a rectangular array A of size $(t_b+1) \times n$ where the $j^{th}, 1 \le j \le t_b+1$ row contains the elements $m_{(j-1)n+1}\,m_{(j-1)n+2}\,\cdots\,m_{(j-1)n+n}$. For each column i of A, $1 \le i \le n$ we do the following: we construct the unique t_b degree polynomial $q_i(x)$ passing through the points $(1, m_i), (2, m_{n+i}), \ldots, (t_b+1, m_{t_b n+i})$ where $m_i, m_{n+i}, \ldots, m_{t_b n+i}$ belong to the ith column A. Then $q_i(x)$ is evaluated at $t_b + t_o + t_f$ points namely, $x = t_b + 2, t_b + 3, \ldots n$ to obtain $c_{1i}, c_{2i}, \ldots, c_{(t_b+t_o+t_f)i}$. Finally, we obtain a square array D of size $n \times n$ containing n^2 elements, where

$$D = \begin{bmatrix} m_1 & m_2 & \cdots & m_i & \cdots & m_n \\ & \cdots & & & & \\ m_{(j-1)n+1} & m_{(j-1)n+2} & \cdots & m_{(j-1)n+i} & \cdots & m_{(j-1)n+n} \\ & \cdots & & \cdots & & \cdots \\ m_{t_b n+1} & m_{t_b n+2} & \cdots & m_{t_b n+i} & \cdots & m_{t_b n+n} \\ c_{11} & c_{12} & \cdots & c_{1i} & \cdots & c_{1n} \\ & \cdots & & \cdots & & \cdots \\ c_{j1} & c_{j2} & \cdots & c_{ji} & \cdots & c_{jn} \\ & \cdots & & \cdots & & \cdots \\ c_{(t_b+t_o+t_f)1} & c_{(t_b+t_o+t_f)2} & \cdots & c_{(t_b+t_o+t_f)i} & \cdots & c_{(t_b+t_o+t_f)n} \end{bmatrix} = \begin{bmatrix} A \\ C \end{bmatrix} \text{ where}$$

C is the sub-matrix of D containing last $t_b + t_o + t_f$ rows. Thus D is the row concatenation of A of size $(t_b+1) \times n$ (containing elements of $\mathbf{M}$) and matrix C, whose elements are obtained from A by **Extrapolation Technique**. We now prove certain properties of the array D.

Lemma 1. *In D, all the n elements of any column can be uniquely generated from any $t_b + 1$ elements of the same column.*

Proof: Without loss of generality, we prove this for i^{th} column of D. The elements in the ith column are $m_i, m_{n+i}, \ldots, m_{t_b n+i}, c_{1i}, c_{2i}, \ldots, c_{ji}, \ldots c_{(t_b+t_o+t_f)i}$. From the construction, the points $(1, m_i), (2, m_{n+i}), \ldots, (t_b + 1, m_{t_b n+i})$, $(t_b + 2, c_{1i}), (t_b + 3, c_{2i}), \ldots, (n, c_{(t_b+t_o+t_f)i})$ lie on a unique t_b degree polynomial $q_i(x)$. Any $t_b + 1$ points uniquely determines $q_i(x)$ and hence the remaining $t_b+t_o+t_f$ points. $\qquad\square$

Lemma 2. *The elements of message $\mathbf{M}$ can be uniquely determined from any $t_b + 1$ rows of D.*

Proof: From the construction of D, the elements of $\mathbf{M}$ are arranged in the first t_b+1 rows. If the first t_b+1 rows are known then the lemma holds trivially. On the other hand, if some other t_b+1 rows are known, then from Lemma 1, i^{th} $1 \leq i \leq n$ column of D can be completely generated from $t_b + 1$ elements of the same column. Hence, knowledge of any $t_b + 1$ rows can reconstruct the whole matrix D and hence the message (first t_b+1 rows of D). □

Lemma 3. *Modification of at most t_b elements along any column of D is detectable.*

Proof: Recall that in D, the points (corresponds to i^{th} column of D) $(1, m_i)$, $(2, m_{n+i}), \ldots, (t_b + 1, m_{t_b n + i}), (t_b + 2, c_{1i}), \ldots, (n, c_{(t_b + t_o + t_f)i})$ lie on a unique t_b degree polynomial $q_i(x)$. Now suppose t_b values are changed in such a manner that they lie on some other t_b degree polynomial $q_i'(x)$ where $q_i(x) \neq q_i'(x)$. Since both $q_i(x)$ and $q_i'(x)$ are of degree t_b, they can match on additional t_b common points. But still there are at least $n - 2t_b = t_o + t_f + 1$ points still passing only the original polynomial $q_i(x)$ (but not through $q_i'(x)$). Hence any attempt to interpolate a t_b degree polynomial passing through the elements of a column (in which at most t_b values has been changed) will clearly indicate that at most t_b values are changed along the column. Hence the lemma holds. □

We are now ready to describe our protocol. Let the set of n wires be denoted as $\mathcal{W} = \{w_1, w_2, \ldots, w_n\}$. Let δ be a bound on the probability that the protocol fails to deliver the correct message. We require the size of the field $\mathbb{F}$ be $\Omega(\frac{Q(n)}{\delta})$, for

---

Protocol PPRMT_Single_Phase - The Single Phase PPRMT Protocol

1. **S** generates a rectangular array D containing n^2 field elements, from the $(t_b + 1) \times n$ elements of message $\mathbf{M}$ using **Extrapolation Technique**. **S** then forms n polynomials $p_j(x), 1 \leq j \leq n$, each of degree $n - 1$ where $p_j(x)$ is formed using the jth row of D as follows: the coefficient of $x^i, 0 \leq i \leq n - 1$ in $p_j(x)$ is the $(i + 1)^{th}$ element of jth row of D.

2. **S** chooses another n^2 field elements at random, say r_{ji}, $1 \leq i, j \leq n$. Over w_j, **S** sends the following to **R**: the polynomial $p_j(x)$ and the n ordered pairs $(r_{ji}, p_i(r_{ji}))$, for $1 \leq i \leq n$. Let $v_{ji} = p_i(r_{ji})$.

3. Let $\mathcal{F}$ denotes the set of wires that delivered nothing and let $\mathcal{B}$ denotes the set of wires that delivered invalid information (like higher degree polynomials etc.). Note that the wires in $\mathcal{B}$ are Byzantine corrupted because omission or fail-stop adversary is not allowed to modify the contents. **R** removes all the wires in $(\mathcal{F} \cup \mathcal{B})$ from $\mathcal{W}$ to work on the remaining wires in $\mathcal{W} \setminus (\mathcal{F} \cup \mathcal{B})$ out of which at most $t_b - |\mathcal{B}|$ could be Byzantine corrupted. Let **R** receives $p_j'(x)$ and (r_{ji}', v_{ji}') $1 \leq i \leq n$ over $w_j \in \mathcal{W} \setminus (\mathcal{F} \cup \mathcal{B})$. We say that w_j *contradicts* w_i if: $v_{ji}' \neq p_i'(r_{ji}')$ where $w_i, w_j \in \mathcal{W} \setminus (\mathcal{F} \cup \mathcal{B})$. Among all the wires in $\mathcal{W} \setminus (\mathcal{F} \cup \mathcal{B})$, **R** checks if there is a wire contradicted by at least $(t_b - |\mathcal{B}|) + 1$ wires. All such wires are Byzantine corrupted and removed (see Lemma 4).

4. To retrieve $\mathbf{M}$, **R** tries to reconstruct the array D as generated originally by **S** as follows: Corresponding to each $w_j \in \mathcal{W} \setminus (\mathcal{F} \cup \mathcal{B})$, which is not removed in step 3, **R** fills the jth row of D in the following manner: coefficient of $x^i, 0 \leq i \leq n - 1$ in $p_j'(x)$ occupies $(i + 1)^{th}$ column in the jth row of D; i.e., the coefficients of $p_j'(x)$ are inserted in j^{th} row of D such that the coefficient of x^i in $p_j'(x)$ occupies $(i + 1)^{th}$ column in the jth row of D.

5. After doing the above step for each $w_j \in \mathcal{W} \setminus (\mathcal{F} \cup \mathcal{B})$, which is not removed in step 3, **R** has at least $t_b + 1$ rows inserted in D (see Lemma 6). **R** then checks the validity of these rows as follows: corresponding to the $i^{th}, 1 \leq i \leq n$ column, **R** checks whether the points corresponding to the inserted elements of ith column lie on a t_b degree polynomial.

6. If the above test fails for at least one column of D, then **R** outputs "FAILURE" and halts. Otherwise, **R** regenerates the complete D correctly and recovers $\mathbf{M}$ from the first $t_b + 1$ rows (see Lemma 6).

some polynomial $Q(n)$, but this is acceptable because complexity of the protocol increases logarithmically with field size.

Lemma 4. *In* **PPRMT_Single_Phase**, *if any $w_j \in \mathcal{W} \setminus (\mathcal{F} \cup \mathcal{B})$ is contradicted by at least $(t_b - |\mathcal{B}|) + 1$ wires, then the polynomial $p_j(x)$ over w_j has been changed by adversary or in effect w_j is faulty.*

Proof: The wires in $\mathcal{B}$ are already identified to be Byzantine corrupted and hence neglected by **R**. Also the wires in $\mathcal{F}$ delivers nothing and hence neglected by **R**. So among the remaining $\mathcal{W} \setminus (\mathcal{F} \cup \mathcal{B})$ wires, at most $(t_b - |\mathcal{B}|)$ could be Byzantine corrupted. Also there cannot be any contradiction between two honest wires and hence any honest wire can be contradicted by at most $(t_b - |\mathcal{B}|)$ wires. Thus if a wire is contradicted by at least $(t_b - |\mathcal{B}|) + 1$ wires then it is faulty. □

Lemma 5. *In the protocol, if the adversary corrupts a polynomial over wire w_j in such a way that w_j is not removed during step 3, then **R** will always be able to detect it at the end of step 5 and outputs "FAILURE".*

Proof: At the beginning of step 5, there are at least $t_b + 1$ rows present in the partially reconstructed D. This follows from the fact there always exist $t_b + 1$ honest wires which will deliver correct polynomials to **R**. As mentioned in Lemma 4, any honest wire can be contradicted by at most $(t_b - |\mathcal{B}|)$ wires and hence is not be removed by **R** during step 4. So the coefficients of the polynomials corresponding to these honest wires will be present in partially reconstructed D.

Now if w_j (which has delivered a faulty polynomial) is not removed during step 3, then during step 4, the coefficients of $p'_j(x)$ are inserted in the jth row of partially reconstructed D. Since $p_j(x) \neq p'_j(x)$, there is at least one coefficient in $p'_j(x)$ which is different from the corresponding coefficient in $p_j(x)$. Let $p_j(x)$ differs from $p'_j(x)$ in the coefficient of x^i. Then $(i+1)^{th}$ column of partially reconstructed D differs from the $(i+1)^{th}$ column of original D at j^{th} position. The proof now follows from Lemma 3. Hence **R** outputs "FAILURE". □

Lemma 6. *In* **PPRMT_Single_Phase**, *if the test in step 5 succeeds for all the n columns of partially constructed D, then **R** will never output "FAILURE" and always recovers **M** correctly.*

Proof: As explained in previous Lemma, at the beginning of step 5, there will be at least $t_b + 1$ rows present in the partially reconstructed D. Now if the test in step 5 succeeds for all the n columns of partially constructed D, it implies that all the rows present in the partially reconstructed D are same as the corresponding rows in the original D. From Lemma 1, **R** will be able to completely regenerate all the n columns of original D. The proof now follows from Lemma 2. It is easy to see that **R** does not outputs "FAILURE" in this case.

Theorem 4. **PPRMT_Single_Phase** *terminates with a non-"FAILURE" output with high probability.*

Proof: Since no honest wire contradicts another honest wire, from Lemma 4, all the wires removed by **R** during step 3 are indeed faulty. We need to show that if a wire is corrupted (the polynomial over the wire is changed), then it will be contradicted by all the honest players with high probability. Let π_{ij} be the probability that a corrupted wire w_j will not be contradicted by a honest wire w_i. This means that the adversary can ensure that $p_j(r_{ij}) = p'_j(r_{ij})$ with a probability of π_{ij}. Since there are only $n-1$ points at which these two polynomials intersect, this allows the adversary to guess the value of r_{ij} with a probability of at least $\frac{\pi_{ij}}{n-1}$. But since r_{ij} was selected uniformly in $\mathbb{F}$, the probability of guessing it is at most $\frac{1}{|\mathbb{F}|}$. Therefore we have $\pi_{ij} \le \frac{n-1}{|\mathbb{F}|}$ for each i, j. Thus the total probability that the adversary can find w_i, w_j such that corrupted wire w_j will not be contradicted by w_i is at most $\sum_{i,j} \pi_{ij} \le \frac{n^2(n-1)}{|\mathbb{F}|}$. Since $\mathbb{F}$ is chosen such that $|\mathbb{F}| \ge \frac{Q(n)}{\delta}$, it follows that the protocol outputs a non-"FAILURE" value with probability $> 1 - \delta$ if we set $Q(n) = n^3$. $\square$

Note. *PPRMT_Single_Phase is a special kind of a probabilistic reliable message transmission protocol where* **R** *actually knows whether he outputs the correct message. But according to our definition of PPRMT, inability of* **R** *to "detect" every occurrence of an error is acceptable. Thus, our protocol has a strictly stronger property than that of necessary.*

Lemma 7. *PPRMT_Single_Phase reliably sends $n(t_b + 1)$ field elements by communicating $O(n^2)$ field elements. In terms of bits, the protocol sends $n(t_b + 1) \log |\mathbb{F}|$ bits by communicating $O(n^2 \log |\mathbb{F}|)$ bits.*

Proof: Over each wire, **S** sends a polynomial of degree $n-1$ and n ordered pair. Thus the total communication complexity is $O(n^2)$. Since each element from field $\mathbb{F}$ can be represented by $\log |\mathbb{F}|$ bits, the communication complexity of the protocol is $O(n^2 log|\mathbb{F}|)$ bits. $\square$

Achieving PPRMT in Constant Factor Overhead in Single Phase

In the presence of Byzantine fault, ℓ field elements can be transmitted by communicating $O(\ell)$ field elements in three phases [9] with perfect reliability. Also, achieving the same in single phase in the presence of Byzantine adversary is impossible [12]. However it is attainable in case of probabilistic reliability. In **PPRMT_Single_Phase**, *if $t_o = t_f = 0$, then $(t_b + 1)n = O(n^2)$ field elements (when $t_o = 0, t_f = 0$, $n = 2t_b + 1$ and so $t_b = O(n)$) can be sent by communicating $O(n^2)$ field elements. Thus, by allowing a small error probability in the reliability we can send ℓ field elements by communicating $O(\ell)$ field elements in only single phase.*

In Theorem 3, substituting $n = 2t_b + t_o + t_f + 1$ and $\ell = n(t_b + 1)$, we find that any single phase PPRMT protocol must communicate $\Omega(n^2)$ elements to send $n(t_b + 1)$ elements. Now, from Lemma 7, the communication complexity of **PPRMT_Single_Phase** is $O(n^2)$. Hence our protocol has **optimal communication complexity**. In terms of bits, **PPRMT_Single_Phase** sends $n(t_b + 1) \log |\mathbb{F}|$ bits by communicating $n^2 \log |\mathbb{F}|$ bits where $\mathbb{F} = \frac{Q(n)}{\delta}$, $Q(n) = n^3$ and $1 - \delta$ is the least probability with which the protocol terminates without "FAILURE". So, our protocol is **bit-optimal**.

Finally, we would like to point out that single phase PPRMT protocols can also be designed using the idea of check vectors proposed by Rabin and Ben-Or [10] for VSS. However, simple extension of their idea does not leads to a **bit-optimal** single phase PPRMT protocol.

3   Multiphase PPSMT Protocol in Undirected Networks

In this section, we provide characterization, lower bound on the communication complexity of any multiphase PPSMT protocol and also design one such protocol whose communication complexity matches with the lower bound.

3.1   Characterization for Multiphase PPSMT Protocol

In the previous section, we have shown how randomization affects the possibility and optimality of PPRMT protocol in the presence of a mixed adversary. We now explore the effect of randomization on the possibility and optimality of PPSMT protocol tolerating a mixed adversary. Our first step towards this exploration is to characterize the possibility of any multiphase PPSMT protocol.

Theorem 5. *Multiphase PPSMT between* **S** *and* **R** *in an undirected network tolerating a mixed adversary characterized by 4-tuple* (t_b, t_o, t_f, t_p) *is possible if and only if the network is* $(t_b + \max(t_b, t_p) + t_o + t_f + 1)$-*(**S,R**)-connected.*

Proof: Necessity: We consider two cases for proving the necessity.

- **Case 1:** $t_p \leq t_b$: In this case, the network is $(2t_b + t_o + t_f + 1)$-(**S,R**) connected which is necessary for PPRMT (Theorem 2) and hence obviously for PPSMT.

- **Case 2:** $t_p > t_b$: Here, the network is $(t_b + t_p + t_o + t_f + 1)$-(**S,R**)-connected. This condition is necessary for PPSMT because, if the network is $(t_b + t_p + t_o + t_f)$-(**S,R**)-connected, then the adversary may strategize to simply block all message

---

Phase I: S to R   Protocol SECURE - A Three Phase PPSMT Protocol
- Along $w_i, 1 \leq i \leq n$, **S** sends to **R** two randomly picked elements ρ_{i1} and ρ_{i2} chosen from $\mathbb{F}$.

Phase II: R to S
- Suppose **R** receives values in syntactically correct form along $n' \leq n$ wires. **R** neglects the remaining $(n - n')$ wires. Let **R** receives ρ'_{i1} and ρ'_{i2} along wire w_i, where w_i is not neglected by **R**.

- **R** chooses uniformly at random an element $K \in \mathbb{F}$. **R** then broadcasts to **S** the following: identities of the $(n - n')$ wires neglected by him, the secret K and the values $(K\rho'_{i1} + \rho'_{i2})$ for all i such that w_i is not neglected by **R**.

Phase III: S to R
- **S** correctly receives the identities of $(n - n')$ wires neglected by **R** during **Phase II** (because irrespective of the value of t_b and t_p, n is at least $2t_b + t_o + t_f + 1$. So any information which is broadcast over n wires will be received correctly). **S** eliminates these wires. **S** also correctly receives K and the values, say $u_i = (K\rho'_{i1} + \rho'_{i2})$ for each i, such that wire w_i is not eliminated by **R**.

- **S** then computes the set H such that $H = \{w_i | u_i = (K\rho_{i1} + \rho_{i2})\}$. Furthermore, **S** calculates the secret key ρ where: $\rho = \sum_{w_i \in H} \rho_{i2}$. **S** then broadcasts the set H and the blinded message $\mathbf{M} \oplus \rho$ to **R**, where **M** is a single field element.

Message Recovery by R
- **R** correctly receives H and computes his version of ρ'. If z' is the blinded message received, **R** outputs $\mathbf{M} = z' \oplus \rho'$.

through $(t_b + t_o + t_f)$ vertex disjoint paths and thereby ensure that every value received by $\mathbf{R}$ is also listened by the adversary.

Sufficiency: Suppose that network is $(t_b + \max(t_b, t_p) + t_o + t_f + 1)$-$(\mathbf{S}, \mathbf{R})$-connected. Then from Menger's theorem [6], there exist at least $n = (t_b + \max(t_b, t_p) + t_o + t_f + 1)$ vertex disjoint paths from $\mathbf{S}$ to $\mathbf{R}$. We model these paths as wires $w_1, w_2, \ldots, w_n$. We design a three phase PPSMT protocol called **SECURE** to securely send a single field element.

It can be shown that with a probability of at least $\left(1 - \frac{1}{|\mathbb{F}|}\right)$, $\rho' = \rho$ and hence $\mathbf{R}$ almost always learns the correct message (Proof is similar to that of the correctness and security of the information-checking protocol of [10]). Since $n = t_b + \max(t_b, t_p) + t_o + t_f + 1$, there exists at least one wire say w_i, which is not controlled by the adversary. So, the corresponding ρ_{i2} is unknown to adversary implying information theoretic security for $\rho = \sum_{w_i \in H} \rho_{i2}$ and hence for **M**. It is easy to see that the communication complexity of **SECURE** is $O(n^2)$. □

MultiPhase PSMT vs MultiPhase PPSMT: *From [7], for any multiphase Perfectly Secure Message Transmission (PSMT) protocol, the network should be $(2t_b + t_o + t_f + t_p + 1)$-$(\mathbf{S}, \mathbf{R})$ connected. Thus, except when either t_b or $t_p = 0$, Theorem 5 shows that allowing a negligible error probability in the reliability of the protocol (without sacrificing the secrecy) significantly helps in the possibility of multiphase secure message transmission protocol.*

Note: *Theorem 5 characterizes multiphase PPSMT protocol. A single phase PPSMT protocol tolerating Byzantine adversary is given in [5]. The characterization, lower bound on the communication complexity and an optimal single phase PPSMT tolerating mixed adversary is given in [8]. The connectivity requirement for single phase PPSMT is more[10] than multiphase PPSMT [8].*

3.2 Lower Bound on Communication Complexity of Multiphase PPSMT Protocol

We now prove the lower bound on the communication complexity of any r-phase ($r \geq 2$) PPSMT protocol which sends ℓ field elements tolerating a mixed adversary (t_b, t_o, t_f, t_p). Let $n \geq t_b + \max(t_b, t_p) + t_o + t_f + 1$.

Theorem 6. *Any r-phase ($r \geq 2$) PPSMT protocol which securely sends ℓ field elements in the presence of a threshold adversary (t_b, t_o, t_f, t_p) needs to communicate at least $\Omega\left(\frac{n\ell}{n-(t_b+t_o+t_f+t_p)}\right)$ field elements.*

Proof: The proof follows from Lemma 8 and Lemma 9, which are proved below.

Lemma 8. *The communication complexity of any multi-phase PPSMT protocol to send a message against an adversary corrupting up to $b(\leq t_b), F(\leq t_f)$ and $P(\leq t_b + t_o + t_p)$ of the wires in Byzantine, Fail-stop and passive manner respectively is not less than the communication complexity of distributing n shares for*

---

[10] In [8], it is shown that for the existence of single phase PPSMT protocol the network should be $2t_b + 2t_o + t_f + t_p + 1$-$(\mathbf{S}, \mathbf{R})$-connected.

the message such that any set of $n - F$ correct shares has full information about the message while any set of P shares has no information.

To prove the lemma, we begin with defining a weaker version of single-phase PPSMT called PPSMT with Error Detection (PPSMTED). We then prove the equivalence of communication complexity of PPSMTED protocol to send message $\mathbf{M}$ and the share complexity of distributing n shares for $\mathbf{M}$ such that any set of $n - F$ correct shares has full information about $\mathbf{M}$ while any set of P shares has no information about $\mathbf{M}$. To prove the aforementioned statement, we first show their equivalence (Claim 1). Finally, we will show the equivalence of single-phase protocol PPSMTED and multiphase PPSMT protocol in terms of communication complexity and also answer the question: why it is weaker than multiphase PPSMT protocol (Claim 3). These two equivalence will prove the desired equivalence as stated in this lemma. Note that b, F and P are bounded by t_b, t_f and $t_b + t_o + t_p$ respectively.

Definition 1. *A single phase PPSMT protocol is called PPSMTED if it satisfies the following:*

1. *If the adversary is passive on all the P ($P \leq t_b + t_o + t_p$ which is the maximum limit on the number of passive adversaries) corrupted wires then $\mathbf{R}$ securely receives the message sent by $\mathbf{S}$.*
2. *If the adversary corrupts information over some b wires ($b \leq t_b$), then $\mathbf{R}$ detects it, and aborts.*
3. *If adversary blocks some $F \leq t_f$ wires, without doing any other modification, then $\mathbf{R}$ recovers message correctly. Else if adversary blocks more than t_f wires or do some modification (or both), then $\mathbf{R}$ aborts.*
4. *The adversary obtains no information about the transmitted message.*

We next show that the properties of PPSMTED protocol for sending message $\mathbf{M}$ is equivalent to the problem of distributing n shares for $\mathbf{M}$ such that any set of $n - F$ correct shares has full information about $\mathbf{M}$ while any set of P shares has no information about the message.

Claim 1. *Let Π be a PPSMTED protocol tolerating an adversary that can corrupt up to any b, F and P of the n wires connecting $\mathbf{S}$ and $\mathbf{R}$ in Byzantine, fail-stop and passive manner respectively. In an execution of Π for sending a message $\mathbf{M}$, the data $s_i, 1 \leq i \leq n$ sent by the $\mathbf{S}$ along wires $w_i, 1 \leq i \leq n$ form n shares for $\mathbf{M}$ such that any set of $n - F$ correct shares has full information about $\mathbf{M}$ while any set of P shares has no information.*

Proof: The fact that any set of P shares have no information about $\mathbf{M}$ follows directly from property 1 and 4 of definition of PPSMTED. We now show that any set of $n - F$ correct shares has full information about $\mathbf{M}$. The proof is by contradiction. For a set of wires $A \subseteq W$, let $Message(\mathbf{M}, A)$, denotes the set of messages sent along the wires in A during the execution of PPSMTED to send $\mathbf{M}$. Now for any set $C, |C| \geq n - F$ of honest wires, $Message(\mathbf{M}, C)$ should uniquely determine the message $\mathbf{M}$. Suppose not, then there exists another message $\mathbf{M}'$ such that $Message(\mathbf{M}, C) = Message(\mathbf{M}', C)$. By definition the fail-stop adversary can block all the messages sent along the F wires not in C. Thus for two different executions of PPSMTED to send two distinct message $\mathbf{M}$ and $\mathbf{M}'$, there exists an adversary strategy such that view of $\mathbf{R}$ at the end of two executions is exactly same. This is a contradiction to the property 3 of PPSMTED protocol Π which outputs the correct message if at most F fail-stop errors take place. $\square$

The above claim also says that the communication complexity of PPSMTED protocol to send $\mathbf{M}$ is same as the share complexity (length of the sum of all shares) of distributing n shares for a message $\mathbf{M}$ such that any set of $n-F$ correct shares has full information about $\mathbf{M}$ while any set of P shares has no information about the message. Now we step forward to show the communication complexity of PPSMTED protocol is the lower bound on the communication complexity of any multiphase PPSMT protocol.

Before that we take a closer look at the execution of any multi-phase PPSMT protocol. $\mathbf{S}$ and $\mathbf{R}$ are modeled as polynomial time Turing machines with access to a random tape. The number of random bits used by the $\mathbf{S}$ and $\mathbf{R}$ are bounded by a polynomial $q(n)$. Let $r_1, r_2 \in \{0,1\}^{q(n)}$ denote the contents of the random tapes of $\mathbf{S}$ and $\mathbf{R}$ respectively. The message $\mathbf{M}$ is an element from the set $\{0,1\}^{p(n)}$, where $p(n)$ is a polynomial. A transcript for an execution of a multiphase PPSMT protocol Π is the concatenation of all the messages sent by $\mathbf{S}$ and $\mathbf{R}$ along all the wires.

Definition 2. *A passive transcript $T(\Pi, \mathbf{M}, r_1, r_2)$ is a transcript for the execution of the multiphase protocol Π with $\mathbf{M}$ as the message to be sent, r_1, r_2 as the contents of the random tapes of sender $\mathbf{S}$ and the receiver $\mathbf{R}$ and the adversary remaining passive throughout the execution. Let $T(\Pi, \mathbf{M}, r_1, r_2, w_i)$ denote the passive transcript restricted to messages exchanged along the wire w_i. When $\Pi, \mathbf{M}, r_1, r_2$ are obvious from the context, we drop them and denote the passive transcript restricted to a wire w_i by T_{w_i}. Similarly, T_B denotes the set of passive transcripts over the set of wires in B.*

Given $(\mathbf{M}, r_1, r_2)$ it is possible for $\mathbf{S}$ to compute $T(\Pi, \mathbf{M}, r_1, r_2)$ by simulating $\mathbf{R}$ with random tape r_2. Similarly given $(\mathbf{M}, r_1, r_2)$ $\mathbf{R}$ can compute $T(\Pi, \mathbf{M}, r_1, r_2)$ by simulating $\mathbf{S}$. Note that although $\mathbf{S}$ and receiver require both r_1, r_2 to generate the transcript, $\mathbf{R}$ requires only r_2 in order to obtain the message $\mathbf{M}$ from the transcript. This is clear since $\mathbf{R}$ does not have access to r_1 during the execution of Π but still can retrieve the message $\mathbf{M}$ from the messages exchanged.

Definition 3. *A transcript T_B, with $n - F \leq |B| \leq n$ is said to be a valid fault-free transcript with respect to $\mathbf{R}$ if there exists random string r_2 and message $\mathbf{M}$ such that protocol Π at $\mathbf{R}$ with r_2 as the contents of the random tape and T_B as the messages exchanged, terminates by outputting the message $\mathbf{M}$.*

Definition 4. *Two transcripts T_B and T'_B, where $n - F \leq B \leq n$ are said to be adversely close if the two transcripts differ only on a set of wires A such that $|A| \leq b + (|B| - (n - F))$. Formally $|\{w_i \in W | T_{w_i} \neq T'_{w_i}\}| \leq b + (|B| - (n - F))$.*

Claim 2. *Two valid fault-free transcripts* $T_B(\Pi, \mathbf{M}, r_1, r_2)$ *and* $T_B(\Pi, \mathbf{M}', r_1', r_2')$ *with two different message inputs* $\mathbf{M}, \mathbf{M}'$, *cannot be adversely close to each other, where* $n - F \leq B \leq n$.

Proof: Suppose two valid fault-free transcripts $T_B(\Pi, \mathbf{M}, r_1, r_2)$ and $T_B(\Pi, \mathbf{M}', r_1', r_2')$ are adversely close, then there is a set of wires A, $|A| \leq b + (|B| - (n - F))$ such that the two transcripts differ only on messages sent along the wires in A. Without loss of generality, assume last $b + (|B| - (n - F))$ wires belong to A with $A = X \circ Y$ where $|X| = b$ and $|Y| = (|B| - (n - F))$. Consider the following two executions of Π where the contents of $\mathbf{S}$'s and $\mathbf{R}$'s random tapes are r_1, r_2 respectively.

• $\mathbf{S}$ wants to send $\mathbf{M}$. $\mathbf{S}$ and $\mathbf{R}$ executes Π while the adversary stop the wires in Y to deliver any message. As $T_{B-Y}(\Pi, \mathbf{M}, r_1, r_2)$ is a valid transcript with respect to $\mathbf{M}$, $\mathbf{R}$ terminates with output $\mathbf{M}$.

• $\mathbf{S}$ wants to send $\mathbf{M}$. $\mathbf{S}$ and $\mathbf{R}$ executes Π. The adversary blocks messages over Y and changes the messages along wires in X such that the view of $\mathbf{S}$ is $T_{B-Y}(\Pi, \mathbf{M}, r_1, r_2)$ but the view of $\mathbf{R}$ is $T_{B-Y}(\Pi, \mathbf{M}', r_1', r_2')$. Since $T_{B-Y}(\Pi, \mathbf{M}', r_1', r_2')$ is a valid transcript with respect to $\mathbf{M}'$, $\mathbf{R}$ will terminate with output $\mathbf{M}'$.

The two scenarios differ only in the adversarial behavior and in the contents of $\mathbf{R}$'s random tape. In both the scenarios $\mathbf{S}$ wanted to send message $\mathbf{M}$. But the message received by receiver $\mathbf{R}$ in the second case is an incorrect message $\mathbf{M}'$. Thus, with only probability $1/2$, $\mathbf{R}$ will output the correct message $\mathbf{M}$. This is a contradiction because Π is a PPSMT protocol.   □

Till now, we have shown that a transcript over at least $n - F$ correct wires allows $\mathbf{R}$ to output $\mathbf{M}$ correctly. We now show how to reduce a multiphase PPSMT protocol into a single phase PPSMTED protocol.

Protocol PPSMTED

• $\mathbf{S}$ computes the passive transcript $T(\Pi, \mathbf{M}, r_1, r_2)$ for some random r_1 and r_2 and sends $T(\Pi, \mathbf{M}, r_1, r_2, w_i)$ to $\mathbf{R}$ along w_i.

• If $\mathbf{R}$ does not receives information through at least $n - F$ wires then $\mathbf{R}$ outputs ERROR and stop. Otherwise, let $\mathbf{R}$ receives information over the set of wires $B = \{w_{i_1}, w_{i_2}, \ldots, w_{i_\alpha}\}$ where $n - F \leq |B| \leq n$. $\mathbf{R}$ concatenates the values received along these wires to obtain a transcript T_B (which may be corrupted along t_b wires) and does the following:

- for each $\mathbf{M} \in \{0, 1\}^{p(n)}$ and $r_2 \in \{0, 1\}^{q(n)}$ do:
    If T_B is a valid transcript with random tape contents r_2 for message $\mathbf{M}$ then output $\mathbf{M}$ and stop.
  Output ERROR.

Claim 3. *The Communication complexity of any multiphase PPSMT protocol* Π *is at least the communication complexity of* **PPSMTED** *protocol. Also* Π *has stronger properties than* **PPSMTED**. *Finally,* **PPSMTED** *does not reveals* $\mathbf{M}$ *to the adversary.*

Proof: The communication complexity of any multiphase PPSMT protocol Π assuming the adversary to be passive during the complete execution is trivially a lower bound for any multiphase PPSMT protocol with corruption in any phases. In **PPSMTED**, **S** communicates the transcript generated by him assuming adversary to be passive throughout the execution of Π to **R**. The cost of communicating such a transcript by **PPSMTED** is same as of Π with the assumption that adversary remain passive throughout the execution of Π. **PPSMTED** is weaker than Π for the following reason: under the passive adversary assumption Π always outputs M but **PPSMTED** does not output M for certain adversarial behavior . But in that case it detects it and aborts.

The message sent along the wire w_i in **PPSMTED** is the concatenation of the messages sent along w_i in an execution of Π. Hence the adversary cannot obtain any information about the message **M**. From Claim 2, we know that valid transcripts of two different messages cannot be adversely close to each other. So irrespective of the actions of the adversary, the transcript received by **R** cannot be a valid transcript for any message other than **M** for any value of r_2. Hence if **R** outputs a message **M** then it is the same message sent by **S**. $\square$

This completes the proof of Lemma 8. We now prove the share complexity of distributing n shares for a message such that any set of $n - F$ correct shares has full information while any set of P shares has no information about the message.

Lemma 9. *The share-complexity (that is the length of the sum of all shares) of distributing n shares for a message of size ℓ field elements from $\mathbb{F}$ such that any set of $n - F$ correct shares has full information about the message while any set of P shares has no information about the message is $\Omega(\frac{n\ell}{(n-F-P)})$.*

Proof: Let X_i denotes the ith share. For any subset $A \subseteq \{1, 2 \ldots n\}$ let X_A denote the set of variables $\{X_i | i \in A\}$. Let **M** be a value drawn uniformly at random from $\mathbb{F}^l$. Then the secret **M** and the shares X_i are random variables. Let $H(X)$ for a random variable denote its entropy. Let $H(X|Y)$ denotes the entropy of X conditional on Y. The conditional entropy measures how much entropy a random variable X has remaining if we have already learned completely the value of a second random variable Y [2]. Since **M** is a value drawn uniformly at random from $\mathbb{F}^\ell$, we have $H(\mathbf{M}) = \ell$. Since any set B consisting of $n - F$ correct shares has full information about **M**, we have $H(\mathbf{M}|X_B) = 0$. Consider any subset $A \subset B$ such that $|A| = P$. Since any set of P shares has no information about **M**, we have $H(\mathbf{M}|X_A) = H(\mathbf{M})$. It is clear that

$$H(\mathbf{M}|X_A) = H(\mathbf{M}|X_A|X_{B-A}) + H(X_{B-A}) \leq H(\mathbf{M}|X_A, X_{B-A})) + H(X_{B-A}) = H(X_{B-A})$$
So $H(\mathbf{M}) \leq H(X_{B-A})$ { since $H(\mathbf{M}|X_A) = H(\mathbf{M})$}

Since $|B| = n - F$ and $|A| = P$, $|B - A| = n - F - P$. So for any set C of size $|B - A| = n - F - P$,

$$H(X_C) \geq H(\mathbf{M}) \Rightarrow \sum_{i \in C} H(X_i) \geq H(\mathbf{M})$$

Since there are $\binom{n}{n-F-P}$ possible subsets of cardinality $n - F - P$, summing the above equation over all possible subsets of cardinality $n - F - P$ we get

$$\sum_C \sum_{i \in C} H(X_i) \geq \binom{n}{n-F-P} H(\mathbf{M})$$

Now in all the possible $\binom{n}{n-F-P}$ subsets of size $n - F - P$, each of the term $H(X_i)$ appears $\binom{n-1}{n-F-P-1}$ times. So

$$\binom{n-1}{n-F-P-1} \sum_{i=1}^{n} H(X_i) \geq \binom{n}{n-F-P} H(\mathbf{M}) \Rightarrow \sum_{i=1}^{n} H(X_i) \geq \frac{n}{n-F-P} H(\mathbf{M})$$

which is equal to $\frac{n\ell}{n-F-P}$. Thus the share-complexity for any $\mathbf{M} \in \mathbb{F}^\ell$ is $\Omega\left(\frac{n\ell}{n-F-P}\right)$.   □

Since $P \leq t_b + t_o + t_b$ and $F \leq t_f$, $\Omega\left(\frac{n\ell}{n-F-P}\right) = \Omega\left(\frac{n\ell}{n-(t_b+t_o+t_f+t_p)}\right)$. Theorem 6 now follows from Lemma 8 and Lemma 9.   □

Note. *In terms of bits, any multiphase PPSMT protocol must communicate $\Omega\left(\frac{n\ell}{n-(t_b+t_o+t_f+t_p)} \log |\mathbb{F}|\right)$ bits to send $\ell \log |\mathbb{F}|$ bits, where $|\mathbb{F}|$ is a function of δ. In the next section, we give a concrete PPSMT protocol satisfying this bound and show how to set $|\mathbb{F}|$ as a function of δ.*

Randomization Helps in Reducing the Communication Complexity of Multiphase Secure Protocols: *In [7], it is shown that any multiphase PSMT protocol has a communication complexity of $\Omega\left(\frac{n\ell}{n-(2t_b+t_o+t_f+t_p)}\right)$ to securely send ℓ field elements. Comparing this bound with Theorem 6, we find that allowing a negligible error probability in reliability (without sacrificing the privacy) significantly reduces the communication complexity of multiphase secure protocol. We support this claim by designing a four phase PPSMT protocol whose total communication complexity matches the bound proved in Theorem 6.*

3.3   Constant Phase Bit Optimal PPSMT Protocol

Here we design a *bit optimal* multiphase PPSMT protocol called **PPSMT_Mixed** tolerating mixed adversary. The protocol terminates in four phases and uses the three phase **SECURE** protocol (described in Theorem 5) as a black-box[11]. The four phase protocol **PPSMT_Mixed** securely sends ℓ field elements by communicating $O(\ell)$ field elements against only Byzantine adversary, thus achieving *secrecy* with *constant* overhead.

---

[11] Since $n = t_b + \max(t_b, t_p) + t_o + t_f + 1$, we can execute **SECURE** protocol as a black-box. We cannot use any single phase PPSMT protocol as a black-box because the connectivity requirement for single phase and multi phase PPSMT are different.

If $t_p \geq t_b$, then the protocol securely sends n^2 field elements by communicating $O(n^3)$ field elements and if $t_b > t_p$, then $(t_b - t_p)n^2$ field elements by communicating $O(n^3)$ field elements. Let, $n = t_b + \max(t_b, t_p) + t_o + t_f + 1$. In the protocol, depending upon whether $t_b \leq t_p$ or $t_p < t_b$, the field size $|\mathbb{F}|$ is set to at least $\frac{3n^2}{\delta}$ or $\frac{4n^4(t_b - t_p)}{\delta t_b}$ respectively, where δ is the error probability of the protocol. Before describing the protocol, we first recall an algorithm from [12].

Consider the following problem: Suppose $\mathbf{S}$ and $\mathbf{R}$ by some means agree on a sequence of n numbers $x = [x_1 x_2 \ldots x_n] \in \mathbb{F}^n$ such that the adversary knows $n - f$ components of x, but the adversary has no information about the other f components of x, however, $\mathbf{S}$ and $\mathbf{R}$ do not necessarily know which values are known to the adversary. The goal is for $\mathbf{S}$ and $\mathbf{R}$ to agree on a sequence of f numbers $y_1 y_2 \ldots y_f \in \mathbb{F}$ such that the adversary has no information about $y_1 y_2 \ldots y_f$. This is achieved by the following algorithm [12]:

Algorithm EXTRAND$_{n,f}(x)$. Let V be a $n \times f$ Vandermonde matrix with members in $\mathbb{F}$. This matrix is published as a part of the protocol specification. $\mathbf{S}$ and $\mathbf{R}$ both locally compute the product $[y_1 \ y_2 \ \cdots \ y_f] = [x_1 \ x_2 \ \cdots \ x_n]V$.

Lemma 10 ([12]). *The adversary gets no information about $[y_1 \ y_2 \ \cdots \ y_f]$ computed in EXTRAND.*

Theorem 7. *By setting $|\mathbb{F}| \geq \frac{3n^2}{\delta}$ (if $t_p \geq t_b$) or $|\mathbb{F}| \geq \frac{4n^4(t_b - t_p)}{\delta t_b}$ (if $t_b > t_p$) the protocol* **PPSMT_Mixed** *securely transmits the message* $\mathbf{M}$ *with an error probability bounded by δ.*

Proof: For better understanding, we first prove the theorem when $t_b > t_p$. So $|\mathbb{F}| \geq \frac{4n^4(t_b - t_p)}{\delta t_b}$. It is evident from the protocol construction that the theorem holds if the following are true:

1. For all $1 \leq i \leq n$, $\rho_i' = \rho_i$ with probability $\geq (1 - \frac{\delta}{4})$.
2. For all $1 \leq i \leq n$, $y_i' = y_i$ with probability $\geq (1 - \frac{\delta}{4})$.
3. If the wire w_i were indeed corrupt (i.e., the n^2 tuple sent over w_i is changed by the adversary), then $w_i \in L_{fault}$ with probability $\geq (1 - \frac{\delta}{4})$.
4. The protocol **PPRMT_Single_Phase** to send the vector d fails with probability of at most $\frac{\delta}{4}$.
5. The adversary learns no (additional) information about the transmitted message $\mathbf{M}$.

The error probability of the protocol depends upon the error probability of the first four events. If each of the above are true, then the protocol's failure probability is bounded by δ. We prove now each of the above four claims separately.

Claim 4. *In* **PPSMT_Mixed**, *for all $1 \leq i \leq n$, $\rho_i' = \rho_i$ with probability $\geq (1 - \frac{\delta}{4})$.*

Protocol PPSMT_Mixed
A Bit Optimal 4-Phase PPSMT Protocol Tolerating Mixed Adversary

The message M is a sequence of n^2 field elements if $t_b \leq t_p$, otherwise it is a sequence of $(t_b - t_p)n^2$ field elements.

Phase I (R to S)

- R selects at random n^3 elements, r_{ij}, $1 \leq i \leq n, 1 \leq j \leq n^2$ from field $\mathbb{F}$. R also randomly selects $\rho_1, \rho_2, \ldots \rho_n$ from $\mathbb{F}$.
- R computes $y_i = \sum_{j=1}^{n^2} \rho_i^j r_{ij}, 1 \leq i \leq n$. Note that ρ_i^j is j^{th} power of ρ_i.
- R sends to S over $w_i, 1 \leq i \leq n$, the n^2 field elements $r_{ij}, 1 \leq j \leq n^2$. R also sends $\rho_i, y_i, 1 \leq i \leq n$ to S using $2n$ **parallel invocations of the three phase SECURE protocol (described in Theorem 5)** as there are total $2n$ elements to send. Hence **Phase I, II** and **Phase III** are used to do $2n$ parallel executions of **SECURE** protocol.

Phase IV (S to R)

- Let S receives $r'_{ij}, 1 \leq j \leq n^2$ along $w_i, 1 \leq i \leq n$. S adds w_i to a list $L_{erasure}$, if S does not receive any information over w_i.
- Let S receives ρ'_i and $y'_i, 1 \leq i \leq n$ after the $2n$ parallel executions of the three phase **SECURE** protocol initiated by R. For each i, such that $w_i \notin L_{erasure}$, S verifies whether $y'_i \overset{?}{=} \sum_{j=1}^{n^2} \rho_i'^j r'_{ij}$. If false, then S adds the wire w_i to the set of faulty wires, denoted by L_{faulty}. S sets $L_{honest} = \mathcal{W} \setminus (L_{faulty} \cup L_{erasure})$. If $t_p \geq t_b$, then S computes a random pad $Z = (z_1, z_2, \ldots, z_{n^2})$ of size n^2 field elements as follows:

$$Z = EXTRAND_{n^2|L_{honest}|, n^2}(r'_{ij}|w_i \in L_{honest})$$

However, if $t_b > t_p$, S computes a random pad Z of length $(t_b - t_p)n^2$ from $n^2|L_{honest}|$ elements using the above method.

- S computes $d = M \oplus Z$. If $t_p \geq t_b$ then d is of size n^2, so S broadcasts d to R. On the other hand, if if $t_b > t_p$ then d consists of $(t_b - t_p)n^2$ field elements and S reliably sends d to R by invoking $\frac{(t_b - t_p)}{t_b} * n$ parallel executions of single phase **PPRMT_Single_Phase** protocol (This is possible because n is at least $2t_b + t_o + t_f + 1$, which is necessary and sufficient for single phase PPRMT. Since **PPRMT_Single_Phase** protocol reliably sends nt_b field elements, d consisting of $(t_b - t_p)n^2$ field elements can be communicated by S by invoking the single phase PPRMT protocol $\frac{(t_b - t_p)}{t_b} * n$ times). S also broadcasts the set L_{faulty} and $L_{erasure}$ to R.

Message recovery by R.

- R correctly receives L_{faulty} and $L_{erasure}$ and sets $L_{honest} = \mathcal{W} \setminus (L_{faulty} \cup L_{erasure})$. R receives d with certainty (probability one) when $t_p \geq t_b$ and with high probability when $t_b > t_p$. If $t_b \leq t_p$, then R computes $Z^R = (z_1, z_2, \ldots, z_{n^2})$ of size n^2 field elements as follows:

$$Z^R = EXTRAND_{n^2|L_{honest}|, n^2}(r_{ij}|w_i \in L_{honest})$$

If $t_b > t_p$, then R computes Z^R of length $(t_b - t_p)n^2$ using the above method and recovers M by computing $M = Z^R \oplus d$.

Proof: In **PPSMT_Mixed**, ρ_i's are sent using n parallel execution of the three phase **SECURE** protocol. From Theorem 5, the error probability of a single execution of **SECURE** protocol is bounded by $\frac{1}{|\mathbb{F}|}$. Hence the total error probability of n parallel executions of **SECURE** to communicate ρ_i, $1 \leq i \leq n$, is bounded by $\frac{n}{|\mathbb{F}|}$. If $|\mathbb{F}| \geq \frac{4n}{\delta}$, then the total error probability of n parallel executions of **SECURE** is bounded by $\frac{\delta}{4}$. Since, $|\mathbb{F}| \geq \frac{4n^4(t_b - t_p)}{\delta t_b} > \frac{4n}{\delta}$, the claim holds. $\qquad \square$

Claim 5. *In **PPSMT_Mixed**, for all $1 \leq i \leq n$, $y'_i = y_i$ with probability $\geq (1 - \frac{\delta}{4})$.*

Proof: Similar to the proof of the above claim. $\qquad \square$

Claim 6. *In* **PPSMT_Mixed**, *if wire* w_i *is corrupted (i.e., at least one of the value* $r_{ij}, 1 \leq j \leq n^2$ *is changed by the adversary) and for all* i, $\rho'_i = \rho_i$ *then* $w_i \in L_{fault}$ *with probability* $\geq (1 - \frac{\delta}{4})$.

Proof: From the security argument of **SECURE** protocol, the adversary gains no information about ρ_i, y_i for all $1 \leq i \leq n$. Assume that adversary has changed the n^2 tuple over some wire w_i and it is not marked as faulty by **S**. This implies that $y_i = \sum_{j=1}^{n^2} \rho_i^j r_{ij} = \sum_{j=1}^{n^2} \rho_i^j r'_{ij} = y'_i$. As inferred by the expression, y_i and y'_i are the y-values (evaluated at $x = \rho_i$) of the polynomials of degree n^2 constructed using $r_{ij}, 1 \leq j \leq n^2$ and $r'_{ij}, 1 \leq j \leq n^2$ as coefficients. Since the polynomials are of degree n^2, there are at most n^2 points of intersection between the two. The point ρ_i is chosen uniformly by **R** in $\mathbb{F}$. Thus, with probability at most $\frac{n^2}{|\mathbb{F}|}$, the protocol fails to detect the faulty wire. In order to bound this error probability by $\frac{\delta}{4}$, we require $|\mathbb{F}|$ to be at least $\frac{4n^2}{\delta}$. Since, $|\mathbb{F}| \geq \frac{4n^4(t_b - t_p)}{\delta t_b} > \frac{4n^2}{\delta}$, the claim holds. □

Claim 7. *In* **PPSMT_Mixed**, *the single phase PPRMT protocol* **PPRMT_Single_Phase** *which is executed parallely* $\frac{n(t_b - t_p)}{t_b}$ *times to reliably send* d, *fails with probability of at most* $\frac{\delta}{4}$.

Proof: In **PPSMT_Mixed**, d is sent during **Phase IV** using $\frac{n(t_b - t_p)}{t_b}$ parallel executions of **PPRMT_Single_Phase** protocol. If δ' is the failure probability of a single execution of **PPRMT_Single_Phase**, the total failure probability to send d is bounded by $\frac{n(t_b - t_p)\delta'}{t_b}$. To obtain $\frac{n(t_b - t_p)\delta'}{t_b} \leq \frac{\delta}{4}$, we require $\delta' \leq \frac{\delta t_b}{4n(t_b - t_p)}$. Now from Theorem 4, if $|\mathbb{F}| = \frac{n^3}{\delta'}$ then the error probability of **PPRMT_Single_Phase** is bounded by δ'. So to bound the error probability of **PPRMT_Single_Phase** by $\delta' \leq \frac{\delta t_b}{4n(t_b - t_p)}$, we require $|\mathbb{F}| \geq \frac{4n^4(t_b - t_p)}{\delta t_b}$ which is true in this case. Hence the claim follows. □

Thus Theorem 7 is true if $t_b > t_p$ and $|\mathbb{F}| \geq \frac{4n^4(t_b - t_p)}{\delta t_b}$. If $t_p \geq t_b$, then **PPSMT_Mixed** will have an error probability of δ if the error probability of each of first three events mentioned in Theorem 7 is bounded by $\frac{\delta}{3}$. This is because 4th event does not occur, as d is broadcast in this case during **Phase IV**, instead of sending by single phase PPRMT. It is easy to check that by setting $|\mathbb{F}| \geq \frac{3n^2}{\delta}$, the theorem holds for $t_b \leq t_p$. □

Note: *From Theorem 7, the field size should be either* $\frac{3n^2}{\delta}$ *or* $\frac{4n^4(t_b - t_p)}{\delta t_b}$. *However, in* **PPSMT_Mixed**, *during* **Phase I**, **R** *needs to select at least* n^3 *random field elements from* $\mathbb{F}$. *So depending upon* δ, *we will set the field size as* $max(n^3, \frac{3n^2}{\delta})$. *Setting field size like this will not affect the working of the protocol.*

Theorem 8. *In* **PPSMT_Mixed**, *the adversary learns no information about the transmitted message* **M**.

Proof: The proof is divided into the following two cases:

1. **Case I: If** $t_p \geq t_b$: In this case, $n = t_b + t_p + t_o + t_f + 1$. In the worst case, the adversary can passively listen the contents over $t_b + t_o + t_p$ wires and block t_f wires. So there will be only one honest wire w_i and hence the adversary will have no information about the n^2 random elements sent over w_i. In this case, **S** generates a random pad of length n^2 and sends **M** containing n^2 field elements, using this pad. The proof follows from the correctness of EXTRAND algorithm.

2. **Case II: If** $t_b > t_p$: In this case, $n = 2t_b + t_o + t_f + 1$. In the worst case, the adversary can passively listen the contents of at most $t_b + t_p + t_o$ wires and block t_f wires. So there are at least $(t_b - t_p)$ honest wires and hence the adversary will have no information about the n^2 random elements sent over these honest wires. In this case, **S** generates a random pad of length $(t_b - t_p)n^2$ and sends **M** containing $(t_b - t_p)n^2$ field elements, using this pad. The proof now follows from the correctness of EXTRAND algorithm. □

Theorem 9. *The communication complexity of* **PPSMT_Mixed** *is* $O(n^3)$.

Proof: During **Phase I**, **R** sends n^2 random field elements over each of the n wires causing a communication complexity of $O(n^3)$. **R** also invokes $2n$ parallel executions of **SECURE** protocol with communication complexity of $O(n^2)$. This incurs total communication overhead of $O(n^3)$. During **Phase IV**, **S** sends d to **R**. If $t_p \geq t_b$, then d will consist of n^2 field elements and hence broadcasting it to **R** incurs a communication complexity of $O(n^3)$. On the other hand, if $t_b > t_p$, d consist of $(t_b - t_p)n^2$ field elements. In this case, **S** will send d by invoking $\frac{(t_b - t_p)}{t_b} * n$ parallel executions of single phase PPRMT protocol. Since, each execution of the single phase PPRMT protocol has a communication complexity of $O(n^2)$, total communication complexity is $O\left(\frac{(t_b - t_p)*n^3}{t_b}\right)$, which is $O(n^3)$. Thus, overall communication complexity of **PPSMT_Mixed** is $O(n^3)$. □

Finally to comment on the communication complexity of **PPSMT_Mixed** in terms of bits, we state the following: **PPSMT_Mixed** sends $(t_b - t_p)n^2 \log |\mathbb{F}|$ (if $t_b > t_p$) or $n^2 \log |\mathbb{F}|$ bits (if $t_b \leq t_p$) by communicating $O(n^3 \log |\mathbb{F}|)$ bits, where $|\mathbb{F}|$ is either $\frac{4n^4(t_b - t_p)}{\delta t_b}$ or $\frac{3n^2}{\delta}$ respectively. From Theorem 6, if $t_b \geq t_p$ (n will $2t_b + t_o + t_f + 1$), then any four phase PPSMT protocol needs to communicate $\Omega(n^3 \log |\mathbb{F}|)$ bits to securely send $(t_b - t_p)n^2 \log |\mathbb{F}|$ bits. Similarly, if $t_p \geq t_b$ (n will be $t_b + t_p + t_o + t_f + 1$), then any four phase PPSMT protocol need to communicate $\Omega(n^3 \log |\mathbb{F}|)$ bits in order to securely send $n^2 \log |\mathbb{F}|$ bits. Since total communication complexity of **PPSMT_Mixed** in both cases is $O(n^3 \log |\mathbb{F}|)$ bits, our protocol is **bit optimal**.

Significance of the Protocol: *In [7], the authors have designed a PSMT protocol achieving optimum communication complexity in* $O(log(t_o + t_f))$ *phases. Our PPSMT protocol achieves optimum communication complexity in* **four phases,**

which shows the power of randomization. However, our protocol does not sacrifice security in any sense for gaining **optimality***.*

Achieving Probabilistic Reliability and Perfect Security with Constant Overhead in Four Phases: In [13], the lower bound on the communication complexity of any multiphase PSMT protocol has been proved to be $\Omega\left(\frac{n\ell}{n-2t_b}\right)$ in the presence of Byzantine adversary. Hence, communicating any message *secretly* with *constant* overhead is *impossible* by *any* PSMT protocol. However protocol **PPSMT_Mixed** achieves this bound. In **PPSMT_Mixed**, if $t_o = t_p = t_f = 0$, then it sends $t_b n^2 = O(n^3)$ field elements in four phases by communicating $O(n^3)$ field elements (when $t_o = t_f = t_p = 0$, $n = 2t_b + 1$ and so $t_b = O(n)$). Thus we get *secrecy* with *constant* overhead in four phases when **PPSMT_Mixed** is executed considering *only* Byzantine adversary. Like **PPRMT_Single_Phase**, **PPSMT_Mixed** is also a *special* kind of a PPSMT protocol in that **R** actually *knows* if the protocol outputs the correct message or not.

4 Conclusion

We have studied the problem of PPRMT and PPSMT in the presence of mixed adversary. The paper shows considerably strong effect of randomization in the *possibility*, *feasibility* and *optimality* of reliable and secure message transmission protocols. We summarize our results in Table 1 and Table 2.

Table 1. Connectivity Requirement for the Existence of Protocol

Model	Single Phase	Multiple Phase
PRMT(Mixed Adversary)	$n \geq 2t_b + t_o + t_f + 1$ [7]	$n \geq 2t_b + t_o + t_f + 1$ [7]
PPRMT(Mixed Adversary)	$n \geq 2t_b + t_o + t_f + 1$, Theorem 2	$n \geq 2t_b + t_o + t_f + 1$, Theorem 2
PSMT(Mixed Adversary)	$n \geq 3t_b + 2t_o + 2t_f + t_p + 1$ [11]	$n \geq 2t_b + t_o + t_f + t_p + 1$ [7]
PPSMT(Mixed Adversary)	$n \geq 2t_b + 2t_o + t_f + t_p + 1$ [8]	$n \geq t_b + \max(t_b, t_p) + t_o + t_f + 1$, Theorem 5

Table 2. Protocols with Optimum Communication Complexity. ℓ is the message size.

Model	Communication Complexity	Number of Phases	Remarks
PRMT (Byzantine Adversary)	$O(\ell)$	3	$\ell = n^2$ [9].
PPRMT (Byzantine Adversary)	$O(\ell)$	1	$\ell = O(n^2)$ (Protocol **PPRMT_Single_Phase** given in this paper by substituting $t_o = t_f = 0$).
PSMT (Byzantine Adversary)	$O\left(\frac{n\ell}{n-3t_b}\right)$	1	$\ell = O(n)$ [11].
	$O\left(\frac{n\ell}{n-2t_b}\right)$	2	Exponential computation [1].
	$O\left(\frac{n\ell}{n-2t_b}\right)$	3	Polynomial computation [9].
PPSMT (Byzantine Adversary)	$O\left(\frac{n\ell}{n-t_b}\right)$	1	$\ell = O(n)$ [8].
	$O(\ell)$	4	$\ell = O(n^3)$ (by substituting $t_o = t_f = t_p = 0$ in **PPSMT_Mixed**)
PSMT (Mixed Adversary)	$O\left(\frac{n\ell}{n-(2t_b+t_o+t_f+t_p)}\right)$	$O(\log(t_o + t_f))$	$\ell = n\log(t_o + t_f)$ [7]
PPSMT (Mixed Adversary)	$O\left(\frac{n\ell}{n-(t_b+t_o+t_f+t_p)}\right)$	4	$\ell = n^2$ or $\ell = (t_b - t_p)n^2$ Protocol **PPSMT_Mixed** given in this paper

References

1. Agarwal, S., Cramer, R., de Haan, R.: Asymptotically optimal two-round perfectly secure message transmission. In: Dwork, C. (ed.) CRYPTO 2006. LNCS, vol. 4117, pp. 394–408. Springer, Heidelberg (2006)
2. Cover, T.H., Thomas, J.A.: Elements of Information Theory. John Wiley & Sons, Chichester (2004)
3. Dolev, D., Dwork, C., Waarts, O., Yung, M.: Perfectly secure message transmission. JACM 40(1), 17–47 (1993)
4. Franklin, M., Wright, R.N.: Secure communication in minimal connectivity models. In: Nyberg, K. (ed.) EUROCRYPT 1998. LNCS, vol. 1403, pp. 346–360. Springer, Heidelberg (1998)
5. Kurosawa, K., Suzuki, K.: Almost secure (1-round, n-channel) message transmission scheme. Cryptology ePrint Archive, Report 2007/076 (2007), http://eprint.iacr.org/
6. Menger, K.: Zur allgemeinen kurventheorie. Fundamenta Mathematicae 10, 96–115 (1927)
7. Patra, A., Choudhary, A., Srinathan, K., Rangan, P.C.: Bit optimal protocols for perfectly reliable and secure message transmission in the presence of mixed adversary. Manuscript
8. Patra, A., Choudhary, A., Srinathan, K., Rangan, P.C.: Does randomization helps in reliable and secure communication. Manuscript
9. Patra, A., Choudhary, A., Srinathan, K., Rangan, C.P.: Constant phase bit optimal protocols for perfectly reliable and secure message transmission. In: Barua, R., Lange, T. (eds.) INDOCRYPT 2006. LNCS, vol. 4329, pp. 221–235. Springer, Heidelberg (2006)
10. Rabin, T., Ben-Or, M.: Verifiable secret sharing and multiparty protocols with honest majority. In: Proc. of twenty-first annual ACM symposium on Theory of computing, pp. 73–85. ACM Press, New York (1989)
11. Srinathan, K.: Secure Distributed Communication. PhD thesis, Indian Institute of Technology Madras (2006)
12. Srinathan, K., Narayanan, A., Rangan, C.P.: Optimal perfectly secure message transmission. In: Franklin, M. (ed.) CRYPTO 2004. LNCS, vol. 3152, pp. 545–561. Springer, Heidelberg (2004)
13. Srinathan, K., Prasad, N.R., Rangan, C.P.: On the optimal communication complexity of multiphase protocols for perfect communication. In: IEEE Symposium on Security and Privacy, pp. 311–320 (2007)
14. Wang, Y., Desmedt, Y.: Secure communication in multicast channels: The answer to Franklin and Wright's question. Journal of Cryptology 14(2), 121–135 (2001)

Secret Swarm Unit
Reactive k−Secret Sharing*
(Extended Abstract)

Shlomi Dolev[1], Limor Lahiani[1], and Moti Yung[2]

[1] Department of Computer Science,
Ben-Gurion University of the Negev, Beer-Sheva, 84105, Israel
{dolev,lahiani}@cs.bgu.ac.il
[2] Department of Computer Science,
Columbia University, New York, and Google, USA
moti@cs.columbia.edu

Abstract. Secret sharing is a basic fundamental cryptographic task. Motivated by the virtual automata abstraction and swarm computing, we investigate an extension of the k-secret sharing scheme, in which the secret components are changed on the fly, independently and without (internal) communication, as a reaction to a global external trigger. The changes are made while maintaining the requirement that k or more secret shares may reveal the secret and no $k - 1$ or fewer reveal the secret.

The application considered is a swarm of mobile processes, each maintaining a share of the secret which may change according to common outside inputs e.g., inputs received by sensors attached to the process.

The proposed schemes support addition and removal of processes from the swarm as well as corruption of a small portion of the processes in the swarm.

Keywords: secret sharing, mobile computing.

1 Introduction

Secret sharing is a basic and fundamental technique [13]. Motivated by the high level of interest in the virtual automata abstraction and swarm computing, e.g., [3,2,1,4,5] we investigate an extension of the k-secret sharing scheme, in which the secret shares are changed on the fly, while maintaining the requirement that k or more shares reveal the secret and no $k - 1$ or fewer reveal the secret.

There is a great interest in pervasive ad hoc and swarm computing [14], and in particular in swarming unmanned aerial vehicles (UAV) [9,4]. A unit of UAVs that collaborate in a mission is more robust than a single UAV that has to complete a mission by itself. This is a known phenomenon in distributed computing where a single point of failure has to be avoided. Replicated memory and state

---

* Partially supported by the Israeli Ministry of Science, Lynne and William Frankel Center for Computer Sciences and the Rita Altura trust chair in Computer Sciences.

K. Srinathan, C. Pandu Rangan, M. Yung (Eds.): Indocrypt 2007, LNCS 4859, pp. 123–137, 2007.

machine abstractions are used as general techniques for capturing the strength of distributed systems in tolerating faults and dynamic changes.

In this work we integrate cryptographic concerns into these abstractions. In particular, we are interested in scenarios in which some of the participants of the swarm are compromised and their secret shares are revealed. We would like the participants to execute a global transition without communicating with each other and therefore without knowing the secret, before or after the transition. Note that secure function computation (e.g., [10]) requires communication whenever inputs should be processed, while we require transition with no internal communication.

Our contributions. We define and present three reactive k-secret solutions. The first solution is based on the Chinese remainder, the second is based on polynomial representation, and the third uses replication of states. In the first solution we allow the addition arithmetic operation as a possible transition, where each share of the secret is modified according to the added value, without collecting the global secret value. The second solution supports both arithmetic addition and multiplication of the secret by a given input. The last solution implements a general I/O automaton, where the transition to the next step is performed according to the input event accepted by the swarm.

To avoid compromising the global secret of the swarm, the participants maintain only a share of the secret. In the Chinese remainder scheme, the participant share of the global value reveals partial information on the secret value. The polynomial based scheme assumes unbounded secret share values, which enable it to ensure that no information is revealed to the swarm members. Note that the shares of the polynomial based solution have number of bits that is approximately the number of bits of the secret, while the *total* number of bits of the shares of the Chinese remainder is approximately the number of bits required to describe the secret. The third solution replicates states of a given automaton and distributes several distinct replicas to each participant in the swarm. The relative majority of the distributed replicas represent the state of the swarm. The participant changes the states of all the replicas it maintains according to the global input arriving to the swarm. In this case, general automaton can be implemented by the swarm, revealing only partial knowledge on the secret of the swarm.

We remark that it is also possible to device Vandermonde matrix based scheme that supports other operations such as the bitwise-xor operation of the secret shares. In this case the secret is masked by a random number, and operations over shares are according to the relevant portions of the global input.

Paper organization. The system settings are described in Section 2. The k-secret addition implementation that is based on the Chinese remainder appears in Section 3. The solution that supports addition and multiplication by a number is the polynomial-based solution presented is Section 4. The I/O automaton implementation appears in Section 5. Finally, conclusions appear in Section 6. Some of the details are omitted from this extended abstract and can be found in [6].

2   Swarm Settings

A *swarm* consists of at least n processes (executed by, say, Unmanned Aerial Vehicles UAVs, mobile-sensors, processors) which receive inputs from the outside environment simultaneously[1]. The swarm as a unit holds a secret, where shares of the secret are distributed among the swarm members in a way that at least k are required to reveal the secret (some of our schemes require more than k), and any less than k shares can not reveal the secret. Yet, in some of our schemes the shares may imply some information regarding the secret, namely, the secret can be guessed with some positive probability < 1 greater than the probability of a uniform guess over the secret domain. We consider both listening adversary and Byzantine adversary, and present different schemes used by the processes to cope with them. We assume that at most $f < k$ of the n processes may be captured and compromised by an adversary. Communication among the processes in the swarm is avoided or performed in a safe land, alternatives of more expensive secure communication techniques are also mentioned.

Reactive k-secret counting. Assume that we have a swarm, which consists of n processes. The task of the swarm is to manage a global value called *counter*, which is updated according to the swarm input events. The value of the counter is the actual secret of the swarm. Each swarm member holds a share of the global counter in a way that any k or more members may reveal the secret with some positive probability Pr_k, yet any less than k members fail to reveal it.

Any event sensed by the processes is modeled as a system input. The swarm receives *inputs* and sends *outputs* to the outside environment. An input to the swarm arrives at all processes simultaneously. The output of the swarm is a function of the swarm state. There are two possible assumptions concerning the swarm output, the first, called *threshold accumulated output*, where the swarm outputs only when at least a predefined number of processes have this output locally. The second means to define the swarm output is based on secure internal communication within the swarm, the communication takes place when the local state of a process indicates that a swarm output is possible[2]. In the sequel we assume the *threshold accumulated output* where the adversary cannot observe outputs below the threshold. Whenever the output is above the threshold, the adversary may observe the swarm output together with the outside environment, and is "surprised" by the non anticipated output of the swarm.

We consider the following input actions, defined for our first solution:

• *set(x):* Sets the secret share with the value x. The value x is distributed in a secure way to processes in the swarm, each process receives a secret share x. This operation is either done in a safe land, or uses encryption techniques.

---

[1] Alternatively, the processes can communicate the inputs to each other by atomic broadcast or other weaker communication primitive.

[2] In this case, one should add "white noise" of constant output computations to mask the actual output computations.

- *step(δ):* Increments (or decrements in case δ is negative) the current counter value by δ. The processes of the swarm independently receive the input δ.
- *regain consistency:* Ensures that the processes carry the current counter value in a consistent manner. We view the execution of this command as an execution in a safe land, where the adversary is not present. The commands are used for reestablishing security level. Recovery and preparation for succeeding obstacles is achieved, by redistributing the counter shares, in order to prepare the swarm to cope with future joins, leaves, and state corruption.

The processes in the swarm communicate, if possible creating additional processes instead of the processes that left, and redistribute the counter value. This is the mechanism used to obtain a proactive security property.

We assume that the number of processes *leaving* the swarm between any two successive *regain consistency* actions, is bounded by lp. The operation taken by a leaving process is essentially an erase of its memory[3].

- *join request:* A process requests to join the swarm.
- *join reply:* A process reply to a join request of another process, by sending the joining process a secret share.

The adversary operations are:

- *listening:* Listening to all communication, but cannot send messages.
- *capture a process:* Remove a process from the swarm and reveal a snapshot of the memory of a process[4]. The adversary can invoke this operation at most $f = k - 1$ times between any two successive global resets of the swarm secret. Reset is implemented by using the set input actions.
- *corrupt process state:* The adversary is capable of changing the state of a process. In this case, the adversary is called a Byzantine adversary. Byzantine adversary also models the case in which transient faults occur, e.g., causing several of the processes not to get the same input sequence. We state for each of our schemes the number of times the adversary can invoke this operation between any two successive regain consistency operations.

3   Reactive k-Secret Counting – The Chinese Remainder Solution

According to the Chinese Remainder Theorem (CRT), any positive integer m is uniquely specified by its residue modulo relatively prime numbers $p_1 < ... < p_l$, where $\prod_{i=1}^{l-1} p_i \leq m < \prod_{i=1}^{l} p_i$ and $p_1 < p_2 < ... < p_l$. We use the CRT for defining the swarm's global counter, denoted by GC, which is the actual swarm secret.

---

[3] One may wish to design a swarm in which the members maintain the population of the swarm, in this case, as an optimization for a mechanism based on secure heart-bits, a leaving process may notify the other members on the fact it is leaving.

[4] In the sequel we assume that a joining process reveals information equivalent to a captured process, though, if it happen that the (listening) adversary is not presented during the join no information is revealed.

Using a Chinese remainder counter. Let $\mathcal{P} = \{p_1, p_2, \ldots, p_l\}$, such that $p_1 < p_2 < \ldots < p_l$, is the set of l relatively prime numbers which defines the global counter GC. The integer values of GC run from 0 to GC_{max}, where $GC_{max} = \prod_{i=1}^{l} p_i - 1$.

A *counter component* is a pair $\langle r_i, p_i \rangle$, where $r_i = GC \bmod p_i$ and $p_i \in \mathcal{P}$. The swarm's global counter GC can be denoted by a sequence of l counter components $\langle \langle r_1, p_1 \rangle, \langle r_2, p_2 \rangle, .., \langle r_l, p_l \rangle \rangle$, the *CRT-representation* of GC, or $\langle r_1, r_2, .., r_l \rangle$, when $\mathcal{P}$ is known. Note that this representation implies that GC can hold up to $\prod_{i=1}^{l} p_i$ distinct values. A *counter share* is simply a set of distinct counter components.

We assume that there is a lower bound p_{min} on the relatively prime numbers in $\mathcal{P}$ such that $p_{min} < p_1 < p_2 < \ldots < p_l$.

For a given $\mathcal{P} = \{p_1, p_2, \ldots, p_l\}$ and $GC = \langle \langle r_1, p_1 \rangle, \langle r_2, p_2 \rangle, .., \langle r_l, p_l \rangle \rangle$, we distribute the counter GC among the n processes in a way that (a) k or more members may reveal the secret with some probability, yet (b) any fewer than k members fail to reveal it.

In order to support simple join input actions, we use the CRT-representation of GC in a way that each process holds a counter share of size $s = \lfloor \frac{l}{k} \rfloor$, namely, s counter components out of l. Let Pr_m denote the probability that all the l components of GC are present in a set of m distinct counter shares, each of size s as specified. We now compute Pr_m. For any $0 \le m < k$ it holds that $Pr_m = 0$, since at least one counter component is missing. For $m \ge k$ it holds that $Pr_m = [1 - (1-p)^m]^l$, where p is the probability of a counter component to be chosen. As the components are chosen with equal probability out of the l components of GC, it holds that $p = \frac{s}{l} \approx \frac{1}{k}$. Assuming k divides l, it holds that $p = \frac{s}{l} = \frac{1}{k}$. The probability that a certain counter component appears in one of the m counter shares is $1 - (1-p)^m$. Hence, the probability that no component is missing is $[1 - (1-p)^m]^l$. Therefore, $Pr_m = [1 - (1-p)^m]^l$. Thus, the expected number m of required partial counters is a function of n, l, and k.

Note that when the GC value is incremented (decremented) by δ, each counter component $\langle r_i, p_i \rangle$ of GC is incremented (decremented) by δ modulo p_i. For example, let $\mathcal{P} = \{2, 3, 5, 7\}$ ($p_1 = 2$, $p_2 = 3$, $p_3 = 5$, $p_4 = 7$), $l = 4$ and $GC = 0$. The CRT-representation of GC is $\langle \langle 0, 2 \rangle, \langle 0, 3 \rangle, \langle 0, 5 \rangle, \langle 0, 7 \rangle \rangle$ or $\langle 0, 0, 0, 0 \rangle$ for the given set of primes $\mathcal{P}$. After incrementing the value of GC by one, it holds that $GC = \langle 1, 1, 1, 1 \rangle$. Incrementing by one again, results in $GC = \langle 0, 2, 2, 2 \rangle$, then $\langle 1, 0, 3, 3 \rangle$, $\langle 0, 1, 4, 4 \rangle$, $\langle 1, 2, 0, 5 \rangle$, and so on.

Next we describe the way the Chinese remainder counter supports the required input actions as appears in Figure 1.

Line-by-line code description. The code in Figure 1 describes input actions of process i. Each process i has a share of s counter components: s relatively prime numbers $primes_i[1..s]$ and s relative residues $residues_i[1..s]$, where $residues_i[j] = GC \bmod primes_i[j] \; \forall j = 1..s$.

Each input action includes a message of the form $\langle type, srcid, destid, parameters \rangle$, where $type$ is the message type indicating the input action type, $srcid$ is the identifier of the source process, $destid$ the identifier of the destination

```
 1  seti(⟨set, srcid, i, share⟩)
 2      for j = 1..s
 3          primesi[j] ⟵ getPrime(share, j)
 4          residuesi[j] ⟵ getResidue(share, j)

 5  stepi(⟨stp, srcid, i, δ⟩)
 6      for j = 1..s do
 7          residuesi[j] ⟵ (residuesi[j] + δ) mod primesi[j]

 8  regainConsistencyRequesti(⟨rgn_rqst, srcid, i⟩)
 9      leaderId ⟵ leaderElection()
10      if leaderId = i then
11          globalCounterComponentsi ⟵ listenAll(⟨rgn_rply, i, j, share⟩)
12          if size(globalCounterComponentsi) < l then
13              globalCounterComponentsi ⟵ initGlobalCounterComponents()
14          for every process id j in the swarm do:
15              share ⟵ randomShare(globalCounterComponents)
16              send(⟨set, i, j, share⟩)
17          globalCounterComponentsi ⟵ ∅
18      else
19          send(⟨rgn_rply, i, leaderId, ⟨primesi[1..s], residuesi[1..s]⟩⟩)

20  regainConsistencyReplyi(⟨rgn_rply, srcid, i, share⟩)
21      if leaderId = i then
22          globalCounterComponentsi ⟵ globalCounterComponentsi ∪ {share}

23  joinRequesti(⟨join_rqst, srcid, i⟩)
24      sentPrimes ⟵ ∅
25      while |sentPrimes| < s do
26          waitingTime ⟵ random([1..maxWaiting(n)])
27          while waitingTime not elapsed do
28              listen(⟨join_rply, i, pid, p, r⟩)
29              sentPrimes = sentPrimes ∪ {p}
30          if |sentPrimes| < s then
31              p′ ⟵ getRandom(primesi \ sentPrimes)
32              r′ ⟵ getAssociatedResidue(p′)
33              send(⟨join_rply, i, srcid, p′, r′⟩)

34  joinReplyi(⟨join_rply, srcid, i, p, r⟩)
35      shareSize ⟵ size(primesi)
36      if shareSize < s then
37          if p ∉ primesi then
38              primesi[shareSize] ⟵ p
39              primesi[shareSize] ⟵ r
```

Fig. 1. Chinese Remainder, program for swarm member i

process and further parameters required for the actions executed as a result of the input action.

- *set:* On *set*, process i receives a message of type *set*, indicating the *set* input action and a counter share *share*, namely a set of s counter components (line 1). Process i sets $primes_i$ and $residues_i$ with the received primes and relative residues of the received counter share *share* (lines 2–4).
- *step:* On *step*, process i receives a message of type *stp*, indicating the *step* input action, and an increment value $δ$, which may be negative (line 5). See a similar technique in [7]. The $δ$ value indicates a positive or negative change in the global counter that affects all the counter shares.

Incrementing (or decrementing) the global counter by $δ$ is done by incrementing (or decrementing) each residue r_{i_j} in the counter share of process i by $δ$

modulo p_{i_j} such that $residues_i[j]$ is set with $(residues_i[j] + \delta) \bmod primes_i[j]$ (lines 6,7).

• *regainConsistencyRequest:* On *regainConsistancyRequest*, the processes are assumed to be in a safe place without the threat of any adversary (alternatively, a global secure function computation technique is used).

Process i receives a message of type *rgn_rqst* (line 8) which triggers a leader election procedure (line 9). Once the leader is elected, it is responsible for distributing the global counter components to the swarm members. If process i is the leader (line 10) it first listens to regain consistency reply messages, initializing the set *globalCounterComponents* with the counter components received from other swarm members (line 11).

If the number of distinct global counter components is smaller than s, i.e., some of the global counter components are missing, then process i initializes *globalCounterComponents* with the set of components calculated by the method *initGlobalCounterComponents()* (lines 12,13). This method sets the values of the global counter GC by setting l distinct primes and l relative residues.

Having set the global counter components, process i (the leader) randomly chooses a share of size s (out of l) components and sends it to a swarm member. Note that there is also a straightforward deterministic way to distribute the shares, or alternatively check the result of the random choice. The random share is chosen with equal probability for every swarm member and sent in a message of type *set* (lines 14-16).

After the shares are sent, the set *globalCounterComponents* is initialized as an empty set, to avoid revealing the counter in case the leader is later compromised (line 17). In case process i is not the leader, it sends its share to the leader (lines 18,19).

• *regainConsistencyReply:* On *regainConsistancyReply*, the processes are assumed to be in a safe place without the threat of any adversary.

Process i receives a message of type *rgn_rply* with the counter share of a process whose identifier is *srcid* (line 20). If process i is the leader, then it adds the components of the received share to its own set of *globalCounterComponents* (lines 21,22).

• *joinRequest:* An input message of type *join_rqst* indicates a request by a new process with identifier *srcid* to join the swarm (line 23). Process i holds a set *sentPrimes* of primes which were sent by other processes in a join reply message. This set is initially empty (line 24).

The join procedure is designed to restrict the shares the (listening) adversary may reveal during the join procedure to the shares assigned to the joining process. Thus, if the number of distinct primes which were sent to the joining process is at least s, then process i should not reply to the join request (line 25).

Otherwise, it sets *waitingTime* with a random period of time, which is a number of time units within the range 1 and $maxWaiting(n)$; where $maxWaiting$ is a function which depends on the number of swarm members n and the time unit size (line 26).

During the random period of time $waitingTime$, process i listens to join replies sent by other processes. Each reply includes a prime p and its relative residue r. While listening, process i adds the sent prime p to the set of primes $sentPrimes$ (lines 27–29). After the $waitingTime$ elapsed, process i checks if at least s distinct counter components were sent back to the joining process (line 30). If not, it randomly chooses a prime number p' out of the primes that appears in its share but not in $sentPrimes$. It then sends back to process $srcid$ a join reply with the random counter component $\langle p', r' \rangle$, where r' is the residue associated with p', namely $r' = GC \bmod p'$ (lines 31–33).

We assume that at most one sender may succeed in sending the reply. If one has failed, process i knows which counter component was successfully sent. Note that the counter components can be encrypted. In that case, the $join\_rqst$ message includes a public key. Otherwise, we regard each join as a process capture by the adversary.

- $joinReply$: Process i receives an input message of type $join\_rply$, which indicates a reply for a join request by a process joining the swarm. The message includes a counter component: Prime p and its relative residue r (line 34).

Process i sets the $shareSize$ with the size of $primes_i$ and indicates the current size of its current share, which should eventually be s (line 35). If $shareSize$ is smaller than s (line 36), then process i should not ignore incoming join replies since it is missing counter components. Process i checks whether the received prime p was already received. If so, process i adds the received counter components by adding p to $primes_i$ and r to $residues_i$ (lines 37–39).

Theorem 1. *In any execution in which the adversary captures at most $k-1$ processes, the probability of the adversary guessing the secret, i.e., guessing the value of the global counter, is bounded by $\frac{1}{p_{min}}$.*

Byzantine adversary and error correcting. We now turn to considering the case of the Byzantine adversary, in which some errors take place, such as input not received by all swarm members. Let m be any positive integer, where $\prod_{i=1}^{l-1} p_i \le m < \prod_{i=1}^{l} p_i$ and $p_1 < p_2 < ... < p_l$. By the CRT, m is uniquely specified by its residues modulo relatively prime numbers $p_1 < ... < p_l$.

The integer m can be represented by $l + l_0$ ($l_0 > 0$) residues modulo relatively prime numbers $p_1 < ... < p_{l+l_0}$. Clearly, this representation is not unique and uses l_0 redundant primes. The integer m can be considered a code word, while the extended representation (using $l + l_0$ primes) yields a natural *error-correcting code* [8].

The error correction is based on the property that for any two integers $m, m' < \prod_{i=1}^{l+l_0} p_i$ the sequences $\{(m \bmod p_1), ..., (m \bmod p_{l+l_0})\}$ and $\{(m' \bmod p_1), ..., (m' \bmod p_{l+l_0})\}$ differ in at least l_0 coordinates.

On the presence of errors, the primes may also be faulty. For that, let us assume that each process keeps the whole set $\mathcal{P}$ instead of only a share of it. Let us also assume that $\mathcal{P}$ is of size $l + l_0$, where l_0 primes are redundant. Under this assumption, we can update the $regainConsistency$ action, so that the processes first agree on $\mathcal{P}$ by a simple majority function and only then agree on the residues

$\langle r_1, r_2, ..., r_{l+l_0} \rangle$ matching the relatively primes $\{p_1 < p_2 < ... < p_{l+l_0}\} = \mathcal{P}$. In that case, the number of Byzantine values or faults, modeled by f, is required to be less than the majority and less than $\frac{l_0}{2}$.

Once $\mathcal{P}$ is agreed on, the received counter components $\langle p_j, r_j \rangle$ where $p_j \notin \mathcal{P}$ are discarded, while the rest of the components are considered candidates to be the real global counter components. The swarm then needs to agree on the residues and again, it is done by a simple majority function executed for every residue out of the $l + l_0$ residues. After the swarm has agreed on the primes and the residues (and in fact, on the counter), the consistency of the counter components can be checked using Mandelbaum's technique [12]. If $f < l_0$ errors have occurred, then they can be detected. Also, the number of errors which can be corrected is $\lfloor l_0/2 \rfloor$.

Chinese remainder solution with single component share. In this case, we have l relatively prime numbers in $\mathcal{P}$ and each secret share is a single pair $\langle r_i, p_i \rangle$, where $p_i \in \mathcal{P}$ and $r_i = GC \bmod p_i$. Using Mandelbaum's technique [12] we may distribute $l \geq k$ shares that represent a value of a counter defined by any k of them. So in case of the listening adversary with no joins, any k will reveal the secret and less than k will not. This distribution also supports the case of Byzantine/Transient faults, where more than k should be read in order to reveal the correct value of the secret. In such a case, a join can be regarded as a transient fault, having the joining process choose its share to be a random value.

4   Reactive k-Secret – Counting/Multiplying Polynomial-Based Solution

Here we consider a global counter, where its value can be multiplied by some factor, as well as increased (decreased) as described in Section 3. The global counter is based on Shamir's (k, n)-threshold scheme [13], according to which, a secret is divided into n shares in a way that any k or more shares can reveal it yet no fewer than k shares can imply any information regarding the secret. Given $k + 1$ points in the 2-dimensional plane $(x_0, y_0), \ldots (x_k, y_k)$, where the values x_i are distinct, there is one and only one polynomial $p(x)$ of degree k such that $p(x_i) = y_i$ for $i = 0..k$. The secret, assumed to be a number S, is encoded by $p(x)$ such that $p(0) = S$. In order to divide the secret S into n shares $S_1, \ldots, S_n$, we first need to construct the polynomial $p(x)$ by picking k random coefficients $a_1, \ldots, a_k$ such that $p(x) = S + a_1 x^1 + a_2 x^2 + \ldots + a_k x^k$. The n shares $S_1, \ldots, S_n$ are pairs of the form $S_i = \langle i, p(i) \rangle$. Given any subset of k shares, $p(x)$ can be found by interpolation and the value of $S = p(0)$ can be calculated.

Our polynomial-based solution encodes the value of the global counter GC, which is the actual secret shared by the swarm members. The global counter GC is represented by a polynomial $p(x) = GC + a_1 x^1 + a_2 x^2 + \ldots + a_l x^l$, where $a_1, \ldots, a_l$ are random. Now, the l counter components are the points $\langle i, p(i) \rangle$ for $i = 1..l$. Instead of each secret share being a single point, a share is a tuple of $s = \lfloor \frac{l}{k} \rfloor$ such points. This way, compromising at most $k - 1$ processes ensures that at least one point is missing and therefore the polynomial $p(x)$ cannot be calculated.

Having each process holding a single point as in Shamir's scheme, implies a complicated join action since it requires collecting all the swarm members' shares and computing the polynomial $p(x)$ in order to calculate a new point for the new joining process. Such action should be avoided under the threat of an adversary. A tuple of $s = \lfloor \frac{l}{k} \rfloor$ points implies a safe join action. The processes share their own points with the new joining process and there is no need to collect all the points.

Lemma 1. *Let $P(x)$ be a polynomial of degree d. Given a set of $d+1$ points $(x_0, y_0), \ldots (x_d, y_d)$, where $P(x_i) = y_i$ for $i = 0, .., d$ and a number δ. The polynomial $Q(x)$, also of degree d, where $Q(x_i) = y_i + \delta$, equals to $P(x) + \delta$.*

Lemma 2. *Let $P(x)$ be a polynomial of degree d. Given a set of $d+1$ points $(x_0, y_0), \ldots (x_d, y_d)$, where $P(x_i) = y_i$ for $i = 0, .., d$ and a number μ. The polynomial $Q(x)$ of degree d, where $Q(x_i) = y_i \cdot \mu$, equals to $P(x) \cdot \mu$.*

According to lemma 1 adding δ to $y_0, \ldots y_l$, where $P(i) = y_i$ for $i = 0, .., l$, results in a new polynomial $Q(x)$ where $Q(x) = P(x) + \delta$. Hence, increasing (decreasing) the second coordinate of the counter shares by δ increases (decreases) the secret by δ as well, since $Q(0) = P(0) + \delta$. Similarly, according to lemma 2 multiplication of the second coordinate in some factor μ implies the multiplication of the secret value in μ.

The code for the polynomial based solution is omitted from this extended abstract and cane be found in [6]. The procedure for each input actions are very similar to the input actions in the Chinese remainder counter, see Figure 1. In this case, the secret shares are tuples of s points given in two arrays $xCoords[1..s]$ and $yCoords[1..s]$, matching the x and y coordinates, respectively. These arrays replace the $primes[1..s]$ and $residues[1..s]$ arrays in the Chinese remainder counter. Another difference is that the step operation has an additional parameter $type$, which defines whether to increment (decrement) or multiply the counter components by δ.

Theorem 2. *In any execution in which the adversary captures at most $k - 1$ processes, the adversary does not reveal any information concerning the secret.*

Byzantine adversary and error correcting. Analogously to the Chinese remainder case, we can design a scheme that is robust to faults. Having n distinct points of the polynomial $p(x)$ of degree l, the Berlekamp-Welch decoder [15] can decode the secret as long as the number of errors e is less than $(n - l)/2$.

Polynomial-based solution with single component share. Using Berlekamp-Welch technique [15] we may distribute $n \geq k$ shares that represent a value of a counter defined by any k of them. So in the case of the listening adversary with no joins, any k will reveal the secret, and fewer than k will not. Similarly to the Chinese remainder solution, this secret share distribution also supports the case of Byzantine/Transient faults, where more than k should be read to reveal the correct value of the secret. Again, a join can be regarded as a transient fault, having the joining process choose its share to be a random value.

5   Virtual Automaton

We would like the swarm members to implement a virtual automaton where the state is not known. Thus, if at most f, where $f < n$, swarm members are compromised, the global state is not known and the swarm task is not revealed.

In this section we present the scheme assuming possible errors, as the error free is a straightforward special case.

We assume that our automaton is modeled as an I/O automaton [11] and described as a five-tuple:

- An action signature $sig(A)$, formally a partition of the set $acts(A)$ of actions into three disjoint sets $in(acts(A))$, $out(acts(A))$ and $int(acts(A))$ of input actions, output actions, and internal actions, respectively. The set of local controlled actions is denoted by $local(A) = out(A) \cup int(A)$.
- A set $states(A)$ of states.
- A nonempty set $start(A) \subseteq states(A)$ of initial states.
- A transition relation $steps(A) : states(A) \times acts(A) \longrightarrow states(A)$, where for every state $s \in states(A)$ and an input action π there is a transition $(s, \pi, s') \in steps(A)$.
- An equivalence relation $part(A)$ partitioning the set $local(A)$ into at most a countable number of equivalence classes.

We assume that the swarm implements a given I/O automaton A. The swarm's *global state* is the current state in the execution of A. Each process i in the swarm holds a tuple $cur\_state_i = \langle s_{i_1}, s_{i_2}, \ldots, s_{i_m} \rangle$ of m distinct states, where $s_{i_j} \in states(A)$ for all $j = 1..m$ and at most one of the m states is the swarm's global state. Formally, the swarm's global state is defined as the state which appears in at least threshold T out of n cur_state tuples ($T \leq n$). If there are more than one such states, then the swarm's global state is a predefined default state.

The *output* of process i is a tuple $out_i = \langle o_{i_1}, o_{i_2}, \ldots, o_{i_m} \rangle$ of m output actions, where $o_{i_j} \in out(acts(A))$ for all $j = 1..m$. The swarm's *global output* is defined as the result of the output action which appears in at least threshold T out of n members' output.

We assume the existence of a devices (sensors, for example) which receives the output of swarm members (maybe in the form of directed laser beams) and thus can be exposed to identify the swarm's global output by a threshold of the members outputs.

We assume an adversary which can compromise at most $f < n$ processes between two successive global reset operations of the swarm's global state. We assume that the adversary knows the automaton A and the threshold T. Therefore, when compromising f processes it can sample the cur_state tuples of the compromised processes and assume that the most common state, i.e., appears as many times in the compromised cur_state tuples, is most likely to be the global state of the swarm.

Consider the case in which $f = 1$ and $T = \lfloor n/2 + 1 \rfloor$. If $|cur\_state| = 1$ (i.e., there is a single state in cur_state), then an adversary which compromises process i, knows the state $s_{i_1} \in cur\_state_i$. The probability that s_{i_1} is the swarm's global state is at least $\frac{T}{n}$ and since T is a lower bound, the probability may reach 1 when all shares are identical. If $|cur\_state| = 2$, then an adversary which compromises process i, knows the state $s_{i_1}, s_{i_2} \in cur\_state_i$. The probability that either of the states s_{i_1} or s_{i_2} is the swarm's global state is at least $\frac{T}{n}$. Since there is no information on which state of the two is the most likely to be the swarm's global state, the only option for the adversary is to arbitrarily choose one of the two states with equal probability. Therefore, the probability of revealing the swarm's global state is at least $\frac{T}{2n}$ and at most $\frac{T}{n}$ in case $|cur\_state| = 2$. Generally, if $|cur\_state| = m$, then the probability of revealing the swarm's global state is at least $\frac{T}{m \cdot n}$, and at most $\frac{T}{(m-1) \cdot n}$ for $f = 1$. As the number of states in cur_state increases, the probability to reveal the swarm's global state decreases.

We consider the following input actions:

- $set(\langle s_{i_1}, \ldots, s_{i_m} \rangle)$: Sets cur_states with the given tuple. The tuples are distributed in a way that at least $T + f + lp$ of them contain the swarm's global state. Thus, even if f shares are corrupted and lp are missing because of the leaving processes, the swarm threshold is respected. Moreover, in order to ensure uniqness of the global state in the presence of corruptions and joins, any other state has less than $T - f$ replicas.
- $step(\delta)$: Emulates a step of the automaton for each of the states in cur_state_i. By the end of the emulation each process has output. Here, δ is any possible input of the simulated automaton.
- $regain\ consistency$: Ensures that there are at least $T + f + lp$ members, whose cur_states tuples include the current state of A. Any other state has less than $T - f$ replicas.
- $join$: A process joins the swarm, and constructs its cur_states tuple by randomly collecting states from other processes. Note that the scheme benefits from smooth joins, since the number f that includes the join operations is taken in consideration while calculating the swarm's global state upon regain consistency operation. That is, a threshold of T is required for a state in order to be the swarm's global state. Therefore, in case swarm members maintain the population of the swarm (updated by joins, leaves and possibly by periodic heartbits) a join may be simply done by sending a join request message, specifying the identifier of the joining process. However, the consistency of the swarm will definitely benefit if shares are uniformly chosen for the newcomers. In this way, if the adversary was not listening during the join procedure, there is high probability that the joining processes will assist in encoding the current secret.

Line-by-line code description. The code in Figure 2 describes input actions of process i. Each process i has an m-tuple cur_state_i of m states in $states(A)$, where at most one of them is the swarm's global state.

- set: On input action set, process i receives a message of type set and an m-tuple of distinct states in $states(A)$ (line 1). It then sets its tuple cur_state_i with the received tuple (line 2).
- $step$: On input action $step$, process i receives a message of type stp and δ, which is an input parameter for the I/O automaton (line 3). For every state s_{i_j} in

```
1   set_i(⟨set, srcid, i, ⟨s_{i_1}, ..., s_{i_m}⟩⟩)
2       cur_state_i ⟵ ⟨s_{i_1}, ..., s_{i_m}⟩

3   step_i(⟨stp, srcid, i, δ⟩)
4       for j = 1..m do
5           ⟨s'_{i_j}, o_{i_j}⟩ ⟵ follow the transaction in steps(A) for s_{i_j} and δ
6       next_state ⟵ ⟨s'_{i_1}, ..., s'_{i_m}⟩
7       output_acts ⟵ ⟨o'_{i_1}, ..., o'_{i_m}⟩
8       executeOutputActions(output_acts)
9       cur_state_i ⟵ next_state

10  regainConsistencyRequest_i(⟨rgn_rqst, srcid, i⟩)
11      leaderId ⟵ leaderElection()
12      if leaderId = i then
13          allStateTuples_i ⟵ collectAllStates()
14          candidates ⟵ mostPopularStates(allStateTuples)
15          if |candidates| == 1 then
16              globalState ⟵ first(candidates)
17          else
18              globalState ⟵ defaultGlobalState
19          distributeStateTuples(globalState)
20          allStateTuples_i ⟵ ∅
21          delete candidates
22      else
23          send(⟨rgn_rply, i, leaderId, cur_state_i, ⟩)

24  regainConsistencyReply_i(⟨rgn_rply, srcid, i, stateTuple⟩)
25      if leaderId = i then
26          allStateTuples_i ⟵ allStateTuples_i ∪ {stateTuple}

27  joinRequest_i(⟨join_rqst, srcid, i⟩)
28      addMember(srcid)
```

Fig. 2. Virtual automaton, program for swarm member i

cur_state_i process i emulates the automaton A by executing a single transaction on s_{i_j} and δ (line 5). As a result, there is a new state s'_{i_j} and an output action o_{i_j}. Process i initializes a tuple $next\_state$ of all the new states s'_{i_j} for all $j = 1..m$ (line 6) and the resulting output actions o'_{i_j} for all $j = 1..m$ (line 7). It then executes the output actions in $output\_acts$ (line 8) and finally, it updates cur_state_i to be the tuple of new states $next\_state$ (line 9).

• $regainConsistencyRequest$: On input action $regainConsistency$ the processes are assumed to be in a safe land with no threat of any adversary. Process i receives a message of type rgn_rqst from process identified by $srcid$ (line 10).

The method $leaderElection()$ returns the process identifier of the elected leader (line 11). If process i is the leader, then it should distribute state tuples using set input actions in a way that at least $T + f$ swarm members have tuples that include the global state and all other states appear no more than T times. Possibly by randomly choosing shares to members, such that the probability for assigning the global state share to a process is equal to, or slightly greater than, $T/n + f/n$ while the probability of any other state to be assigned to a process is the same (smaller) probability.

First, the leader collects all the state tuples (line 13) and then executes the method $mostPopularStates()$ in order to find the candidates to be the swarm's global state (lines 14). If there is a single candidate (line 15), then it is the global state and *globalState* is set with the first (and only) state in *candidates* (line 16). In case there is more than one candidate (line 17), the leader sets *globalState* with a predefined default global state (line 18).

The leader then distributes the state tuples (line 19) and deletes both the collected tuples *allStateTuples* and the candidates for the global state *candidates* (lines 20,21). If process i is not the leader, then it sends its cur_state_i tuple to the leader (lines 22,23).

• *regainConsistencyReply:* On input action *regainConsistencyReply* the processes are also assumed to be in a safe land. Process i receives a message of type rgn_rply, which is a part of the regain consistency procedure. The message includes the identifier *srcid* of the sender and the sender's state tuple (line 24). If process i is the leader (line 25), then it adds the received tuple to the set $allStateTuples_i$ of already received tuples. Otherwise, it ignores the message.

• *joinRequest:* On input action, *joinRequest* process i receives a message of type $join\_rqst$ from a process identified by *srcid*, which is asking to join the swarm (line 27). Process i executes the method $addMember(srcid)$, which adds *srcid*, the identifier of the joining process, to its population list of processes in the swarm.

6   Conclusions

We have presented three (in fact four, including the Vandermonde matrix based scheme) approaches for reactive k-secret sharing that require no internal communication to perform a transition.

The two first solutions maybe combined as part of the reactive automaton to define share of the state, for example to enable an output of the automaton whenever a share value of the counter is prime. Thus the operator of the swarm may control the output of each process by manipulating the counter value, e.g., making sure the counter secret shares are never prime, until a sufficient number and combination of events occurs.

We believe that such a distributed manipulation of information without communicating the secret shares, that is secure even from the secret holders, should be further investigated. At last, the similarity in usage of Mandelbaum and Berlekamp-Welch techniques may call for arithmetic generalization of the concepts.

References

1. Dolev, S., Gilbert, S., Lahiani, L., Lynch, N., Nolte, T.: Virtual Stationary Automata for Mobile Networks. In: Anderson, J.H., Prencipe, G., Wattenhofer, R. (eds.) OPODIS 2005. LNCS, vol. 3974, Springer, Heidelberg (2006) Also invited paper in Forty-Third Annual Allerton Conference on Communication, Control, and Computing. Also, Brief announcement. In: PODC 2005. Proc. of the 24th Annual ACM Symp. on Principles of Distributed Computing, p. 323 (2005) Technical Report MIT-LCS-TR-979, Massachusetts Institute of Technology (2005)

2. Dolev, S., Gilbert, S., Lynch, A.N., Schiller, E., Shvartsman, A., Welch, J.: Virtual Mobile Nodes for Mobile Ad Hoc Networks. In: DISC 2004. International Conference on Principles of DIStributed Computing, pp. 230–244 (2004) Also Brief announcement. In: PODC 2004. Proc. of the 23th Annual ACM Symp. on Principles of Distributed Computing (2004)
3. Dolev, S., Gilbert, S., Lynch, N.A., Shvartsman, A., Welch, J.: GeoQuorum: Implementing Atomic Memory in Ad Hoc Networks. Distributed Computing 18(2), 125–155 (2003)
4. Dolev, S., Gilbert, S., Schiller, E., Shvartsman, A., Welch, J.: Autonomous Virtual Mobile Nodes. In: DIALM/POMC 2005. Third ACM/SIGMOBILE Workshop on Foundations of Mobile Computing, pp. 62–69 (2005) Brief announcement. In: SPAA 2005. Proc. of the 17th International Conference on Parallelism in Algorithms and Architectures, p. 215 (2005) Technical Report MIT-LCS-TR-992, Massachusetts Institute of Technology (2005)
5. Dolev, S., Lahiani, L., Lynch, N., Nolte, T.: Self-Stabilizing Mobile Location Management and Message Routing. In: Tixeuil, S., Herman, T. (eds.) SSS 2005. LNCS, vol. 3764, pp. 96–112. Springer, Heidelberg (2005)
6. Dolev, S., Lahiani, L., Yung, M.: Technical Report TR-#2007-12, Department of Computer Science, Ben-Gurion University of the Negev (2007)
7. Dolev, S., Welch, L.J.: Self-Stabilizing Clock Synchronization in the Presence of Byzantine Faults. Journal of the ACM 51(5), 780–799 (2004)
8. Goldrich, O., Ron, D., Sudan, M.: Chinese Remaindering with Errors. In: Proc. of 31st STOC. ACM (1999)
9. Kivelevich, E., Gurfil, P.: UAV Flock Taxonomy and Mission Execution Performance. In: Proc. of the 45th Israeli Conference on Aerospace Sciences (2005)
10. Kilian, J., Kushilevitz, E., Micali, S., Ostrovsky, R.: Reducibility and Completeness In Multi-Party Private Computations. In: FOCS 1994. Proceedings of Thirty-fifth Annual IEEE Symposium on the Foundations of Computer Science, Journal version in SIAM J. Comput. 29(4), 1189-1208 (2000)
11. Lynch, N., Tuttle, M.: An introduction to Input/Output automata, Centrum voor Wiskunde en Informatica, Amsterdam, The Netherlands, 2(3), 219–246 (September 1989) Also Tech. Memo MIT/LCS/TM-373
12. Mandelbaum, D.: On a Class of Arithmetic and a Decoding Algorithm. IEEE Transactions on Information Theory 21(1), 85–88 (1976)
13. Shamir, A.: How to Share a Secret. CACM 22(11), 612–613 (1979)
14. Weiser, M.: The Computer for the 21th Century. Scientific American (September 1991)
15. Welch, L., Berlekamp, E.R.: Error Correcting for Algebraic Block Codes, U.S. Patent 4633470 (September 1983)

New Formulae for Efficient Elliptic Curve Arithmetic

Huseyin Hisil, Gary Carter, and Ed Dawson

Information Security Institute, Queensland University of Technology
{h.hisil, g.carter, e.dawson}@qut.edu.au

Abstract. This paper is on efficient implementation techniques of Elliptic Curve Cryptography. In particular, we improve timings[1] for Jacobi-quartic (3M+4S) and Hessian (7M+1S or 3M+6S) doubling operations. We provide a faster mixed-addition (7M+3S+1d) on modified Jacobi-quartic coordinates. We introduce tripling formulae for Jacobi-quartic (4M+11S+2d), Jacobi-intersection (4M+10S+5d or 7M+7S+3d), Edwards (9M+4S) and Hessian (8M+6S+1d) forms. We show that Hessian tripling costs 6M+4C+1d for Hessian curves defined over a field of characteristic 3. We discuss an alternative way of choosing the base point in successive squaring based scalar multiplication algorithms. Using this technique, we improve the latest mixed-addition formulae for Jacobi-intersection (10M+2S+1d), Hessian (5M+6S) and Edwards (9M+1S+1d+4a) forms. We discuss the significance of these optimizations for elliptic curve cryptography.

Keywords: Elliptic curve, efficient point multiplication, doubling, tripling, DBNS.

1 Introduction

One of the main challenges in elliptic curve cryptography is to perform scalar multiplication efficiently. In the last decade, much effort has been spent in representing the elliptic curves in special forms which permit faster point doubling and addition. In particular,

- Cohen, Miyaji and Ono [1] showed fast implementation in Weierstrass form on Jacobian coordinates.
- Smart [2], Joye and Quisquater [3], Liardet and Smart [4], Billet and Joye [5] showed ways of doing point multiplication to resist side channel attacks.
- Doche, Icart and Kohel [6] introduced the fastest doubling[2] and tripling in Weierstrass form on two different families of curves.

---

[1] **M:** Field multiplication, **S:** Field squaring, **C:** Field cubing on characteristic 3 fields, **d:** Multiplication by a curve constant. **a:** Addition. For simplicity in our analysis, we fix 1S=0.8M, 1C=0.1M, 1d=0M.

[2] With the improvements of Bernstein, Birkner, Lange and Peters [7,8].

K. Srinathan, C. Pandu Rangan, M. Yung (Eds.): Indocrypt 2007, LNCS 4859, pp. 138–151, 2007.
© Springer-Verlag Berlin Heidelberg 2007

- Negre [9], Kim, Kim and Choe [10], Smart [11] investigated the case of efficient arithmetic for low odd characteristic curves.
- Dimitrov, Imbert and Mishra [12] showed the first efficient inversion free tripling formula on Jacobian coordinates. Meloni [13] showed how special addition chains can be used in point multiplication.
- Avanzi, Dimitrov, Doche and Sica [14] and Doche and Imbert [15] provided an extended way of using DBNS[3] in elliptic curve cryptography.
- Edwards [16] introduced a new representation of elliptic curves. Bernstein and Lange [17,18] showed the importance of this new system for providing fast arithmetic and side channel resistance. They have also built a database [7] of explicit formulae that are reported in the literature together with their own optimizations. Our work greatly relies on the formulae reported in this database.

This paper is composed of several optimizations regarding elliptic curve arithmetic operations. We improve some of the previously reported elliptic curve group operations namely Jacobi-quartic doubling, Jacobi-quartic mixed addition and Hessian doubling. As well, we introduce elliptic curve point tripling formulae for Jacobi-quartic, Jacobi-intersection, Hessian and Edwards forms. We introduce a technique for successive squaring based point multiplication algorithms which speeds-up mixed addition in some forms. This technique enables faster mixed-addition for Jacobi-intersection, Hessian and Edwards curves. The optimizations in this paper are solely efficiency oriented. Therefore, these results do not cover side channel resistance. Some immediate outcomes are;

- Jacobi-quartic form becomes competitive in efficiency oriented applications. For instance, successive squaring based point multiplications on modified/ extended Jacobi-quartic coordinates can be performed faster than standard[4] Edwards coordinates for all S/M values.
- The tripling formulae that are introduced provide a wider background for new comparisons on DBNS based point multiplications in different systems.
- Point multiplication in Hessian form can be performed faster than Jacobian form whenever S/M is near to 1.
- Hessian tripling formula enables efficient implementation of DBNS based point multiplication algorithms with Hessian curves defined over fields of characteristic 3.

This paper is organized as follows. We show faster doubling formulae in Section 2. We introduce new tripling formulae in Section 3. We provide a faster Jacobi-quartic mixed addition in Section 4. We describe an alternative strategy for the selection of base point for point multiplication in Section 5. We draw our conclusions in Section 6.

---

[3] DBNS: Double Base Number System.
[4] Very recently, Bernstein and Lange [7] developed the inverted-Edwards coordinates which requires less memory and bandwidth.

2   New Doubling Formulae

We observe that output coordinates of an inversion-free formula can be represented alternatively by selecting congruent elements from $K[E]$, i.e. the coordinate ring of E over K. In other words, an inversion-free formula (which is originally derived from its affine version) can be modified using the curve equation. We do not report the detail for finding all new explicit formulae since the process is composed of fairly tedious steps. Nevertheless, a step-by-step derivation of doubling a formula for Jacobi-quartic form is given in the Appendix as an example. Tripling formulae in Section 3 can be derived building on the same ideas.

Let K be a field with $\operatorname{char}(K) \neq 2, 3$. An elliptic curve in Jacobi-quartic form [5] is defined by $E(K)\colon y^2 = x^4 + 2ax^2 + 1$ where $a^2 - 1$ is nonzero. The identity element is the point $(0, 1)$. We do not give background material on curves since it is well documented in the literature. Our optimization leads to the following strategy for inversion-free Jacobi-quartic doubling. (See the Appendix).

$$X_3 = Y_1((X_1 - Z_1)^2 - (X_1^2 + Z_1^2)), \qquad Z_3 = (X_1^2 + Z_1^2)(X_1^2 - Z_1^2),$$

$$Y_3 = 2(Y_1(X_1^2 + Z_1^2))^2 - (X_3^2 + Z_3^2).$$

The operation count shows that Jacobi-quartic doubling costs **3M+4S** (previous best [7], 1M+9S) when the points are represented on the modified coordinates, $(X_1 : Y_1 : Z_1 : X_1^2 : Z_1^2)$. This redundant representation will also be used to improve the mixed-addition for Jacobi-quartic form in Section 4. The operations can be scheduled as follows. We denote the input cache registers as (X_1^2) and (Z_1^2). Re-caching is done on (X_3^2) and (Z_3^2).

$X_3 \leftarrow X_1 + Z_1$	$Y_3 \leftarrow Z_3 \times Y_1$	$Y_3 \leftarrow Y_3^2$	$(Z_3^2) \leftarrow Z_3^2$
$X_3 \leftarrow X_3^2$	$X_3 \leftarrow Y_3 - X_3$	$Y_3 \leftarrow 2 * Y_3$	$Y_3 \leftarrow Y_3 - (Z_3^2)$
$X_3 \leftarrow X_3 \times Y_1$	$R_0 \leftarrow (X_1^2) - (Z_1^2)$	$(X_3^2) \leftarrow X_3^2$	
$Z_3 \leftarrow (X_1^2) + (Z_1^2)$	$Z_3 \leftarrow Z_3 \times R_0$	$Y_3 \leftarrow Y_3 - (X_3^2)$	

Most applications overwrite (X_1^2) and (Z_1^2). In this case, temporary register R_0 can be replaced with (X_1^2) or (Z_1^2). In addition, (X_3^2) and (Z_3^2) can be the same registers as (X_1^2) and (Z_1^2), respectively. This scheduling method uses 6 additions and 1 multiplication by 2. If the caching is performed more redundantly as $(X_1 : Y_1 : Z_1 : X_1^2 : Z_1^2 : (X_1^2 + Z_1^2))$, it is possible to save one addition and to avoid the use of R_0.

Let K be a field with $\operatorname{char}(K) \neq 2$. An elliptic curve in Hessian form [2] is defined by $E(K)\colon x^3 + y^3 + 1 = cxy$ where $(c/3)^3 - 1$ is nonzero. The identity element is the point at infinity. The cost of inversion-free doubling was reported as 6M+3S with respect to the following formula where each coordinate is cubed and used in the obvious way [7].

$$X_3 = Y_1(Z_1^3 - X_1^3), \quad Y_3 = X_1(Y_1^3 - Z_1^3), \quad Z_3 = Z_1(X_1^3 - Y_1^3).$$

However, the same formula costs **7M+1S** when the following strategy is used.

$$A \leftarrow X_1^2, \quad B \leftarrow Y_1(X_1 + Y_1), \quad C \leftarrow A + B, \quad D \leftarrow Z_1(Z_1 + X_1),$$
$$E \leftarrow A + D, \quad F \leftarrow C(X_1 - Y_1), \quad G \leftarrow E(Z_1 - X_1),$$
$$X_3 \leftarrow GY_1, \quad Y_3 \leftarrow -(F + G)X_1, \quad Z_3 \leftarrow FZ_1.$$

Furthermore, it is possible to save 1 reduction by delaying the reduction steps after computing $X_1^2, Y_1(X_1 + Y_1)$ and $Z_1(Z_1 + X_1)$ whenever desired. Lazy reductions effect the timings when arbitrary moduli are used for field reductions. If fast reduction moduli (such as NIST primes) are used then this advantage vanishes. Note, we do not include this optimization in our complexity analysis. The operations for Hessian doubling can be scheduled as follows.

$R_0 \leftarrow X_1^2$	$R_2 \leftarrow Z_1 \times R_2$	$R_0 \leftarrow R_1 \times R_0$	$X_3 \leftarrow R_1 \times Y_1$
$R_1 \leftarrow X_1 + Y_1$	$R_2 \leftarrow R_0 + R_2$	$Z_3 \leftarrow R_0 \times Z_1$	$R_2 \leftarrow -(R_0 + R_1)$
$R_1 \leftarrow Y_1 \times R_1$	$R_1 \leftarrow R_0 + R_1$	$R_1 \leftarrow Z_1 - X_1$	$Y_3 \leftarrow R_2 \times X_1$
$R_2 \leftarrow Z_1 + X_1$	$R_0 \leftarrow X_1 - Y_1$	$R_1 \leftarrow R_2 \times R_1$	

An alternative layout is as follows.

$$X_3 = (((X_1 + Y_1)^2 - (X_1^2 + Y_1^2)) - ((Y_1 + Z_1)^2 - (Y_1^2 + Z_1^2))) \cdot$$
$$(((X_1 + Z_1)^2 - (X_1^2 + Z_1^2)) + 2(X_1^2 + Z_1^2))$$
$$Y_3 = (((X_1 + Z_1)^2 - (X_1^2 + Z_1^2)) - ((X_1 + Y_1)^2 - (X_1^2 + Y_1^2))) \cdot$$
$$(((Y_1 + Z_1)^2 - (Y_1^2 + Z_1^2)) + 2(Y_1^2 + Z_1^2))$$
$$Z_3 = (((Y_1 + Z_1)^2 - (Y_1^2 + Z_1^2)) - ((X_1 + Z_1)^2 - (X_1^2 + Z_1^2))) \cdot$$
$$(((X_1 + Y_1)^2 - (X_1^2 + Y_1^2)) + 2(X_1^2 + Y_1^2))$$

This strategy costs **3M+6S**. There are no lazy reduction possibilities. It requires more additions and more temporary registers. However, it will be faster whenever $1S < 0.8M$. It is known that $1S \approx 0.66M$ when fast reduction moduli are used. The operations can be scheduled as follows.

$R_0 \leftarrow X_1^2$	$R_0 \leftarrow R_0^2$	$R_2 \leftarrow R_2^2$	$Y_3 \leftarrow R_0 - R_1$
$R_1 \leftarrow Y_1^2$	$R_0 \leftarrow R_0 - R_3$	$R_4 \leftarrow 2 * R_4$	$R_5 \leftarrow R_2 + R_5$
$R_2 \leftarrow Z_1^2$	$R_1 \leftarrow X_1 + Z_1$	$R_2 \leftarrow R_2 - R_5$	$Y_3 \leftarrow Y_3 \times R_5$
$R_3 \leftarrow R_0 + R_1$	$R_1 \leftarrow R_1^2$	$R_5 \leftarrow 2 * R_5$	$Z_3 \leftarrow R_1 - R_2$
$R_4 \leftarrow R_0 + R_2$	$R_1 \leftarrow R_1 - R_4$	$X_3 \leftarrow R_2 - R_0$	$R_0 \leftarrow R_0 + R_3$
$R_5 \leftarrow R_1 + R_2$	$R_2 \leftarrow Y_1 + Z_1$	$R_4 \leftarrow R_1 + R_4$	$Z_3 \leftarrow Z_3 \times R_0$
$R_0 \leftarrow X_1 + Y_1$	$R_3 \leftarrow 2 * R_3$	$X_3 \leftarrow X_3 \times R_4$	

The comparison of doubling costs in different systems is depicted in Table 1.

Table 1. Cost comparison of elliptic curve point doubling operations in different coordinate systems. The bold values are the old and the new timings that are explained in this section. We assume 1S=0.8M.

System	Cost analysis	Total
Hessian (OLD) [3]	6M + 3S	**8.4M**
Jacobi-quartic (OLD) [5]	1M + 9S + 1d	**8.2M**
Hessian (NEW-1)	7M + 1S	**7.8M**
Hessian (NEW-2)	3M + 6S	**7.8M**
Doche/Icart/Kohel(3) [7,6]	2M + 7S + 2d	7.6M
Jacobian [1,7]	1M + 8S + 1d	7.4M
Jacobian, $a = -3$ [1,7]	3M + 5S	7.0M
Jacobi-intersections [7,4]	3M + 4S	6.2M
Inverted Edwards [7,8]	3M + 4S + 1d	6.2M
Edwards [7,17,18]	3M + 4S	6.2M
Jacobi-quartic (NEW)	3M + 4S	**6.2M**
Doche/Icart/Kohel(2) [7,6]	2M + 5S + 2d	6.0M

3 New Tripling Formulae

Since DBNS based point multiplication algorithms [12,14,15] have been introduced, there has been a demand for fast tripling formulae. We introduce tripling formulae for Jacobi-quartic, Hessian, Jacobi-intersection and Edwards forms in this section. Tripling formulae can be derived by the composition of doubling and addition formulae. However, a straight forward derivation yields expensive expressions. Nevertheless, it is possible to do simplifications using the curve equation. To the best of our knowledge, no algorithm is known to guarantee the best strategy. Therefore, one can expect further improvements in these formulae in the future. At least, there should be multiplication/squaring tradeoffs.

Following the same notation in Section 2, we introduce Jacobi-quartic tripling. The formula is as follows.

$$X_3 = X_1(X_1^8 - 6X_1^4Z_1^4 - 8aX_1^2Z_1^6 - 3Z_1^8)$$
$$Y_3 = Y_1(X_1^{16} + 8aX_1^{14}Z_1^2 + 28X_1^{12}Z_1^4 + 56aX_1^{10}Z_1^6 + 6X_1^8Z_1^8 +$$
$$64a^2X_1^8Z_1^8 + 56aX_1^6Z_1^{10} + 28X_1^4Z_1^{12} + 8aX_1^2Z_1^{14} + Z_1^{16})$$
$$Z_3 = Z_1(3X_1^8 + 8aX_1^6Z_1^2 + 6X_1^4Z_1^4 - Z_1^8)$$

The terms can be organized as follows.

$$A \leftarrow (X_1^2)^2, \quad B \leftarrow (Z_1^2)^2, \quad C \leftarrow 2(((X_1^2) + (Z_1^2))^2 - (A + B)),$$

$$D \leftarrow (a^2 - 1)C^2, \quad E \leftarrow 4(A - B), \quad F \leftarrow 2(A + B) + aC, \quad G \leftarrow E^2,$$

$$H \leftarrow F^2, \quad J \leftarrow (E + F)^2 - (G + H), \quad K \leftarrow 2(H - D),$$

$$X_3 \leftarrow X_1(J - K), \quad Y_3 \leftarrow Y_1(K^2 - 4GD), \quad Z_3 \leftarrow Z_1(J + K),$$

Jacobi-quartic tripling costs 4M+10S+2d in the modified Jacobi-quartic coordinates. The formulae that will take advantage of the fast Jacobi-quartic addition formula [19] require the addition of XY coordinate to modified coordinates

$$(X_1 : Y_1 : Z_1 : X_1^2 : Z_1^2 : (X_1^2 + Z_1^2)).$$

This modification fixes the complexity to **4M+11S+2d**. We refer the reader to EFD [7] for compatible versions developed by Bernstein and Lange.

Temporary registers R_5 and R_6 can be replaced with volatile cache registers (X_1^2) and (Z_1^2). As well, (X_3^2) and (X_3^2) can be the same registers as (X_1^2) and (Z_1^2), respectively.

In the same fashion, we introduce Hessian tripling. We follow the same notation for Hessian curves that is given in Section 2. Set $k = c^{-1}$. One can treat k as the curve constant confidently since addition and doubling formulae do not depend on c. An efficient tripling can be performed using the following formula.

$$X_3 = X_1^3(Y_1^3 - Z_1^3)(Y_1^3 - Z_1^3) + Y_1^3(X_1^3 - Y_1^3)(X_1^3 - Z_1^3)$$
$$Y_3 = Y_1^3(X_1^3 - Z_1^3)(X_1^3 - Z_1^3) - X_1^3(X_1^3 - Y_1^3)(Y_1^3 - Z_1^3)$$
$$Z_3 = k(X_1^3 + Y_1^3 + Z_1^3)((X_1^3 - Y_1^3)^2 + (X_1^3 - Z_1^3)(Y_1^3 - Z_1^3))$$

The operations are organized as follows.

$$A \leftarrow X_1^3, \quad B \leftarrow Y_1^3, \quad C \leftarrow Z_1^3, \quad D \leftarrow A - B, \quad E \leftarrow A - C,$$

$$H \leftarrow D^2, \quad J \leftarrow E^2, \quad K \leftarrow F^2, \quad X_3 \leftarrow 2AK - B(K - H - J),$$

$$Y_3 \leftarrow 2BJ - A(J - H - K), \quad Z_3 \leftarrow k(A + B + C)(J + H + K).$$

This formula costs **8M+6S+1d**. Furthermore, there exists 2 lazy reduction points. (First, delay reduction when computing AK and $B(K - H - J)$, then delay reduction when computing BJ and $A(J - H - K)$). If the Hessian curve is defined over a field of characteristic 3, the tripling formula simplifies to the following. Note, it is enough to choose a nonzero k in this case.

$$X_3 = (X_1(Y_1 - Z_1)(Y_1 - Z_1) + Y_1(X_1 - Y_1)(X_1 - Z_1))^3$$
$$Y_3 = (Y_1(X_1 - Z_1)(X_1 - Z_1) - X_1(X_1 - Y_1)(Y_1 - Z_1))^3$$
$$Z_3 = k((X_1 + Y_1 + Z_1)^3)^3$$

It is easy to see that this formula costs **6M+4C+1d**. (Reuse $X_1(Y_1 - Z_1)$ and $Y_1(X_1 - Z_1)$). Furthermore, 2 additional lazy reductions can be done in the computation of X_3 and Y_3. It is interesting to note that the cost of 5P=2P+3P is less than a point addition. Recently, Kim, Kim and Choe [10] introduced 4M+5C+2d tripling formula in Jacobian/ML coordinates which is faster than the tripling introduced here.

Next, we introduce the tripling formula for Jacobi-intersection form [4]. Let K be a field with char$(K) \neq 2,3$ and let $a \in K$ with $a(1 - a) \neq 0$. The elliptic curve in Jacobi-intersection form is the set of points which satisfy the equations

$s^2 + c^2 = 1$ and $as^2 + d^2 = 1$ simultaneously. The identity element is the point $(0, 1, 1)$. The inversion-free tripling formula is as follows. With $k = a - 1$,

$$S_3 = S_1(k(kS_1^8 + 6S_1^4C_1^4 + 4S_1^2C_1^6) - 4S_1^2C_1^6 - 3C_1^8)$$
$$C_3 = C_1(k(3kS_1^8 + 4kS_1^6C_1^2 - 4S_1^6C_1^2 - 6S_1^4C_1^4) - C_1^8)$$
$$D_3 = D_1(k(-kS_1^8 + 4S_1^6C_1^2 + 6S_1^4C_1^4 + 4S_1^2C_1^6) - C_1^8)$$
$$T_3 = T_1(k(-kS_1^8 - 4S_1^6C_1^2 + 6S_1^4C_1^4) - 4S_1^2C_1^6 - C_1^8)$$

The operations can be organized as follows.

$$E \leftarrow S_1^2, \quad F \leftarrow C_1^2, \quad G \leftarrow E^2, \quad H \leftarrow F^2, \quad J \leftarrow G^2, \quad K \leftarrow H^2,$$

$$L \leftarrow ((E+F)^2 - H - G), \quad M \leftarrow L^2, \quad N \leftarrow (G+L)^2 - J - M,$$

$$P \leftarrow (H+L)^2 - K - M, \quad R \leftarrow 2k^2J, \quad S \leftarrow 2kN,$$

$$T \leftarrow 3kM, \quad U \leftarrow 2P, \quad V \leftarrow 2K, \quad W \leftarrow (k+1)U, \quad Y \leftarrow (k+1)S,$$

$$S_3 \leftarrow S_1((R-V)+(T+W)-2(U+V)), \quad C_3 \leftarrow C_1((R-V)-(T-Y)+2(R-S)),$$

$$D_3 \leftarrow D_1((T+W)-(R-S)-(U+V)), \quad T_3 \leftarrow T_1((T-Y)-(R-S)-(U+V)).$$

This formula costs **4M+10S+5d**. An alternative strategy costs **7M+7S+3d**. Here is the alternative organization of operations.

$$E \leftarrow S_1^2, \quad F \leftarrow C_1^2, \quad G \leftarrow E^2, \quad H \leftarrow F^2, \quad J \leftarrow 2H, \quad K \leftarrow 2J,$$

$$L \leftarrow (2F+E)^2 - G - K, \quad M \leftarrow kG, \quad N \leftarrow K+J, \quad P \leftarrow M^2,$$

$$R \leftarrow NM, \quad U \leftarrow ML, \quad V \leftarrow H^2, \quad W \leftarrow HL,$$

$$S_3 \leftarrow S_1(R+kW+2(P-V)-W-P-V), \quad C_3 \leftarrow C_1(2(P-V)-U+P+V-R+kU),$$

$$D_3 \leftarrow D_1(U-P-V+R+kW), \quad T_3 \leftarrow T_1(R-kU-W-P-V).$$

Finally, we introduce the tripling formula for Edwards curves [16,17,18]. Let K be a field with $\text{char}(K) \neq 2$ and let $c, d \in K$ with $cd(1 - c^4d) \neq 0$. Then, the Edwards curve, $(x^2 + y^2) = c^2(1 + dx^2y^2)$, is birationally equivalent to an elliptic curve [17,18]. The identity element is the point $(0, c)$. The same formula was independently developed by Bernstein, Birkner, Lange and Peters [8]. Edwards tripling costs **9M+4S**. For further results, we refer the reader to Bernstein, Birkner, Lange and Peters [8]. The inversion-free tripling formula is as follows.

$$X_3 = X_1(X_1^4 + 2X_1^2Y_1^2 - 4c^2Y_1^2Z_1^2 + Y_1^4)(X_1^4 - 2X_1^2Y_1^2 + 4c^2Y_1^2Z_1^2 - 3Y_1^4)$$
$$Y_3 = Y_1(X_1^4 + 2X_1^2Y_1^2 - 4c^2X_1^2Z_1^2 + Y_1^4)(3X_1^4 + 2X_1^2Y_1^2 - 4c^2X_1^2Z_1^2 - Y_1^4)$$
$$Z_3 = Z_1(X_1^4 - 2X_1^2Y_1^2 + 4c^2Y_1^2Z_1^2 - 3Y_1^4)(3X_1^4 + 2X_1^2Y_1^2 - 4c^2X_1^2Z_1^2 - Y_1^4)$$

The cost comparison of tripling formulae in different systems is depicted in Table 2. (Also see Table 3 in the Appendix).

Table 2. Cost comparison of elliptic curve point tripling formulae in different coordinate systems. The bold lines correspond to the complexities of the formulae that are introduced in this section. We assume 1S=0.8M.

System	Tripling Cost	Total
Jacobian [12,7]	5M+10S+1d	13M
Hessian	8M+6S+1d	**12.8M**
Jacobi-quartic	4M+11S+2d	**12.8M**
Jacobi-intersection-2	7M+7S+3d	**12.6M**
Jacobian, $a = -3$ [12,7]	7M+7S	12.6M
Edwards [7,8]	9M+4S+1d	**12.2M**
Inverted Edwards [7,8]	9M+4S+1d	12.2M
Jacobi-intersection-1	4M+10S+5d	**12M**
Doche/Icart/Kohel(3) [6]	6M+6S+2d	10.8M
Hessian, char= 3	6M+4C+2d	**6.4M**

4   Mixed-Addition for Modified Jacobi-Quartic Coordinates

Following the outline on modified Jacobi-quartic doubling (see Section 2), we provide a mixed-addition which is faster than the previous best. The updated formula [19,7] is as follows.

$$X_3 = (Y_1 + (X_1 + Z_1)^2 - (X_1^2 + Z_1^2))(2X_2 + Y_2) -$$
$$2X_2((X_1 + Z_1)^2 - (X_1^2 + Z_1^2)) - (Y_1Y_2)$$
$$Y_3 = 4X_2((X_1 + Z_1)^2 - (X_1^2 + Z_1^2))(X_1^2 + a(Z_1^2 + X_1^2X_2^2) + Z_1^2X_2^2) +$$
$$4(Z_1^2 + X_1^2X_2^2)(Y_1Y_2)$$
$$Z_3 = 2(Z_1^2 - X_1^2X_2^2)$$

The operations can be organized as follows.

$$A \leftarrow (X_1^2) + (Z_1^2), \quad B \leftarrow (X_1 + Z_1)^2 - A, \quad C \leftarrow B + Y_1, \quad D \leftarrow (X_1^2)(X_2^2),$$

$$E \leftarrow 2BX_2, \quad F \leftarrow (Z_1^2) + D, \quad G \leftarrow 2E, \quad H \leftarrow Y_1Y_2,$$

$$X_3 \leftarrow C(2X_2 + Y_2) - E - H, \quad Y_3 \leftarrow 4FH + ((Z_1^2)(X_2^2) + aF + (X_1^2))G,$$

$$Z_3 \leftarrow 2((Z_1^2) - D), \quad (X_3^2) \leftarrow X_3^2, \quad (Z_3^2) \leftarrow Z_3^2.$$

This formula costs **7M+3S+1d** (previous best, 8M+3S+1d). Let C_0 and C_1 be static registers. $C_0 \leftarrow X_2^2$ and $C_1 \leftarrow 2X_2 + Y_2$ are precomputed and stored permanently. If C_1 is not used, an extra addition and a multiplication by 2 is to be performed for each mixed-addition. The operations can be scheduled as follows.

The formulae that will take the advantage of the fast Jacobi-quartic addition formula that is described by Duquesne [19] require the addition of XY coordinate to modified coordinates,

$$(X_1 : Y_1 : Z_1 : X_1^2 : Z_1^2 : (X_1^2 + Z_1^2)).$$

$$
\begin{array}{c|c|c|c}
X_3 \leftarrow X_1 + Z_1 & Z_3 \leftarrow (Z_1^2) - R_1 & R_1 \leftarrow R_1 + (Z_1^2) & R_1 \leftarrow R_1 \times R_0 \\
X_3 \leftarrow X_3^2 & Z_3 \leftarrow 2 * Z_3 & Y_3 \leftarrow Y_3 \times R_1 & R_1 \leftarrow 2 * R_1 \\
X_3 \leftarrow X_3 - (X_1^2) & X_3 \leftarrow X_3 \times C_1 & Y_3 \leftarrow 4 * Y_3 & Y_3 \leftarrow Y_3 + R_1 \\
R_0 \leftarrow X_3 - (Z_1^2) & X_3 \leftarrow X_3 - Y_3 & R_1 \leftarrow a * R_1 & (X_3^2) \leftarrow X_3^2 \\
X_3 \leftarrow R_0 + Y_1 & R_0 \leftarrow R_0 \times X_2 & R_1 \leftarrow R_1 + (X_1^2) & (Z_3^2) \leftarrow Z_3^2 \\
Y_3 \leftarrow Y_1 \times Y_2 & R_0 \leftarrow 2 * R_0 & R_2 \leftarrow (Z_1^2) \times C_0 & \\
R_1 \leftarrow (X_1^2) \times C_0 & X_3 \leftarrow X_3 - R_0 & R_1 \leftarrow R_1 + R_2 &
\end{array}
$$

We refer the reader to EFD [7] for compatible versions developed by Bernstein and Lange. This coordinate system is named as extended Jacobi-quartic coordinates.

5   Alternative Base Points

In this section, we introduce a technique that is useful for successive squaring based point multiplication algorithms. Our technique improves the mixed-addition timings reported in the literature. We show how an affine point can be represented alternatively in its projective version. Point addition with these alternative points is faster for some of the forms. We will abuse the terminology and call this type of addition as mixed-addition too since these points require the same amount of storage as affine points and they are kept fixed during the point multiplication.

We follow the same notation for Jacobi-intersection curves in Section 3. Let $(S_1 : C_1 : D_1 : T_1)$ and (s_2, c_2, d_2) with $s_2 \neq 0$ be two points to be added. We can observe that representing the base point (s_2, c_2, d_2) as $(1 : (c_2/s_2) : (d_2/s_2) : (1/s_2))$ leads to a faster formulation. We rename this new representation as $(1 : C_2 : D_2 : T_2)$. With this setup we have,

$$
\begin{aligned}
S_3 &= (T_1 C_2 + D_1)(C_1 T_2 + S_1 D_2) - T_1 C_2 C_1 T_2 - D_1 S_1 D_2 \\
C_3 &= T_1 C_2 C_1 T_2 - D_1 S_1 D_2 \\
D_3 &= T_1 D_1 T_2 D_2 - a S_1 C_1 C_2 \\
T_3 &= (T_1 C_2)^2 + D_1^2
\end{aligned}
$$

This formula is from Liardet and Smart [4]. The operations can be organized as follows.

$$
E \leftarrow T_1 C_2, \quad F \leftarrow S_1 D_2, \quad G \leftarrow C_1 T_2, \quad H \leftarrow EG, \quad J \leftarrow D_1 F,
$$

$$
S_3 \leftarrow (E + D_1)(G + F) - H - J, \quad C_3 \leftarrow H - J,
$$

$$
D_3 \leftarrow T_1 D_1 (T_2 D_2) - a S_1 C_1 C_2, \quad T_3 \leftarrow E^2 + D_1^2,
$$

If $(T_2 D_2)$ is cached permanently, this formula costs **10M+2S+1d** (previous best, 11M+2S+1d). The cost of computing the alternative base point $(1 : C_2 : D_2 : T_2)$ can be omitted if it is directly selected as the base point itself.

We follow the same notation for Hessian curves in Section 2. Let $(X_1 : Y_1 : Z_1)$ and (x_2, y_2) with $x_2 \neq 0$ be two points to be added. We can observe that representing the base point (x_2, y_2) as $(1 : (y_2/x_2) : (1/x_2))$ leads to a faster formulation. We rename this new representation as $(1 : Y_2 : Z_2)$. Then, we have

$$X_3 = Y_1^2 Z_2 - X_1 Z_1 Y_2^2$$
$$Y_3 = X_1^2 (Y_2 Z_2) - Y_1 Z_1$$
$$Z_3 = Z_1^2 Y_2 - X_1 Y_1 Z_2^2$$

This formula costs **5M+6S**. The operations can be organized as follows where $S_0 \leftarrow Y_2^2$, $S_1 = Z_2^2$ and $S_2 \leftarrow 2(Y_2 + Z_2)^2 - S_0 - S_1$ are cached permanently.

$$A \leftarrow X_1^2, \quad B \leftarrow Y_1^2, \quad C \leftarrow Z_1^2, \quad D \leftarrow A + B, \quad E \leftarrow A + C, \quad F \leftarrow B + C,$$

$$G \leftarrow (X_1 + Y_1)^2 - D, \quad H \leftarrow (X_1 + Z_1)^2 - E, \quad J \leftarrow (Y_1 + Z_1)^2 - F,$$

$$X_3 \leftarrow 2BZ_2 - HS_0, \quad Y_3 \leftarrow AS_2 - J, \quad Z_3 \leftarrow 2CY_2 - GS_1.$$

The same idea works for Edwards curves reducing the number of additions in the mixed-addition formula. We follow the same notation for Edwards curves in Section 3. The Edwards mixed-addition formula that is described by Bernstein and Lange [17,18,7] costs 9M+1S+1d+7a. Let $(X_1 : Y_1 : Z_1)$ and (x_2, y_2) with $x_2 \neq 0$ be two points to be added. We can observe that representing the base point (x_2, y_2) as $(1 : (y_2/x_2) : (1/x_2))$ leads to a slightly faster formulation. We rename this new representation as $(1 : Y_2 : Z_2)$. With this setup we have,

$$X_3 = (X_1 Y_2 + Y_1)(Z_1 Z_2)((Z_1 Z_2)^2 - d(X_1 Y_1 Y_2))$$
$$Y_3 = (Y_1 Y_2 - X_1)(Z_1 Z_2)((Z_1 Z_2)^2 + d(X_1 Y_1 Y_2))$$
$$Z_3 = c((Z_1 Z_2)^2 + d(X_1 Y_1 Y_2))((Z_1 Z_2)^2 - d(X_1 Y_1 Y_2))$$

The operation count shows that the alternative Edwards mixed-addition costs **9M+1S+1d+4a**. This formula invokes 3 fewer field additions. Note, the curve parameter c can always be made 1. In this case, multiplication by c is eliminated naturally. The operations for Edwards mixed-addition can be scheduled as follows.

$R_0 \leftarrow X_1 \times Y_2$	$Y_3 \leftarrow Y_3 - X_1$	$R_1 \leftarrow d * R_1$	$X_3 \leftarrow X_3 \times R_0$
$R_0 \leftarrow R_0 + Y_1$	$Z_3 \leftarrow Z_1 \times Z_2$	$Z_3 \leftarrow Z_3^2$	$Y_3 \leftarrow Y_3 \times Z_3$
$Y_3 \leftarrow Y_1 \times Y_2$	$X_3 \leftarrow R_0 \times Z_3$	$R_0 \leftarrow Z_3 - R_1$	$Z_3 \leftarrow Z_3 \times R_0$
$R_1 \leftarrow Y_3 \times X_1$	$Y_3 \leftarrow Y_3 \times Z_3$	$Z_3 \leftarrow Z_3 + R_1$	$Z_3 \leftarrow c * Z_3$

6   Conclusion

We provided several optimizations for doing arithmetic on some special elliptic curve representations. In particular, we have improved the group operations of the Jacobi-quartic form which was initially recommended for providing side channel resistance. With our improvements, Jacobi-quartics became one of the fastest special curves in the speed ranking. For instance, successive squaring based point multiplication can be performed faster than standard Edwards coordinates for all possible scenarios because both coordinates shares the same complexity for doubling and Jacobi-quartics is faster in addition and mixed-addition. Staying in the same context, if the curve constants are large, extended Jacobi-quartic coordinates provide better timings than inverted-Edwards coordinates.

We have developed tripling formulae for Jacobi-quartic, Jacobi-intersection, Hessian and Edwards forms. These tripling formulae provide a wider background for studying DBNS based applications.

Hessian curves were initially used for providing side channel resistance. We improved Hessian doubling and mixed-addition formulae. With these improvements, point multiplication in Hessian form can be performed faster than Jacobian form if S/M is near to 1. In addition, we showed that the tripling can be performed very efficiently in characteristic 3 case. This improvement enables efficient implementation of DBNS based point multiplication with Hessian (char=3) curves.

We described how the mixed-additions can be done faster in Jacobi-intersection, Hessian and Edwards forms.

One should expect further results in the near future. For example, not all tripling formulae have been developed for all known systems yet. The quintupling formulae are also likely to appear for various forms shortly. Furthermore, the formulae that we introduced might be further improved in time.

Acknowledgements

The authors wish to thank Daniel Bernstein and Tanja Lange for announcing our formulae on EFD [7]. This study would not have been possible without their support and freely available scripts. The authors also wish to thank Christophe Doche for his corrections and suggestions on the preprint version of this paper.

References

1. Cohen, H., Miyaji, A., Ono, T.: Efficient elliptic curve exponentiation using mixed coordinates. In: Ohta, K., Pei, D. (eds.) ASIACRYPT 1998. LNCS, vol. 1514, pp. 51–65. Springer, Heidelberg (1998)
2. Smart, N.P.: The Hessian form of an elliptic curve. In: Koç, Ç.K., Naccache, D., Paar, C. (eds.) CHES 2001. LNCS, vol. 2162, pp. 118–125. Springer, Heidelberg (2001)
3. Joye, M., Quisquater, J.J.: Hessian elliptic curves and side-channel attacks. In: Koç, Ç.K., Naccache, D., Paar, C. (eds.) CHES 2001. LNCS, vol. 2162, pp. 402–410. Springer, Heidelberg (2001)
4. Liardet, P.Y., Smart, N.P.: Preventing SPA/DPA in ECC systems using the Jacobi form. In: Koç, Ç.K., Naccache, D., Paar, C. (eds.) CHES 2001. LNCS, vol. 2162, pp. 391–401. Springer, Heidelberg (2001)
5. Billet, O., Joye, M.: The Jacobi model of an elliptic curve and side-channel analysis. In: Fossorier, M.P.C., Høholdt, T., Poli, A. (eds.) AAECC. LNCS, vol. 2643, pp. 34–42. Springer, Heidelberg (2003)

6. Doche, C., Icart, T., Kohel, D.R.: Efficient scalar multiplication by isogeny decompositions. In: Yung, M., Dodis, Y., Kiayias, A., Malkin, T.G. (eds.) PKC 2006. LNCS, vol. 3958, pp. 191–206. Springer, Heidelberg (2006)

7. Bernstein, D.J., Lange, T.: Explicit-formulas database (2007), Accessible through: http://hyperelliptic.org/EFD

8. Bernstein, D.J., Birkner, P., Lange, T., Peters, C.: Optimizing double-base elliptic-curve single-scalar multiplication. In: INDOCRYPT. LNCS, Springer, Heidelberg (2007)

9. Negre, C.: Scalar multiplication on elliptic curves defined over fields of small odd characteristic. In: Maitra, S., Madhavan, C.E.V., Venkatesan, R. (eds.) INDOCRYPT 2005. LNCS, vol. 3797, pp. 389–402. Springer, Heidelberg (2005)

10. Kim, K.H., Kim, S.I., Choe, J.S.: New fast algorithms for arithmetic on elliptic curves over fields of characteristic three. Cryptology ePrint Archive, Report, 2007/179 (2007), http://eprint.iacr.org/

11. Smart, N.P., Westwood, E.J.: Point multiplication on ordinary elliptic curves over fields of characteristic three. Applicable Algebra in Engineering, Communication and Computing 13(6), 485–497 (2003)

12. Dimitrov, V.S., Imbert, L., Mishra, P.K.: Efficient and secure elliptic curve point multiplication using double-base chains. In: Roy, B. (ed.) ASIACRYPT 2005. LNCS, vol. 3788, pp. 59–78. Springer, Heidelberg (2005)

13. Meloni, N.: Fast and secure elliptic curve scalar multiplication over prime fields using special addition chains. Cryptology ePrint Archive, Report, 2006/216 (2006), http://eprint.iacr.org/

14. Avanzi, R.M., Dimitrov, V., Doche, C., Sica, F.: Extending scalar multiplication using double bases. In: Lai, X., Chen, K. (eds.) ASIACRYPT 2006. LNCS, vol. 4284, pp. 130–144. Springer, Heidelberg (2006)

15. Doche, C., Imbert, L.: Extended double-base number system with applications to elliptic curve cryptography. In: Barua, R., Lange, T. (eds.) INDOCRYPT 2006. LNCS, vol. 4329, pp. 335–348. Springer, Heidelberg (2006)

16. Edwards, H.M.: A normal form for elliptic curves. Bulletin of the AMS 44(3), 393–422 (2007)

17. Bernstein, D.J., Lange, T.: Faster addition and doubling on elliptic curves. Cryptology ePrint Archive, Report, 2007/286 (2007), http://eprint.iacr.org/

18. Bernstein, D.J., Lange, T.: Faster addition and doubling on elliptic curves. In: ASIACRYPT 2007. LNCS, vol. 4833, pp. 29–50. Springer, Heidelberg (2007)

19. Duquesne, S.: Improving the arithmetic of elliptic curves in the Jacobi model. Inf. Process. Lett. 104(3), 101–105 (2007)

Appendix

We give a step by step derivation of the new doubling formula for Jacobi-quartic form. The original formula, described by Billet and Joye [5], is as follows.

$$X_3 = 2X_1Y_1Z_1$$
$$Y_3 = 2aX_1^2Z_1^6 + 4X_1^4Z_1^4 + Y_1^2Z_1^4 + 2aX_1^6Z_1^2 + X_1^4Y_1^2$$
$$Z_3 = Z_1^4 - X_1^4$$

Step 1: Modify the point $(X_3 : Y_3 : Z_3)$ to $(-X_3 : Y_3 : -Z_3)$. These two points correspond to the same affine point.

$$X_3 = -2X_1Y_1Z_1$$
$$Y_3 = 2aX_1^2Z_1^6 + 4X_1^4Z_1^4 + Y_1^2Z_1^4 + 2aX_1^6Z_1^2 + X_1^4Y_1^2$$
$$Z_3 = X_1^4 - Z_1^4$$

Step 2: Organize X_3 and Z_3. Here, Y_3 should be computed after X_3 and Z_3.

$$X_3 = Y_1((X_1^2 + Z_1^2)) - Y_1(X_1 + Z_1)^2$$
$$Z_3 = (X_1^2 + Z_1^2)(X_1^2 - Z_1^2)$$
$$Y_3 = 2aX_1^2Z_1^6 + 4X_1^4Z_1^4 + Y_1^2Z_1^4 + 2aX_1^6Z_1^2 + X_1^4Y_1^2$$

Step 3: Use the curve equation, $Y_1^2 = X_1^4 + 2aX_1^2Z_1^2 + Z_1^4$, to find a suitable polynomial representation for Y_3.

$$X_3 = Y_1((X_1^2 + Z_1^2)) - Y_1(X_1 + Z_1)^2$$

$$Z_3 = (X_1^2 + Z_1^2)(X_1^2 - Z_1^2)$$

$$
\begin{aligned}
Y_3 &= 2aX_1^2Z_1^6 + 4X_1^4Z_1^4 + Y_1^2Z_1^4 + 2aX_1^6Z_1^2 + X_1^4Y_1^2 \\
&= 2aX_1^2Z_1^6 + 4X_1^4Z_1^4 + 2aX_1^6Z_1^2 + (X_1^4 + Z_1^4)Y_1^2 \\
&\equiv 2aX_1^2Z_1^6 + 4X_1^4Z_1^4 + 2aX_1^6Z_1^2 + (X_1^4 + Z_1^4)(Z_1^4 + 2aX_1^2Z_1^2 + X_1^4) \\
&\equiv Z_1^8 + 4aX_1^2Z_1^6 + 6X_1^4Z_1^4 + 4aX_1^6Z_1^2 + X_1^8 \\
&\equiv 2(Y_1Z_1^2 + X_1^2Y_1)^2 - (Z_1^8 - 2X_1^4Z_1^4 + X_1^8) - (4X_1^2Y_1^2Z_1^2) \\
&\equiv 2(Y_1(X_1^2 + Z_1^2))^2 - X_3^2 - Z_3^2
\end{aligned}
$$

Table 3. This table contains the best speeds in different systems. The rows are sorted with respect to doubling costs (then mixed-addition). Total complexity is computed for commonly accepted ratios; $1S=0.8M$ and $1d=0M$. $\mathbf{I}$ is used to describe the cost of a field inversion. The bold values are our contributions. The underlined values are the fastest speeds. The double lines are used for the alternative formulae.

System	Double	Total	Triple	Total	Add	Total	Mixed-Add	Total
Doche/Icart/Kohel(3) [6, 7]	4M+5S+2d	8M	6M+6S+2d	10.8M	11M+6S+1d	15.8M	7M+4S+1d	10.2M
Hessian [3]	7M+1S 3M+6S	**7.8M** **7.8M**	8M+6S+1d	**12.8M**	12M	12M	10M 5M+6S	10M **9.8M**
Jacobian [1, 12, 7]	1M+8S+1d	7.4M	5M+10S+1d	13M	10M+4S	13.2M	7M+4S	10.2M
Jacobian, $a = -3$ [1, 12, 7]	3M+5S	7M	7M+7S	12.6M	10M+4S	13.2M	7M+4S	10.2M
Jacobi-intersection [4, 7]	3M+4S	6.2M	4M+10S+5d 7M+7S+3d	**12M** 12.6M	13M+2S+1d	14.6M	11M+2S+1d 10M+2S+1d	12.6M **11.6M**
Edwards [7, 17, 18]	3M+4S	6.2M	9M+4S 7M+7S	12.2M 12.6M	10M+1S+1d	10.8M	9M+1S+1d+7a	9.8M
Extended Jacobi-quartic [5, 19, 7]	3M+4S	**6.2M**	4M+11S+2d	12.8M	8M+3S+1d	10.4M	9M+1S+1d+4a 7M+3S+1d	**9.8M** **9.4M**
Inverted Edwards [7, 8]	3M+4S+1d	6.2M	9M+4S+1d	12.2M	9M+1S+1d	9.8M	8M+1S+1d	8.8M
Doche/Icart/Kohel(2) [6, 7]	2M+5S+2d	6M	-	-	12M+5S	16M	8M+4S+1d	11.2M
Hessian, char= 3 [11]	3M+2C	3.2M	6M+4C+1d	**6.4M**	12M	12M	9M+1C	9.1M

A Graph Theoretic Analysis of Double Base Number Systems

Pradeep Kumar Mishra and Vassil Dimitrov

University of Calgary,
Calgary, AB, Canada
pradeep@math.ucalgary.ca, dimitrov@atips.ca

Abstract. Double base number systems (DBNS) provide an elegant way to represent numbers. These representations also have many interesting and useful properties, which have been exploited to find many applications in Cryptography and Signal Processing. In the current article we present a scheme to represent numbers in double (and multi-) base format by combinatorial objects like graphs and diagraphs. The combinatorial representation leads to proof of some interesting results about the double and multibase representation of integers. These proofs are based on simple combinatorial arguments. In this article we have provided a graph theoretic proof of the recurrence relation satisfied by the number of double base representations of a given integer. The result has been further generalized to more than 2 bases. Also, we have uncovered some interesting properties of the sequence representing the number of double base representation of a positive integer n. It is expected that the combinatorial representation can serve as a tool for a better understanding of the double (and multi-) base number systems.

Keywords: Double base number system, DBNS-graphs, MB-graphs.

1 Introduction

For last couple of years, there have been many papers emphasizing the use of double base number system (DBNS) in cryptography ([1,2,7,9,10,11,14,17,18]). In [6] and [15], authors have discussed elliptic curve scalar multiplication using a representation of the scalar in more than one bases. Double base number system, first time proposed in [13], is a non-traditional way of representing numbers. Unlike traditional systems, which use only one radix to represent numbers, DBNS uses 2 radii to represent a number. For example, if 2 and 3 are used as the radii, an integer n is expressed as sum of terms like $\pm 2^{b_i} 3^{t_i}$, where b_i, t_i are integers. Recently, in [17] an elliptic curve scalar multiplication scheme has been presented which uses 3 bases to represent the scalar. The proposed algorithm performs even better than its 2 base counterparts. This indicates that the DBNS can be easily generalized to more than 2 bases, which will greatly enhance their applicability to real life situations. The current article is devoted to analyse and explore some interesting propeties of double (and multi) base number system using combinatorial and graph theoretic arguments.

K. Srinathan, C. Pandu Rangan, M. Yung (Eds.): Indocrypt 2007, LNCS 4859, pp. 152–166, 2007.

Graphs are very interesting combinatorial objects widely used in discrete mathematics and computer science. In the current article we will represent a number in DBNS by means of a bipartite graph or a diagraph. We will prove some interesting results about DBNS representations using simple combinatorial arguments on these objects. Usual arithmetic operations like addition, multiplication can now be described by graph theoretic operations. This representation may be of interest to people working in various areas of computer science.

Two most interesting properties of DBNS are: (i) sparsity and (ii) redundancies. Sparsity means that a number can be represented as a sum of very few terms of the form $\pm 2^{b_i} 3^{t_i}$. This is important in cryptographic applications like exponentiation or scalar multiplication. In fact, it is the number of point addition operation needed to compute the scalar multiplication. In [2], it has been proved that in certain DBNS representations, the number of addition could be sublinear in the size of the scalar. Redundancy means that such representation is not unique. Redundancy implies that one can choose a particular representation a given number depending upon the application in which it is used. Also, these properties raise some interesting questions about the DBNS representations. Given a number n, what is the shortest DBNS representation for n requiring the minimum number of summands? Such representations are called *cannonical* representations. A number can have several cannonical representations. Computing a cannonical representation is again a very difficult computational problem. Also, given an integer n, exactly how many DBNS representationcan help in finding solutions to these problems.

In the current article, we tackle the problem of redundancy. Let p_n represent the number of DBNS representation of the integer n. Then the sequence p_n satisifies an interesting recurrence relation. In the current article, we will provide a graph theoretic proof of the recurrence relation. The relation can be proved using other mathematical tools. In [10], authors have provided one proof using generating functions. However, the proof can not be extended to more than 2 bases. The beauty of the proof given in this article is that it can be generalized to any number of radii. In the current article we have also provided a general version of the recurrence relation and proved it using a special type of graphs and combinatorial arguments.

2 Background: Double Base Number System

For the last decade, a new number representation scheme has been a subject of intensive studies, due to its applicability to digital signal processing and cryptography. This number representation is called double base number system. We start with the following definition from B.M.M. de Weger [8].

Definition 1. *Given a set P of primes, a P-integer is a positive integer all of whose prime divisors are in P.*

The double base number system is a representation scheme in which every positive integer k is represented as the sum or difference of $\{2,3\}$-integers (i.e., numbers of the form $2^b 3^t$) as

$$k = \sum_{i=1}^{m} s_i \, 2^{b_i} 3^{t_i}, \quad \text{with } s_i \in \{-1, 1\}, \text{ and } b_i, t_i \geq 0. \tag{1}$$

The term 2-integer is also used for terms of the form $2^b 3^t$. This number representation scheme is highly redundant. If one considers the DBNS with only positive signs ($s_i = 1$), then it is seen that, 10 has exactly five different DBNS representations, 100 has exactly 402 different DBNS representations and 1000 has exactly 1 295 579 different DBNS representations. Probably, the most important theoretical result about the double base number system is the following theorem from [12].

Theorem 1. *Every positive integer k can be represented as the sum of at most*
$O \left(\dfrac{\log k}{\log \log k} \right)$ *$\{2, 3\}$-integers.*

□

The proof is based on Baker's theory of linear forms of logarithms and more specifically on a result by R. Tijdeman [21].

Table 1. Table indicating sparseness of DBNS representation

range of n	maximum #2-integers in the canonical represention of n
$1 \leq n \leq 22$	2
$23 \leq n \leq 431$	3
$432 \leq n \leq 18, 431$	4
$18432 \leq n \leq 3\,448\,733$	5

Some of these representations are of special interest, most notably the ones that require the minimal number of $\{2, 3\}$-integers; i.e., an integer can be represented as the sum of m terms ($\{2, 3\}$-integers), but cannot be represented as the sum of $m - 1$ or less. These are called *canonical* representations. Even such representations are not unique for numbers greater than 8. For example, 10 has two canonical representations, 2+8, 1+9. In Table 1, we present some numerical figures to demonstrate sparseness of DBNS.

Finding one of the canonical DBNS representations, especially for very large integers, seems to be a very difficult task. One can apply a greedy algorithm to find a fairly sparse representation very quickly: given $k > 0$, find the largest number of the form $z = 2^b 3^t$ less than or equal to k, and apply the same procedure with $k - z$ until reaching zero. The greedy algorithm returns *near* canonical solutions, but not the real canonical ones. A small example is 41. Greedy returns $36 + 4 + 1$, a 3-term representation, where as the canonical solution is $32 + 9$. However, greedy algorithm is easy to implement and it guarantees a representation satisfying the asymptotic bound given by Theorem 1 (see [12]).

3   Graphical Representation of Numbers: The DBNS Graphs

We can represent natural numbers by means of a special type of bipartite graphs, we call *DBNS-graphs*. Let $V_1 = \{1, 2, 2^2, \cdots\}$ and let $V_2 = \{1, 3, 3^2, \cdots\}$ be two sets of vertices. A DBNS-graph is a bipartite graph whose vertex set is $V_1 \bigcup V_2$ and the set of edges is a subset of $\{(2^a, 3^b) : a \geq 0, b \geq 0\}$. In practice, we will take V_1 and V_2 to be finite sets.

Let n be a natural number and let $n = 2^{a_1} 3^{b_1} + \cdots + 2^{a_k} 3^{b_k}$ be a DBNS representation of n. We can represent n by a DBNS-graph D_n defined as follows: Let $\bar{a} = max_{1 \leq i \leq k}\{a_i\}$ and let $\bar{b} = max_{1 \leq i \leq k}\{b_i\}$. Then the vertex set of D_n is $V = \{1, 2, \cdots 2^{\bar{a}}\} \bigcup \{1, 3, \cdots, 3^{\bar{b}}\}$ and the edge set is $E = \{(2^{a_1}, 3^{b_1}), \cdots, (2^{a_k}, 3^{b_k})\}$. Due to redundancy of the DBNS, for every natural number n, there are several DBNS graphs representing n. We can represent 0 by the null bipartite graph. Thus, the null graph is also a DBNS-graph.

Fig. 1. A DBNS-graph representing $33 = 2^3 3^1 + 2^0 3^2$ and a DBNS-digraph representing $80 = 2^3 3^2 + 2^2 3^1 - 2^1 3^0 - 2^0 3^1 + 2^0 3^0$

To accommodate negative integers, we can use bipartite digraphs. We use the following convention: the arcs of the type $(2^i, 3^j)$ are taken to be positive (represent the summand $2^i 3^j$ in the DBNS representation of the number) and the arcs of type $(3^j, 2^i)$ to be negative (represent $-2^i 3^j$). We use this convention throughout this paper, although one can use the other sign convention too. Thus any integer, positive, negative or zero can be represented by a DBNS-digraph. If n is represented by a DBNS digraph D_n, then $-n$ can be represented by the digraph $-D_n$, obtained from D_n by just reversing the directions of the arcs of D_n. In Figure 1, a DBNS-graph representing 33 and a DBNS-digraph representing 80 have been shown. Same types of graphs can be used to represent numbers with more than 2 bases. We defer the discussion of such graphs to Section 4 and concentrate on 2 bases here, 2 and 3 only.

3.1   Some Special DBNS-Graphs

It is simple to see that the binary and ternary representation are special cases of DBNS. In fact, if we restrict the vertex set to V_2 to be $\{3^0 = 1\}$ only, then we get the binary number system. The DBNS-digraphs with this restriction will represent the signed binary system with both positive and negative coefficients. The NAF representation [20] is a further restriction, in which no two consecutive vertices in V_1 (like 2^a and 2^{a+1}) are of positive degree.

If we impose the restriction $V_1 = \{1, 2\}$ on the DBNS-graphs (resp digraph), the representations obtained are the (resp signed) ternary representations.

In [9], the authors use a special type of DBNS representation, in which the binary and ternary indices form two monotonic sequences. Such representations have DBNS-graphs with non-intersecting edges.

It is an interesting question to see which numbers are represented by complete DBNS-graphs. A DBNS-graph is complete if its vertex set is $V_1 = \{1, 2, \cdots, 2^m\}$ $\bigcup \{1, 3, \cdots, 3^n\}$ and it contains all the edges $(2^i, 3^j), 0 \leq i \leq m, 0 \leq j \leq n$. It is simple to see that these graphs represent the numbers

$$(1 + 2 + \cdots + 2^m)(1 + 3 + \cdots + 3^n) = \frac{(2^{m+1} - 1)(3^{n+1} - 1)}{2}$$

For $m = n = 0$, the corresponding complete DBNS-graph represents 1, for $m = n = 1$ it represents 12 and so on. If we take $m = 0$, and allow n to vary, the complete DBNS-graphs so obtained will represent the numbers $1, 4, 13, 40, \cdots$. On the other hand if we take $n = 0$ and allow m to vary, then we get the numbers 1, 3, 7, 15, 31, $\cdots$, which are the binary numbers $(1)_2, (11)_2, (111)_2, (1111)_2, \cdots$.

3.2   Operations on DBNS-Graphs

We assume that all DBNS-graphs are simple, i.e. without any parallel edges. If during any graph theoretic operation on a DBNS-graph a pair of parallel edges appear in it, we use the following rule to resolve the parallel edges:

D-Rule: If some operation on a DBNS-graph D_n creates a pair of parallel edges between the vertices 2^a and 3^b, then replace the pair of edges by a single edge between 2^{a+1} and 3^b. The rationale behind the rule is trivial. If the DBNS representation of a number has $2^a 3^b + 2^a 3^b$ then we can replace these two terms by a single term $2^{a+1} 3^b$.

In case of digraphs, the D-rule has to be slightly modified:

Modified D-Rule: If there are parallel edges between the vertices 2^a and 3^b then

1. If the edges are in opposite directions (i.e. edges are $(2^a, 3^b)$ and $(3^b, 2^a)$), just eliminate them.
2. If they are in the direction $(2^a, 3^b)$, then replace them by $(2^{a+1}, 3^b)$.
3. If they are in the direction $(3^b, 2^a)$, then replace them by $(3^b, 2^{a+1})$.

Digression: NAF representation [20] is extensively used in cryptography due to the fact that it is the signed binary representation of an integer with minimal

Hamming weight [5]. In the last Section we have indicated that NAF representations are DBNS-digraphs with $V_2 = \{1\}$ with the restriction that no two consecutive vertices in V_1 are of degree one. We can obtain the NAF representation of a number n with the following simple operations. Let D_n be the DBNS-graph representing the binary expansion of n. Then starting from the vertex $1 \in V_1$, see if any two consecutive vertices are of degree 1. If such a pair of vertices found with edges $(2^a, 3^0)$ and $(2^{a+1}, 3^0)$, then replace the edges by the arcs $(2^{a+2}, 3^0)$ and $(3^0, 2^a)$. This may lead to parallel edges. If so, remove them using modified D-rule. Repeat the process till the last vertex in V_1 is reached.

Note that, if both the last two vertices (say, 2^{m-1} and 2^m) in V_1 have degree 1, then one has to extend V_1 to one more vertex (namely 2^{m+1}). This explains, why the length NAF representation of n is at most 1 more than that of the binary representation.

We define the following simple operations on a DBNS-graph.

SUCC: If D_n is a DBNS-graph, then $SUCC(D_n)$ is the graph obtained by adding the edge $(2^0, 3^0) = (1, 1)$ to D_n. Note that addition of this new edge may introduce a pair of parallel edges in D_n. In that case the parallel edge has to be avoided using the D-rule, may be once or more than once. Also, note that $SUCC(D_n)$ is a DBNS-graph representing $n + 1$.

RT-operation: Let D_n be a DBNS-graph. Then $RT(D_n)$ is a graph obtained by replacing each edge $(2^a, 3^b) \in D_n$ by $(2^a, 3^{b+1})$. Note that RT operation always creates a isolated vertex at 3^0. RT stands for right-twist. The operation twists the graph in the right side. It is simple to show that if the graph D_n stands for the number n, then $RT(D_n)$ stands for $3n$.

Similarly we can define the **LT-operation**, where LT stands for left-twist and it is the inverse operation of RT. If in D_n, the vertex 3^0 is an isolated vertex, then $LT(D_n)$ is the graph obtained by replacing each edge $(2^a, 3^b)$ of D_n by $(2^a, 3^{b-1})$. $LT(D_n)$ is undefined if the vertex 3^0 is not an isolated vertex in D_n. It is obvious that if the graph D_n stands for n, then $LT(D_n)$ stands for $n/3$.

Also, we define the following notation. If S is a set of DBNS-graphs and X is one of the above operations, then by $X(S)$, we mean the set of graphs obtained by applying operation X to each member of the set S, provided such application is possible, otherwise $X(S)$ is undefined. For example, $SUCC(S) = \{SUCC(G) : G \in S\}$.

Due to high redundancy of DBNS, every integer $n > 3$ can be represented by several DBNS-graphs. Let $\mathcal{S}_n$ represent the set of all DBNS-graphs representing n. Clearly, $\mathcal{S}_1$ is $\{D_1\} = \{(2^0, 3^0)\}$. Also, $\mathcal{S}_2 = SUCC(\mathcal{S}_1) = \{D_2\} = \{(2^1, 3^0)\}$. We know, 1 and 2 have unique DBNS representations. These representations are given by the singleton sets $\mathcal{S}_1$ and $\mathcal{S}_2$ respectively.

What is $\mathcal{S}_3$? We know 3 has 2 DBNS representations, namely $1 + 2$ and 3. So $SUCC(\mathcal{S}_2)$ is a proper subset of $\mathcal{S}_3$. It is simple to see that $\mathcal{S}_3 = SUCC(\mathcal{S}_2) \bigcup RT(\mathcal{S}_1)$. That is the second graph representing 3 can be obtained by applying RT-operation to the graph representing 1. In fact, we have the following general theorem.

Theorem 2. *For any positive integer* n,

$$
\begin{aligned}
\mathcal{S}_n &= SUCC(\mathcal{S}_{n-1}) \bigcup RT(\mathcal{S}_{n/3}) \qquad &\text{if } 3 \mid n \\
&= SUCC(\mathcal{S}_{n-1}) \qquad &\text{otherwise}
\end{aligned}
\tag{2}
$$

Proof: We use induction to prove the theorem. Clearly, it is true for $n = 1, 2, 3$. Let it be true for all integers less than n. Now let us consider the case of n. Obviously, the set in the right hand side is a subset of the set in the left hand side. We need to prove the other inclusion only. Let $D \in \mathcal{S}_n$. We wish to show that D is in the set in the RHS. If D has the edge $1 = (2^0, 3^0)$, then removing this edge from D, we get a member D' of $\mathcal{S}_{n-1}$. Hence $D = SUCC(D')$ is in the sets in RHS. If D does not contain the edge 1, let it contain some edge $2^j = (2^j, 3^0)$. Let us consider the graph $D'' \in \mathcal{S}_{n-1}$ obtained from D by removing the edge 2^j and introducing the edges $1 = (2^0, 3^0), 2 = (2^1, 3^0), \cdots, 2^{j-1} = (2^{j-1}, 3^0)$. Clearly, $D'' \in \mathcal{S}_{n-1}$ and $SUCC(D'') = D$, hence D is in the set in right-hand side also. Suppose, D has no edge of the form $2^j = (2^j, 3^0)$. Then, 3^0 must be an isolated vertex in D. Then n must be a multiple of 3. Let us consider the graph $D''' = LT(D)$. It is in $\mathcal{S}_{n/3}$ and $D = RT(D''')$. Hence, in this case also D belongs to the set in the RHS. This completes the proof. $\qquad\square$

Theorem 2 has two implications. It gives us a methodology to compute the sets $\mathcal{S}_n$ iteratively from $\mathcal{S}_{n-1}$ and $\mathcal{S}_{n/3}$ (if n is a multiple of 3). If these latter sets are unknown, then we can start from $\mathcal{S}_1$ and compute all $\mathcal{S}_i$ upto n to obtain $\mathcal{S}_n$. Another important implication of this theorem is the following corollary.

Corollary 1. *For any positive integer* n, *let* $P(n)$ *denote the number of distinct DBNS representation of* n. *Then* $P(1) = 1$ *and for* $n > 1$, $P(n)$ *satisfies the following recurrence relation:*

$$
\begin{aligned}
P(n) &= P(n-1) + P(n/3) \qquad &\text{if } 3 \mid n \\
&= P(n-1) \qquad &\text{otherwise}
\end{aligned}
\tag{3}
$$

Proof: Clearly, $|\mathcal{S}_j| = P(j)$ for all $j \geq 1$. We only need to prove that there is no duplicate in the sets $SUCC(\mathcal{S}_{n-1}) \bigcup RT(\mathcal{S}_{n/3})$ if $3|n$ or in $SUCC(\mathcal{S}_{n-1})$ if n is not a multiple of 3. Clearly SUCC operation on $\mathcal{S}_{n-1}$ can not generate any duplicate. Also RT on $\mathcal{S}_{n/3}$ can not generate any duplicate. The only question is, can any element of $SUCC(\mathcal{S}_{n-1})$ be equal to one in $RT(\mathcal{S}_{n/3})$? The answer is obviously no. Because SUCC operation adds one edge $(2^0, 3^0)$ and eliminates any parallel edges using the D-rule. So SUCC operation always adds one edge of the type $(2^i, 3^0)$. But, the RT operation creates an isolated vertex at 3^0. Hence no DBNS-graph generated by an SUCC operation can be equal to any graph generated by an RT operation. This completes the proof. $\qquad\square$

We note that, the recurrence (3) has been extensively studied in connection with partition of integers for last six decades. In this article we have established the connection of the recurrence relation with the number of double base representation of a positive integer n. Unfortunately, the recurrence can not be solved as an explicit function in n. There has been many attempts for approximate solutions.

Note that the above results can be generalized to any base $\{2, s\}$. Taking powers of 2 and powers of s in the vertex sets of DBNS graphs and redefining the RT operation, one can prove that,

Corollary 2. *For any positive integer n, let $P_s(n)$ denote the number of distinct DBNS representation of n using the bases 2 and s. Then $P_s(1) = 1$ and for $n > 1$, $P_s(n)$ satisfies the following recurrence relation:*

$$
\begin{aligned}
P_s(n) &= P_s(n-1) + P_s(n/s) & \text{if } s \mid n \\
&= P_s(n-1) & \text{otherwise}
\end{aligned}
\tag{4}
$$

$\square$

It is worth mentioning here that the authors have also found alternative proofs of Equation 3 and 4 using generating functions and Mehler's functional equation [16]. However, those proofs can not be generalized to more than two bases. The proofs provided in this article using the DBNS graphs can be easily extended to more than 2 bases, which has been dealt with in the next section. This justifies the use of the graph theoretic representation.

4   Generalization to More Than 2 Bases

In this section we will generalize the results obtained in last section to more than two bases. Let us first consider the simple case of 3 bases, namely, 2, 3 and 5. A graph representing an integer in three or more bases will be called a multi-base graph (MB-graph). An MB-graph, like a DBNS-graph, is a bipartite graph, with the usual vertex sets $V_1 = \{2^0, 2^1, \cdots\}$ and $V_2 = \{3^0, 3^1, \cdots\}$. To accommodate each of the base elements other than 2 and 3, we will add an attribute to the edges. For example in the case of bases 2, 3 and 5, the edges will have one attribute. This attribute can be something like colour and we use an integer variable to represent it. We will refer to the value of this attribute variable as the *intensity* of the attribute or simply intensity of the edge. If a particular representation has k bases, then the corresponding MB-graphs will have edges with $k - 2$ attributes.

In an MB-graph with bases 2, 3 and 5, an edge of intensity c joining two vertices 2^a and 3^b will represent the summand $2^a 3^b 5^c$ in the multi-base representation of an integer. We can represent such an edge by an ordered triple, $(2^a, 3^b, c)$, where the third component is the intensity of the edge. Note that if in an MB-graph all the edges are of intensity 0, then it is a DBNS-graph. Two edges with different intensities between the same pair of vertices will not be treated as parallel edges. In other words we will allow parallel edges between a pair of vertices if the edges are of different intensity (e.g. colour). If the parallel edges are of same intensity then they are to be eliminated by D-rule.

We define a new operation σ on MB-graphs as follows. Let D be a MB-graph. Then $\sigma(D)$ is the graph obtained from D by increasing intensity of each of its edges by 1. For example, if D has edges $\{(2^{a_1}, 3^{b_1}, c_1), (2^{a_2}, 3^{b_2}, c_2), \cdots\}$ then $\sigma(D)$ has edges $\{(2^{a_1}, 3^{b_1}, c_1 + 1), (2^{a_2}, 3^{b_2}, c_2 + 1), \cdots\}$. Clearly, if D represents n, then $\sigma(D)$ represents $5n$.

Let $\mathcal{T}_n$ denote the set of all MB-graphs representing an integer n using the bases 2, 3 and 5 and let $Q(n)$ be the cardinality of the set $\mathcal{T}_n$. Then obviously $\mathcal{S}_n = \mathcal{T}_n$ and $P(n) = Q(n)$ for $n = 1, 2, 3, 4$. To illustrate how $Q(n)$ differs from $P(n)$ for $n \geq 5$, let us define $\mathcal{U}_n$ and $R(n)$ as follows. Let $\mathcal{U}_n$ be the set of MB-graphs representing n such that each edge in each of the graphs in $\mathcal{U}_n$ has intensity at least one. Let $R(n)$ denote the size of $\mathcal{U}_n$. In other words, $R(n)$ is the number of representation of n using terms of the form $2^a 3^b 5^c$, where $c \geq 1$. Clearly, each of these terms is a multiple of 5. So, n has such a representation if and only if $5 \mid n$. Thus we have the following result:

For any $n \geq 5$, we have

$$
\begin{aligned}
\mathcal{U}_n &= \sigma(\mathcal{T}_{n/5}) \quad && if \;\; 5 \mid n \\
&= \phi && if \;\; 5 \nmid n
\end{aligned}
\tag{5}
$$

Hence, taking cardinalities in both sides, we obtain,

$$
\begin{aligned}
R(n) &= Q(n/5) \quad && if \;\; 5 \mid n \\
&= 0 && if \;\; 5 \nmid n
\end{aligned}
\tag{6}
$$

We define a binary operation $\oplus$ as follows. Let $\mathcal{A}$ and $\mathcal{B}$ be two sets of graphs. Then,

$$
\mathcal{A} \oplus \mathcal{B} = \{G_1 \textstyle\bigcup G_2 \mid G_1 \in \mathcal{A} \wedge G_2 \in \mathcal{B}\}
$$

Let us now look at the graphs in $\mathcal{T}_n$. Clearly, $\mathcal{S}_n \subset \mathcal{T}_n$. What else are there in $\mathcal{T}_n$? Let $k = \lfloor n/5 \rfloor$. Let $i \leq k$ and let us consider the graphs in the set

$$
\mathcal{S}_{n-5i} \oplus \mathcal{U}_{5i}
$$

Clearly each graph in this set represents n. Hence $\mathcal{S}_{n-5i} \oplus \mathcal{U}_{5i} \subset \mathcal{T}_n$. So,

$$
\mathcal{S}_n \bigcup (\bigcup_{i=0}^{k} \mathcal{S}_{n-5i} \oplus \mathcal{U}_{5i}) \subset \mathcal{T}_n
$$

Conversely, let M_n be a multi-base graph in $\mathcal{T}_n$. If the intensity of each of its edges is zero, then it is a DBNS-graph in $\mathcal{S}_n$. Otherwise, the non-DBNS component of M_n (i.e. the set of edges with nonzero intensities) can be 5 or 10 or $\cdots$ or $5k$. If it is $5i$ for some $1 \leq i \leq k$, then $M_n \in \mathcal{S}_{n-5i} \oplus \mathcal{U}_{5i}$. Thus the other inclusion is also true. Thus we have proved the following theorem:

Theorem 3. *For any positive integer* n,

$$\mathcal{T}_n = \mathcal{S}_n \bigcup (\mathcal{S}_{n-5} \oplus \mathcal{U}_5) \bigcup \cdots \bigcup (\mathcal{S}_{n-5k} \oplus \mathcal{U}_{5k})$$

$$= \bigcup_{i=1}^{k} (\mathcal{S}_{n-5i} \oplus \mathcal{U}_{5i}) \tag{7}$$

$$= \bigcup_{i=1}^{k} (\mathcal{S}_{n-5i} \oplus \sigma(\mathcal{T}_i)) \qquad (using Eq(\ 5))$$

where $k = \lfloor n/5 \rfloor$ *and* $\mathcal{U}_0 = \mathcal{T}_0 =$ *the set containing the null MB-graph only.* □

As the graphs in $\mathcal{S}_{n-5i} \oplus \mathcal{U}_{5i}$ have non-DBNS component $5i$, $\mathcal{S}_{n-5i} \oplus \mathcal{U}_{5i}$, $0 \le i \le k$ is a union of pairwise disjoint sets. Taking the cardinality of the sets in Equation 7 we get,

Corollary 3. *Let* $Q(n)$ *be the number of multibase expansion of an integer* n *using a bases 2, 3 and 5. Then* $Q(1) = 1$ *and*

$$Q(n) = P(n) + P(n-5)R(5) + \cdots + P(n-5k)R(5k)$$
$$= P(n) + P(n-5)Q(1) + \cdots + P(n-5k)Q(k) \quad (using Eq(\ 6)) \tag{8}$$
$$= \Sigma_{i=0}^{k} P(n-5i)Q(i)$$

where $k = \lfloor n/5 \rfloor$. □

In other words, is $P(x), Q(x)$ and $R(x)$ are the generating functions of $P(n)$, $Q(n)$ and $R(n)$ respectively, then we have,

$$Q(x) = P(x) \oplus R(x)$$

where $\oplus$ is the convolution operator.

Let us now consider the more general case, i.e. the case of any number of bases. We choose our bases from the set of primes $\{2, 3, 5, 7, \cdots\}$. Let B_k be the set of first k primes. Let $\mathcal{S}_n^{(k)}$ be the set of multi-base graphs representing n using the set of bases B_k. Let $P^{(k)}(n)$ be the number of multi-base representation of n using the base set B_k. The correspondence between this new notation and the older one is:

$$\mathcal{S}_n = \mathcal{S}_n^{(2)}$$
$$P(n) = P^{(2)}(n)$$
$$\mathcal{T}_n = \mathcal{S}_n^{(3)}$$
$$Q(n) = P^{(3)}(n)$$

Moreover, we have now these notations for the binary representations also. The number of binary representation of an integer n is $P^{(1)}(n)$, which is 1 for all n.

The following theorem can be proved by induction on l.

Theorem 4. *For any positive integer n,*

$$\mathcal{S}_n^{(l)} = \mathcal{S}_n^{(l-1)} \bigcup (\mathcal{S}_{n-b_l}^{(l-1)} \oplus \mathcal{S}_1^{(l)}) \bigcup \cdots \bigcup (\mathcal{S}_{n-b_l k}^{(l-1)} \oplus \mathcal{S}_k^{(l)}) \qquad (9)$$

where b_l is the l-th prime base and $k = \lfloor n/b_l \rfloor$.
  Also,

$$P^{(l)}(n) = P^{(l-1)}(n) + P^{(l-1)}(n - b_l)P^{(l)}(1) + \cdots + P^{(l-1)}(n - b_l k)P^{(l)}(k) \qquad (10)$$
$$= \Sigma_{i=0}^{k} P^{(l-1)}(n - 5i)P^{(l)}(i)$$

□

For example, if we use 4 bases, namely, the base set $B_4 = \{2, 3, 5, 7\}$, then the sequence $P^{(4)}(n)$ of number of multi-base representations of an integer n using B_4, satisfies the following recurrence relation:

$$P^{(4)}(n) = P^{(3)}(n) + P^{(3)}(n - 7)P^{(4)}(1) + \cdots + P^{(3)}(n - 7k)P^{(4)}(k)$$
$$= \Sigma_{i=0}^{k} P^{(3)}(n - 7i)P^{(4)}(i) \qquad (11)$$

where $k = \lfloor n/7 \rfloor$.

We have carried out numerous experiments using the above relations. The number of representations of n grows very fast in the number of base elements. For example 100 has 402 DBNS representation (base 2 and 3), 8425 representations using the bases 2, 3 and 5 and has 43777 representations using the bases 2, 3, 5, and 7. The number of representations for some values of $P^{(l)}(n)$ for $l = 2, 3, 4$ for various n have been given in Table 2. This gives some idea about the degree of redundancy of multi-base representations.

Table 2. Values of $P(n)$, $P_5(n)$, $P^{(3)}(n)$ and $P^{(4)}(n)$ for some small values of n

n	$P(n)$	$P_5(n)$	$P^{(3)}(n)$	$P^{(4)}(n)$
10	5	3	8	10
20	12	5	32	48
50	72	18	489	1266
100	402	55	8425	43777
150	1296	119	63446	586862
200	3027	223	316557	4827147
300	11820	569	4016749	142196718

5   A Partition of $\mathcal{S}_n$

Let us consider the DBNS-graphs in $\mathcal{S}_n$. We can partition $\mathcal{S}_n$ into the following subsets: Let $\mathcal{S}_n^i$ be the set of DBNS-graphs in $\mathcal{S}_n$, which are obtained by application of i, $i \geq 0$, RT-operations on graphs of the sets $\mathcal{S}_j$, $j < n$. In $\mathcal{S}_1$, we have only one graph, which is not obtained by any RT-operation. Hence $\mathcal{S}_1^0 = \mathcal{S}_1$

Table 3. Sizes of the sets $\mathcal{S}_n^i$

n	0-RT	1-RT	2-RT	3-RT	n	0-RT	1-RT	2-RT	3-RT	4-RT
1	1				45	1	15	35	12	
3	1	1			48	1	16	40	15	
6	1	2			51	1	17	45	18	
9	1	3	1		54	1	18	51	23	
12	1	4	2		57	1	19	57	28	
15	1	5	3		60	1	20	63	33	
18	1	6	5		63	1	21	70	40	
21	1	7	7		66	1	22	77	47	
24	1	8	9		69	1	23	84	54	
27	1	9	12	1	72	1	24	92	63	
30	1	10	15	2	75	1	25	100	72	
33	1	11	18	3	78	1	26	108	81	
36	1	12	22	5	81	1	27	117	93	1
39	1	13	26	7	84	1	28	126	105	2
42	1	14	30	9	87	1	29	135	117	3

and $\mathcal{S}_1^i = \phi$ for all $i \geq 1$. Similarly for $\mathcal{S}_2$. In $\mathcal{S}_3$, there are two graphs, one is obtained by $SUCC(\mathcal{S}_2)$ and other is obtained by $RT(\mathcal{S}_1)$. Hence $|\mathcal{S}_3^0| = |\mathcal{S}_3^1| = 1$ and $\mathcal{S}_3^i = \phi$ for all $i > 1$.

In Table 3 we have tabulated the size of these sets for various values of n. Note that, by Theorem 2, the size of $\mathcal{S}_n$ remains the same for $n = 3k, 3k+1, 3k+2$. Hence we do not list these sets for all values of n, but only for multiples of 3. Also, in Table 3, the first column represents n. The second column represents the size of $\mathcal{S}_n^0$, the third column represents $\mathcal{S}_n^1$ and so on. To be brief, these columns have been entitled as 0-RT (2nd column), 1-RT (3rd), etc.

Observe that in $\mathcal{S}_1$ there is only one element, namely $\{(2^0, 3^0)\}$. The other $\mathcal{S}_n$ are generated from $\mathcal{S}_1$ by application of SUCC and RT operations. If we repeatedly apply SUCC operation on $\mathcal{S}_1$ (and no RT-operation), each time we get only one graph. Hence, $\mathcal{S}_n^0$ will always contain only one graph. This graph represents the unique binary representation of n. This is the reason why all the entries in the second column (0-RT) of Table 3 are 1's.

The third column in Table 3 represents the size of the subset $\mathcal{S}_n^1$ of in $\mathcal{S}_n$. $\mathcal{S}_n^1$ contains the graphs, which are obtained by just one application of RT-operation. Note that $\mathcal{S}_n^1$ contains $SUCC(\mathcal{S}_{n-1}^1)$. If n is a multiple of 3, then it also contains $RT(\mathcal{S}_{n/3}^0)$, which is a singleton set. Hence the size of $\mathcal{S}_n^1$ increases by 1 each time n increases by 3. Thus, we have, $|\mathcal{S}_n^1| = \lfloor n/3 \rfloor$.

The fourth column of Table 3 gives the sizes of $\mathcal{S}_n^2$. Obviously, $|\mathcal{S}_n^2| = 0$ for $n = 1, \cdots, 8$. We have $|\mathcal{S}_9^2| = 1$. Also, $|\mathcal{S}_{10}^2| = |SUCC(\mathcal{S}_9^2)| = 1$ and $|\mathcal{S}_{11}^2| = |SUCC(\mathcal{S}_{10}^2)| = 1$. But $|\mathcal{S}_{12}^2| = |SUCC(\mathcal{S}_{11}^2)| + |RT(\mathcal{S}_4^1)| = 2$. When n is a multiple of 3, $|\mathcal{S}_n^2| = |SUCC(\mathcal{S}_{n-1}^2)| + |RT(\mathcal{S}_{n/3}^1)|$. Hence, the size of $\mathcal{S}_n^2$ is given by the sequence $1, 2, 3, 5, 7, 9, 12, 15, 18, 22, \ldots$ for $n = 9, 12, 15, 18, 21, \ldots$.

Let us see how Table 3 can be constructed without actually constructing the sets $\mathcal{S}_n^i$. Let $A_{i,j}$ denote the entry in the ith row and jth column of Table 3.

The first column which represents n, contains 1 and then multiples of 3 only. We have,

- $A_{i,2} = 1$, $\forall\, i$. Also, $A_{1,j} = 0$, $\forall\, j \geq 3$.
- For $i > 1$ and $j > 2$, we have $A_{ij} = A_{i-1,j} + A_{1+\lfloor(i-1)/3\rfloor, j-1}$

If we add all elements in the kth row of this table, starting from the second column, we get $|\mathcal{S}_k|$, i.e. the numbers of DBNS-graph representing k. In the last section, we had denoted it by $P(k)$. Thus we get an explicit value for $P(k)$ if we add kth row of Table 3 (except the first column entry).

6   Integer Arithmetic Using DBNS-Graphs

Arithmetic operation on integers can be carried out using DBNS-graphs (or DBNS-digraphs). For example, addition of two integers can be carried out by taking the union of the DBNS-graphs representing them. Ofcourse, such union may introduce parallel edges, which are to be removed using the D-rule or modified D-rule (for digraphs).

More formally, let n_1 and n_2 be two integers represented by the DBNS-graphs D_{n_1} and D_{n_2} respectively. Let vertex set of D_{n_1} be $V_1 = \{1, 2, \cdots, 2^{a_1}\} \bigcup \{1, 3, \cdots 3^{b_1}\}$ and edge set be $E_1 = \{(2^{a_{i_1}}, 3^{b_{i_1}}), \cdots, (2^{a_{i_l}}, 3^{b_{i_l}})\}$. Let vertex set of D_{n_2} be $V_2 = \{1, 2, \cdots, 2^{a_2}\} \bigcup \{1, 3, \cdots 3^{b_2}\}$ and edge set be $E_2 = \{(2^{a_{j_1}}, 3^{b_{j_1}}), \cdots, (2^{a_{j_m}}, 3^{b_{j_m}})\}$. Let $max\{a_1, a_2\} = a$ and $max\{b_1, b_2\} = b$. Then, the sum $D_{n_1} + D_{n_2}$ of D_{n_1} and D_{n_2} is the graph on the vertex set $V_1 \bigcup V_2 = \{1, 2, \cdots, 2^a\} \bigcup \{1, 3, \cdots 3^b\}$ and edge set is $E_1 \bigcup E_2$, subject to D-rule.

Subtraction can be carried out very similarly. For subtraction we have to use DBNS-digrahs. It has been mentioned earlier that if the DBNS-digraph D_{n_2} represents n_2, then the DBNS-digraph $-D_{n_2}$, obtained by reversing the orientation of edges of D_{n_2} represents $-n_2$. Hence, we define $D_{n_1} - D_{n_2}$ as $D_{n_1} + (-D_{n_2})$. Clearly, $D_{n_1} - D_{n_2}$ represents the number $n_1 - n_2$.

Multiplication is slightly tricky. We define the following operation. For integers c and d, we define an operation $SH_{c,d}$ on the edges of a DBNS-graph as follows. Let $e = (2^a, 3^b)$ be an edge. We define,

$$SH_{c,d}(e) = SH_{c,d}((2^a, 3^b)) = (2^{a+c}, 3^{b+d})$$

If E is an set of edges, then we define $SH_{c,d}(E) = \{SH_{c,d}(e) : e \in E\}$. Now we define multiplication of two DBNS-graphs as follows: If the vertex sets of D_{n_1} and D_{n_2} are as given above, then the vertex set of the product graph $D_{n_1} * D_{n_2}$ is $V = \{1, 2, \cdots, 2^{a_1+a_2}\} \bigcup \{1, 3, \cdots 3^{b_1+b_2}\}$ and the edge set is $SH_{a_{i_1}, b_{i_1}}(E_2) \bigcup \cdots \bigcup SH_{a_{i_l}, b_{i_l}}(E_2)$ subject to D-rule.

Division is a restrictive operation. Given any two integers n_1 and n_2 we can not always divide n_1 by n_2. This restriction escalates in case of DBNS-graphs. We can divide D_{n_1} by D_{n_2} only if

1. D_{n_2} has only one edge $(2^c, 3^d)$, i.e. n_2 is only a DBNS term $2^c 3^d$.

2. D_{n_1} has a *nice* representation, i.e., if edge set of D_{n_1} is $E_1 = \{(2^{a_{i_1}}, 3^{b_{i_1}}),$ $\cdots, (2^{a_{i_l}}, 3^{b_{i_l}})\}$, then $a_{i_k} > c, b_{i_k} > d, \forall k, 1 \leq k \leq l$. Note that this is not true for all DBNS-graph representing n_1 even if $n_2 | n_1$.

If these two conditions are met, then a DBNS-graph representing n_1/n_2 has the vertex set $V' = \{1, 2, \cdots, 2^{a_1 - c}\} \bigcup \{1, 3, \cdots 3^{b_1 - d}\}$ and edge set $SH_{-c, -d}(E_1)$.

DBNS numbers can be represented in the computer by the adjacency matrix or adjacency list of the corresponding DBNS-graph. Also, arithmetic operations on the numbers can be carried out by matrix operation on the adjacency matrices of corresponding DBNS graphs. The scheme of representing an integer by means of the adjacency matrix of a DBNS-graph does not seem to be efficient in memory. But as the DBNS representation is very sparse, these matrices will be very sparse too. Also, the number of basic computational operations, like XOR (for addition) and Shift (for multiplication) per arithmetic operation will be very few. This may lead to efficient integer arithmetic.

Without going for details, which is more or less trivial, we mention here that the integer arithmetic defined above can be implemented in a computer using adjacency matrices of the DBNS-graphs. The addition operation is a generalization of the usual XORing algorithm used for integer addition in case of binary representation of the numbers. The multiplication algorithm can be seen as a generalization of the traditional shift and add algorithm for integer multiplication. It can be an interesting work to calculate complexity of these new algorithms and compare them with the existing ones.

7   Conclusion

In the current article, we have proposed a graph theoretic representation of integers using double and multi-base number system. The representation can be a powerful tool to study the structure of these system of representation. These number representations are highly redundant. We have proposed and proved some interesting relations satisfied by the number of double/multi- base representation of an integer n. Most of the proofs are based on simple graph theoretic arguments.

References

1. Avanzi, R.M., Sica, F.: Scalar Multiplication on Koblitz Curves using Double Bases. Available at http://eprint.iacr.org/2006/067.pdf
2. Avanzi, R.M., Dimitrov, V., Doche, C., Sica, F.: Extending Scalar Multiplication to Double Bases. In: Lai, X., Chen, K. (eds.) ASIACRYPT 2006. LNCS, vol. 4284, pp. 130–144. Springer, Heidelberg (2006)
3. Berth é, V., Imbert, L.: On converting numbers to the double-base number system. In: Luk, F.T. (ed.) Advanced Signal Processing Algorithms, Architecture and Implementations XIV, Proceedings of SPIE, vol. 5559, pp. 70–78. SPIE, San Jose, CA (2004)

4. Ciet, M., Lauter, K., Joye, M., Montgomery, P.L.: Trading inversions for multiplications in elliptic curve cryptography. Designs, Codes and Cryptography 39(2), 189–206 (2006)
5. Bosma, W.: Signed bits and fast exponentiation. J. Theor. Nombres Bordeaux 13, 27–41 (2001)
6. Ciet, M., Lauter, K., Joye, M., Montgomery, P.L.: Trading inversions for multiplications in elliptic curve cryptography. Designs, Codes and Cryptography 39(2), 189–206 (2006)
7. Ciet, M., Sica, F.: An Analysis of Double Base Number Systems and a Sublinear Scalar Multiplication Algorithm. In: Dawson, E., Vaudenay, S. (eds.) Mycrypt 2005. LNCS, vol. 3715, pp. 171–182. Springer, Heidelberg (2005)
8. de Weger, B.M.M.: Algorithms for Diophantine equations of CWI Tracts. In: Centrum voor Wiskunde en Informatica, Amsterdam (1989)
9. Dimitrov, V.S., Imbert, L., Mishra, P.K.: Efficient and secure elliptic curve point multiplication using double-base chains. In: Roy, B. (ed.) ASIACRYPT 2005. LNCS, vol. 3788, pp. 59–78. Springer, Heidelberg (2005)
10. Dimitrov, V.S., Imbert, L., Mishra, P.K.: The Double Base Number System and Its Applications to Elliptic Curve Cryptography. Research Report LIRMM #06032 (May 2006)
11. Dimitrov, V., Järvinen, K.U., Jacobson, M.J., Chan, W.F., Huang, Z.: FPGA Implementation of Point Multiplication on Koblitz Curves Using Kleinian Integers. In: Goubin, L., Matsui, M. (eds.) CHES 2006. LNCS, vol. 4249, pp. 445–459. Springer, Heidelberg (2006)
12. Dimitrov, V.S., Jullien, G.A., Miller, W.C.: An algorithm for modular exponentiation. Information Processing Letters 66(3), 155–159 (1998)
13. Dimitrov, V.S., Jullien, G.A., Miller, W.C.: Theory and applications of the double-base number system. IEEE Transactions on Computers 48(10), 1098–1106 (1999)
14. Doche, C., Imbert, L.: Extended Double-Base Number System with Applications to Elliptic Curve Cryptography. In: Barua, R., Lange, T. (eds.) INDOCRYPT 2006. LNCS, vol. 4329, pp. 335–348. Springer, Heidelberg (2006)
15. Doche, C., Icart, T., Kohel, D.: Efficient Scalar Multiplication by Isogeny Decompositions. In: Yung, M., Dodis, Y., Kiayias, A., Malkin, T.G. (eds.) PKC 2006. LNCS, vol. 3958, pp. 191–206. Springer, Heidelberg (2006)
16. Mahler, K.: On a Special Functional Equation. J of London Math Soc. 15, 115–123 (1939)
17. Mishra, P.K., Dimitrov, V.: Efficient Quintuple Formulas and Efficient Elliptic Curve Scalar Multiplication using Multibase Number Representation. In ISC (to appear, 2007)
18. Mishra, P.K., Dimitrov, V.: WIndow-based Elliptic CUrve Scalar Multiplication Using Double Base Number Representation. In Inscrypt (to appear, 2007)
19. Pennington, W.B.: On Mahler's partition problem. Annals of Math. 57, 531–546 (1953)
20. Reitwiesner, G.: Binary Arithmetic. Adv. Comput. 1, 231–308 (1962)
21. Tijdeman, R.: On the maximal distance between integers composed of small primes. Compositio Mathematica 28, 159–162 (1974)

Optimizing Double-Base Elliptic-Curve Single-Scalar Multiplication*

Daniel J. Bernstein[1], Peter Birkner[2], Tanja Lange[2], and Christiane Peters[2]

[1] Department of Mathematics, Statistics, and Computer Science (M/C 249)
University of Illinois at Chicago, Chicago, IL 60607–7045, USA
djb@cr.yp.to
[2] Department of Mathematics and Computer Science
Technische Universiteit Eindhoven, P.O. Box 513, 5600 MB Eindhoven, The Netherlands
p.birkner@tue.nl, tanja@hyperelliptic.org, c.p.peters@tue.nl

Abstract. This paper analyzes the best speeds that can be obtained for single-scalar multiplication with variable base point by combining a huge range of options:

- many choices of coordinate systems and formulas for individual group operations, including new formulas for tripling on Edwards curves;
- double-base chains with many different doubling/tripling ratios, including standard base-2 chains as an extreme case;
- many precomputation strategies, going beyond Dimitrov, Imbert, Mishra (Asiacrypt 2005) and Doche and Imbert (Indocrypt 2006).

The analysis takes account of speedups such as $\mathbf{S} - \mathbf{M}$ tradeoffs and includes recent advances such as inverted Edwards coordinates.

The main conclusions are as follows. Optimized precomputations and triplings save time for single-scalar multiplication in Jacobian coordinates, Hessian curves, and tripling-oriented Doche/Icart/Kohel curves. However, even faster single-scalar multiplication is possible in Jacobi intersections, Edwards curves, extended Jacobi-quartic coordinates, and inverted Edwards coordinates, thanks to extremely fast doublings and additions; there is no evidence that double-base chains are worthwhile for the fastest curves. Inverted Edwards coordinates are the speed leader.

Keywords: Edwards curves, double-base number systems, double-base chains, addition chains, scalar multiplication, tripling, quintupling.

1 Introduction

Double-base number systems have been suggested as a way to speed up scalar multiplication on elliptic curves. The idea is to expand a positive integer n as a sum of very few terms $c_i 2^{a_i} 3^{b_i}$ with $c_i = 1$ or $c_i = -1$, and thus to express

---

* Permanent ID of this document: d721c86c47e3b56834ded945c814b5e0. Date of this document: 2007.10.03. This work has been supported in part by the European Commission through the IST Programme under Contract IST–2002–507932 ECRYPT.

K. Srinathan, C. Pandu Rangan, M. Yung (Eds.): Indocrypt 2007, LNCS 4859, pp. 167–182, 2007.

a scalar multiple nP as a sum of very few points $c_i 2^{a_i} 3^{b_i} P$. Unfortunately, the time to add these points is only one facet of the time to compute nP; computing the points in the first place requires many doublings and triplings. Minimizing the number of additions is minimizing the wrong cost measure.

At Asiacrypt 2005, Dimitrov, Imbert, and Mishra [9] introduced double-base chains $\sum c_i 2^{a_i} 3^{b_i}$, where again $c_i = 1$ or $-c_i = 1$, with the new restrictions $a_1 \geq a_2 \geq a_3 \geq \cdots$ and $b_1 \geq b_2 \geq b_3 \geq \cdots$ allowing a Horner-like evaluation of nP with only a_1 doublings and only b_1 triplings. But the new restrictions introduced by double-base chains substantially increase the number of additions.

At Indocrypt 2006, Doche and Imbert [11] improved double-base chains by introducing an analogue of signed-sliding-window methods, keeping the restrictions $a_1 \geq a_2 \geq a_3 \geq \cdots$ and $b_1 \geq b_2 \geq b_3 \geq \cdots$ but allowing c_i and $-c_i$ to be chosen from a coefficient set S larger than $\{1\}$, leading to shorter chains and thus fewer additions. Doche and Imbert studied in detail the sets $\{1\}$, $\{1, 2, 3, 4, 9\}$, $\{1, 2, \ldots, 2^4, 3, \ldots, 3^4\}$, $\{1, 5, 7\}$, $\{1, 5, 7, 11, 13, 17, 19, 23, 25\}$. For each set they counted (experimentally) the number of additions, doublings, and triplings in their chains, compared these to the number of additions and doublings in the standard single-base sliding-window methods, and concluded that these new double-base chains save time in scalar multiplication. However, there are several reasons to question this conclusion:

- The comparison ignores the cost of precomputing all the cP for $c \in S$. These costs are generally lower for single-base chains, and are incurred for every scalar multiplication (unless P is reused, in which case there are much faster scalar-multiplication methods).
- The comparison relies on obsolete addition formulas. For example, [11] uses mixed-addition formulas that take $8\mathbf{M} + 3\mathbf{S}$: i.e., 8 field multiplications and 3 squarings. Faster formulas are known, taking only $7\mathbf{M} + 4\mathbf{S}$; this speedup has a larger benefit for single-base chains than for double-base chains.
- The comparison relies on obsolete curve shapes. For example, [11] uses doubling formulas that take $4\mathbf{M} + 6\mathbf{S}$, but the standard choice $a_4 = -3$ improves Jacobian-coordinate doubling to $3\mathbf{M} + 5\mathbf{S}$, again making single-base chains more attractive. Recent work has produced extremely fast doubling and addition formulas for several non-Jacobian curve shapes.

In this paper we carry out a much more comprehensive comparison of elliptic-curve scalar-multiplication methods. We analyze a much wider variety of coordinate systems, including the most recent innovations in curve shapes and the most recent speedups in addition formulas; see Section 3. In particular, we include Edwards curves in our comparison; in Section 2 we introduce new fast tripling formulas for Edwards curves, and in the appendix we introduce quintupling formulas. Our graphs include the obsolete addition formulas for Jacobian coordinates ("Std-Jac" and "Std-Jac-3") to show how striking the advantage of better group operations is. We account for the cost of precomputations, and we account for the difference in speeds between addition, readdition, and mixed addition. We include more choices of chain parameters, and in particular identify better choices of S for double-base chains. We cover additional exponent lengths of interest in cryptographic applications.

We find, as in [11], that double-base chains achieve significant improvements for curves in Jacobian coordinates and for tripling-oriented Doche-Icart-Kohel curves; computing scalar multiples with the $\{2,3\}$-double-base chains is faster than with the best known single-base chains. For integers of bit-length ℓ about 0.22ℓ triplings and 0.65ℓ doublings are optimal for curves in Jacobian coordinates; for the Doche-Icart-Kohel curves the optimum is about 0.29ℓ triplings and 0.54ℓ doublings. For Hessian curves we find similar results; the optimum is about 0.25ℓ triplings and 0.6ℓ doublings.

On the other hand, for Edwards curves it turns out that the optimum for base-$\{2,3\}$ chains uses very few triplings. This makes the usefulness of double-base chains for Edwards curves questionable. The same result holds for $\{2,5\}$-double-base chains. Based on our results we recommend traditional single-base chains for implementors using Edwards curves. Similar conclusions apply to Jacobi intersections, extended Jacobi-quartic coordinates, and inverted Edwards coordinates.

In the competition between coordinate systems, inverted Edwards coordinates are the current leader, followed closely by extended Jacobi-quartic coordinates and standard Edwards coordinates, and then by Jacobi intersections. Jacobian coordinates with $a_4 = -3$, despite double-base chains and all the other speedups we consider, are slower than Jacobi intersections. Tripling-oriented Doche/Icart/Kohel curves are competitive with Jacobian coordinates—but not nearly as impressive as they seemed in [11]. For the full comparison see Section 5.

2   Edwards Curves

Edwards [14] introduced a new form for elliptic curves over fields of characteristic different from 2 and showed that – after an appropriate field extension – every elliptic curve can be transformed to this normal form. Throughout this paper we focus on fields k of characteristic at least 5. We now briefly review arithmetic on Edwards curves and then develop new tripling formulas. Hisil, Carter, and Dawson independently developed essentially the same tripling formulas; see [16].

Background on Edwards curves. We present Edwards curves in the slightly generalized version due to Bernstein and Lange [5]. An *elliptic curve in Edwards form*, or simply *Edwards curve*, over a field k is given by an equation

$$x^2 + y^2 = 1 + dx^2y^2, \qquad \text{where } d \in k \setminus \{0,1\}.$$

Two points (x_1, y_1) and (x_2, y_2) are added according to the *Edwards addition law*

$$(x_1, y_1), (x_2, y_2) \mapsto \left(\frac{x_1y_2 + y_1x_2}{1 + dx_1x_2y_1y_2}, \frac{y_1y_2 - x_1x_2}{1 - dx_1x_2y_1y_2} \right). \tag{2.1}$$

The neutral element of this addition is $(0, 1)$. The inverse of any point (x_1, y_1) on E is $(-x_1, y_1)$. Doubling can be performed with exactly the same formula as addition. If d is not a square in k the addition law is complete, i.e., it is defined for all pairs of input points on the Edwards curve over k and the result gives the sum of the input points.

Bernstein and Lange [5] study Edwards curves for cryptographic applications and give efficient explicit formulas for the group operations. To avoid inversions they work with the homogenized equation in which a point $(X_1 : Y_1 : Z_1)$ corresponds to the affine point $(X_1/Z_1, Y_1/Z_1)$ on the Edwards curve. Their newer paper [4] proposes using $(X_1 : Y_1 : Z_1)$ to represent $(Z_1/X_1, Z_1/Y_1)$. These *inverted Edwards coordinates* save $1\mathbf{M}$ in addition. For contrast we refer to the former as *standard Edwards coordinates*.

Doubling on Edwards curves. If both inputs are known to be equal the result of the addition can be obtained using fewer field operations. We briefly describe how the the special formulas for doubling were derived from the general addition law (2.1) in [5]. The same approach will help in tripling.

Since (x_1, y_1) is on the Edwards curve one can substitute the coefficient d by $(x_1^2 + y_1^2 - 1)/(x_1^2 y_1^2)$ as follows:

$$(x_1, y_1), (x_1, y_1) \mapsto \left(\frac{2x_1 y_1}{1 + dx_1^2 y_1^2}, \frac{y_1^2 - x_1^2}{1 - dx_1^2 y_1^2} \right) = \left(\frac{2x_1 y_1}{x_1^2 + y_1^2}, \frac{y_1^2 - x_1^2}{2 - (x_1^2 + y_1^2)} \right).$$

This reduces the degree of the denominator from 4 to 2 which is reflected in faster doublings.

Tripling on Edwards curves. One can triple a point by first doubling it and then adding the result to itself. By applying the curve equation as in doubling we obtain

$$3(x_1, y_1) = \left(\frac{((x_1^2 + y_1^2)^2 - (2y_1)^2)}{4(x_1^2 - 1)x_1^2 - (x_1^2 - y_1^2)^2} x_1, \frac{((x_1^2 + y_1^2)^2 - (2x_1)^2)}{-4(y_1^2 - 1)y_1^2 + (x_1^2 - y_1^2)^2} y_1 \right).$$

We present two sets of formulas to do this operation in standard Edwards coordinates. The first one costs $9\mathbf{M} + 4\mathbf{S}$ while the second needs $7\mathbf{M} + 7\mathbf{S}$. If the $\mathbf{S}/\mathbf{M}$ ratio is very small, specifically below $2/3$, then the second set is better while for larger ratios the first one is to be preferred.

The explicit formulas were verified to produce the 3-fold of the input point $(X_1 : Y_1 : Z_1)$ by symbolically computing $3(X_1 : Y_1 : Z_1)$ using the addition and doubling formulas from [5] and comparing it with $(X_3 : Y_3 : Z_3)$.

Here are our $9\mathbf{M} + 4\mathbf{S}$ formulas for tripling:

$$A = X_1^2; \; B = Y_1^2; \; C = (2Z_1)^2; \; D = A + B; \; E = D^2; \; F = 2D \cdot (A - B);$$
$$G = E - B \cdot C; \; H = E - A \cdot C; \; I = F + H; \; J = F - G;$$
$$X_3 = G \cdot J \cdot X_1; \; Y_3 = H \cdot I \cdot Y_1; \; Z_3 = I \cdot J \cdot Z_1.$$

Here are our $7\mathbf{M} + 7\mathbf{S}$ formulas for tripling:

$$A = X_1^2; \; B = Y_1^2; \; C = Z_1^2; \; D = A + B; \; E = D^2; \; F = 2D \cdot (A - B);$$
$$K = 4C; \; L = E - B \cdot K; \; M = E - A \cdot K; \; N = F + M; \; O = N^2; \; P = F - L;$$
$$X_3 = 2L \cdot P \cdot X_1; \; Y_3 = M \cdot ((N + Y_1)^2 - O - B); \; Z_3 = P \cdot ((N + Z_1)^2 - O - C).$$

Appendix A contains formulas for quintupling on Edwards curves.

3   Fast Addition on Elliptic Curves

There is a vast literature on elliptic curves. See [12,6,15] for overviews of efficient group operations on elliptic curves, and [5, Section 6] for an analysis of scalar-multiplication performance without triplings.

Those overviews are not a satisfactory starting point for our analysis, because they do not include the most recent improvements in curve shapes and in addition formulas. Fortunately, all of the latest improvements have been collected into the Bernstein/Lange "Explicit-Formulas Database" (EFD) [3], with Magma scripts verifying the correctness of the formulas. For example, this database now includes our tripling formulas, the tripling formulas from [4] (modeled after ours) for inverted Edwards coordinates, and the formulas from [16] for other systems.

Counting operations. In Section 5 we assume $\mathbf{S} = 0.8\mathbf{M}$, but in this section we record the costs separately. We ignore costs of the cheaper field operations such as field additions, field subtractions, and field doublings.

We also ignore the costs of multiplications by curve parameters (for example, d in Edwards form). We assume that curves are sensibly selected with small parameters so that these multiplications are easy.

Jacobian coordinates. Recall that k is assumed to be a field of characteristic at least 5. Every elliptic curve over k can then be written in Weierstrass form $E : y^2 = x^3 + a_4 x + a_6$, $a_4, a_6 \in k$, where $f(x) = x^3 + a_4 x + a_6$ is squarefree. The set $E(k)$ of k-rational points of E is the set of tuples (x_1, y_1) satisfying the equation together with a point P_∞ at infinity.

The most popular representation of an affine point $(x_1, y_1) \in E(k)$ is as *Jacobian coordinates* $(X_1 : Y_1 : Z_1)$ satisfying $Y_1^2 = X_1^3 + a_4 X_1 Z_1^2 + a_6 Z_1^6$. An *addition* of generic points $(X_1 : Y_1 : Z_1)$ and $(X_2 : Y_2 : Z_2)$ in Jacobian coordinates costs $11\mathbf{M} + 5\mathbf{S}$. A *readdition*—i.e., an addition where $(X_2 : Y_2 : Z_2)$ has been added before—costs $10\mathbf{M} + 4\mathbf{S}$, because Z_2^2 and Z_2^3 can be cached and reused. A *mixed addition*—i.e., an addition where Z_2 is known to be 1—costs $7\mathbf{M} + 4\mathbf{S}$. A *doubling*—i.e., an addition where $(X_1 : Y_1 : Z_1)$ and $(X_2 : Y_2 : Z_2)$ are known to be equal—costs $1\mathbf{M} + 8\mathbf{S}$. A *tripling* costs $5\mathbf{M} + 10\mathbf{S}$.

If $a_4 = -3$ then the cost for doubling changes to $3\mathbf{M} + 5\mathbf{S}$ and that for tripling to $7\mathbf{M} + 7\mathbf{S}$. Not every curve can be transformed to allow $a_4 = -3$ but important examples such as the NIST curves [18] make this choice. We refer to this case as Jacobian-3.

Most of the literature presents slower formulas producing the same output, and correspondingly reports higher costs for arithmetic in Jacobian coordinates. See, for example, the P1363 standards [18] and the aforementioned overviews. We include the slower formulas in our experiments to simplify the comparison of our results to previous results in [11] and [9] and to emphasize the importance of using faster formulas. We refer to the slower formulas as Std-Jac and Std-Jac-3.

More coordinate systems. Several other representations of elliptic curves have attracted attention because they offer faster group operations or extra features such as unified addition formulas that also work for doublings. Some of these representations can be reached through isomorphic transformation for any curve in Weierstrass form while others require, for example, a point of order 4. Our analysis includes all of the curve shapes listed in the following table:

Curve shape	ADD	reADD	mADD	DBL	TRI
3DIK	$11M + 6S$	$10M + 6S$	$7M + 4S$	$2M + 7S$	$6M + 6S$
Edwards	$10M + 1S$	$10M + 1S$	$9M + 1S$	$3M + 4S$	$9M + 4S$
ExtJQuartic	$8M + 3S$	$8M + 3S$	$7M + 3S$	$3M + 4S$	$4M + 11S$
Hessian	$12M + 0S$	$12M + 0S$	$10M + 0S$	$7M + 1S$	$8M + 6S$
InvEdwards	$9M + 1S$	$9M + 1S$	$8M + 1S$	$3M + 4S$	$9M + 4S$
JacIntersect	$13M + 2S$	$13M + 2S$	$11M + 2S$	$3M + 4S$	$4M + 10S$
Jacobian	$11M + 5S$	$10M + 4S$	$7M + 4S$	$1M + 8S$	$5M + 10S$
Jacobian-3	$11M + 5S$	$10M + 4S$	$7M + 4S$	$3M + 5S$	$7M + 7S$
Std-Jac	$12M + 4S$	$11M + 3S$	$8M + 3S$	$3M + 6S$	$9M + 6S$
Std-Jac-3	$12M + 4S$	$11M + 3S$	$8M + 3S$	$4M + 4S$	$9M + 6S$

The speeds listed here, and the speeds used in our analysis, are the current speeds in EFD.

"ExtJQuartic" and "Hessian" and "JacIntersect" refer to the latest addition formulas for Jacobi quartics $Y^2 = X^4 + 2aX^2Z^2 + Z^4$, Hessian curves $X^3 + Y^3 + Z^3 = 3dXYZ$, and Jacobi intersections $S^2 + C^2 = T^2, aS^2 + D^2 = T^2$. EFD takes account of the improvements in [13] and [16].

"3DIK" is an abbreviation for "tripling-oriented Doche-Icart-Kohel curves," the curves $Y^2 = X^3 + a(X + Z^2)^2 Z^2$ introduced last year in [10]. (The same paper also introduces doubling-oriented curves that do not have fast additions or triplings and that are omitted from our comparison.) We note that [10] states incorrect formulas for doubling. The corrected and faster formulas are:

$$B = X_1^2; \ C = 2A \cdot Z_1 2 \cdot (X_1 + Z_1 2); \ D = 3(B + C); \ E = Y_1^2; \ F = E^2;$$
$$Z_3 = (Y_1 + Z_1)^2 - E - Z_1 2; \ G = 2((X_1 + E)^2 - B - F);$$
$$X_3 = D^2 - 3A \cdot Z_3^2 - 2G; \ Y_3 = D \cdot (G - X_3) - 8F;$$

which are now also included in the EFD.

4   Background: Double-Base Chains for Single-Scalar Multiplication

This section reviews the previous state of the art in double-base chains for computing nP given P.

The non-windowing case. The "base-2" equation

$$314159P$$
$$= 2(2(2(2(2(2(2(2(2(2(2(2(2(2(2(2(2(2(P)) + P)) - P))) + P) + P)) - P))) + P) + P)))) - P$$

can be viewed as an algorithm to compute $314159P$, starting from P, with a chain of 18 doublings and 8 additions of P; here we count subtractions as additions. One can express this chain more concisely—with an implicit application of Horner's rule—as

$$314159P = 2^{18}P + 2^{16}P - 2^{14}P + 2^{11}P + 2^{10}P - 2^8P + 2^5P + 2^4P - 2^0P.$$

The slightly more complicated "double-base-2-and-3" equation

$$314159P = 2^{15}3^2P + 2^{11}3^2P + 2^83^1P + 2^43^1P - 2^03^0P$$
$$= 3(2(2(2(2(2(2(2(2(3(2(2(2(2(2(2(2(P)))) + P)))) + P)))) + P))))) - P$$

can be viewed as a better algorithm to compute $314159P$, starting from P, with a chain of 2 triplings, 15 doublings, and 4 additions of P. If 1 tripling has the same cost as 1 doubling and 1 addition then this chain has the same cost as 17 doublings and 6 additions which is fewer operations than the 18 doublings and 8 additions of P needed in the base-2 expansion. One can object to this comparison by pointing out that adding mP for $m > 1$ is more expensive than adding P—typically P is provided in affine form, allowing a mixed addition of P, while mP requires a more expensive non-mixed addition—so a tripling is more expensive than a doubling and an addition of P. But this objection is amply answered by dedicated tripling formulas that are *less* expensive than a doubling and an addition. See Sections 2 and 3 for references and the new tripling formulas for Edwards curves.

Double-base chains were introduced by Dimitrov, Imbert, and Mishra in a paper [9] at Asiacrypt 2005. There were several previous "double-base number system" papers expanding nP in various ways as $\sum c_i 2^{a_i} 3^{b_i} P$ with $c_i \in \{-1, 1\}$; the critical advance in [9] was to require $a_1 \geq a_2 \geq a_3 \geq \cdots$ and $b_1 \geq b_2 \geq b_3 \geq \cdots$, allowing a straightforward chain of doublings and triplings without the expensive backtracking that plagued previous papers.

Issues in comparing single bases to double bases. One can object that taking advantage of triplings requires considerable extra effort in finding a double-base chain for nP: finding the integer $2^a 3^b$ closest to n, for a range of several b's, is clearly more difficult than finding the integer 2^a closest to n. Perhaps this objection will be answered someday by an optimized algorithm that finds a double-base chain in less time than is saved by applying that chain. We rely on a simpler answer: we focus on cryptographic applications in which the same n is used many times (as in [8, Section 3]), allowing the chain for n to be constructed just once and then reused. Our current software has not been heavily optimized but takes under a millisecond to compute an expansion of a cryptographic-size integer n.

A more troubling objection is that the simple base-2 chains described above were obsolete long before the advent of double-base chains. Typical speed-oriented elliptic-curve software instead uses "sliding window" base-2 chains that use marginally more temporary storage but considerably fewer additions—see below. Even if double-base chains are faster than obsolete base-2 chains, there is no reason to believe that they are faster than state-of-the-art sliding-window base-2 chains. This objection is partly answered by an analogous improvement to double-base chains—see below—but the literature does not contain a careful comparison of optimized double-base chains to optimized single-base chains.

The sliding-windows case. The "sliding-windows base-2" equation

$$314159P = 2^{16}5P - 2^{11}7P + 2^83P + 2^43P - 2^0P$$
$$= 2(2(2(2(2(2(2(2(2(2(2(2(2(2(2(5P))))) - 7P))) + 3P)))) + 3P)))) - P$$

can be viewed as an algorithm to compute $314159P$, starting from $\{P, 3P, 5P, 7P\}$, with a chain of 16 doublings and 4 additions. It can therefore be viewed as an algorithm to compute $314159P$, starting from P, with 17 doublings and 7 additions; this operation count includes the obvious chain of 1 doubling and 3 additions to produce $2P, 3P, 5P, 7P$ from P.

The idea of starting with $\{P, 2P, 3P, 4P, \ldots, (2^w - 1)P\}$ ("fixed length-w windows") was introduced by Brauer long ago in [7]. By optimizing the choice of w as a function of the bitlength ℓ, Brauer showed that one can compute nP for an ℓ-bit integer n using $\approx \ell$ doublings and at most $\approx \ell/\lg \ell$ additions (even without subtractions). The idea to start with $\{P, 2P, 3P, 5P, 7P, \ldots, (2^w - 1)P\}$ ("sliding length-w windows") was introduced by Thurber in [20], saving some additions. For comparison, the simple base-2 chains considered earlier use $\approx \ell$ doublings and $\approx \ell/3$ additions (on average; as many as $\ell/2$ in the worst case). The benefit of windows increases slowly with ℓ.

Doche and Imbert, in their paper [11] at Indocrypt 2006, introduced an analogous improvement to double-base chains. Example: The "sliding-windows double-base-2-and-3" equation

$$314159P = 2^{12}3^33P - 2^73^35P - 2^43^17P - 2^03^0P$$
$$= 3(2(2(2(2(3(3(2(2(2(2(2(2(2(2(3P))))) - 5P))))) - 7P))))) - P$$

can be viewed as an algorithm to compute $314159P$, starting from $\{P, 3P, 5P, 7P\}$, with a chain of 3 triplings, 12 doublings, and 3 additions. It can therefore be viewed as an algorithm to compute $314159P$, starting from P, with 3 triplings, 13 doublings, and 6 additions.

Doche and Imbert state an algorithm to compute double-base chains for arbitrary coefficient sets S containing 1. In their experiments they focus on sets of the form $\{1, 2, 3, 2^2, 3^2, \ldots, 2^k, 3^k\}$ or sets of odd integers co-prime to 3. In this paper we study several coefficient sets including all sets considered in [11] and additional sets such as $\{P, 2P, 3P, 5P, 7P\}$.

Computing a chain. Finding the chain $314159 = 2^{18} + 2^{16} - 2^{14} + 2^{11} + 2^{10} - 2^8 + 2^5 + 2^4 - 2^0$ is a simple matter of finding the closest power of 2 to 314159, namely $2^{18} = 262144$; then finding the closest power of 2 to the difference $|314159 - 262144| = 52015$, namely $2^{16} = 65536$; and so on.

Similarly, by inspecting the first few bits of a nonzero integer n one can easily see which of the integers

$$
\begin{array}{lllll}
\pm 1, & \pm 2, & \pm 2^2, & \pm 2^3, & \pm 2^4, & \ldots \\
\pm 3, & \pm 2 \cdot 3, & \pm 2^2 3, & \pm 2^3 3, & \pm 2^4 3, & \ldots \\
\pm 5, & \pm 2 \cdot 5, & \pm 2^2 5, & \pm 2^3 5, & \pm 2^4 5, & \ldots \\
\pm 7, & \pm 2 \cdot 7, & \pm 2^2 7, & \pm 2^3 7, & \pm 2^4 7, & \ldots
\end{array}
$$

is closest to n. By subtracting that integer from n and repeating the same process one expands n into Thurber's base-2 sliding-window chain $\sum_i c_i 2^{a_i}$ with $c_i \in \{-7, -5, -3, -1, 1, 3, 5, 7\}$ and $a_1 > a_2 > a_3 > \cdots$. For example, $2^{16} \cdot 5 = 327680$ is closest to 314159; $-2^{11} \cdot 7 = -14336$ is closest to $314159 - 327680 = -13521$; continuing in the same way one finds the chain $314159 = 2^{16}5P - 2^{11}7P + 2^8 3P + 2^4 3P - 2^0 P$ shown above. Similar comments apply to sets other than $\{1, 3, 5, 7\}$.

Dimitrov, Imbert, and Mishra in [9, Section 3] proposed a similar algorithm to find double-base chains with $c_i \in \{-1, 1\}$; Doche and Imbert in [11, Section 3.2] generalized the algorithm to allow a wider range of c_i. For example, given n and the set $\{1, 3, 5, 7\}$, the Doche-Imbert algorithm finds the product $c_1 2^{a_1} 3^{b_1}$ closest to n, with $c_1 \in \{-7, -5, -3, -1, 1, 3, 5, 7\}$, subject to limits on a_1 and b_1; it then finds the product $c_2 2^{a_2} 3^{b_2}$ closest to $n - c_1 2^{a_1} 3^{b_1}$, with $c_2 \subset \{7, 5, 3, 1, 1, 3, 5, 7\}$, subject to the chain conditions $a_1 \geq a_2$ and $b_1 \geq b_2$; continuing in this way it expands n as $\sum_i c_i 2^{a_i} 3^{b_i}$ with $c_i \in \{-7, -5, -3, -1, 1, 3, 5, 7\}$, $a_1 \geq a_2 \geq \cdots$, and $b_1 \geq b_2 \geq \cdots$.

(The algorithm statements in [9] and [11] are ambiguous on the occasions that n is equally close to two or more products $c 2^a 3^b$. Which (c, a, b) is chosen? In our new experiments, when several $c 2^a 3^b$ are equally close to n, we choose the first (c, b, a) in lexicographic order: we prioritize a small c, then a small b, then a small a.)

The worst-case and average-case chain lengths produced by this double-base algorithm are difficult to analyze mathematically. However, the average chain length for all n's can be estimated with high confidence as the average chain length seen for a large number of n's. Dimitrov, Imbert, and Mishra used 10000 integers n for each of their data points; Doche and Imbert used 1000; our new experiments use 10000. We also plan to compute variances but have not yet done so.

5   New Results

This section describes the experiments that we carried out and the multiplication counts that we achieved. The results of the experiments are presented as a table and a series of graphs.

Parameter space. Our experiments included several bit sizes ℓ, namely 160, 200, 256, 300, 400, and 500. The choices $200, 300, 400, 500$ were used in [11] and we include them to ease comparison. The choices 160 and 256 are common in cryptographic applications.

Our experiments included the eight curve shapes described in Section 3: 3DIK, Edwards, ExtJQuartic, Hessian, InvEdwards, JacIntersect, Jacobian, and Jacobian-3. For comparison with previous results, and to show the importance of optimized curve formulas, we also carried out experiments for Std-Jac and Std-Jac-3.

Our experiments included many choices of the parameter a_0 in [11, Algorithm 1]. The largest power of 2 allowed in the algorithm is 2^{a_0}, and the largest power of 3 allowed in the algorithm is 3^{b_0} where $b_0 = \lceil (\ell - a_0)/\lg 3 \rceil$. Specifically, we tried each $a_0 \in \{0, 10, 20, \dots, 10\lfloor \ell/10 \rfloor\}$. This matches the experiments reported in [11] for $\ell = 200$. We also tried all integers a_0 between 0.95ℓ and 1.00ℓ.

Our experiments included several coefficient sets S, i.e., sets of coefficients c allowed in $c2^a3^b$: the set $\{1\}$ used in [9]; the sets $\{1, 2, 3\}$, $\{1, 2, 3, 4, 8, 9, 16, 27, 81\}$, $\{1, 5, 7\}$, $\{1, 5, 7, 11, 13, 17, 19, 23, 25\}$ appearing in the graphs in [11, Appendix B] with labels "$(1, 1)$" and "$(4, 4)$" and "S_2" and "S_8"; and the sets $\{1, 2, 3, 4, 9\}$, $\{1, 2, 3, 4, 8, 9, 27\}$, $\{1, 5\}$, $\{1, 5, 7, 11\}$, $\{1, 5, 7, 11, 13\}$, $\{1, 5, 7, 11, 13, 17, 19\}$ appearing in the tables in [11, Appendix B]. We also included the sets $\{1, 2, 3, 5\}$, $\{1, 2, 3, 5, 7\}$, and so on through $\{1, 2, 3, 5, 7, 9, 11, 13, 15, 17, 19, 23, 25\}$; these sets are standard in the base-2 context but do not seem to have been included in previous double-base experiments.

(We have considered additional sets such as $\{1, 2, 3, 4, 5, 7, 9\}$. Multiples of 6 are not worthwhile but we see some potential for coefficients $4, 8, 10, \dots$ in the context of 3DIK coordinates. However, those sets are not yet included in our experiments.)

We used straightforward combinations of additions, doublings, and triplings for the initial computation of cP for each $c \in S$. We have considered, but not yet included in our experiments, the use of quintuplings, merged operations, etc. for this computation. Reader beware: as mentioned in Section 1, the costs of these computations are ignored in [11], allowing arbitrarily large sets S for free and allowing arbitrarily small costs of computing nP; the costs in [11] thus become increasingly inconsistent with the costs in this paper (and in reality) as S grows.

We follow the standard (although debatable) practice of counting $\mathbf{S} = 0.8\mathbf{M}$ and disregarding other field operations. We caution the reader that other weightings of field operations can easily change the order of two systems with similar levels of performance.

Experiments and results. There are 8236 combinations of ℓ, a_0, and S described above. For each combination, we

- generated 10000 uniform random integers $n \in \{0, 1, \dots, 2^\ell - 1\}$,
- converted each integer into a chain as specified by a_0 and S,
- checked that the chain indeed computed n starting the chain from 1, and
- counted the number of triplings, doublings, additions, readditions, and mixed additions for those 10000 choices of n.

We converted the results into multiplication counts for 3DIK, Edwards, ExtJQuartic, Hessian, InvEdwards, JacIntersect, Jacobian, Jacobian-3, Std-Jac, and Std-Jac-3, obtaining a cost for each of the 82360 combinations of ℓ, curve shape, a_0, and S.

Figure 1 shows, for each ℓ (horizontal axis) and each curve shape, the minimum cost per bit obtained when a_0 and S are chosen optimally. The implementor can easily read off the ranking of coordinate systems from this graph. Table 1 displays the same information in tabular form, along with the choices of a_0 and S.

There is no unique optimal choice of a_0 and S for every curve shape which gives rise to the fastest computation of a given ℓ-bit integer. For example, using Jacobian coordinates the best result is achieved by precomputing odd coefficients

Table 1. Optimal parameters for each curve shape and each ℓ

ℓ	Curve shape	Mults	Mults/ℓ	a_0	a_0/ℓ	S
160	3DIK	1502.393800	9.389961	80	0.5	$\{1\}$
200	3DIK	1879.200960	9.396005	100	0.5	$\{1, 2, 3, 5, 7\}$
256	3DIK	2393.193800	9.348413	130	0.51	$\{1, 2, 3, 5, \ldots, 13\}$
300	3DIK	2794.431020	9.314770	160	0.53	$\{1, 2, 3, 5, \ldots, 13\}$
400	3DIK	3706.581360	9.266453	210	0.53	$\{1, 2, 3, 5, \ldots, 13\}$
500	3DIK	4615.646620	9.231293	270	0.54	$\{1, 2, 3, 5, \ldots, 17\}$
160	Edwards	1322.911120	8.268194	156	0.97	$\{1, 2, 3, 5, \ldots, 13\}$
200	Edwards	1642.867360	8.214337	196	0.98	$\{1, 2, 3, 5, \ldots, 15\}$
256	Edwards	2089.695120	8.162872	252	0.98	$\{1, 2, 3, 5, \ldots, 15\}$
300	Edwards	2440.611880	8.135373	296	0.99	$\{1, 2, 3, 5, \ldots, 15\}$
400	Edwards	3224.251900	8.060630	394	0.98	$\{1, 2, 3, 5, \ldots, 25\}$
500	Edwards	4005.977080	8.011954	496	0.99	$\{1, 2, 3, 5, \ldots, 25\}$
160	ExtJQuartic	1310.995340	8.193721	156	0.97	$\{1, 2, 3, 5, \ldots, 13\}$
200	ExtJQuartic	1628.386660	8.141933	196	0.98	$\{1, 2, 3, 5, \ldots, 15\}$
256	ExtJQuartic	2071.217580	8.090694	253	0.99	$\{1, 2, 3, 5, \ldots, 15\}$
300	ExtJQuartic	2419.026660	8.063422	299	1	$\{1, 2, 3, 5, \ldots, 21\}$
400	ExtJQuartic	3196.304940	7.990762	399	1	$\{1, 2, 3, 5, \ldots, 25\}$
500	ExtJQuartic	3972.191800	7.944384	499	1	$\{1, 2, 3, 5, \ldots, 25\}$
160	Hessian	1560.487660	9.753048	100	0.62	$\{1, 2, 3, 5, \ldots, 13\}$
200	Hessian	1939.682780	9.698414	120	0.6	$\{1, 2, 3, 5, \ldots, 13\}$
256	Hessian	2470.643200	9.650950	150	0.59	$\{1, 2, 3, 5, \ldots, 13\}$
300	Hessian	2888.322160	9.627741	170	0.57	$\{1, 2, 3, 5, \ldots, 13\}$
400	Hessian	3831.321760	9.578304	240	0.6	$\{1, 2, 3, 5, \ldots, 17\}$
500	Hessian	4772.497740	9.544995	300	0.6	$\{1, 2, 3, 5, \ldots, 19\}$
160	InvEdwards	1290.333920	8.064587	156	0.97	$\{1, 2, 3, 5, \ldots, 13\}$
200	InvEdwards	1603.737760	8.018689	196	0.98	$\{1, 2, 3, 5, \ldots, 15\}$
256	InvEdwards	2041.223320	7.973529	252	0.98	$\{1, 2, 3, 5, \ldots, 15\}$
300	InvEdwards	2384.817880	7.949393	296	0.99	$\{1, 2, 3, 5, \ldots, 15\}$
400	InvEdwards	3152.991660	7.882479	399	1	$\{1, 2, 3, 5, \ldots, 25\}$
500	InvEdwards	3919.645880	7.839292	496	0.99	$\{1, 2, 3, 5, \ldots, 25\}$
160	JacIntersect	1438.808960	8.992556	150	0.94	$\{1, 2, 3, 5, \ldots, 13\}$
200	JacIntersect	1784.742200	8.923711	190	0.95	$\{1, 2, 3, 5, \ldots, 15\}$
256	JacIntersect	2266.135540	8.852092	246	0.96	$\{1, 2, 3, 5, \ldots, 15\}$
300	JacIntersect	2644.233000	8.814110	290	0.97	$\{1, 2, 3, 5, \ldots, 15\}$
400	JacIntersect	3486.773860	8.716935	394	0.98	$\{1, 2, 3, 5, \ldots, 25\}$
500	JacIntersect	4324.718620	8.649437	492	0.98	$\{1, 2, 3, 5, \ldots, 25\}$
160	Jacobian	1558.405080	9.740032	100	0.62	$\{1, 2, 3, 5, \ldots, 13\}$
200	Jacobian	1937.129960	9.685650	130	0.65	$\{1, 2, 3, 5, \ldots, 13\}$
256	Jacobian	2466.150480	9.633400	160	0.62	$\{1, 2, 3, 5, \ldots, 13\}$
300	Jacobian	2882.657400	9.608858	180	0.6	$\{1, 2, 3, 5, \ldots, 13\}$
400	Jacobian	3819.041260	9.547603	250	0.62	$\{1, 2, 3, 5, \ldots, 17\}$
500	Jacobian	4755.197420	9.510395	310	0.62	$\{1, 2, 3, 5, \ldots, 19\}$
160	Jacobian-3	1504.260200	9.401626	100	0.62	$\{1, 2, 3, 5, \ldots, 13\}$
200	Jacobian-3	1868.530560	9.342653	130	0.65	$\{1, 2, 3, 5, \ldots, 13\}$
256	Jacobian-3	2378.956000	9.292797	160	0.62	$\{1, 2, 3, 5, \ldots, 13\}$
300	Jacobian-3	2779.917220	9.266391	200	0.67	$\{1, 2, 3, 5, \ldots, 17\}$
400	Jacobian-3	3681.754460	9.204386	260	0.65	$\{1, 2, 3, 5, \ldots, 17\}$
500	Jacobian-3	4583.527180	9.167054	330	0.66	$\{1, 2, 3, 5, \ldots, 21\}$

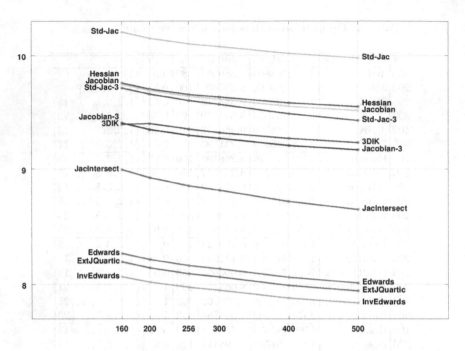

Fig. 1. Multiplications per bit (all bits, all shapes)

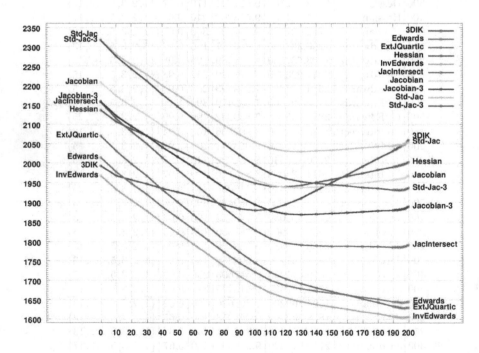

Fig. 2. Importance of doubling/tripling ratio (200 bits, all shapes)

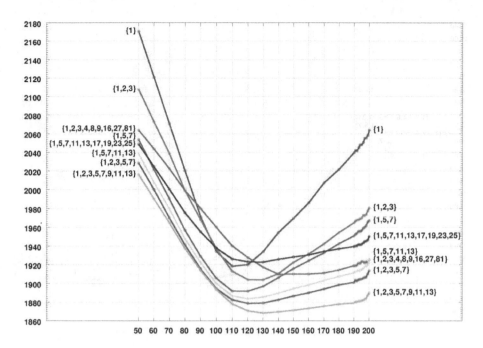

Fig. 3. Importance of parameter choices (200 bits, Jacobian-3)

up to 13 for an integer of bit length at most 300. For 400-bit integers the optimum uses $S = \{1, 2, 3, 5, \ldots, 17\}$ and in the 500-bit case also 19 is included.

None of the optimal results for $\ell \geq 200$ uses a set of precomputed points discussed in [9] or [11]. Independent of the ratio between doubling and tripling cost and the cost of doubling the optimal coefficient sets were those used in (fractional) sliding-window methods, i.e. the sets $\{1, 2, 3, 5, \ldots\}$.

Figure 2 shows, for each a_0 (horizontal axis) and each curve shape, the cost for $\ell = 200$ when S is chosen optimally. This graph demonstrates the importance of choosing the right bounds for a_0 and b_0 depending on the ratio of the doubling/tripling costs. We refer to Table 1 for the best choices of a_0 and S for each curve shape.

The fastest systems are Edwards, ExtJQuartic, and InvEdwards. They need the lowest number of multiplications for values of a_0 very close to ℓ. These systems are using larger sets of precomputations than slower systems such as Jacobian-3 or Jacobian, and fewer triplings. The faster systems all come with particularly fast addition laws, making the precomputations less costly, and particularly fast doublings, making triplings less attractive. This means that currently double-base chains offer no or very little advantage for the fastest systems. See [5] for a detailed description of single-base scalar multiplication on Edwards curves.

Not every curve can be represented by one of these fast systems. For curves in Jacobian coordinates values of a_0 around 0.6ℓ seem optimal and produce significantly faster scalar multiplication than single-base representations.

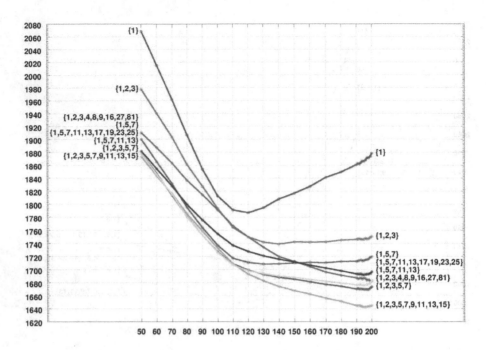

Fig. 4. Importance of parameter choices (200 bits, Edwards)

Figure 3 shows, for a smaller range of a_0 (horizontal axis) and each choice of S, the cost for Jacobian-3 coordinates for $\ell = 200$. This graph demonstrates several interesting interactions between the doubling/tripling ratio, the choice of S, and the final results. Figure 4 is a similar graph for Edwards curves. The optimal scalar-multiplication method in that graph uses $a_0 \approx 195$ with coefficients in the set $\pm\{1, 2, 3, 5, 7, 11, 13, 15\}$. The penalty for using standard single-base sliding-window methods is negligible.

References

1. Avanzi, R., Cohen, H., Doche, C., Frey, G., Lange, T., Nguyen, K., Vercauteren, F.: The Handbook of Elliptic and Hyperelliptic Curve Cryptography. CRC, Boca Raton, USA (2005)
2. Barua, R., Lange, T. (eds.): INDOCRYPT 2006. LNCS, vol. 4329. Springer, Heidelberg (2006)
3. Bernstein, D.J., Lange, T.: Explicit-formulas database, http://www.hyperelliptic.org/EFD
4. Bernstein, D.J., Lange, T.: Inverted Edwards coordinates. In: AAECC 2007 (to appear, 2007)
5. Bernstein, D.J., Lange, T.: Faster addition and doubling on elliptic curves. In: Asiacrypt 2007 [17], pp. 29–50 (2007), http://cr.yp.to/newelliptic/

6. Blake, I.F., Seroussi, G., Smart, N.P.: Elliptic curves in cryptography. London Mathematical Society Lecture Note Series, vol. 265. Cambridge University Press, Cambridge (1999)
7. Brauer, A.: On addition chains. Bulletin of the American Mathematical Society 45, 736–739 (1939)
8. Diffie, W., Hellman, M.E.: New directions in cryptography. IEEE Transactions on Information Theory 22(6), 644–654 (1976)
9. Dimitrov, V., Imbert, L., Mishra, P.K.: Efficient and secure elliptic curve point multiplication using double-base chains. In: ASIACRYPT 2005 [19], pp. 59–78 (2005)
10. Doche, C., Icart, T., Kohel, D.R.: Efficient scalar multiplication by isogeny decompositions. In: PKC 2006 [21], pp. 191–206 (2006)
11. Doche, C., Imbert, L.: Extended double-base number system with applications to elliptic curve cryptography. In: Indocrypt 2006 [2], pp. 335–348 (2006)
12. Doche, C., Lange, T.: Arithmetic of Elliptic Curves, Ch. 13 in [1], pp. 267–302. CRC Press, Boca Raton, USA (2005)
13. Duquesne, S.: Improving the arithmetic of elliptic curves in the Jacobi model. Information Processing Letters 104, 101–105 (2007)
14. Edwards, H.M.: A normal form for elliptic curves. Bulletin of the American Mathematical Society 44, 393–422 (2007),
    http://www.ams.org/bull/2007-44-03/S0273-0979-07-01153-6/home.html
15. Hankerson, D., Menezes, A.J., Vanstone, S.A.: Guide to elliptic curve cryptography. Springer, Berlin (2003)
16. Hisil, H., Carter, G., Dawson, E.: New formulae for efficient elliptic curve arithmetic. In: Indocrypt 2007. LNCS, vol. 4859, pp. 138–151. Springer, Heidelberg (2007)
17. Kurosawa, K. (ed.): Advances in cryptology–ASIACRYPT 2007. LNCS, vol. 4833. Springer, Heidelberg (2007)
18. IEEE P1363. Standard specifications for public key cryptography. IEEE (2000)
19. Roy, B. (ed.): ASIACRYPT 2005. LNCS, vol. 3788. Springer, Heidelberg (2005)
20. Thurber, E.G.: On addition chains $l(mn) \leq l(n) - b$ and lower bounds for $c(r)$. Duke Mathematical Journal 40, 907–913 (1973)
21. Yung, M., Dodis, Y., Kiayias, A., Malkin, T. (eds.): PKC 2006. LNCS, vol. 3958. Springer, Heidelberg (2006)

A   Appendix: Quintupling on Edwards Curves

In this section we give formulas to compute the 5-fold of a point on an Edwards curve. We present two different versions which lead to the same result but have a different complexity. The first version needs $17\mathbf{M} + 7\mathbf{S}$ while version 2 needs $14\mathbf{M} + 11\mathbf{S}$.

The question which version to choose for a specific platform can be answered by looking at the $\mathbf{S}/\mathbf{M}$-ratio. If $\mathbf{S}/\mathbf{M} = 0.75$ both sets of formulas have the same complexity, namely $17\mathbf{M} + 5.25\mathbf{M} = 14\mathbf{M} + 8.25\mathbf{M} = 22.25\mathbf{M}$. For a $\mathbf{S}/\mathbf{M}$-ratio less than 0.75 one should use version 2; if the $\mathbf{S}/\mathbf{M}$-ratio is greater 0.75, then version 1 of the quintupling formulas achieves best performance.

Both versions were verified to produce the 5-fold of the input point $(X_1 : Y_1 : Z_1)$ by symbolically computing $5(X_1 : Y_1 : Z_1)$ using the addition and doubling formulas from [5] and comparing it with $(X_5 : Y_5 : Z_5)$.

It is interesting to note that the formulas do not have minmal degree. The new variables X_5, Y_5, Z_5 have total degree 33 in the initial variables even though one would expect expect degree 5^2. Indeed, X_5, Y_5, Z_5 are all divisible by the degree 8 polynomial $((X_1^2 - Y_1^2)^2 + 4Y_1^2(Z_1^2 - Y_1^2))((X_1^2 - Y_1^2)^2 + 4X_1^2(Z_1^2 - X_1^2))$. Minimizing the number of operations lead to better results for our extended polynomials.

The 17M + 7S-formulas for quintupling:

$$A = X_1^2;\ B = Y_1^2;\ C = Z_1^2;\ D = A + B;$$
$$E = 2C - D;\ F = D \cdot (B - A);\ G = E \cdot ((X_1 + Y_1)^2 - D);$$
$$H = F^2;\ I = G^2;\ J = H + I;\ K = H - I;\ L = J \cdot K;$$
$$M = D \cdot E;\ N = E \cdot F;\ O = 2M^2 - J;\ P = 4N \cdot O;$$
$$Q = 4K \cdot N \cdot (D - C);\ R = O \cdot J;\ S = R + Q;\ T = R - Q;$$
$$X_5 = X_1 \cdot (L + B \cdot P) \cdot T;\ Y_5 = Y_1 \cdot (L - A \cdot P) \cdot S;\ Z_5 = Z_1 \cdot S \cdot T;$$

The 14M + 11S-formulas for quintupling:

$$A = X_1^2;\ B = Y_1^2;\ C = Z_1^2;\ D = A + B;\ E = 2C - D;\ F = A^2;\ G = B^2;\ H = F + G;$$
$$I = D^2 - H;\ J = E^2;\ K = G - F;\ L = K^2;\ M = 2I \cdot J;\ N = L + M;\ O = L - M;$$
$$P = N \cdot O;\ Q = (E + K)^2 - J - L;\ R = ((D + E)^2 - J - H - I)^2 - 2N;$$
$$S = Q \cdot R;\ T = 4Q \cdot O \cdot (D - C);\ U = R \cdot N;\ V = U + T;\ W = U - T;$$
$$X_5 = 2X_1 \cdot (P + B \cdot S) \cdot W;\ Y_5 = 2Y_1 \cdot (P - A \cdot S) \cdot V;\ Z_5 = Z_1 \cdot V \cdot W.$$

Note, that only variables $A \ldots E$ have the same values in the two versions.

Transitive Signatures from Braid Groups

Licheng Wang[1], Zhenfu Cao[2], Shihui Zheng[1], Xiaofang Huang[1],
and Yixian Yang[1]

[1] Information Security Center, State Key Laboratory of Networking and Switching
Technology, Beijing University of Posts and Telecommunications,
Beijing 100876, P.R. China
[2] Dept. Computer Science and Engineering, Shanghai Jiao Tong University,
Shanghai 200240, P.R. China

Abstract. Transitive signature is an interesting primitive due to Micali
and Rivest. During the past years, many constructions of transitive signatures have been proposed based on various assumptions. In this paper,
we provide the first construction of transitive signature schemes by using
braid groups. In the random oracle model, our proposals are proved to
be transitively unforgeable against adaptively chosen message attack under the assumption of the intractability of one-more matching conjugate
problem (OM-MCP) over braid groups. Moreover, the proposed schemes
are invulnerable to currently known quantum attacks.

Keywords: Transitive signature, braid group, one-more matching conjugate problem, provable security, random oracle model.

1 Introduction

1.1 Primitive of Transitive Signature and Related Constructions

The primitive of transitive signature was firstly proposed by Micali and Rivest[25]
in 2002. Transitive signature aims to dynamically build an authenticated graph,
edge by edge[4]. In other words, the message space of a transitive signature
scheme is the set of *all* potential edges of the authenticated graph. The signer,
having secret key tsk and public key tpk, can at any time pick a pair i, j of
nodes and create a signature for (i, j), thereby adding edge (i, j) to the graph[4].
The most remarkable characteristic of transitive signature is the newly conceived
property — *composability*, which requires that given a signature of an edge (i, j)
and a signature of an edge (j, k), anyone in possession of the public key can
create a signature of the edge (i, k). Security of transitive signature asks that
without the secret key tsk, it should be hard to create a valid signature of
edge (i, j) unless i, j are connected by a path whose edges have been explicitly
authenticated by the signer[4].

In 2002, Micali and Rivest[25] proposed the first transitive signature scheme
based on discrete logarithm (DL) assumption; In the end of this paper, they also
conceived another transitive signature scheme based on RSA assumption. The
former was proved unforgeable against adaptive chosen-message attacks, while

K. Srinathan, C. Pandu Rangan, M. Yung (Eds.): Indocrypt 2007, LNCS 4859, pp. 183–196, 2007.
© Springer-Verlag Berlin Heidelberg 2007

the latter was merely proved to be secure under non-adaptive chosen-message attacks[1]. Shortly afterwards, Bellare and Neven[4] proposed new transitive signature schemes based on factoring and RSA assumptions. Transitive signature based on bilinear maps was also proposed in 2004 [28]. In 2005, Bellare and Neven[5] gave a good survey on research of transitive signatures and proposed some new schemes, as well as some new proofs. Their paper shows that transitive signature schemes can be constructed based on lots of assumptions, such as RSA assumption, one-more RSA-inversion assumption, factoring assumption, DL assumption, one-more DL assumption and one-more gap Diffie-Hellman assumption.

Meanwhile, research on transitive signatures also made progress along another parallel direction. The original primitive introduced in [25] was defined over a graph G which was modeled under general scenario, i.e., without the limitation that G should be undirected. The implementation in [25], however, was merely suitable for signing data with undirected graph structure. The subsequent works in [4,5,35] were all concerned with undirected graphs. Finding a transitive signature scheme for a directed graph had been remained open until the published of [21], in which a transitive signature scheme for a directed tree was proposed. Since the directed tree is a special case of the directed graph, the proposed scheme in [21] was a partial solution for the open problem. The authors of [21] also showed that a transitive signature scheme for the undirected graph can be constructed from a bundling homomorphism. This means that the transitive signature scheme for the undirected graph is closely related with a fail-stop signature scheme[21]. Unfortunately, the scheme in [21] was proved insecure about a year later [34]. In 2005, Huang et al.[14] proposed an efficient directed transitive signature scheme, in which the basic technique in [25] was improved to suite for *chains*, and then signing a directed transitive binary relation G by combining the proposed scheme with the undirected transitive signature schemes whenever G can be partitioned into several chains such that these chains can be connected by some small equivalence relations. Most recently, Yi[33] also proposed a RSA-based directed transitive signature scheme for directed graphs of which transitive reductions are directed trees. As far as we know, Huang's method [14] and Yi's method[33] are the nearest, though not perfect, answers for directed transitive signatures.

1.2 Background of Braid-Based Public Key Cryptography

Today, most public key cryptosystems that remain unbroken are based on the perceived difficulty of certain problems in particular large finite (abelian) groups. The theoretical foundations for these cyrptosystems lie in the intractability of problems closer to number theory than group theory[24]. On quantum computer, most of these problems turned out to be efficiently solved by algorithms due to Shor[29], Kitaev[17] and Proos-Zalka [27]. Although practical quantum computers are as least 10 years away, their potential will soon create distrust in current

---

[1] Later, it was also proved unforgeable against adaptive chosen-message attacks under a stronger assumption—one-more RSA-inversion assumption[5].

cryptographic methods[22]. In order to enrich cryptography as well as not to put all eggs in one basket[22], there have been many attempts to develop alternative public key cryptography (PKC) based on different kinds of problems [2,19,22,24].

With this background, some non-abelian groups attracted many researchers' attention. The most popular group lies in this area is braid group. In 1999, Anshel et al.[2] proposed an algebraic method for PKC. In their pioneering work, braid groups go upon the stage of modern cryptography. Also, Ko et al.[19] published their paper on braid-based PKC in 2000. From then, the subject has met with a quick success [1,7,18,23]. However, from 2001 to 2003, repeated cryptanalytic success[8,15,16,26] also diminished the initial optimism on the subject significantly. In 2004, Dehornoy[9] gave a good survey on the state of the subject, and suggested that significant research is still needed to reach a definite conclusion on cryptographic potential of braid groups. Most recently, Ko et al.[20] also analyzed the current attacks and proposed new methods for generating secure keys for braid cryptography.

1.3   Contributions and Organizations

Just as claimed in [4], it is standard practice in cryptography to seek new and alternative realizations of primitives of potential interest, both to provide firmer theoretical foundations for the existence of the primitive by basing it on alternative conjectured hard problems and to obtain performance improvements. In order to accomplish both of these objectives, we present new transitive signature schemes by using braid groups. In the random oracle model, the proposed schemes are proved to be transitively unforgeable against adaptively chosen message attack assuming that the one-more matching conjugate problem over braid groups is intractable. Of course, our schemes are only suite for undirected graphs. Therefore, the possible improvements on our construction include at least three aspects: The first is to design braid-based transitive signatures for directed graph; The second is to remove the necessary of "one-more" flavor assumption; And the third is to prove the security of our design in the standard model.

The rest of the paper is organized as follows: Necessary preliminaries on braid cryptography are presented in Section 2; The basic notations and models of transitive signatures are given in Section 3; Then we propose two braid-based transitive signature schemes in Section 4, and give performance analysis and security level evaluation in Section 5. Finally, concluding remarks are presented in Section 6. Meanwhile, all corresponding proofs are arranged in appendices.

2   Preliminaries

2.1   Braid Group and Related Cryptographic Assumptions

The n-braid group B_n is presented by the Artin generators $\sigma_1, \cdots, \sigma_{n-1}$ and relations $\sigma_i\sigma_j = \sigma_j\sigma_i$ for $|i - j| > 1$ and $\sigma_i\sigma_j\sigma_i = \sigma_j\sigma_i\sigma_j$ for $|i - j| = 1$. Two braids x, y are *conjugate*, written $x \sim y$, if $y = axa^{-1}$ for some $a \in B_n$. The conjugator search problem (CSP) defined as: Find $a \in B_n$ such that $y = axa^{-1}$ for a given instance $(x, y) \in B_n \times B_n$ with $x \sim y$. So far, most PKC using braid groups are based on variants of CSP. Although there are some algorithms for solving CSP in braid groups [10,11,12,13], none of them has been proved that can solve CSP in polynomial time (with respect to the braid index n). According to [9] and[20], most of current attacks against braid-based PKC take advantage of the way the keys are generated and at present we see no serious reason for doubting that braid groups are and will remain a promising platform for cryptography[9].

In [18], Ko et al. introduced the concept of *CSP-hard pair*. Suppose that S_1 and S_2 are two subgroups of B_n, a pair $(x, x') \in S_1 \times S_2$ is said to be *CSP-hard* if $x \sim x'$ and CSP is intractable in the group B_n for the instance (x, x'). Clearly, if (x, x') is CSP-hard, so is (x', x)[18]. In fact, we think it is more meaningful to define similar concept as a special sampling algorithm than to define CSP-hardness for one particular pair. Therefore, we say that a *CSP-hard pair generator* K_{csp} is a probabilistic polynomial-time algorithm defined as follows:

- $\mathsf{K}_{csp}(n)$: On inputs the security parameter n, outputs a triple $(p, q, w) \in B_n^3$, where $q = wpw^{-1}$ and the LCF-lengths of p, q and w are bounded by $\mathcal{O}(n^2)$, while finding a conjugator for the pair (p, q) is intractable.

In general, given a triple (p, q, w) generated by CSP-hard pair generated K_{csp} and a random braid $c \in B_n$, it is difficult to produce $wpcw^{-1}$ without knowing w. This kind of problem is defined as conjugate adjoin problem (CAP), which is also seems as hard as CSP[30].

As for how to construct such a CSP-hard pair generator, please refer to [9],[18] and[20]. In particular in [20], Ko et al. proposed several alternate ways of generating hard instances of the conjugacy problem for use braid cryptography. Choosing good parameters and CSP-hard pair generators are non-trivial problems outside the scope of this paper.

Enlightened by the idea of "One-More-RSA-Inversion Problems" [6], Wang et al. [31] defined the so-called One-More Matching Conjugate Problem (OM-MCP) as follows.

Suppose that the braid index n is viewed as the security parameter. An om-mcp attacker is a probabilistic polynomial-time algorithm $\mathcal{A}$ that gets input p, q and has access to two oracles: the matching conjugate oracle $\mathcal{O}_{mc}(\cdot)$ and the challenge oracle $\mathcal{O}_{ch}()$. The attacker $\mathcal{A}$ *wins* the game if it succeeds in matching conjugates with all $\eta(n)$ braids output by the challenge oracle, but submits strictly less than $\eta(n)$ queries to the matching conjugate oracle, where $\eta : \mathbb{N} \to \mathbb{N}$ is arbitrary polynomials over $\mathbb{N}$. More formally, $\mathcal{A}$ is invoked in the following experiment.

Experiment $\mathbf{Exp}_{\mathsf{K}_{csp}, \mathcal{A}}^{om-mcp}(n)$

$$(p, q, w) \xleftarrow{\$} \mathsf{K}_{csp}(n); k \leftarrow 0; l \leftarrow 0;$$
$$(r_1, \cdots, r_{k'}) \xleftarrow{\$} \mathcal{A}^{\mathcal{O}_{mc}, \mathcal{O}_{ch}}(p, q, n);$$

If $k' = k$ and $l < k$ and $\forall\, i = 1, \cdots, k : (r_i \sim c_i) \wedge (qr_i \sim pc_i)$
Then return 1 else return 0

where the oracles are defined as

Oracle $\mathcal{O}_{mc}(b)$	Oracle $\mathcal{O}_{ch}()$
$l \leftarrow l + 1;$	$k \leftarrow k + 1; c_k \xleftarrow{\$} B_n;$
Return wbw^{-1}	Return c_k

The om-mcp advantage of $\mathcal{A}$, denoted by $\mathbf{Adv}_{\mathrm{K}_{csp},\mathcal{A}}^{om-mcp}(n)$, is the probability that the above experiment returns 1, taken over the coins of K_{csp}, the coins of $\mathcal{A}$, and the coins used by the challenge oracle across its invocations. The *one-more matching conjugate assumption* says that the one-more matching conjugate problem associated to K_{csp} is hard, i.e., the function $\mathbf{Adv}_{\mathrm{K}_{csp},\mathcal{A}}^{om-mcp}(n)$ is negligible with respect to the security parameter n for all probabilistic polynomial-time adversaries $\mathcal{A}$.

Note that in one-more type experiments, the adversaries are not permitted to choose challenges by themselves. But they can submit queries on their own choices. Only if the number of the submitted queries is strictly less than the number of challenges that they answered correctly, they won the experiments. Please refer to [6] and[31] for more details about this issue.

2.2 Notations and Definitions

Let $\mathbb{N} = \{1, 2, \cdots\}$ denote the set of positive integers. The notation $x \xleftarrow{\$} S$ denotes that x is selected randomly from set S. If $\mathcal{A}$ is a randomized algorithm, then the notation $x \xleftarrow{\$} \mathcal{A}^{\mathcal{O}_1, \cdots, \mathcal{O}_m}(a_1, a_2, \cdots, a_n)$ denotes that x is assigned the outcome of the experiment of running $\mathcal{A}$, which is permitted to access the oracles $\mathcal{O}_1, \cdots, \mathcal{O}_m$, on inputs $a_1, a_2, \cdots, a_n$.

A graph $G = \langle V, E \rangle$ has a finite set $V \subseteq \mathbb{N}$ of vertices and a finite set $E \subseteq V \times V$ of edges. We write an edge between u and v as the pair (u, v) in any case, whether the graph is directed or undirected. G's *transitive closure* is the graph $\mathbf{cl}(G) = \langle V, \mathbf{cl}(E) \rangle$, where $(u, v) \in \mathbf{cl}(E)$ if and only if there is a path from u to v in G. Considering that E is a binary relation defined over the set V, its transitive closure $\mathbf{cl}(E)$ can be calculated efficiently by using Warshall algorithm and its improved variants[32].

Definition 1 (Transitive Signature[25,33]). *A transitive signature scheme $\mathcal{TS}$ is defined by four polynomial-time algorithms as follows:*

- **TKG**, *the key generation algorithm, is a probabilistic algorithm that takes 1^k as input and returns a pair (tpk, tsk), where k is viewed as the security parameter.*
- **TSign**, *the signature algorithm, is a deterministic or probabilistic algorithm that takes as inputs the secret key tsk and an edge (i, j), and returns an original signature σ_{ij} of (i, j) relative to tsk, where $i, j \in N$.*

- **TVf**, *the verification algorithm, is a deterministic algorithm that takes as inputs tpk, an edge (i, j), and a candidate signature σ_{ij}, and returns either 1 or 0. If the output is 1, σ_{ij} is said to be valid or pass the verification.*
- **Comp**, *the composition algorithm, is a deterministic algorithm that takes as inputs tpk, two edges (i, j) and (j, k), and corresponding signatures σ_{ij} and σ_{jk}, and returns either a composed signature σ_{ik} of edge (i, k) or $\perp$ to indicate failure.*

For a transitive signature scheme, formulating its correctness is indeed a deliberate task[5]. Briefly, a signature σ $(\neq \perp)$ is said to be

- *original*, if it is one of the output of the signing algorithm **TSign**;
- *valid*, if it passes the verification;
- *legitimate*, if it is either original or one of output of the composition algorithm **Comp**.

In this paper, we follow the concept in [5] and define the correctness only for legitimate signatures. In Fig.1, we describe an experiment $\mathbf{Exp}_{TS,A}^{correct}(k)$, by which the correctness of transitive signature is defined. This kind of method for defining correctness of transitive signatures was firstly introduced by Bellare et al.[4] and widely used in [5,33,35].

In this experiment, the boolean value *Legit* is set to **false** if A ever makes an illegitimate oracle query. And the boolean value *NotOK* is set to **true** if one of the following events happens:

- A signature is claimed to be original, but cannot pass the verification.
- A signature is composed legitimately, but cannot pass the verification.
- A signature is composed legitimately, and passes the verification, but is different from the original signature the signer would sign.

We assume that A wins in this experiment if the output is **true**. Then, the correctness of TS can be formally defined as follows.

Definition 2 (Correctness). *A transitive signature scheme TS is correct if, for any k, any (even computationally unbounded) attacker A wins the experiment $\mathbf{Exp}_{TS,A}^{correct}(k)$ with the probability 0.*

In order to give a rigid definition of the security of a transitive signature, we employing the method used in [33]. For a transitive signature scheme TS, we associate it to any polynomial-time forger $\mathcal{F}$ (called tu-cma[2] adversary) and security parameter $k \in \mathbb{N}$ the experiment $\mathbf{Exp}_{TS,\mathcal{F}}^{tu-cma}(k)$ of Fig.2, which provides $\mathcal{F}$ with input tpk and an oracle **TSign**$(tsk, \cdot, \cdot)$. Meanwhile, the oracle is assumed to maintain states. The advantage of $\mathcal{F}$ in this experiment is defined by

$$\mathbf{Adv}_{TS,\mathcal{F}}^{tu-cma}(k) = \mathbf{Pr}[\mathbf{Exp}_{TS,\mathcal{F}}^{tu-cma}(k) = 1]. \tag{1}$$

Then, the security of TS can be formally defined as follows.

---

[2] Abbr. of "transitive unforgeable under adaptive chosen-message attack".

Experiment $\mathbf{Exp}_{TS,A}^{correct}(k)$

1: $(tpk, tsk) \xleftarrow{\$} \mathbf{TKG}(1^k)$;
2: $S \leftarrow \emptyset;\ Legit \leftarrow \mathtt{true};\ NotOK \leftarrow \mathtt{false}$;
3: Run $\mathcal{A}$ with its oracles until it halts, replying to its oracle queries as follows:
4: **if** $\mathcal{A}$ makes **TSign** query on i, j **then**
5:    **if** $i = j$ **then**
6:       $Legit \leftarrow \mathtt{false}$;
7:    **else**
8:       Let σ be the output of the **TSign** oracle;
9:       $S \leftarrow S \cup \{((i,j), \sigma)\}$;
10:      **if** $\mathbf{TVf}(tpk, i, j, \sigma) = 0$ **then**
11:         $NotOK \leftarrow \mathtt{true}$;
12:      **end if**
13:   **end if**
14: **end if**
15: **if** $\mathcal{A}$ makes **Comp** query on $i, j, k, \sigma_1, \sigma_2$ **then**
16:    **if** $\mathbf{TVf}(tpk, i, j, \sigma_1) = 0$ or $\mathbf{TVf}(tpk, j, k, \sigma_2) = 0$ or $|\{i, j, k\}| < 3$ **then**
17:       $Legit \leftarrow \mathtt{false}$;
18:    **else**
19:       Let σ be output of the **Comp** oracle;
20:       $S \leftarrow S \cup \{((j,k), \sigma)\}$;
21:       **if** $\mathbf{TVf}(tpk, i, k, \sigma) = 0$ **then**
22:          $NotOK \leftarrow \mathtt{true}$;
23:       **end if**
24:       $\tau \leftarrow \mathbf{TSign}(tsk, i, k)$;
25:       **if** $\sigma \neq \tau$ **then**
26:          $NotOK \leftarrow \mathtt{true}$;
27:       **end if**
28:    **end if**
29: **end if**
30: When $\mathcal{A}$ halts, output $(Legit \wedge NotOK)$ and halt.

Fig. 1. Experiment for Correctness Definition

Experiment $\mathbf{Exp}_{TS,A}^{tu-cma}(k)$

1: $(tpk, tsk) \xleftarrow{\$} \mathbf{TKG}(1^k)$;
2: $(i', j', \sigma') \xleftarrow{\$} \mathcal{F}^{\mathbf{TSign}(tsk, \cdot, \cdot)}(tpk)$;
3: $S = \{(i, j, \sigma) :$ whenever σ is output by $\mathbf{TSign}(tsk, i, j)$ and given to $\mathcal{F}\}$;
4: $E = \{(i, j) : \exists (i, j, \sigma) \in S\}$;
5: **if** $(i', j') \in \mathbf{cl}(E)$ or $\mathbf{TVf}(tpk, i', j', \sigma') = 0$ **then**
6:    return 0;
7: **else**
8:    return 1;
9: **end if**

Fig. 2. Experiment for Security Definition

Definition 3 (Security). *A transitive signature scheme $\mathcal{TS}$ is unforgeable against adaptive chosen-message attacks if the advantage function $\mathbf{Adv}_{\mathcal{TS},\mathcal{F}}^{tu-cma}(k)$ is negligible (with respect to the security parameter k) for any attacker $\mathcal{F}$ whose running time is polynomial in k.*

Remark 1. Many transitive signature schemes use the node certification paradigm, i.e., containing a underlying standard signature schemes SS=($SKG, SSign, SVf$) for creating node certificates [4,25,28,33,35]. Although Micali and Rivest[25] gave concrete description for SS in their pioneering paper, almost all subsequent researchers provided only abstract definitions for SS[4,28,33]. On the one hand, the involved underlying standard signatures are a significant factor in the cost of the transitive signature schemes. On the other hand, Bellare et al.[5] showed that for many schemes, node certificates can be eliminated by specifying the public label of a node i as the output of a hash function applied to i. Furthermore, the edge label provides an "implicit authentication" of the associated node label that allows us to prove that the revised scheme is transitively unforgeable under adaptive chosen-message attacks, in a model where the hash function is random oracle[5]. In fact, the semantic of "transitivity" is fully manifested by the edge authentication process, but little concerning with the node certificates. Therefore, in this paper, we directly omit the description of the underlying node authentication signature scheme and pay more attention to edge authentication signature schemes. If someone requests a underlying signature scheme for our proposed braid-based transitive signature schemes, then any standard signature scheme that is existentially unforgeable against adaptively chosen message attack (EUF-CMA) is a good candidate. For example, he can employ the braid-based signature schemes SCSS, TCSS or ECSS (See [31] for more details).

3   Transitive Signatures from Braid Groups

Now, we try to sketch out two transitive signature schemes by using braid groups. Suppose that $G = (V, E)$ is the given authentication graph, where $V \subset N$ and $E \subset V \times V$ are node set and edge set, respectively. Let $H : V \to B_n$ be an one-way and collision-resistant hash function which maps a given node in V to a braid in B_n (See [9,18,19] for detail of how to construct the desired hash functions by using braids).

The first braid-based transitive signature scheme, denoted by $\mathcal{BBTS}1$ consists of the following four algorithms:

- **TKG**(n): Suppose that n is the system security parameter. Let

$$(p,q,w) \xleftarrow{\$} \mathrm{K}_{csp}(n) \qquad (2)$$

  and return the public key $tpk = (p,q)$ and the private key $tsk = w$, where $p, q, w \in B_n$ and (p,q) is a CSP-hard pair.
- **TSign**(w,i,j): For a given edge $(i,j) \in E$, its signature is a braid

$$\sigma_{ij} = wb_ib_j^{-1}w^{-1}, \qquad (3)$$

where $b_i = H(i), b_j = H(j)$. Here, we assume without loss of generality that $i < j$. If this is not the case, one can swap i and j.

- **TVf**$(p, q, i, j, \sigma_{ij})$: A signature σ_{ij} for an edge $(i, j) \in E$ is valid (or acceptable equivalently) if and only if

$$(\sigma_{ij} \sim b_i b_j^{-1}) \wedge (q \cdot \sigma_{ij} \sim p \cdot b_i b_j^{-1}) \tag{4}$$

holds, where $b_i = H(i), b_j = H(j)$.

- **Comp**$(p, q, i, j, k, \sigma_{ij}, \sigma_{jk})$: Given two signatures σ_{ij} (for the edge $(i, j) \in E$) and σ_{jk} (for the edge $(j, k) \in E$), if

$$\mathbf{Tvf}(p, q, i, j, \sigma_{ij}) = 0 \text{ or } \mathbf{Tvf}(p, q, j, k, \sigma_{jk}) = 0, \tag{5}$$

then return $\perp$ as an indication of failure; Otherwise, the signature on the edge (i, k) would be a braid

$$\sigma_{ik} = \sigma_{ij} \cdot \sigma_{jk}. \tag{6}$$

Here, we again assume that $i < j < k$. If this is not the case, one can resort them.

Remark 2. In **TSign** algorithm, if we do not consider wether $i < j$ hold, then we have $\sigma_{ij} \neq \sigma_{ji}$ in general. But this fact never means that we reach a directed transitive signature scheme, since one can get σ_{ji} easily by computing the inverse braid of σ_{ij}.

Before proving the correctness and the security of the above scheme, let us describe another braid-based transitive signature scheme, denoted by $\mathcal{BBTS}2$:

- **TKG**(n): Suppose that n is the system security parameter. Let

$$(p, q, w) \xleftarrow{\$} K_{csp}(n). \tag{7}$$

Pick another complex braid $c \in B_n$ and compute $d = wpcw^{-1}$. Then, return the public key $tpk = (p, q, c; d)$ and the private key $tsk = w$, where $p, q, c, d, w \in B_n$ and both (p, q) and (pc, d) are CSP-hard pairs.

- **TSign**(w, i, j): For a given edge $(i, j) \in E$, its signature is a braid

$$\sigma_{ij} = wpb_i b_j^{-1} cw^{-1}, \tag{8}$$

where $b_i = H(i), b_j = H(j)$. We also assume that $i < j$.

- **TVf**$(p, q, c, i, j, \sigma_{ij})$: A signature σ_{ij} for an edge $(i, j) \in E$ is valid (or acceptable equivalently) if and only if

$$(\sigma_{ij} \sim pb_i b_j^{-1} c) \wedge (q^{-1} \cdot \sigma_{ij} \sim b_i b_j^{-1} c), \tag{9}$$

where $b_i = H(i), b_j = H(j)$.

- **Comp**$(p, q, c, d, i, j, k, \sigma_{ij}, \sigma_{jk})$: Given two signatures σ_{ij} (for the edge $(i, j) \in E$) and σ_{jk} (for the edge $(j, k) \in E$), if

$$\mathbf{Tvf}(p, q, c, i, j, \sigma_{ij}) = 0 \text{ or } \mathbf{Tvf}(p, q, c, j, k, \sigma_{jk}) = 0, \tag{10}$$

then return $\perp$ as an indication of failure; Otherwise, the signature on the edge (i, k) would be a braid

$$\sigma_{ik} = \sigma_{ij} \cdot d^{-1} \cdot \sigma_{jk}. \tag{11}$$

We also assume that $i < j < k$.

Remark 3. To some extend, $\mathcal{BBTS}2$ is a simple variant of $\mathcal{BBTS}1$. On the one hand, $\mathcal{BBTS}2$ has a bit less efficient than $\mathcal{BBTS}1$: In **Tvf** and **Comp**, $\mathcal{BBTS}2$ needs an additional multiplication (Note that both wp and cw^{-1} can be pre-computed). On the other hand, $\mathcal{BBTS}2$ has a remarkable new trait: one cannot perform the composition work without possession of d or w. We know that w is the private key to which nobody, except the signer himself, can fetch. However, we can send d to a semi-trusted party who will fulfil the composition work. Note that d is not involved into the verification process. This does not violate with the standard definition of transitive signature, since d is also a component of public key tpk.

Now, it is time to consider the correctness and the security of the proposed schemes. For space limitation, we just list the results as follows. All corresponding proofs would be found in the journal version.

Theorem 1. *Both the proposed braid-based transitive signature schemes, $\mathcal{BBTS}1$ and $\mathcal{BBTS}2$ meet the correctness requirement of Definition 2.*

Theorem 2. *In the random oracle model, the first braid-based transitive signature scheme $\mathcal{BBTS}1$ is transitively unforgeable against adaptively chosen message attack assuming that the one-more matching conjugate problem (OM-MCP) is intractable. More specifically, suppose that there is a tu-cma forger $\mathcal{F}$ that makes at most q_h hash queries and at most q_s signing queries, and finally breaks $\mathcal{BBTS}1$ with a non-negligible advantage ϵ within time t, then there exists an om-mcp attacker $\mathcal{A}$ can win the one-more matching conjugate experiment with the probability at least ϵ' within the time t', where*

$$\epsilon' = \epsilon, \tag{12}$$

and

$$t' = t + t_s \cdot q_s + t_h \cdot q_h + t_{mc} \cdot (|V| - 1), \tag{13}$$

where $|V|$ is the number of node of authentication graph, while t_s, t_h and t_{mc} are time for answering a signing oracle query, a hash oracle query and a matching conjugate oracle query, respectively.

Theorem 3. *The second braid-based transitive signature scheme $\mathcal{BBTS}2$ is transitively unforgeable against adaptively chosen message attack assuming that the first braid-based transitive signature scheme $\mathcal{BBTS}1$ is secure in the same sense.*

4 Security Level, Performance Evaluation and Parameters Suggestion

Since the birth of braid group cryptography, it has fascinated many cryptologists with its hard problems and efficient algorithms for parameter generation and group operations[20]. In [18], the upper bound of canonical length for working braids, denoted by l, is very small. Ko et al. set $l = 3$ and $n = 20, 24, 28$ respectively. Their evaluation results[3] say that key generation time and signing time are about 22ms, while verifying time is ranging from 30ms to 60ms. This performance seems *not bad*. However, we think setting $l = 3$ is maybe insecure. According to Myasnikov et al.'s suggestions in [26], setting $l = n^2$ may be a better choice. Of course, with a larger l, the performance of braid operations has to be re-evaluated.

In 2001, Cha et al. [7] gave an efficient implementation of all required operations of braid groups for cryptographic applications. The complexities of all these operations are bounded by $\mathcal{O}(l^2 n \log n)$. In this paper, we use the braid index n as the sole parameter and adopt the setting of $l = n^2$. Thus, the complexities of all operations in our proposals are bounded by $\mathcal{O}(n^5 \log n)$.

According to [18], the security level of the proposed schemes against currently known heuristic attacks is about $(n/4)^{n(n-1)/4}$. Taking the complexity $\mathcal{O}(n^5 \log n)$ into consideration, this seems *a shade worse* than RSA-based transitive signature schemes[5], in which the complexities are bounded by $\mathcal{O}(k^2 \log k)$ and the security level is evaluated with $2^{\mathcal{O}(k)}$ (where k is the bit-length of the RSA modular). However, we know that at current secure RSA modular requires that k is at least 1024, then $\mathcal{O}(k^2 \log k) \approx \mathcal{O}(10^7)$; As for braid-based scheme, if we set $n = 16$, then $(n/4)^{n(n-1)/4} = 2^{120}$, which overwhelms current conventional attacks, and $\mathcal{O}(n^5 \log n) \approx \mathcal{O}(10^6)$, which says that in practice, braid-based schemes maybe are *a shade better* than those RSA-based schemes in [5].

According to [19], the security level of braid-based schemes against brute force attacks is at least $(\lfloor \frac{n-1}{2} \rfloor!)^l$, where n and l are the index of the braid group and the upper bound of the canonical length of the working braids, respectively. Thus, when the setting $l = n^2$ is adopted, the security level of braid-based schemes can be evaluated by $\exp(\mathcal{O}(n^3 \log n))$, according to Stirling's approximation. Apparently, this result shows that braid-based schemes have *huge* potential advantages in security level against brute force attacks over RSA-based ones. For example, if we set $n = 16$ and $k = 1024$ for the braid-based schemes and RSA-based ones respectively, then we have $\exp(n^3 \log n) = \exp(2^{14}) > 2^{16384} \gg 2^k = 2^{1024}$.

In addition, braid-based schemes are invulnerable to currently known quantum attacks. At present, quantum algorithms fall roughly into three categories[3]: quantum search algorithms, hidden subgroup finding techniques, and hybrid quantum algorithms. None of these algorithms can solve the conjugator search problem over braid groups in polynomial time. In 2003, Michael Anshel[3] suggested: by employing Lawrence-Krammer representation[8] with suitable choice of braid group parameters, the number of elementary field operations required to

---

[3] Their experiment was done on a computer with a Pentium III 866MHz processor[18].

solve the associated systems of linear equations overwhelms current conventional attacks and provides a challenge for quantum cryptanalysis.

5   Conclusions

In this paper, we design two transitive signature schemes by using braid groups. As far as we known, this is the first attempt toward braid-based transitive signatures. In the random oracle model, our proposals are proved to be transitively unforgeable against adaptively chosen message attack assume under the assumption of the intractability of one-more matching conjugate problem (OM-MCP) over braid group. Performance analysis and security evaluation show that our proposals have potential advantages over RSA-based transitive signatures. Moreover, the proposed schemes are invulnerable to currently known quantum attacks.

Acknowledgments

This work is supported by Sony (China) research laboratory, National Natural Science Foundation of China under grant Nos. 90604022 and 60673098, and National Basic Research Program of China (973 Program) under grant No. 2007CB310704.

The author Cao is also supported by National Natural Science Foundation of China under grant Nos: 60673079, 60773086 and 60572155.

References

1. Anshel, I., Anshel, M., Fisher, B., Goldfeld, D.: New Key Agreement Protocols in Braid Group Cryptography. In: Naccache, D. (ed.) CT-RSA 2001. LNCS, vol. 2020, pp. 13–27. Springer, Heidelberg (2001)
2. Anshel, I., Anshel, M., Goldfeld, D.: An algebraic method for public-key cryptography. Math. Research Letters 6, 287–291 (1999)
3. Anshel, M.: Braid Group Cryptography and Quantum Cryptoanalysis. In: 8th International Wigner Symposium, May 27-30, 2003, GSUC-CUNY 365 Fifth Avenue, NY, NY 10016, USA (2003)
4. Bellare, M., Neven, G.: Transitive signaures based on factoring and RSA. In: Zheng, Y. (ed.) ASIACRYPT 2002. LNCS, vol. 2501, pp. 397–414. Springer, Heidelberg (2002)
5. Bellare, M., Neven, G.: Transitive signatures: New schemes and proofs. IEEE Transactions on Information Theory 51(6), 2133–2151 (2005)
6. Bellare, M., Namprempre, C., Pointcheval, D., Semanko, M.: The One-More-RSA-Inversion Problems and the Security of Chaum's Blind Signature Scheme. Journal of Cryptology 16(3), 185–215 (2003)
7. Cha, J.C., Ko, K.H., Lee, S.J., Han, J.W., Cheon, J.H., et al.: An efficient implementation of braid groups. In: Boyd, C. (ed.) ASIACRYPT 2001. LNCS, vol. 2248, pp. 144–156. Springer, Heidelberg (2001)

8. Cheon, J.H., Jun, B.: A Polynomial Time Algorithm for the Braid Diffie-Hellman Conjugacy Problem. In: Boneh, D. (ed.) CRYPTO 2003. LNCS, vol. 2729, pp. 212–225. Springer, Heidelberg (2003)
9. Dehornoy, P.: Braid-based cryptography. Contemp. Math., Amer. Math. Soc. 360, 5–33 (2004)
10. Elrifai, E., Morton, H.R.: Algorithms for positive braids. Quart. J. Math. Oxford Ser. 45(2), 479–497 (1994)
11. Franco, N., Gonzales-Menses, J.: Conjugacy problem for braid groups and garside groups. Journal of Algebra 266, 112–132 (2003)
12. Gebhardt, V.: A new approach to the conjugacy problem in garside groups. Journal of Algebra 292, 282–302 (2005)
13. Gonzales-Meneses, J.: Improving an algorithm to solve the multiple simultaneous conjugacy problems in braid groups, Preprint, math.GT/0212150 (2002)
14. Huang, Z.-J., Hao, Y.-H., Wang, Y.-M., Chen, K.-F.: Efficient directed transitive signature scheme. Acta Electronica Sinica 33(8), 1497–1501 (2005)
15. Hughes, J.: The left SSS attack on Ko-Lee-Cheon-Han-Kang-Park key agreement scheme in B45, Rump session Crypto (2000)
16. Hughes, J.: A linear algebraic attack on the AAFG1 braid group cryptosystem. In: Batten, L.M., Seberry, J. (eds.) ACISP 2002. LNCS, vol. 2384, pp. 176–189. Springer, Heidelberg (2002)
17. Kitaev, A.: Quantum measurements and the abelian stabilizer problem. Preprint, quant-ph/9511026 (1995)
18. Ko, K.H., Choi, D.H., Cho, M.S., Lee, J.W.: New signature scheme using conjugacy problem (preprint 2002), http://eprint.iacr.org/2002/168
19. Ko, K.H., Lee, S.J., Cheon, J.H., Han, J.W.: New public-key cryptosystem using braid groups. In: Bellare, M. (ed.) CRYPTO 2000. LNCS, vol. 1880, pp. 166–183. Springer, Heidelberg (2000)
20. Ko, K.H., Lee, J.W., Thomas, T.: Towards generating secure keys for braid cryptography, Designs, Codes and Cryptography (to appear, 2007)
21. Kuwakado, H., Tanaka, H.: Transitive Signature Scheme for Directed Trees. IEICE Transactions on Fundamentals of Electronics, Communications and Computer Sciences E86-A(5), 1120–1126 (2003)
22. Lee, E.: Braig groups in cryptography. IEICE Trans. Fundamentals E87-A(5), 986–992 (2004)
23. Lee, E., Lee, S.-J., Hahn, S.-G.: Pseudorandomness from Braid Groups. In: Kilian, J. (ed.) CRYPTO 2001. LNCS, vol. 2139, pp. 486–502. Springer, Heidelberg (2001)
24. Magliveras, S., Stinson, D., van Trung, T.: New approaches to designing public key cryptosystems using one-way functions and trapdoors in finite groups. Journal of Cryptography 15, 285–297 (2002)
25. Micali, S., Rivest, R.L.: Transitive signaure schemes. In: Preneel, B. (ed.) CT-RSA 2002. LNCS, vol. 2271, pp. 236–243. Springer, Heidelberg (2002)
26. Myasnikov, A., Shpilrain, V., Ushakov, A., Practical, A.: Attack on a Braid Group Based Cryptographic Protocol. In: Shoup, V. (ed.) CRYPTO 2005. LNCS, vol. 3621, pp. 86–96. Springer, Heidelberg (2005)
27. Proos, J., Zalka, C.: Shors discrete logarithm quantum algorithm for elliptic curves. Quantum Information and Computation 3, 317–344 (2003)
28. Shahandashti, S.F., Salmasizadeh, M., Mohajeri, J.: A provably secure short transitive signature scheme from bilinear group Pairs. In: Blundo, C., Cimato, S. (eds.) SCN 2004. LNCS, vol. 3352, pp. 60–76. Springer, Heidelberg (2005)
29. Shor, P.: Polynomail-time algorithms for prime factorization and discrete logarithms on a quantum computer. SIAM J. Comput. 5, 1484–1509 (1997)

30. Wang, L.: PhD. Disseration. Shanghai Jiao Tong University (June 2007)
31. Wang, L., Cao, Z., Zeng, P., Li, X.: One-more matching conjugate problem and security of braid-based signatures. In: ASIACCS 2007, pp. 295–301. ACM, New York (2007)
32. Warren Jr., Henry, S.: A modification of Warshall's algorithm for the transitive closure of binary relations. Communications of the ACM 18(4), 218–220 (1975)
33. Yi, X.: Directed transitive signature scheme. In: Abe, M. (ed.) CT-RSA 2007. LNCS, vol. 4377, pp. 129–144. Springer, Heidelberg (2006)
34. Yi, X., Tan, C.-H., Okamoto, E.: Security of Kuwakado-Tanaka transitive signature scheme for directed trees. IEICE Transactions on Fundamentals of Electronics, Communications and Computer Sciences E87-A(4), 955–957 (2004)
35. Zhu, H.: Model for undirected transitive signatures. IEE Proceedings: Communications 151(4), 312–315 (2004)

Proxy Re-signature Schemes Without Random Oracles*

Jun Shao, Zhenfu Cao**, Licheng Wang, and Xiaohui Liang

Department of Computer Science and Engineering
Shanghai Jiao Tong University
200240, Shanghai, P.R. China
chn.junshao@gmail.com, zfcao@cs.sjtu.edu.cn, wanglc@sjtu.edu.cn,
liangxh127@sjtu.edu.cn

Abstract. To construct a suitable and secure proxy re-signature scheme is not an easy job, up to now, there exist only three schemes, one is proposed by Blaze *et al.* [6] at EUROCRYPT 1998, and the others are proposed by Ateniese and Hohenberger [2] at ACM CCS 2005. However, none of these schemes is proved in the standard model (i.e., do not rely on the random oracle heuristic). In this paper, based on Waters' approach [20], we first propose a multi-use bidirectional proxy re-signature scheme, denoted as S_{mb}, which is existentially unforgeable in the standard model. And then, we extend S_{mb} to be a multi-use bidirectional ID-based proxy re-signature scheme, denoted by S_{id-mb}, which is also existentially unforgeable in the standard model. Both of these two proposed schemes are computationally efficient, and their security bases on the Computational Diffie-Hellman (CDH) assumption.

Keywords: proxy re-signature, standard model, ID-based, bilinear maps, existential unforgeability.

1   Introduction

Proxy re-signature schemes, introduced by Blaze, Bleumer, and Strauss [6], and formalized later by Ateniese and Hohenberger [2], allow a semi-trusted proxy to transform a delegatee's signature into a delegator's signature on the same message by using some additional information. The proxy, however, cannot generate arbitrary signatures on behalf of either the delegatee or the delegator. Generally speaking, a proxy re-signature scheme has eight desirable properties [2], though none of existing schemes satisfies all properties, see Table 1.

1. Unidirectional: In an *unidirectional* scheme, a re-signature key allows the proxy to transform A's signature to B's but not vice versa. In a *bidirectional*

---

* Supported by National Natural Science Foundation of China, No. 60673079 and No. 60572155, Research Fund for the Doctoral Program of Higher Education, No. 20060248008.
** Corresponding author.

K. Srinathan, C. Pandu Rangan, M. Yung (Eds.): Indocrypt 2007, LNCS 4859, pp. 197–209, 2007.

scheme, on the other hand, the re-signature key allows the proxy to transform A's signature to B's as well as B's signature to A's.

2. **Multi-use:** In a *multi-use* scheme, a transformed signature can be re-transformed again by the proxy. In a *single-use* scheme, the proxy can transform only the signatures that have not been transformed.

3. **Private Proxy:** The re-signature key can be kept secret by the proxy in a *private proxy* scheme, but can be recomputed by observing the proxy passively in a *public proxy* scheme.

4. **Transparent:** In a *transparent* scheme, a signature on the same message signed by the delegator is computationally indistinguishable from a signature transformed by a proxy.

5. **Key-Optimal:** In a *key-optimal* scheme, a user is required to protect and store only a small constant amount of secrets no matter how many signature delegations the user gives or accepts.

6. **Non-interactive:** The delegatee is not required to participate in a delegation process.

7. **Non-transitive:** A re-signing right cannot be re-delegated by the proxy alone.

8. **Temporary:** A re-signing right is temporary.

Table 1. The properties that the existing proxy re-signature schemes and ours satisfy

Property	BBS [6]	S_{bi} [2]	S_{uni} [2]	S_{mb}	S_{id-mb}
1.	No	No	Yes	No	No
2.	Yes	Yes	No	Yes	Yes
3.	No	Yes	No	Yes	Yes
4.	Yes	Yes	Yes	Yes	Yes
5.	Yes	Yes	Yes	Yes	Yes
6.	No	No	Yes	No	No
7.	No	No	Yes	No	No
8.	No	No	Yes	No	No

Due to the transformation function, proxy re-signature schemes are very useful and can be applied in many applications, including simplifying key management [6], providing a proof for a path that has been taken, managing group signatures, simplifying certificate management [2], constructing a Digital Rights Management (DRM) interoperable system [19]. However, as mentioned in [2], "Finding suitable and secure proxy re-signature schemes required a substantial effort. Natural extensions of several standard signatures were susceptible to the sort of problems." To our best knowledge, there are only three proxy re-signature schemes, the first one is a *bidirectional, multi-use,* and *public proxy* scheme, proposed by Blaze, Bleumer and Strauss at Eurocrypt 1998 [6], and the left two are both proposed by Ateniese and Hohenberger at ACM CCS 2005 [2]. One of them is a *multi-use bidirectional* scheme, and the other is a *single-use unidirectional* scheme.

However, there exist two disadvantages in the above three schemes.

- All of these three schemes are only proven secure in the random oracle model, i.e., the proof of security relies on the random oracle heuristic. However, it has been shown that some schemes are proven secure in the random oracle model, but are trivially insecure under any instantiation of the oracle [9,5]. Up to now, there are many signatures proven secure in the standard model, such as [10,12,3,4,20,21]. It is natural to ask whether we can construct a new proxy re-signature scheme which can be proved in the standard model.
- The public keys in these three schemes are arbitrary strings unrelated to their owner's identity. A certificate issued by an authority is needed to bind the public key to its owner's identity before the public key is used by others. This creates complexity of certificate management, though proxy re-signature schemes can be used to simplify certificate management. A natural solution to this disadvantage is to apply ID-based cryptography [17]. In ID-based cryptography, a user's unique ID such as an email address is also the user's public key. The corresponding private key is computed from the public key by a Private Key Generator (PKG) who has the knowledge of a master secret. As a result, complexity of certificate management can be eliminated. We can use the method in [11] to convert any proxy re-signature into an ID-based proxy re-signature. However, as mentioned in [15], this method expands the size of signature, and increases the complexity of verification. We hope that we get an ID-based proxy re-signature by a direct construction.

In this paper, we attempt to propose a new proxy re-signature scheme which recovers the above two disadvantages.

1.1 Our Contribution

In this paper, based on Waters' approach [20], we first propose the first proxy re-signature scheme which is existentially unforgeable in the standard model, we denote it as S_{mb}. S_{mb} satisfies bidirectional, multi-use, private proxy, transparent properties. And then we proposed the first ID-based proxy re-signature which is existentially unforgeable in the standard model, we denote it as S_{id-mb}. S_{id-mb} also satisfies bidirectional, multi-use, private proxy, transparent properties. Actually, S_{id-mb} can be considered as an ID-based extension of S_{mb}. As the schemes in [20], both of our proposed schemes are constructed in bilinear groups, and proven secure under the Computational Diffie-Hellman (CDH) assumption. The only drawback of our proposed schemes is the relatively large size of its public parameters inheriting from Waters' approach [20]. However, we can use the techniques of Naccache [14] and Sarkar and Chatterjee [16] to reduce the size of the public parameters.

1.2 Paper Organization

The remaining paper is organized as follows. In Section 2, we review the definitions of (ID-based) proxy re-signatures and their security. And then we present

S_{mb}, S_{id-mb} and their security proofs in Section 3. Finally, We conclude the paper in Section 4.

2   Definitions

The security notions in this section are all for existential unforgeablility under an adaptive chosen message (and identity) attack. That is, a valid forgery should be a valid signature on a new message, which is not signed by the singer before. These security models can be easily extended to cover strong unforgeability [1], where a valid forgery should be a valid signature which is not computed by the signer. However, our concrete schemes do not enjoy security in this stronger sense, since an adversary can easily modify existing signatures into new signatures on same message.

2.1   Bidirectional Proxy Re-signature

In this subsection, we briefly review the definitions about bidirectional proxy re-signatures. The security notion in this subsection is for existential unforgeability under an adaptive chosen message attack, which is weaker than that in [2]. We refer the reader to [2] for details.

Definition 2.1. *A bidirectional proxy re-signature scheme is a tuple of (possibly probabilistic) polynomial time algorithms* (KeyGen, ReKey, Sign, ReSign, Verify), *where:*

- (KeyGen, Sign, Verify) are the same as those in the standard digital signatures[1].
- On input (sk_A, sk_B), the re-signature key generation algorithm, ReKey, outputs a key $rk_{A \leftrightarrow B}$ for the proxy, where sk_A and sk_B are the secret key of A and B, respectively.
- On input $rk_{A \leftrightarrow B}$, a public key pk_A, a message m, and a signature σ, the re-signature function, ReSign, outputs a new signature σ' on message m corresponding to pk_B, if Verify$(pk_A, m, \sigma) = 1$ and $\bot$ otherwise.

Correctness. For any message m in the message space and any key pairs $(pk, sk), (pk', sk') \leftarrow$ KeyGen(1^k), let $\sigma =$ Sign(sk, m) and $rk \leftarrow$ ReKey(sk, sk'). Then the following two conditions must hold:

$$\text{Verify}(pk, m, \sigma) = 1 \quad \text{and} \quad \text{Verify}(pk', m, \text{ReSign}(rk, pk, m, \sigma)) = 1.$$

Unlike the security notion in [2], we define security for bidirectional proxy re-signature schemes by the following game between a challenger and an adversary: (Note that we adopt the method in [8] to define the security notion of bidirectional proxy re-encryption schemes: static corruption, i.e., in this security notion, the adversary has to determine the corrupted parties before the computation starts, and it does not allow adaptive corruption of proxies between corrupted and uncorrupted parties.)

---

[1] For the definition of standard digital signatures, we refer the reader to [13].

Queries. The adversary adaptively makes a number of different queries to the challenger. Each query can be one of the following.

- Uncorrupted Key Generation $\mathcal{O}_{UKeyGen}$: Obtain a new key pair as $(pk, sk) \leftarrow \mathsf{KeyGen}(1^k)$. The adversary is given pk.
- Corrupted Key Generation $\mathcal{O}_{CKeyGen}$: Obtain a new key pair as $(pk, sk) \leftarrow \mathsf{KeyGen}(1^k)$. The adversary is given pk and sk.
- Re-Signature key Generation $\mathcal{O}_{ReKey}$: On input (pk, pk') by the adversary, where pk, pk' were generated before by KeyGen, return the re-signature key $rk_{pk \leftrightarrow pk'} = \mathsf{ReKey}(sk, sk')$, where sk, sk' are the secret keys that correspond to pk, pk'. Like the security notion in [8], here, we also require that both pk and pk' are corrupted, or both are uncorrupted.
- Re-signature $\mathcal{O}_{ReSign}$: On input (pk, pk', m, σ), where pk, pk' were generated before by KeyGen. The adversary is given the re-signed signature $\sigma' = \mathsf{ReSign}(\mathsf{ReKey}(sk, sk'), pk, m, \sigma)$, where sk, sk' are the secret keys that correspond to pk, pk'.
- Signature $\mathcal{O}_{Sign}$: On input a public key pk, a message m, where pk was generated before by KeyGen. The adversary is given the corresponding signature $\sigma = \mathsf{Sign}(sk, m)$, where sk is the secret key that correspond to pk.

Forgery. The adversary outputs a message m^*, a public key pk^*, and a string σ^*. The adversary succeeds if the following hold true:

1. $\mathsf{Verify}(pk^*, m^*, \sigma^*) = 1$.
2. pk^* is not from $\mathcal{O}_{CKeyGen}$.
3. (pk^*, m^*) is not a query to $\mathcal{O}_{Sign}$.
4. $(\Diamond, pk^*, m^*, \blacklozenge)$ is not a query to $\mathcal{O}_{ReSign}$, where $\Diamond$ denotes any public key, and $\blacklozenge$ denotes any signature.

The advantage of an adversary $\mathcal{A}$ in the above game is defined to be $\mathrm{Adv}_{\mathcal{A}} = \Pr[\mathcal{A} \text{ succeeds}]$, where the probability is taken over all coin tosses made by the challenger and the adversary.

2.2 Bidirectional ID-Based Proxy Re-signature

Definition 2.2 (Bidirectional ID-based Proxy Re-Signature). *A Bidirectional ID-based proxy re-signature scheme* $\mathcal{S}$ *consists of the following six random algorithms:* Setup, Extract, ReKey, Sign, ReSign, *and* Verify *where:*

- (Setup, Extract, Sign, Verify) are the same as those in a standard ID-based signature[2].
- On input (d_A, d_B), the re-signature key generation algorithm, ReKey, outputs a key $rk_{A \leftrightarrow B}$ for the proxy, where d_A (d_B) is A's (B's) secret key.
- On input $rk_{A \leftrightarrow B}$, an identity ID_A, a message m, and a signature σ, the re-signature algorithm, ReSign, outputs a new signature σ' on message m corresponding to ID_B, if $\mathsf{Verify}(ID_A, m, \sigma) = 1$ and $\perp$ otherwise.

---

[2] For the definition of ID-based signatures, we refer the reader to [17].

Correctness: This is the same as that in standard proxy re-signature schemes. The following property must be satisfied for the correctness of a proxy re-signature: For any message m in the message space and any two key pairs (ID_A, d_A), and (ID_B, d_B), let $\sigma_A = \text{Sign}(d_A, m)$ and $rk_{A \leftrightarrow B} \leftarrow \text{Rekey}(d_A, d_B)$, the following two equations must hold:

$$\text{Verify}(ID_A, m, \sigma_A) = 1, \text{ and } \text{Verify}(ID_B, m, \text{ReSign}(rk_{A \leftrightarrow B}, ID_A, m, \sigma_A)) = 1.$$

We also define the security notion of bidirectional ID-based proxy re-signature with static corruption by a game between a challenger and an adversary.

Setup. The challenger runs Setup and obtains both the public parameters *params* and the master secret mk. The adversary is given *params* but the master secret mk is kept by the challenger.

Queries. The adversary adaptively makes a number of different queries to the challenger. Each query can be one of the following.

- Extract oracle for corrupted parties $O_{Extract}$: On input an identity ID by the adversary, the challenger responds by running $\text{Extract}(mk, ID)$. and sends the resulting private key d_{ID} to the adversary.

- Re-Signature key Generation $\mathcal{O}_{ReKey}$: On input (ID_A, ID_B) by the adversary, the challenger returns the re-signature key

$$rk_{A \leftrightarrow B} = \text{ReKey}(\text{Extract}(mk, ID_A), \text{Extract}(mk, ID_B)).$$

Here, we also require that both ID_A and ID_B are corrupted, or both are uncorrupted.

- Re-signature $\mathcal{O}_{ReSign}$: On input (ID_A, ID_B, m, σ), the adversary is given the re-signed signature

$$\sigma' = \text{ReSign}(\text{ReKey}(\text{Extract}(mk, ID_A), \text{Extract}(mk, ID_B)), ID_A, m, \sigma).$$

- Signature $\mathcal{O}_{Sign}$: On input an identity ID, a message m. The adversary is given the corresponding signature $\sigma = \text{Sign}(\text{Extract}(mk, ID), m)$.

Forgery. The adversary outputs a message m^*, an identity ID^*, and a string σ^*. The adversary succeeds if the following hold:

1. $\text{Verify}(pk^*, m^*, \sigma^*) = 1$.
2. ID^* is uncorrupted.
3. (ID^*, m^*) is not a query to $\mathcal{O}_{Sign}$.
4. $(\Diamond, ID^*, m^*, \blacklozenge)$ is not a query to $\mathcal{O}_{ReSign}$, where $\Diamond$ denotes any identity, and $\blacklozenge$ denotes any signature.

The advantage of an adversary $\mathcal{A}$ in the above game is defined to be $\text{Adv}_{\mathcal{A}} = \Pr[\mathcal{A} \text{ succeeds}]$, where the probability is taken over all coin tosses made by the challenger and the adversary.

2.3   Bilinear Maps

In this subsection, we briefly review definitions about bilinear maps and bilinear map groups, which follow that in [7].

1. $\mathbb{G}_1$ and $\mathbb{G}_2$ are two (multiplicative) cyclic groups of prime order p;
2. g is a generator of $\mathbb{G}_1$;
3. e is a bilinear map $e : \mathbb{G}_1 \times \mathbb{G}_1 \to \mathbb{G}_2$.

Let $\mathbb{G}_1$ and $\mathbb{G}_2$ be two groups as above. An *admissible bilinear map* is a map $e : \mathbb{G}_1 \times \mathbb{G}_1 \to \mathbb{G}_2$ with the following properties:

1. *Identity:* For all $P \in \mathbb{G}_1$, $e(P, P) = 1$;
2. *Alternation:* For all $P, Q \in \mathbb{G}_1$, $e(P, Q) = e(Q, P)^{-1}$;
3. *Bilinearity:* For all $P, Q, R \in \mathbb{G}_1$, $e(P \cdot Q, R) = e(P, R) \cdot e(Q, R)$ and $e(P, Q \cdot R) = e(P, Q) \cdot e(P, R)$.
4. *Non-degeneracy:* If $e(P, Q) = 1$ for all $Q \in G_1$, then $P = \mathcal{O}$, where $\mathcal{O}$ is a point at infinity.

We say that $\mathbb{G}_1$ is a bilinear group if the group action in $\mathbb{G}_1$ can be computed efficiently and there exists a group $\mathbb{G}_2$ and an efficiently computable bilinear map as above.

2.4   The Computational Diffie-Hellman Assumption (CDH)

Computational Diffie-Hellman Problem. Let $\mathbb{G}$ be a group of prime order p and let g be a generator of $\mathbb{G}$. The CDH problem is as follows: Given $\langle g, g^a, g^b \rangle$ for some $a, b \in \mathbb{Z}_p^*$ compute g^{ab}. An algorithm $\mathcal{A}$ has advantage ε in solving CDH in $\mathbb{G}$ if

$$\Pr[A(g, g^a, g^b) = g^{ab}] \geq \varepsilon$$

where the probability is over the random choice of a, b in $\mathbb{Z}_p^*$, the random choice of $g \in \mathbb{G}^*$, and the random bits of $\mathcal{A}$.

Definition 2.3. *We say that the (ε, t)-CDH assumption holds in $\mathbb{G}$ if no t-time algorithm has advantage at least ε in solving the CDH problem in $\mathbb{G}$.*

3   Bidirectional Proxy Re-signature Schemes

3.1 S_{mb}: Multi-use Bidirectional Scheme

We now present a new multi-use bidirectional proxy re-signature scheme, denoted as S_{mb}, using the signature scheme due to Waters [20]. This scheme requires a bilinear map, as discussed in Section 2. We assume that the messages can be represented as bit strings of length n_m, which is unrelated to p. We can achieve this by a collision-resistant hash function $H : \{0, 1\}^* \to \{0, 1\}^{n_m}$.

KeyGen: On input the security parameter 1^k, it chooses two groups $\mathbb{G}_1$ and $\mathbb{G}_2$ of prime order $p = \Theta(2^k)$, such that an admissible pairing $e : \mathbb{G}_1 \times \mathbb{G}_1 \to \mathbb{G}_2$ can be constructed and chooses a generator g of $\mathbb{G}_1$. Furthermore, it selects a random a from $\mathbb{Z}_p$, and $n_m + 2$ random number $(g_2, u', u_1, \cdots, u_{n_m})$ from $\mathbb{G}_1$, and output the key pair $pk = g_1 = g^a$ and $sk = a$, the public parameters $(\mathbb{G}_1, \mathbb{G}_2, e, g_2, u', u_1, \cdots, u_{n_m})$.

ReKey: On input two secret keys $sk_A = a$, $sk_B = b$, output the re-signature key $rk_{A \to B} = b/a \bmod p$.

   (Note that we make use of the same method and assumptions in [2] to get the re-signature key, we refer the reader to [2][Section 3.3] for details.)

Sign: On input a secret key $sk = a$ and a n_m-bit message m, output $\sigma = (\mathfrak{A}, \mathfrak{B}) = (g_2^a \cdot w^r, g^r)$, where r is chosen randomly from $\mathbb{Z}_p$, and $w = u' \cdot \prod_{i \in \mathcal{U}} u_i$, $\mathcal{U} \subset \{1, \ldots, n_m\}$ is the set of indices i such that $m[i] = 1$, and $m[i]$ is the i-th bit of m.

ReSign: On input a re-signature key $rk_{A \to B}$, a public key pk_A, a signature σ_A, and a n_m-bit message m, check that $\mathtt{Verify}(pk_A, m, \sigma_A) = 1$. If σ_A does not verify, output $\perp$; otherwise, output $\sigma_B = \sigma_A^{rk_{A \to B}} = (g_2^b \cdot w^{rb/a}, g^{rb/a}) = (g_2^b w^{r'}, g^{r'})$, where $r' = rb/a \bmod p$.

Verify: On input a public key pk, a n_m-bit message m, and a purported signature $\sigma = (\mathfrak{A}, \mathfrak{B})$, output 1, if $e(pk, g_2)e(\mathfrak{B}, w) = e(\mathfrak{A}, g)$ and 0 otherwise.

Theorem 3.1 (Security of S_{mb}). *In the standard model, bidirectional proxy re-signature scheme S_{mb} is correct and existentially unforgeable under the Computational Diffie-Hellman (CDH) assumption in $\mathbb{G}_1$; that is, for random $g \in \mathbb{G}_1$, and $x, y \in \mathbb{Z}_p^*$, given (g, g^x, g^y), it is hard to compute g^{xy}.*

Proof. The correctness property is easily observable. We show security following the approaches in [20,15], especially the one in [15].

   If there exists an adversary $\mathcal{A}$ that can break the above proxy re-signature scheme with non-negligible probability ε in time t after making at most q_S sign queries, q_{RS} resign queries, q_K (un)corrupted key queries, and q_{RK} rekey queries, then there also exists an adversary $\mathcal{B}$ that can solve the CDH problem in $\mathbb{G}_1$ with probability $\frac{\varepsilon}{4(q_S + q_{RS})(n_m + 1)}$ in time $t + O((q_S + q_{RS})n_m \rho + (q_S + q_{RS} + q_K)\tau)$, where ρ and τ are the time for a multiplication and an exponentiation in $\mathbb{G}_1$, respectively.

   On input (g, g^a, g^b), the CDH adversary $\mathcal{B}$ simulates a bidirectional proxy re-signature security game for $\mathcal{A}$ as follows:

   To prepare the simulation, $\mathcal{B}$ first sets $l_m = 2(q_S + q_{RS})$, and randomly chooses a number k_m, such that $0 \leq k_m \leq n_m$, and $l_m(n_m + 1) < p$. $\mathcal{B}$ then chooses $n_m + 1$ random numbers x', $x_i (i = 1, \ldots, n_m)$ from $\mathbb{Z}_{l_m}$. Lastly, $\mathcal{B}$ chooses $n_m + 1$ random numbers y', $y_i (i = 1, \ldots, n_m)$ from $\mathbb{Z}_p$.

   To make expression simpler, we use the following notations:

$$F(m) = x' + \sum_{i \in \mathcal{U}} x_i - l_m k_m \quad \text{and} \quad J(m) = y' + \sum_{i \in \mathcal{U}} y_i.$$

Now, $\mathcal{B}$ sets the public parameters:

$$g_2 = g^b, \ u' = g_2^{x' - l_m k_m} g^{y'}, \ u_i = g_2^{x_i} g^{y_i} (1 \le i \le n_m).$$

Note that for any message m, there exists the following equation:

$$w = u' \prod_{i \in \mathcal{U}} u_i = g_2^{F(m)} g^{J(m)}.$$

Queries: $\mathcal{B}$ builds the following oracles:

$\mathcal{O}_{UKeyGen}$: $\mathcal{B}$ chooses a random $x_i \in Z_p^*$, and outputs $pk_i = (g^a)^{x_i}$.

$\mathcal{O}_{CKeyGen}$: $\mathcal{B}$ chooses a random $x_i \in Z_p^*$, and outputs $(pk_i, sk_i) = (g^{x_i}, x_i)$.

$\mathcal{O}_{Sign}$: On input (pk_i, m), if pk_i is corrupted, $\mathcal{B}$ returns the signature $\sigma = (g_2^{x_j} w^r, g^r)$, where $w = u' \prod_{i \in \mathcal{U}} u_i$. Otherwise, $\mathcal{B}$ performs as follows.

- If $F(m) \not\equiv 0 \bmod p$, $\mathcal{B}$ picks a random $r \in \mathbb{Z}_p$ and computes the signature as,

$$\sigma = (g_1^{-J(m)/F(m)} (u' \prod_{i \in \mathcal{U}} u_i)^r, g_1^{-1/F(m)} g^r).$$

For $\tilde{r} = r - a/F(m)$, we have that

$$g_1^{-J(m)/F(m)} (u' \textstyle\prod_{i \in \mathcal{U}} u_i)^r$$
$$= g_1^{-J(m)/F(m)} (g^{J(m)} g_2^{F(m)})^r$$
$$= g_2^a (g_2^{F(m)} g^{J(m)})^{-a/F(m)} (g^{J(m)} g_2^{F(m)})^r$$
$$= g_2^a (g_2^{F(m)} g^{J(m)})^{r - a/F(m)}$$
$$= g^{ab} (u' \textstyle\prod_{i=1}^{n} u_i^{m_i})^{\tilde{r}},$$

and

$$g_1^{-1/F(m)} g^r = g^{r - a/F(m)}$$
$$= g^{\tilde{r}},$$

which shows that σ has the correct signature as in the actual scheme.

- If $F(m) \equiv 0 \pmod p$, $\mathcal{B}$ is unable to compute the signature σ and must abort the simulation.

$\mathcal{O}_{ReKey}$: On input (pk_i, pk_j), if pk_i and pk_j are both corrupted or both uncorrupted, $\mathcal{B}$ returns $rk_{i \to j} = (x_j/x_i) \bmod p$; else, this input is illegal.

$\mathcal{O}_{ReSign}$: On input (pk_i, pk_j, m, σ). If $\texttt{Verify}(pk_i, m, \sigma) \ne 1$, $\mathcal{B}$ outputs $\bot$. Otherwise, $\mathcal{B}$ does:

- If pk_i and pk_j are both corrupted or both uncorrupted, output

$$\texttt{ReSign}(\mathcal{O}_{ReKey}(pk_i, pk_j), pk_i, m, \sigma).$$

- else, output $\mathcal{O}_{Sign}(pk_j, m)$.

Forgery: If $\mathcal{B}$ does not abort as a consequence of one of the queries above, $\mathcal{A}$ will, with probability at least ε, return a message m^* and a valid forgery

$\sigma^* = (\mathfrak{A}^*, \mathfrak{B}^*)$ on m^*. If $F(m^*) \not\equiv 0 \bmod p$, $\mathcal{B}$ aborts. Otherwise, the forgery must be of the form, for some $r^* \in \mathbb{Z}_p$,

$$
\begin{aligned}
\sigma^* &= (g^{ab}(u' \textstyle\prod_{i \in \mathcal{U}} u_i)^{r^*}, g^{r^*}) \\
&= (g^{ab}(g_2^{F(m^*)} g^{J(m^*)})^{r^*}, g^{r^*}) \\
&= (g^{ab+J(m^*)r^*}, g^{r^*}) \\
&= (\mathfrak{A}^*, \mathfrak{B}^*).
\end{aligned}
$$

To solve the CDH instance, $\mathcal{B}$ outputs $(\mathfrak{A}^*) \cdot (\mathfrak{B}^*)^{-J(m^*)} = g^{ab}$.

To conclude, we bound the probability that $\mathcal{B}$ completes the simulation without aborting. For the simulation to complete without aborting, we require that all sign and resign queries on a message m have $F(m) \not\equiv 0 \bmod p$, and that $F(m^*) \equiv 0 \bmod p$.

Let $m_1, \ldots, m_{q_Q}$ be the messages appearing in sign queries or resign queries not involving the message m^*. Clearly, $q_Q \leq q_S + q_{RS}$. We define the events E_i, E_i', and E^* as:

$$
E_i : F(m_i) \not\equiv 0 \bmod p, \ \ E_i' : F(m_i) \not\equiv 0 \bmod l_m, \ \ E^* : F(m^*) \equiv 0 \bmod p.
$$

The probability of $\mathcal{B}$ not aborting is $\Pr[\neg abort] \geq \Pr[\bigwedge_{i=1}^{q_Q} E_i \wedge E^* \wedge E]$. It is easy to see that the events $(\bigwedge_{i=1}^{q_Q} E_i)$, E^*, and E are independent, and $\Pr[E] = 1/q_K$.

From $l_m(n_m + 1) < p$ and x' and $x_i (i = 1, \ldots, n_m)$ are all from $\mathbb{Z}_{l_m}$, we have $0 \leq l_m k_m < p$ and $0 \leq x' + \prod_{i \in \mathcal{U}} x_i < p$. Then it is easy to see that $F(m) \equiv 0 \bmod p$ implies that $F(m) \equiv 0 \bmod l_m$. We can get that $F(m) \not\equiv 0 \bmod l_m$ implies that $F(m) \not\equiv 0 \bmod p$. Hence, we have: $\Pr[E_i] \geq \Pr[E_i']$,

$$
\begin{aligned}
&\Pr[E^*] \\
&= \Pr[F(m^*) \equiv 0 \bmod p \wedge F(m^*) \equiv 0 \bmod l_m] \\
&= \Pr[F(m^*) \equiv 0 \bmod l_m] \\
&\quad \Pr[F(m^*) \equiv 0 \bmod p | F(m^*) \equiv 0 \bmod l_m] \\
&= \tfrac{1}{l_m} \tfrac{1}{n_m+1}
\end{aligned}
$$

and

$$
\begin{aligned}
\Pr[\textstyle\bigwedge_{i=1}^{q_Q} E_i] &\geq \Pr[\textstyle\bigwedge_{i=1}^{q_Q} E_i'] \\
&= 1 - \Pr[\textstyle\bigvee_{i=1}^{q_Q} \neg E_i'] \\
&\geq 1 - \textstyle\sum_{i=1}^{q_Q} \Pr[\neg E_i'] \\
&= 1 - \tfrac{q_Q}{l_m} \\
&\geq 1 - \tfrac{q_S + q_{RS}}{l_m}.
\end{aligned}
$$

and $l_m = 2(q_S + q_{RS})$ as in the simulation.

Hence, we get that

$$
\begin{aligned}
&\Pr[\neg abort] \\
&\geq \Pr[\textstyle\bigwedge_{i=1}^{q_Q} E_i] \Pr[E^*] \\
&\geq \tfrac{1}{l_m(n_m+1)} \cdot (1 - \tfrac{q_S + q_{RS}}{l_m}) \\
&\geq \tfrac{1}{2(q_S + q_{RS})(n_m+1)} \cdot \tfrac{1}{2} \\
&= \tfrac{1}{4(q_S + q_{RS})(n_m+1)}
\end{aligned}
$$

Since there are $O(n_m)$ and $O(n_m)$ multiplications in sign queries and resign queries, respectively, and $O(1)$, $O(1)$, and $O(1)$ exponentiations in sign queries, resign queries and (un)corrupted key queries, respectively, hence the time complexity of $\mathcal{B}$ is $t + O((q_S + q_{RS})n_m\rho + (q_S + q_{RS} + q_K)\tau)$.

Thus, the theorem follows. $\qquad\qquad\qquad\qquad\qquad\qquad\qquad\qquad\qquad\qquad$ □

Discussion of Scheme S_{mb}: This scheme is transparent, since the signature from Sign algorithm is the same of that from ReSign algorithm. This fact also implies that this scheme is multi-use. Furthermore, it is easy to see that $rk_{A \to B} = 1/rk_{B \to A}$, which shows the scheme is bidirectional. Last, since each user just stores one signing key, the scheme is also key optimal.

3.2 S_{id-mb}: *ID*-Based *Multi-use Bidirectional* Scheme

In this subsection, we will extend S_{mb} to an ID-based multi-use bidirectional scheme, denoted as S_{id-mb}. The scheme is consisted of six algorithms. In the following we assume that all identities and messages are n_{id}-bit and n_m-bit strings, respectively. We can achieve this by applying two collision-resistant hash functions, $H_{id}: \{0,1\}^* \to \{0,1\}^{n_{id}}$, and $H_m: \{0,1\}^* \to \{0,1\}^{n_m}$.

Setup: On input the security parameter 1^k, it chooses groups $\mathbb{G}_1$ and $\mathbb{G}_2$ of prime order $p = \Theta(2^k)$, such that an admissible pairing $e: \mathbb{G}_1 \times \mathbb{G}_1 \to \mathbb{G}_2$ can be constructed and pick a generator g of $\mathbb{G}_1$. Furthermore, choose a random number α from $\mathbb{Z}_p$, compute $g_1 = g^\alpha$, and then choose u', u_i ($i = 1, \cdots, n_{id}$), v', and v_i ($i = 1, \cdots, n_m$) from $\mathbb{G}_1$.

   The public parameters are $(\mathbb{G}_1, \mathbb{G}_2, e, g, g_1, g_2, u', u_i(i = 1, \cdots, n_{id}), v', v_i(i = 1, \cdots, n_m))$ and the master secret key is α.

Extract: On input an n_{id}-bit identity ID, output the corresponding private key d_{id},

$$d_{id} = (d_{id}^{(1)}, d_{id}^{(2)}) = (g_2^\alpha (u' \prod_{i \in \mathcal{U}} u_i)^{r_{id}}, g^{r_{id}}),$$

   where r_{id} is a random number from $\mathbb{Z}_p$, $\mathcal{U} \subset \{1, \cdots, n_{id}\}$ is the set of indices i such that $u[i] = 1$, and $u[i]$ is the i-th bit of ID.

Rekey: On input two private keys $d_A = (d_A^{(1)}, d_A^{(2)})$ and $d_B = (d_B^{(1)}, d_B^{(2)})$, output the re-signature key

$$rk_{A \to B} = \frac{d_B}{d_A} = (\frac{d_B^{(1)}}{d_A^{(1)}}, \frac{d_B^{(2)}}{d_A^{(2)}}).$$

   (Note that we make use of the same method and assumptions in [2] to get the re-signature key.)

Sign: On input a private key $d_{id} = (d_{id}^{(1)}, d_{id}^{(2)})$ and a n_m-bit message m, output

$$\sigma = (\mathfrak{A}, \mathfrak{B}, \mathfrak{C}) = (d_{id}^{(1)}(v' \prod_{i \in \mathcal{V}} v_i)^{r_m}, d_{id}^{(2)}, g^{r_m}),$$

   where r_m is a random number from $\mathbb{Z}_p$, $\mathcal{V} \subset \{1, \cdots, n_m\}$ is the set of indices i such that $m[i] = 1$, and $m[i]$ is the i-th bit of m.

ReSign: On input a re-signature key $rk_{A \to B} = (\frac{d_B^{(1)}}{d_A^{(1)}}, \frac{d_B^{(2)}}{d_A^{(2)}})$, an n_{id}-bit identity ID_A, a signature σ_A, and an n_m-bit message, check that $\texttt{Verify}(ID_A, m, \sigma_A) = 1$. If $\sigma_A = (\mathfrak{A}_A, \mathfrak{B}_A, \mathfrak{C}_A)$ does not verify, output $\perp$; otherwise, output

$$\sigma_B = (\mathfrak{A}_A \cdot \frac{d_B^{(1)}}{d_A^{(1)}} \cdot (v' \prod_{i \in \mathcal{V}} v_i)^{\Delta r}, \mathfrak{B}_A \frac{d_B^{(2)}}{d_A^{(2)}}, \mathfrak{C}_A \cdot g^{\Delta r})$$
$$= (d_B^{(1)} (v' \prod_{i \in \mathcal{V}} v_i)^{r_m + \Delta r}, d_B^{(2)}, g^{r_m + \Delta r}),$$

where Δr is a random number from $\mathbb{Z}_p$.

Verify: On input an n_{id}-bit identity ID, an n_m-bit message m, and a purported signature $\sigma = (\mathfrak{A}, \mathfrak{B}, \mathfrak{C})$, output 1, if

$$e(\mathfrak{A}, g) = e(g_2, g_1) e(u' \prod_{i \in \mathcal{U}} u_i, \mathfrak{B}) e(v' \prod_{i \in \mathcal{V}} v_i, \mathfrak{C})$$

and 0 otherwise.

Theorem 3.2 (Security of S_{id-mb}). *In the standard model, ID-based bidirectional proxy re-signature scheme S_{id-mb} is correct and existentially unforgeable under the Computational Diffie-Hellman (CDH) assumption in $\mathbb{G}_1$; that is, for random $g \in \mathbb{G}_1$, and $x, y \in \mathbb{Z}_p^*$, given (g, g^x, g^y), it is hard to compute g^{xy}.*

Due to the limited space, we give the proof of above theorem in the full version [18].

Discussion of Scheme S_{id-mb}: As S_{mb}, S_{id-mb} is bidirectional, multi-use, transparent, and key optimal.

4   Conclusions

We have presented the first two proxy re-signature schemes which are proven secure in the standard model. Especially, the second one is an ID-based proxy re-signature scheme. Both of them are computational efficient, only two exponentiations in $\mathbb{G}_1$ in Sign and ReSign algorithms. However, their public parameters' size is relatively large. We can make a tradeoff between the public parameters' size and the security reduction by using the techniques of Naccache [14] and Sarkar and Chatterjee [16] to reduce its size. Note that, our proposals are only proven secure with static corruption not the adaptive corruption, we left it as the future work.

References

1. An, J.H., Dodis, Y., Rabin, T.: On the Security of Joint Signature and Encrption. In: Knudsen, L.R. (ed.) EUROCRYPT 2002. LNCS, vol. 2332, pp. 83–107. Springer, Heidelberg (2002)
2. Ateniese, G., Hohenberger, S.: Proxy Re-Signatures: New Definitions, Algorithms, and Applications. In: ACM CCS 2005, pp. 310–319 (2005)

3. Boneh, D., Boyen, X.: Short signatures without random oracles. In: Cachin, C., Camenisch, J.L. (eds.) EUROCRYPT 2004. LNCS, vol. 3027, pp. 56–73. Springer, Heidelberg (2004)
4. Boneh, D., Boyen, X.: Secure Identity Based Encryption Without Random Oracles. In: Cachin, C., Camenisch, J.L. (eds.) EUROCRYPT 2004. LNCS, vol. 3027, pp. 443–459. Springer, Heidelberg (2004)
5. Bellare, M., Boldyreva, A., Palacio, A.: An uninstantiable random-oracle-model scheme for a hybrid-encryption problem. In: Cachin, C., Camenisch, J.L. (eds.) EUROCRYPT 2004. LNCS, vol. 3027, pp. 171–188. Springer, Heidelberg (2004)
6. Blaze, M., Bleumer, G., Strauss, M.: Divertible protocols and atomic proxy cryptography. In: Nyberg, K. (ed.) EUROCRYPT 1998. LNCS, vol. 1403, pp. 127–144. Springer, Heidelberg (1998)
7. Boneh, D., Franklin, M.: Identity-based encryption from the Weil pairing. SIAM Journal of Computing 32(3), 586–615 (2003)
8. Canetti, R., Hohenberger, S.: Chosen-ciphertext secure proxy re-encryption. Cryptology ePrint Archieve: Report 2007/171(2007)
9. Canetti, R., Goldreich, O., Halevi, S.: The random oracle methodology, revisited. In: STOC 1998, pp. 209–218 (1998)
10. Cramer, R., Shoup, V.: Signature schemes based on the strong RSA assumption. ACM TISSEC 3(3), 161–185 (2000)
11. Galindo, D., Herranz, J., Kiltz, E.: On the Generic Construction of Identity-Based Signatures with Additional Properties. In: Lai, X., Chen, K. (eds.) ASIACRYPT 2006. LNCS, vol. 4284, pp. 178–193. Springer, Heidelberg (2006)
12. Gennaro, R., Halevi, S., Rabin, T.: Secure hash-and-sign signatures without the random oracle. In: Stern, J. (ed.) EUROCRYPT 1999. LNCS, vol. 1592, pp. 123–139. Springer, Heidelberg (1999)
13. Goldwasser, S., Micali, S., Rivest, R.L.: A digital signature scheme secure against adaptive chosen-message attacks. SIAM Journal of Computing 17(2), 281–308 (1988)
14. Naccache, D.: Secure and Practical Identity-based encryption. Cryptology ePrint Archive, Report 2005/369
15. Paterson, K.G., Schuldt, J.C.N.: Efficient Identity-based Signatures Secure in the Standard Model. In: Batten, L.M., Safavi-Naini, R. (eds.) ACISP 2006. LNCS, vol. 4058, pp. 207–222. Springer, Heidelberg (2006)
16. Sarkar, P., Chatterjee, S.: Trading time for space: Towards an efficient IBE scheme with short(er) public parameters in the standard model. In: Won, D.H., Kim, S. (eds.) ICISC 2005. LNCS, vol. 3935, pp. 424–440. Springer, Heidelberg (2006)
17. Shamir, A.: Identity-based cryptosystems and signature schemes. In: Blakely, G.R., Chaum, D. (eds.) CRYPTO 1984. LNCS, vol. 196, pp. 47–53. Springer, Heidelberg (1985)
18. Shao, J., Cao, Z., Wang, L., Liang, X.: Proxy Re-signature Scheme without Random Oracles. Cryptology ePrint Archive, Report (2007)
19. Taban, G., Cárdenas, A.A., Gligor, V.D.: Towards a Secure and Interoperable DRM Architecture. In: ACM DRM 2006, pp. 69–78 (2006)
20. Waters, B.: Efficient Identity-based Encryption Without Random Oracles. In: Cramer, R.J.F. (ed.) EUROCRYPT 2005. LNCS, vol. 3494, pp. 114–127. Springer, Heidelberg (2005)
21. Zhang, F., Chen, X., Susilo, W., Mu., Y.: A New Short Signature Scheme without Random Oracles from Bilinear Pairings. In: Nguyen, P.Q. (ed.) VIETCRYPT 2006. LNCS, vol. 4341, pp. 67–80. Springer, Heidelberg (2006)

First-Order Differential Power Analysis on the Duplication Method

Guillaume Fumaroli[1], Emmanuel Mayer[2], and Renaud Dubois[1]

[1] Thales Communications
160 Boulevard de Valmy – BP 82
92704 Colombes cedex – France
firstname.lastname@fr.thalesgroup.com
[2] DGA/CELAR
BP 57419
35174 Bruz cedex – France
firstname.lastname@dga.defense.gouv.fr

Abstract. Cryptographic embedded systems are vulnerable to Differential Power Analysis (DPA). In particular, the S-boxes of a block cipher are known to be the most sensitive parts with respect to this very kind of attack. While many sound countermeasures have been proposed to withstand this weakness, most of them are too costly to be adopted in real-life implementations of cryptographic algorithms. In this paper, we focus on a widely adopted lightweight variation on the well-known Duplication Method. While it is known that this design is vulnerable to higher-order DPA attacks, we show that it can also be efficiently broken by first-order DPA attacks. Finally, we point out ad hoc costless countermeasures that circumvent our attacks.

Keywords: Side-channel analysis, differential power analysis, zero attack, spectral analysis.

1 Introduction

The formal definition of an algorithm along with its inputs and outputs are conventionally considered to be the only elements available to an attacker in traditional cryptography. This assumption becomes unfortunately inacurrate in most applications where the attacker also has access to the physical implementation of the algorithm. Sensitive information stored in electronic components can indeed be obtained from passive or active side-channels such as timing of operation, power consumption or fault injection.

In 1998, Kocher et al. introduce the framework of Differential Power Analysis (DPA) [8]. Since then, DPA has been extensively studied in the cryptographic community. DPA leverages statistical properties in the side-channel leakage of targeted implementations, for example electrical consumption or electromagnetic radiation, in order to obtain information about processed sensitive data. Both symmetric and asymmetric embedded cryptographic systems are known to be potentially vulnerable to this kind of attack.

K. Srinathan, C. Pandu Rangan, M. Yung (Eds.): Indocrypt 2007, LNCS 4859, pp. 210–223, 2007.

Most countermeasures against DPA attacks consist in masking all the intermediate data processed by the device and rewriting its operations in the masked domain [6,3,1,10]. In this paper, we analyse a widely adopted lightweight variation on the so-called Duplication Method that is suggested in [6]. While it is known that masking countermeasures are still vulnerable to higher order attacks [9], we show that the design presented in [6] can be efficiently broken by *first-order* DPA attacks.

The remainder of the paper is organized as follows. Section 2 presents the general Duplication Method and its lightweight variation in the context of block ciphers. Section 3 introduces the analytical canvas used in the sequel. Section 4 and 5 present and analyse our attacks. Section 6 describes adapted low-cost countermeasures to these attacks. Section 7 concludes this contribution.

2   The Duplication Method

Goubin et al. [6] suggest that any intermediate variable $y \in \mathfrak{S}'$ occuring in a cryptographic algorithm be split in its actual implementation as k variables $y_1, y_2, \ldots, y_k \in \mathfrak{S}'$ allowing to reconstruct y. If we assume that the variables y_1, $y_2, \ldots$ or y_k are uniformly distributed, the knowledge of at most $k-1$ values in the set $\{y_1, \ldots, y_k\}$ at each computation gives no information about y. Hence, the countermeasure withstands any order $k-1$ attack, where at most $k-1$ variables in the set $\{y_1, \ldots, y_k\}$ can be obtained by the attacker.

In the context of block ciphers, the variables y_i are randomly picked along with the property $y = y_1 \oplus y_2 \oplus \cdots \oplus y_k$ where "$\oplus$" denotes the usual exclusive-or operator.

The overall transformation achieved by a block cipher can be seen as the composition of several affine and non-affine transformations.

Let us first consider that y is the input to some affine transformation $y \mapsto Ay \oplus b$. Such a transformation can be adapted in a straightforward way by applying A and adding b_i to each y_i where $b = b_1 \oplus b_2 \oplus \cdots \oplus b_k$, i.e. by replacing the original transformation $y \mapsto Ay \oplus b$ with

$$(y_1, \ldots, y_k) \mapsto (Ay_1 \oplus b_1, Ay_2 \oplus b_2, \ldots, Ay_k \oplus b_k),$$

leading to a computation overhead linear in k. Notice that it is as secure but more efficient to directly add b to one of the variables y_i, say y_1. Hence, the transformation $y \mapsto Ay \oplus b$ is actually adapted as

$$(y_1, \ldots, y_k) \mapsto (Ay_1 \oplus b, Ay_2, \ldots, Ay_k).$$

Let us now consider that $X \in \{0,1\}^m$ is the input to some S-box $S : \{0,1\}^m \to \{0,1\}^n$. The most generic adaptation then consists in replacing the original mapping $y \mapsto S(y)$ with

$$(y_1, \ldots, y_k) \mapsto (S_1(y_1, \ldots, y_k), \ldots, S_k(y_1, \ldots, y_k)),$$

where $S_i : (\{0,1\}^m)^k \to \{0,1\}^n$, $1 \leq i \leq k-1$, are randomly picked invertible S-boxes and the S-box S_k is computed such that

$$\bigoplus_{i=1}^k S_i(y_1, \ldots, y_k) = S(y)$$

for all y. Clearly, this approach requires to generate and store $kn2^{km}$ bits. This represents a computation and space overhead that exceeds the capacity of most embedded devices even for a small k. Goubin et al. thus suggest a lightweight variation in which a bijective function $\varphi : \{0,1\}^m \to \{0,1\}^m$ and invertible S-boxes $S'_i : \{0,1\}^m \to \{0,1\}^m$, $1 \leq i \leq k-1$, are randomly selected and the S-box S'_k is computed such that

$$\bigoplus_{i=1}^k S'_i(y) = S \circ \varphi^{-1}(y)$$

for all y. The original transformation then becomes

$$(y_1, \ldots, y_k) \mapsto (S'_1(\varphi(y_1 \oplus \cdots \oplus y_k)), \ldots, S'_k(\varphi(y_1 \oplus \cdots \oplus y_k)).$$

This leads to a computation and space overhead linear in k.

An important restriction on φ is that $\varphi(y)$ can be computed without explicitly recombining the variable y even partially, i.e. without having to compute a variable of the form $\bigoplus_{i \in I} y_i$ where $I \subseteq \{1, \ldots, k\}$ and $\#I \geq 2$.

In order to fulfil the former restriction, the authors put forward to choose functions φ in the set $\mathcal{F}$ of linear or quadratic bijective functions. For example, if $\mathcal{F}$ is the set of linear function, $\varphi(y)$ can be computed as $\varphi(y) = \bigoplus_{i=1}^k \varphi(y_i)$. A similar relation also allows to efficiently compute $\varphi(y)$ without recombining y when $\mathcal{F}$ is a particular subset of the set of quadratic functions as pointed out in [6].

3   An Analytical Canvas for DPA Attacks

This section introduces the formalism and some general results that will be used in the sequel of the paper.

3.1   First-Order DPA Attacks

We use the formalism of Joye et al. in [7]. Let $I(x, s) \in \mathfrak{S}'$ denote an intermediate variable manipulated by the device at time τ, that only depends on known data $x \in \mathfrak{S}$ and on a small portion of the secret data $s \in \mathcal{S}$. For each possible value $\hat{s} \in \mathcal{S}$ for s, the attacker determines two sets

$$\mathfrak{S}_b(\hat{s}) = \{x \mid g(I(x, \hat{s})) = b\} \quad \text{for } b \in \{0, 1\} \tag{1}$$

where g is an appropriate *boolean selection function*.

With $\mathscr{C}(t)$ denoting the power consumption of the device at time t, and $\langle\cdot\rangle$ denoting the average operator, the attacker evaluates the first order DPA trace

$$\Delta(\hat{s},t) = \langle\mathscr{C}(t)\rangle_{x\in\mathfrak{S}_1(\hat{s})} - \langle\mathscr{C}(t)\rangle_{x\in\mathfrak{S}_0(\hat{s})}\,. \tag{2}$$

Finally, the attacker retains all key hypothesis $\hat{s}\in\mathcal{S}$ for which the graph of $\Delta(\hat{s},t)$ contains a significant peak. Indeed, for the right key hypothesis $\hat{s}=s$, the partition of the set $\mathscr{C}(t)$ is meaningful to the actual power consumption of the device at time τ, resulting in a peak in the corresponding DPA trace around that instant.

3.2 Soundness of a First-Order DPA Attack

In this paper, the soundness of first-order DPA attacks is proven in a rigorous analytical way. For this, the expected magnitude of the DPA trace is evaluated at time τ for all possible key hypothesis, i.e. the quantity $\mathrm{E}(\Delta(\hat{s},\tau))$ are computed for each $\hat{s}\in\mathcal{S}$. The attack is successful if $\mathrm{E}(\Delta(s,\tau))$ significantly exceeds $\mathrm{E}(\Delta(\hat{s},\tau))$ for almost all $\hat{s}\neq s$.

Let us now derive some useful formulae to evaluate $\mathrm{E}(\Delta(\hat{s},\tau))$. Our analysis is set in the *Hamming weight model*:

$$\mathscr{C}(\tau) = \alpha\,\mathsf{H}(\mathrm{I}(x,s)) + \beta$$

with α,β denoting some fixed devide-dependent parameters, and $\mathsf{H}(\cdot)$ denoting the Hamming weight operator.

Let $\mathfrak{S}_0'(\hat{s})$ and $\mathfrak{S}_1'(\hat{s})$ denote the actual set to which $\mathrm{I}(x,s)$ belongs respectively when $x\in\mathfrak{S}_0(\hat{s})$ and $x\in\mathfrak{S}_1(\hat{s})$.

We have the following result:

Proposition 1. *Let $\mathrm{I}(x,s)$ be uniformly distributed over its definition set $\mathfrak{S}'$. Let $\varepsilon_b(\hat{s}) = \mathrm{E}(\mathsf{H}(\mathrm{I}(x,s))) - \mathrm{E}(\mathsf{H}(\mathrm{I}(x,s)))\mid\mathrm{I}(x,s)\in\mathfrak{S}_b'(\hat{s}))$ denote the difference between the expected Hamming weight of $\mathrm{I}(x,s)$ when $x\in\mathfrak{S}$, and the expected Hamming weight of $\mathrm{I}(x,s)$ when $x\in\mathfrak{S}_b$.*

The expected value of the DPA trace at time τ under key hypothesis $\hat{s}$ is given by

$$\mathrm{E}(\Delta(\hat{s},\tau)) = \alpha\left(1 + \frac{\mathrm{P}(\mathrm{I}(x,s)\in\mathfrak{S}_0'(\hat{s}))}{\mathrm{P}(\mathrm{I}(x,s)\in\mathfrak{S}_1'(\hat{s}))}\right)\varepsilon_0(\hat{s}).$$

Proof. Let $\mu = \mathrm{E}(\mathsf{H}(\mathrm{I}(x,s)))$, $\mu_b(\hat{s}) = \mathrm{E}(\mathsf{H}(\mathrm{I}(x,s))\mid\mathrm{I}(x,s)\in\mathfrak{S}_b'(\hat{s}))$, $\varepsilon_b(\hat{s}) = \mu - \mu_b(\hat{s})$, $p_b(\hat{s}) = \mathrm{P}(\mathrm{I}(x,s)\in\mathfrak{S}_b'(\hat{s}))$.

Since

$$\mu = \frac{\sum_{b\in\{0,1\}} p_b(\hat{s})\mu_b(\hat{s})}{\sum_{b\in\{0,1\}} p_b(\hat{s})},$$

we have

$$\varepsilon_b(\hat{s}) = -\frac{p_{1-b}(\hat{s})}{p_b(\hat{s})}\,\varepsilon_{1-b}(\hat{s}) \quad\text{for all } b\in\{0,1\}.$$

Then,

$$
\begin{aligned}
\mathrm{E}(\Delta(\hat{s}, \tau)) &= \langle \mathscr{C}(t) \rangle_{x \in \mathfrak{S}_1(\hat{s})} - \langle \mathscr{C}(t) \rangle_{x \in \mathfrak{S}_0(\hat{s})} \\
&= (\alpha \, \mathrm{E}(\mathrm{H}(\mathrm{I}(x, s)) \mid \mathrm{I}(x, s) \in \mathfrak{S}_1'(\hat{s})) + \beta) \\
&\quad - (\alpha \, \mathrm{E}(\mathrm{H}(\mathrm{I}(x, s)) \mid \mathrm{I}(x, s) \in \mathfrak{S}_0'(\hat{s})) + \beta) \\
&= \alpha \, (\varepsilon_0(\hat{s}) - \varepsilon_1(\hat{s})) \\
&= \alpha \left(1 + \frac{p_0(\hat{s})}{p_1(\hat{s})} \right) \varepsilon_0(\hat{s}). \qquad \square
\end{aligned}
$$

Corollary 1. *Let $I(x, s)$ be uniformly distibuted over $\mathfrak{S}'$, and $\mathfrak{S}_0'(\hat{s}) \cap \mathfrak{S}_1'(\hat{s}) = \emptyset$. The expected value of the DPA trace at time τ under key hypothesis $\hat{s}$ becomes*

$$
\mathrm{E}(\Delta(\hat{s}, \tau)) = \frac{\alpha \, \varepsilon_0(\hat{s})}{\mathrm{P}(\mathrm{I}(x, s) \in \mathfrak{S}_1'(\hat{s}))}.
$$

Remark 1. Corollary 1 can generally only be applied for the right key hypothesis $\hat{s} = s$. Indeed, in general, if $\hat{s} \neq s$, we have

$$
\mathfrak{S}_0'(\hat{s}) \cap \mathfrak{S}_1'(\hat{s}) \neq \emptyset.
$$

4   Zero Attack When φ Is Variable

Although hardcoding is more efficient, cryptographic implementations in which φ is dynamically regenerated are common in software. It is indeed often believed that variable elements provide more security than constant ones. In [6], the authors suggest that φ be chosen in a set $\mathcal{F}$ of bijective linear functions or quadratic functions without affine parts. In this case, we point out an attack that efficiently recovers the first subkey by short slices of m bits. Our attack is based on ideas similar to that of the *zero attack* against the multiplicative masking of AES [5]. We here exploit the property that for all $f \in \mathcal{F}$, $\varphi(0) = 0$, i.e. 0 is not masked by the countermeasure.

Let us assume that we target the input of a specific S-box in the substitution layer of the first round. Let $x \in \mathfrak{S}$ and $s \in \mathcal{S}$ respectively denote the input bits and the first subkey bits corresponding to this S-box. We target the variable $I(x, s) = \varphi(x \oplus s) \in \mathfrak{S}'$ manipulated by the implementation at time τ. The associated selection function g is defined by $g(\mathrm{I}(x, \hat{s})) = 0$ if $\varphi^{-1}(\mathrm{I}(x, \hat{s})) = 0$, and $g(\mathrm{I}(x, \hat{s})) = 1$ otherwise.

Let us now formaly describe the attack protocol:

1. Prepare a set of power consumption traces $\mathscr{C}(t)$ that contains at least N traces for each possible value for $x \in \mathfrak{S}$. Let $\langle I(x, s) \rangle$ denote the mean value of $I(x, s)$ over N executions for a fixed x. The constant N should be large enough so that for λ close to zero, say $\lambda = 10^{-m}$, we have

$$
\mathrm{P}\left(|\langle I(x, s) \rangle - \mathrm{E}(\varphi(x \oplus s))| > \lambda \right) \to 0
$$

when x is fixed and φ is randomly uniformly chosen in $\mathcal{F}$;

2. Compute the DPA traces $\Delta(\hat{s}, t)$ for each $\hat{s} \in \mathcal{S}$;
3. Create a set $\hat{\mathcal{S}}_\star$ of all $\hat{s} \in \mathcal{S}$ such that the DPA trace $\Delta(\hat{s}, t)$ contains a significant peak;
4. Return the set $\hat{\mathcal{S}}_\star$.

With the notations of the formal canvas presented in section 3, it turns out that:

Proposition 2. *The DPA traces computed in step 2 of the attack protocol are such that for all $s \neq \hat{s}$,*

$$\mathrm{E}(\Delta(\hat{s}, \tau)) = -\alpha \frac{m}{2(2^m - 1)(1 - 2^{-m})},$$

and

$$\mathrm{E}(\Delta(s, \tau)) = \alpha \frac{m}{2(1 - 2^{-m})}.$$

Proof. See Appendix A.1.

With the former result, we have $\mathrm{E}(\Delta(\hat{s}, \tau)) = -(\mathrm{E}(\Delta(s, \tau)))/(2^m - 1)$ for all $\hat{s} \neq s$. Hence, for all typical m, we have $\#\hat{\mathcal{S}}_\star = 1$ so that the right key hypothesis can be immediately identified.

5   Attack When φ Is Constant

This second attack addresses the case when φ is constant. Before we describe the actual attack protocol, we first introduce an attack based on the same idea in a simplified setting.

Let us consider a DES implementation featuring the countermeasure described in section 2 with φ a simple permutation of its input bits. In this setting, the implementation can be broken by a classical first order DPA attack, as if it was not protected. For some key hypothesis, let us predict the value of a bit b at the output of an S-box in the first round. Since b is masked, the power consumption of the device is entirely decorrelated from the actual value of b so that no peak appears in the DPA trace at this point. However, let us further examine the successive transformations that are applied to b between the first and the second substitution layers. b is combined with one bit of the second subkey that is unknown but constant. Hence, depending on the actual value of this second subkey bit, b is either always inverted or left unchanged, which is strictly equivalent with respect to DPA. b also passes through the DES diffusion layers which are almost simple bit permutations such that it is just translated somewhere else in the DES current state. Finally, b passes through φ which translates it but also unmasks its value. Thus, at this point, the power consumption is correlated to the value of b. In this simplified context, the expected DPA peak is only delayed compared to an unprotected implementation, which is transparent to an attacker.

Let us now describe the actual attack protocol that is based exactly on the same idea. We attempt to fix k bits among m at the input of φ in the second round. By examining the diffusion layer, we can determine the bits at the output of the first substitution layer that are related to targeted bits at the input of φ in the second round. Without loss of generality, we may attempt to fix the all-zero pattern at the output of the first substitution layer which fixes an unknown k-bit pattern ζ_* at the input of φ for the right key hypothesis. Clearly, the size of the key hypothesis required to fix the all-zero pattern at the output of the first substitution layer increases with k. It also increases with the diffusion layer complexity. As an example, one bit at the input of φ in the second round depends on the output of one S-box with DES, and on the output of four S-boxes with AES.

Let w denote the m-bit vector of weight k, with ones in the positions that are fixed at the input of φ. Let $\delta_w(v)$ denote the k-bit vector composed of the k bits $v[i]$ such that $w[i] = 1$. With the notations of section 3, the targeted intermediate variable $\mathrm{I}(x,s)$ is the output of φ in the second round. The associated selection function g is defined as $g(\mathrm{I}(x,\hat{s})) = 0$ if $\delta_w(\varphi^{-1}(\mathrm{I}(x,\hat{s}))) = \zeta_*$, and $g(\mathrm{I}(x,\hat{s})) = 1$ otherwise.

Let us now formaly state the attack protocol:

1. Prepare a set of power consumption traces $\mathscr{C}(t)$;
2. Pick the next element w in the ordered list (w_i) of m-bit vectors, where $\mathrm{H}(w_{i_1}) \leq \mathrm{H}(w_{i_2})$ whenever $i_1 \leq i_2$;
3. Compute the DPA traces $\Delta(\hat{s}, t)$ for each $\hat{s} \in \mathcal{S}$, where $\mathcal{S}$ is a set depending on w containing all possibles values for the secret bits involved in the computation of $g(\mathrm{I}(x,s))$;
4. Create a set $\hat{\mathcal{S}}_*$ of all $\hat{s} \in \mathcal{S}$ such that the DPA trace $\Delta(\hat{s}, t)$ contains a significant peak;
5. If $\hat{\mathcal{S}}_* \neq \emptyset$, stop the process and return the set $\hat{\mathcal{S}}_*$;
6. Continue in step 2.

In the following proposition, $\widetilde{\mathrm{Cor}}(\varphi, \ell_w)$ denotes the correlation between φ and ℓ_w, formally stated in the following definition. ℓ_w denotes the linear boolean function $x \mapsto x \cdot w$.

Definition 1 (Correlation). *The usual correlation of the boolean functions f, g:* $\{0,1\}^m \to \{0,1\}$ *is defined as*

$$\mathrm{Cor}(f,g) = \sum_{x \in \{0,1\}^m} (-1)^{f(x)+g(x)}.$$

If $\mathrm{Cor}(f,g) \neq 0$, f *is said to be correlated to* g.

By extension, the correlation between the vectorial boolean function $\varphi : \{0,1\}^m \mapsto \{0,1\}^m$ *and the boolean function* g *is here defined as*

$$\widetilde{\mathrm{Cor}}(\varphi, g) = \sum_{i=1}^{m} \mathrm{Cor}(\varphi_i, g),$$

where φ_i *denotes the i-th component of* φ. *Again, if* $\widetilde{\mathrm{Cor}}(\varphi, g) \neq 0$, φ *is said to be correlated to* g.

Proposition 3. *The DPA traces computed in step 3 of the attack protocol are such that for all $s \neq \hat{s}$,*

$$\mathrm{E}(\Delta(\hat{s}, \tau)) = 0,$$

and

$$\mathrm{E}(\Delta(s, \tau)) = \alpha \, \frac{(-1)^{\mathsf{H}(\varsigma_\star)}}{(1 - 2^{-\,\mathsf{H}(w)})2^{m+1}} \, \widetilde{\mathrm{Cor}}(\varphi, \ell_w)$$

where ℓ_w denotes the linear boolean function $x \mapsto x \cdot w$.

Proof. See Appendix A.2.

As a direct consequence of Proposition 3, the protocol successfully stops in step 5 as soon as φ is correlated to the linear boolean function ℓ_w. In this case, $\#\hat{\mathcal{S}}_\star = 1$ so that the right key hypothesis can be immediately identified.

6   Countermeasures

The attack presented in section 5 would become computationally infeasible if φ was chosen in a set of functions $\mathcal{F}$ with a high resiliency order. However, let us recall that it is also necessary that the output of φ can be computed without recombining its input. It remains as an open problem to find a set $\mathcal{F}$ that fulfils these two conditions and is also sufficiently large to withstand exhaustive search.

As for the attack we presented in section 4, it could be efficiently avoided by picking φ in a set $\mathcal{F}$ of affine functions, rather than in the set of linear or quadratic functions as proposed by Goubin et al. Indeed, in this setting, for all $z \in \{0,1\}^m$, $\varphi(z)$ is uniformly distributed over $\{0,1\}^m$ so that $\mathrm{E}(\mathsf{H}(\varphi(z))) = m/2$. As a consequence, no DPA trace will contain a peak, even under the right hypothesis.

We are left with finding an algorithm that efficiently generates random affine functions of the form $Ax + b$, where $A \in \mathrm{GL}_m(\mathrm{GF}(2))$ is an element of the linear group over $\mathrm{GF}(2)$ of dimension m and b is a binary vector with n components. It is easy to create a vector b from a sequence of random bits, but the generation of a uniform distribution over $\mathrm{GL}_m(\mathrm{GF}(2))$ remains an open problem. As it is observed in [4], on the one hand, one could choose a theoretically sound algorithm with a complexity that is far too high to be practical, and on the other hand, one could choose a simple heuristic for random elements as the *product replacement* algorithm [2] which is known to have a bias.

Let us explain a variant of the *product replacement* algorithm over $G = \mathrm{GL}_m(\mathrm{GF}(2))$. Let us consider the two classical generators of G:

$$g_1 = \begin{bmatrix} 1 & 1 & 0 & \cdots & 0 & 0 & 0 \\ 0 & 1 & 0 & \cdots & 0 & 0 & 0 \\ 0 & 0 & 1 & & 0 & 0 & 0 \\ \vdots & \vdots & & & \vdots & \vdots \\ 0 & 0 & 0 & & 1 & 0 & 0 \\ 0 & 0 & 0 & \cdots & 0 & 1 & 0 \\ 0 & 0 & 0 & \cdots & 0 & 0 & 1 \end{bmatrix} , \quad g_2 = \begin{bmatrix} 0 & 0 & 0 & \cdots & 0 & 0 & 1 \\ 1 & 0 & 0 & \cdots & 0 & 0 & 0 \\ 0 & 1 & 0 & \cdots & 0 & 0 & 0 \\ \vdots & \vdots & & & \vdots & \vdots & \vdots \\ 0 & 0 & 0 & & 0 & 0 & 0 \\ 0 & 0 & 0 & & 1 & 0 & 0 \\ 0 & 0 & 0 & \cdots & 0 & 1 & 0 \end{bmatrix} .$$

Let A be an accumulator $(A \in G)$ and S be an array of s elements of G $(s \geq 2)$. We initialize A with the identity of G and cyclically S with g_1 and g_2. The basic generation step consists in choosing two random elements of S $(S_i$ and S_j where $(i, j) \in \{1, 2, \ldots, s\}^2)$. The accumulator A is replaced by $A \times S_i$ and S_i becomes $S_i \times S_j$. After a preprocessing step executing this basic operation a number of times σ_1, we generate random elements by reading the value of A after each single step.

One obvious disadvantage of this technique is that the returned elements are not independent of each other. For example, if a sequence of elements is generated, then a consecutive quadriple of the form X, Xg_i, Xg_j and Xg_ig_j will occur in the sequence with a probability of order $1/s^3$. This drawback can be circumvented by outputing the accumulator every σ_2 steps.

In the context of DPA countermeasures, it suffices to consider $s = 2$ and $\sigma_1 = \sigma_2 = 1$, so that we can generate a new φ function with only one multiplication matrix. This ensures a good enough variation of φ. From a theoretical point of view, it would be preferable to choose $s = \sigma_1 = \sigma_2 = \log_2(G)$, as explained in [4] in order to reach a uniform distribution over G, but of course it is really much more costly.

7   Conclusion

This paper focuses on a lightweight variation on the Duplication Method implementing a secret function $\varphi \in \mathcal{F}$ to mask the input of an S-Box. We present novel first order DPA attacks that break this countermeasure whether φ is fixed or variable when the set $\mathcal{F}$ is chosen as pointed out in [6]. Finally, we suggest new sets $\mathcal{F}$ that allow the implementation to circumvent our attacks.

Acknowledgments

The authors would like to thank Éric Garrido and Philippe Painchault for helpful discussions about spectral analysis, as well as David Lefranc and anonymous referees for useful comments.

References

1. Akkar, M.L., Giraud, C.: An Implementation of DES and AES Secure Against Some Attacks. In: Koç, Ç.K., Naccache, D., Paar, C. (eds.) CHES 2001. LNCS, vol. 2162, pp. 309–318. Springer, Heidelberg (2001)
2. Celler, F., Leedham-Green, C.R., Murray, S.H., Niemeyer, A.C., O'Brien, E.A.: Generating random element of a finite group. Comm. Algebra 23(13), 4931–4948 (1995)
3. Chari, S., Jutla, C., Rao, J., Rohatgi, P.: Toward Sound Approaches to Counteract Power-Analysis Attacks. In: Wiener, M.J. (ed.) CRYPTO 1999. LNCS, vol. 1666, pp. 398–412. Springer, Heidelberg (1999)

4. Cooperman, G.: Towards a practical, theoretically sound algorithm for random generation in finite groups (2002)
5. Golić, J.D., Tymen, C.: Multiplicative Masking and Power Analysis of AES. In: Kaliski Jr., B.S., Koç, Ç.K., Paar, C. (eds.) CHES 2002. LNCS, vol. 2523, pp. 198–212. Springer, Heidelberg (2002)
6. Goubin, L., Patarin, J.: DES and Differential Power Analysis – The "Duplication" Method. In: Koç, Ç.K., Paar, C. (eds.) CHES 1999. LNCS, vol. 1717, pp. 158–172. Springer, Heidelberg (1999)
7. Joye, M., Paillier, P., Schoenmakers, B.: On Second-Order Differential Power Analysis. In: Rao, J.R., Sunar, B. (eds.) CHES 2005. LNCS, vol. 3659, pp. 293–308. Springer, Heidelberg (2005)
8. Kocher, P., Jaffe, J., Jun, B.: Differential Power Analysis. In: Wiener, M.J. (ed.) CRYPTO 1999. LNCS, vol. 1666, pp. 388–397. Springer, Heidelberg (1999)
9. Messerges, T.S.: Using Second-Order Power Analysis to Attack DPA Resistant Software. In: Paar, C., Koç, Ç.K. (eds.) CHES 2000. LNCS, vol. 1965, pp. 238–251. Springer, Heidelberg (2000)
10. Oswald, E., Mangard, S., Pramstaller, N., Rijmen, V.: A Side-Channel Analysis Description of the AES S-Box. In: Gilbert, H., Handschuh, H. (eds.) FSE 2005. LNCS, vol. 3557, pp. 413–423. Springer, Heidelberg (2005)

A   Proofs

A.1   Proof of Proposition 2

Let us first prove the following lemma.

Lemma 1. *With $\mathcal{F}$ denoting the set of bijective functions over $\{0,1\}^m$ that are linear or quadratic in their input bits, and φ denoting a random variable uniformly distributed over $\mathcal{F}$,*

$$
P(\varphi(z) = y) = \begin{cases} 1 & \text{if } \{z,y\} = \{0\}, \\ 0 & \text{if } \{z,y\} = \{0, \xi \neq 0\}, \\ \frac{1}{2^m - 1} & \text{otherwise,} \end{cases}
$$

where $z, y \in \{0,1\}^m$ are fixed, and the probability is taken over the possible choices for φ in $\mathcal{F}$.

Proof. Cases $\{z,y\} = \{0\}$ and $\{z,y\} = \{0, \xi \neq 0\}$ are trivial. We here give a proof for the case $z \neq 0$ and $y \neq 0$.

Let $z, y \in \mathrm{GF}(2)^m \setminus \{0\}$ and consider the mapping

$$
f_{(z,y)} : \begin{vmatrix} \mathcal{F} & \to & \mathrm{GF}(2)^m \\ \varphi & \mapsto & \varphi(z) \oplus y \end{vmatrix}
$$

According to the rank theorem,

$$
\#\{\varphi \in \mathcal{F} : \varphi(z) \oplus y = 0\} = \#\mathcal{F} / \# \mathrm{Im}(f_{(z,y)}). \tag{3}
$$

Since φ is uniformly distributed over $\mathcal{F}$, (3) can be restated as

$$P(\varphi(z) = y) = \frac{1}{\#\operatorname{Im}(f_{(z,y)})}.$$

We are left with proving that $\#\operatorname{Im}(f_{(z,y)}) = 2^m - 1$.

First, we prove the result for $\mathcal{F} = GL_m(GF(2))$. If $y, z \in GF(2)^m \setminus \{0\}$, there is an invertible linear φ such that $\varphi(z) = y$. Just consider the matrices $M_z, M_y \in GL_m(GF(2))$ with $(M_z)_{*1} = z$ and $(M_y)_{*1} = y$, and let $\varphi = M_z M_x^{-1}$. So, we have $\operatorname{Im}(\varphi \mapsto \varphi(z)) = GF(2)^m \setminus \{0\}$ and thus $\operatorname{Im}(\varphi \mapsto \varphi(z) \oplus y) = GF(2)^m \setminus \{y\}$, i.e. $\#\operatorname{Im}(f_{(z,y)}) = 2^m - 1$.

Now, let $\mathcal{F}$ denote the set of bijective functions quadratic in their input bits. Since any linear function φ can be viewed as a quadratic function, we actually also prove that $\#\operatorname{Im}(f_{(z,y)}) = 2^m - 1$ in this case. □

With Lemma 1, we have

$$P(H(\varphi(z)) = w) = \begin{cases} 1 & \text{if } \{z, w\} = \{0\}, \\ 0 & \text{if } \{z, w\} = \{0, \xi \neq 0\}, \\ \frac{1}{2^m - 1}\binom{n}{w} & \text{otherwise,} \end{cases}$$

and thus

$$E(H(\varphi(z))) = \begin{cases} 0 & \text{if } z = 0, \\ \frac{m2^{m-1}}{2^m - 1} & \text{otherwise.} \end{cases} \tag{4}$$

With the notations of section 3, we have

$$\mathfrak{S}_b'(\hat{s}) = \{I(x, s) \mid g(I(x, \hat{s})) = b\} \quad \text{for } b \in \{0, 1\}, \tag{5}$$

where $g(I(x, \hat{s})) = 0$ if $x \oplus \hat{s} = 0$ and $g(I(x, \hat{s})) = 1$ otherwise.

From (5) and by definition of g, $\mathfrak{S}_0'(\hat{s}) = \{I(x, s) \mid x \oplus \hat{s} = 0\} = \{I(x, s) \mid x \oplus s = \delta_{s\hat{s}}\}$ where $\delta_{s\hat{s}} = s \oplus \hat{s}$. Hence, we have

$$E(H(I(x, s)) \mid I(x, s) \in \mathfrak{S}_0'(\hat{s})) = E(\varphi(x \oplus s) \mid x \oplus s = \delta_{s\hat{s}}). \tag{6}$$

With (4) and (6), we have

$$E(H(I(x, s)) \mid I(x, s) \in \mathfrak{S}_0'(\hat{s})) = \begin{cases} 0 & \text{if } \hat{s} = s, \\ \frac{m2^{m-1}}{2^m - 1} & \text{otherwise,} \end{cases} \tag{7}$$

since $\delta_{s\hat{s}} = 0$ if $\hat{s} = s$, and $\delta_{s\hat{s}} \neq 0$ otherwise.

Let us recall that $\varepsilon_0(\hat{s}) = E(H(I(x, s))) - E(H(I(x, s)) \mid I(x, y) \in \mathfrak{S}_0')$. If we assume that x is uniformly distributed over $\mathfrak{S}$, then $I(x, s)$ can also be assumed uniformly distributed over $\mathfrak{S}'$, so that

$$\varepsilon_0(\hat{s}) = \frac{m}{2} - E(H(I(x, s)) \mid I(x, y) \in \mathfrak{S}_0'). \tag{8}$$

From (7) and (8), we get

$$\varepsilon_0(\hat{s}) = \begin{cases} \frac{m}{2} & \text{if } \hat{s} = s, \\ -\frac{m}{2(2^m - 1)} & \text{otherwise.} \end{cases} \tag{9}$$

Since for all $\hat{s}$, $\mathfrak{S}'_0(\hat{s}) \cap \mathfrak{S}'_1(\hat{s}) = \emptyset$,

$$P(I(x, s) \in \mathfrak{S}'_1(\hat{s})) = 1 - 2^{-m}. \tag{10}$$

From (9) and (10), we finally obtain with Corollary 1 that

$$E(\Delta(\hat{s}, \tau)) = \begin{cases} \alpha \, \frac{m}{2(1-2^{-m})} & \text{if } \hat{s} = s, \\ -\alpha \, \frac{m}{2(2^m-1)(1-2^{-m})} & \text{otherwise.} \end{cases}$$

A.2   Proof of Proposition 3

A.2.1   Case $\hat{s} = s$

Let $x_{[\sigma]}$ denote the m-bit vector defined by $(x_{[\sigma]})_i = x_{\sigma(i)}$ for all i. Let $\bar{u} = (u \, \| \, 0)_{[\sigma]}$ for $u \in \{0,1\}^k$. Let $\ell_{\bar{u}}$ denote the linear boolean function $x \mapsto x \cdot \bar{u}$. Let $\alpha_{(n)}$ denote the n-bit vector such that $(\alpha_{(n)})_i = \alpha$ for all i. Let $\preceq$ denote the product order, i.e. $x \preceq y$ if and only if $x_i \leq y_i$ for all i.

Lemma 2. *Let $Z = (Z_1 \, \| \, Z_2)_{[\sigma]}$ be uniformly distributed over $\{0,1\}^m$, and $Y = \varphi(Z)$. For all $\zeta \in \{0,1\}^k$, $E(H(Y) \mid Z_1 = \zeta) = E(H(Y)) - \varepsilon$, with $\varepsilon = \left(\sum_{u \preceq 1_{(k)}} (-1)^{u \cdot \zeta} \, \widetilde{\mathrm{Cor}}(\varphi, \ell_{\bar{u}}) \right) / 2^{m+1}$.*

Proof. We have

$$E(H(Y) \mid Z_1 = \zeta) = E(H(\varphi((Z_1 \, \| \, Z_2)_{[\sigma]})) \mid Z_1 = \zeta)$$
$$= E(H(\varphi((\zeta \, \| \, Z_2)_{[\sigma]})))$$
$$= E\left(\sum_{i=1}^m \varphi_i((\zeta \, \| \, Z_2)_{[\sigma]}) \right).$$

Since $\varphi_i = \frac{1}{2}(1 - (-1)^{\varphi_i})$, we get

$$E(H(Y) \mid Z_1 = \zeta) = \frac{m}{2} - \frac{1}{2} E\left(\sum_{i=1}^m (-1)^{\varphi_i}((\zeta \, \| \, Z_2)_{[\sigma]}) \right).$$

Let us now introduce the notion of Walsh transform.

Definition 2 (Walsh Transform). *The Walsh transform of a function $f : \{0,1\}^m \to \{0,1\}$ is given by*

$$\hat{f}(w) = \sum_{x \in \{0,1\}^m} (-1)^{w \cdot x} f(x). \tag{11}$$

for all $w \in \{0,1\}^m$.

By writting $(-1)^{\varphi_i}$ in terms of its Walsh transform coefficients, we obtain

$$E(H(Y) \mid Z_1 = \zeta)$$

$$= \frac{m}{2} - \frac{1}{2} E\left(\sum_{i=1}^{m} \frac{1}{2^m} \sum_{(u \,\|\, v)_{[\sigma]}} (-1)^{u \cdot \zeta \cdot v \cdot Z_2} \widehat{(-1)^{\varphi_i}}((u \,\|\, v)_{[\sigma]}) \right)$$

$$= \frac{m}{2} - \frac{1}{2^{m+1}} \sum_{(u \,\|\, v)_{[\sigma]}} (-1)^{u \cdot \zeta} \sum_{i=1}^{m} \widehat{(-1)^{\varphi_i}}((u \,\|\, v)_{[\sigma]}) \, E\left((-1)^{v \cdot Z_2}\right).$$

Since Z is uniformly distributed over $\mathcal{Z}$, we have $E\left((-1)^{v \cdot Z_2}\right) = 0$ if $v \neq 0$, $E\left((-1)^{v \cdot Z_2}\right) = 1$ if $v = 0$, and $E(H(Y)) = \frac{m}{2}$. Hence,

$$E(H(Y) \mid Z_1 = \zeta) = E(H(Y)) - \frac{1}{2^{m+1}} \sum_{u \in \{0,1\}^k} (-1)^{u \cdot \zeta} \sum_{i=1}^{m} \widehat{(-1)^{\varphi_i}}(\bar{u}).$$

By definition, $\mathrm{Cor}(\varphi_i, \ell_{\bar{u}}) = \widehat{(-1)^{\varphi_i}}(\bar{u})$. Hence,

$$E(H(Y) \mid Z_1 = \zeta) = E(H(Y)) - \frac{1}{2^{m+1}} \sum_{u \in \{0,1\}^k} (-1)^{u \cdot \zeta} \widetilde{\mathrm{Cor}}(\varphi, \ell_{\bar{u}}). \qquad \square$$

Now let σ denote the fixed bijection such that $w = (1_{(k)} \,\|\, 0_{(m-k)})_{[\sigma]}$. Lemma 2 can be restated as

$$E(H(I(x,s)) \mid g(I(x,s)) = 0) = E(H(I(x,s))) - \varepsilon_0(s)$$

with $\varepsilon_0(s) = \left(\sum_{\bar{u} \preceq w} (-1)^{u \cdot \zeta_*} \widetilde{\mathrm{Cor}}(\varphi, \ell_{\bar{u}}) \right) / 2^{m+1}$.

Since the protocol did not successfully stop for any element w_i tested before w, it means that no DPA peak – and thus no corresponding bias – happened for these elements. In our attack protocol, by assumption, any element w_i tested before w is such that $H(w_i) \leq H(w)$. In particular, any element $\bar{u} \neq w$ such that $\bar{u} \preceq w$ must have been tested before w. Hence, for all such element $\bar{u}$, we must have $(-1)^{u \cdot \zeta_*} \widetilde{\mathrm{Cor}}(\varphi, \ell_{\bar{u}}) = 0$. Hence, we have

$$\varepsilon_0(s) = \frac{1}{2^{m+1}} (-1)^{1_{(k)} \cdot \zeta_*} \widetilde{\mathrm{Cor}}(\varphi, w) = \frac{(-1)^{H(\zeta_*) \bmod 2}}{2^{m+1}} \widetilde{\mathrm{Cor}}(\varphi, w). \qquad (12)$$

Since Y is uniformly distributed over $\mathfrak{S}' = \{0,1\}^m$, we have

$$P(Y \in \mathfrak{S}'_1(s)) = 1 - P(Y \in \mathfrak{S}'_0(s)) = 1 - 2^{-k}. \qquad (13)$$

Finally, with (12), (13) and Corollary 1, we obtain the expected result.

A.2.2   Case $\hat{s} \neq s$

Let us assume that the S-boxes have ad hoc cryptographic properties. Since the key hypothesis $\hat{s}$ is wrong in this case, there are k' bits in the k-bit pattern that is supposed to be set to ζ_* that are actually uniformly distributed over the set $\{0,1\}^{k'}$, with $k' \geq 1$. Moreover, since the protocol did not successfully stop before this k-bit pattern was tested, it means that no DPA peak – and thus no corresponding bias – happened for any pattern whose length is strictly less than k. In particular, the output of φ is decorrelated from the $k - k'$ bits of the k-bit pattern that are fixed despite the wrong key hypothesis. As a consequence, we must have $\mathfrak{S}'_b = \mathfrak{S}'$ for all $b \in \{0,1\}$.

Hence, whenever $\hat{s} \neq s$, we must have

$$P(I(x,s) \in \mathfrak{S}'_b(\hat{s})) = P(I(x,s) \in \mathfrak{S}') = 1 \quad \text{for all } b \in \{0,1\}, \tag{14}$$

and

$$\begin{aligned}
\varepsilon_0(\hat{s}) &= E(H(I(x,y))) - E(H(I(x,y)) \mid I(x,y) \in \mathfrak{S}'_0(\hat{s})) \\
&= E(H(I(x,y))) - E(H(I(x,y)) \mid I(x,y) \in \mathfrak{S}') \\
&= 0.
\end{aligned} \tag{15}$$

From (14), (15) and Proposition 1, we obtain $E(\Delta_K(t)) = 0$ as expected.

Solving Discrete Logarithms from Partial Knowledge of the Key

K. Gopalakrishnan[1,*], Nicolas Thériault[2,**], and Chui Zhi Yao[3]

[1] Department of Computer Science, East Carolina University, Greenville, NC 27858
[2] Instituto de Matemática y Física, Universidad de Talca, Casilla 747, Talca, Chile
[3] Department of Mathematics, University of California - Riverside, CA 92521

Abstract. For elliptic curve based cryptosystems, the discrete logarithm problem must be hard to solve. But even when this is true from a mathematical point of view, side-channel attacks could be used to reveal information about the key if proper countermeasures are not used. In this paper, we study the difficulty of the discrete logarithm problem when partial information about the key is revealed by side channel attacks. We provide algorithms to solve the discrete logarithm problem for generic groups with partial knowledge of the key which are considerably better than using a square-root attack on the whole key or doing an exhaustive search using the extra information, under two different scenarios. In the first scenario, we assume that a sequence of contiguous bits of the key is revealed. In the second scenario, we assume that partial information on the "Square and Multiply Chain" is revealed.

Keywords: Discrete Logarithm Problem, Generic Groups, Side Channel Attacks.

1 Introduction

The discrete logarithm problem (DLP) is an important problem in modern cryptography. The security of various cryptosystems and protocols (such as Diffie-Hellman key exchange protocol, ElGamal cryptosystem, ElGamal signature scheme, DSA, cryptosystems and signature schemes based on elliptic and hyperelliptic curves) relies on the presumed computational difficulty of solving the discrete logarithm problem. For a survey of the discrete logarithm problem, the reader is referred to [13].

However, even if the DLP is indeed difficult to solve, one has to take other aspects into account in practical implementations. If proper countermeasures are not used, side-channel attacks could be used to reveal partial information about the key. In this paper, we address the problem of how to utilize the partial information effectively when solving the DLP.

---

* This work was done in parts while the author was at the Institute of Pure and Applied Mathematics, UCLA.
** This work was done in parts while the author was at the Fields Institute, Toronto, Canada.

K. Srinathan, C. Pandu Rangan, M. Yung (Eds.): Indocrypt 2007, LNCS 4859, pp. 224–237, 2007.

When one wants to break a system based on DLP, one can of course, ignore the partial information revealed by side channel attacks and simply use a generic algorithm. Alternatively, one can use an exhaustive search using the partial information made available to us. The primary question that we address in this paper is whether we can do something in between? i.e., can we use the partial information revealed by side channel attacks in an intelligent way to break the system?

In some cases, side channel attacks could reveal a string of contiguous bits of the secret key. In this situation, it is always possible to adapt Shank's baby-step giant-step algorithm [19] to perform the search in the remaining possible keyspace; However the memory requirements could make this approach impractical. For example, if 100 bits remain to be identified, computing to the order of 2^{50} group operations can be considered reasonable, but handling (and storing) a table of 2^{50} entries is much more problematic. To avoid this issue, we need a different algorithm, not necessarily deterministic, which has a lower memory requirement.

A number of papers address the question when a large number of observations are available [6,11,12,9]. When only one observation is possible, probabilistic algorithms are known, but they usually assumed that the known bitstring is either in the most or the least significant bits of the key [15,16,23]. In Section 3, we look at what happens when a contiguous sequence of bits is known somewhere in the middle of the binary representation of the key.

In most cases, side channel attacks will reveal information on the square and multiply chain (see the beginning of Section 4 for a definition), and not the bitstring. Extracting the key from partial information on the square and multiply chain requires different approaches than those used when some of the bits are known. In this situation, no "fast" algorithm is known, no matter what the memory requirement is, hence any "fast" algorithm can be considered an improvement.

If uniform formulas are used for the group arithmetic (see [2,1] for example), then a side channel attack will reveal the hamming weight of the key, but not the position of the nonzero bits. If the hamming weight is low enough, fast algorithms are available [22,3], although they can be slower than general searches if the hamming weight is even moderately high. If the field arithmetic is not secured as well, some parts of the square and multiply chain may also be leaked [24,21].

In that situation, no algorithm was known that could improve on the exhaustive search from the partial information, or a search based solely on the hamming weight (note that the two approaches are not compatible). In Section 4, we will show how to significantly improve on the exhaustive search in this context.

2   Background

First, we define the discrete logarithm problem as follows: Let G be a cyclic group of prime order p. Let g be a generator of G. Given $\beta \in G$, determine $\alpha \in \{0, 1, 2, \ldots, p - 1\}$ such that $g^\alpha = \beta$. Here, g and p are public information known to everybody. Although our description is in terms of a multiplicative group, all of the arguments in this paper are essentially identical when applied to additive groups (for example groups coming from elliptic curves).

It is also possible to define the DLP on groups whose order n is not a prime number. However, one could then use the well known technique due to Pohlig and Hellman [14], and reduce the problem to a number of DLPs in groups of order p, where p runs through all the prime factors of n. Hence, without any loss of generality, we will focus on the case when the order of the group is a prime number.

2.1   Generic Algorithms for Solving DLP

In this paper, we only consider *generic algorithms* for solving the DLP. A generic algorithm for solving the DLP is an algorithm that does not exploit the structure of the underlying group in solving the DLP. As a consequence, this algorithm could be used to solve the DLP in any group.

In a generic algorithm, we want to think of the group as though it is presented by means of a black box. More specifically, each group element has a unique encoding or labeling and we have an *oracle* (a black box) which is capable of doing the following things:

- Given the encoding of two elements g and h of the group, the oracle can compute their product $g * h$, in unit time.
- Given the encoding of two elements g and h, the oracle can decide whether $g = h$, in unit time.
- Given the encoding of an element g, the oracle can compute any given power of g (including the inverse g^{-1}), in time $O(\log p)$.

We also note that in some groups, for example those coming from elliptic curves, the inverse operation can be performed in unit time. The time complexity of a generic algorithm is determined by counting the number of times it needs access to the black box.

There are a few well-known generic algorithms to solve the DLP. The baby-step giant-step method due to Shanks [19] is a deterministic generic algorithm that can solve the DLP in time $O(p^{1/2} \log p)$ using space $O(p^{1/2} \log p)$. This algorithm is based on a time-memory trade off technique. The rho method due to Pollard [15,16] is a probabilistic generic algorithm that can solve the DLP in expected running time $O(p^{1/2})$ (under certain assumptions) using only $O(\log p)$ amount of space (requiring the storage of a constant number of group elements), and is based on the birthday paradox. The space efficiency of this algorithm makes it more attractive in comparison to Shanks' method. For an excellent survey of the state of the art in these two methods, the reader is referred to [23].

Victor Shoup [20] established a lower bound of $\Omega(p^{1/2})$ on the time complexity of any probabilistic (and therefore on any deterministic) generic algorithm that can solve the DLP. Hence, both the baby-step giant-step method and the rho method are essentially optimal algorithms with respect to their time complexity and can only be improved in terms of constant factors.

It should be noted that the fastest known algorithms that can solve the DLP for most elliptic curve groups are generic algorithms and thus of exponential complexity (note that the size of the input is $\log p$, whereas the algorithm has $O(p^{1/2})$ complexity). In contrast, subexponential algorithms exist for the factoring problem which is the basis for the RSA cryptosystem. As a consequence, cryptographers believe that elliptic curve based cryptosystems can provide better security (provided the curves and the parameters are chosen appropriately). This is the reason for increasing interest in elliptic curve based cryptography.

2.2   Side Channel Attacks

A side channel attack (on a cryptosystem or a signature scheme) is an attack that focuses on the physical implementation of the algorithm as opposed to the specification of the algorithm. By observing an implementation being executed, an attacker can make correlations between the events that occur in the processor and the data being processed. The first well known versions of side channel attack were based on timing [7] and power [8] analysis (and more recently EM analysis [4,17]).

In timing analysis based attacks, an attacker uses the execution timings to infer about the flow of control and thus about the key. For example, an implementation might take longer to run if a conditional branch is taken than if it is not taken. If the branch is taken or not depending on a bit of the secret key, then the attacker might work out the corresponding bit of the secret key.

In power analysis based attacks, an attacker uses the amount of power consumed by a processor to infer what operations are being performed and thus about the key (EM based attacks use a similar approach on the electromagnetic trace produced by the processor). For example, a multiplication operation would have a distinct power usage profile and will differ considerably from the power usage profile of a squaring operation. The attacker can then use that knowledge to break the system.

The interesting thing about side channel attacks is that they do not contradict the mathematical security provided by the system (even assuming the underlying computational problems are provably difficult to solve) but they simply bypass it.

3   Scenario I – Contiguous Bits of the Key Is Revealed

In this section, we deal with the scenario where the partial information revealed is a sequence of contiguous bits of the key.

Let G be a cyclic group of prime order p and let g be a generator of G. Given $\beta \in G$, recall that the Discrete Logarithm Problem (DLP) is to determine $\alpha \in \{0, 1, 2, \ldots, p - 1\}$ such that $g^{\alpha} = \beta$. In this section, we assume that a sequence of contiguous bits of α is revealed by side channel attacks. Although a variation of the baby-step giant-step method is always possible, we are looking for an algorithm with memory requirements similar to the rho method, i.e. of size $O(\log p)$.

There are three possible cases to consider; the sequence of contiguous bits known may be in the left part, right part or somewhere in the middle (i.e. located away from the extremities, but not necessarily centered). The first two cases are known results and can be found in Appendices A and B. The third case does not appear to have received as much attention, and a new method to approach it is presented below.

3.1   Case III – Middle Part

Let us assume that we have some positive integers M and N such that we can write α in the form

$$\alpha = \alpha_1 MN + \alpha_2 M + \alpha_3 \tag{1}$$

where $0 \leq \alpha_2 < N$ is known and with $0 \leq \alpha_3 < M$. We also assume that $0 \leq \alpha_1 < p/MN$, i.e. that α is reduced modulo p. Note that α_1 is really bounded by $\left\lfloor \frac{p-\alpha_2 M}{MN} \right\rfloor$, but the error introduced is insignificant (a difference of at most 1 on the bound, which vanishes in the O-notation), and it makes the analysis easier to read. In terms of exhaustive search, we have to search through a set of size p/N ($\lfloor (p - \alpha_2 M)/N \rfloor$ to be exact), which requires $O(p/N)$ calls to the oracle, whereas a generic attack on the whole group would require $O(p^{1/2})$ calls to the oracle.

In practice we may be more interested in the case where M and N are powers of 2 – i.e. where $N = 2^{l_2}$ and $M = 2^{l_3}$, so the first l_1 and the last l_3 bits are unknown (we assume that $l_1 + l_2 + l_3$ is the bitlength of the key) – but the arguments presented here will hold for any positive integers N and M.

For now, let us assume that we are given an integer r, $0 < r < p$, such that we can write rMN as $kp + s$ with $|s| < p/2$. We will discuss how to choose r in a few paragraphs. Multiplying both sides of Equation (1) by r, we get

$$\begin{aligned} r\alpha &= r\alpha_1 MN + r\alpha_2 M + r\alpha_3 \\ &= \alpha_1 kp + s\alpha_1 + r\alpha_2 M + r\alpha_3 \\ &= \alpha_1 kp + r\alpha_2 M + \alpha', \end{aligned} \tag{2}$$

where $\alpha' = s\alpha_1 + r\alpha_3$. Raising g to both sides of Equation (2), we get

$$\begin{aligned} g^{\alpha r} &= g^{\alpha_1 kp + r\alpha_2 M + \alpha'} \\ (g^\alpha)^r &= (g^p)^{\alpha_1 k} g^{r\alpha_2 M} g^{\alpha'} \\ \beta^r &= g^{r\alpha_2 M} g^{\alpha'}. \end{aligned} \tag{3}$$

Denoting $\left(\beta \times g^{-\alpha_2 M}\right)^r$ by β', Equation (3) can be written in the form $\beta' = g^{\alpha'}$. Note that β' can be computed from β as r, α_2, and M are known. We can then view determining α' as solving a DLP. When s is positive, $\alpha' = \alpha_3 r + \alpha_1 s$ must be in the interval

$$\left[0, r(M - 1) + s\left(\frac{p}{MN} - 1\right) \right],$$

on which we can use Pollard's kangaroo method. Similarly, if s is negative we must consider the interval

$$\left[s\left(\frac{p}{MN} - 1 \right), r(M-1) \right].$$

In both cases we can restrict the value of α' to an interval of length $rM + |s|\frac{p}{MN}$.

To minimize the cost of the kangaroo method, we must therefore choose $r > 0$ to minimize the value of

$$rM + |s|\frac{p}{MN} \tag{4}$$

under the condition $s \equiv rMN \bmod p$.

Although it is not possible in general to choose r (and s) such that (4) is of the form $O(p/N)$, some situations are more favorable than others.

A perfect example (although a rather unlikely one) occurs when working in the bitstring of a Mersenne prime $p = 2^l - 1$ (with $N = 2^{l_2}$ and $M = 2^{l_3}$), in which case we can choose $r = 2^{l_1} = 2^{l-l_2-l_3}$ and $s = 1$ and we get an interval of length $O(2^{l_1+l_3})$. Similarly, if the difference between p and 2^l is of size $O(2^{l_3})$, we can set $r = 2^{l_1}$ and $s = 2^l - p$ and obtain an interval of length $O(2^{l_1+l_3})$. In both of these situations, using Pollard's kangaroo method would allow us to compute α' in time $O(2^{(l_1+l_3)/2}) = O(\sqrt{p/N})$.

Unfortunately, such an optimal choice of r (and s) is impossible in general. We will now consider how to choose r in order to minimize the range of possible values for $\alpha_1 s + \alpha_3 r$. To do this, we will determine a value T ($0 < T < p$) such that we can ensure both $\alpha_3 r < T$ and $\alpha_1|s| < T$ by choosing r carefully.

We first consider the inequality $\alpha_1|s| < T$. Replacing s by $rMN - kp$ (from the definition of s) and bounding α_1 by $\frac{p}{MN}$ gives us

$$\frac{p}{MN}|rMN - kp| < T$$

and a few simple manipulations turn the inequality into

$$\left| \frac{MN}{p} - \frac{k}{r} \right| < \frac{\left(\frac{MNT}{p^2} \right)}{r}.$$

Thinking of $\frac{MN}{p}$ as the real number γ, and $\frac{MNT}{p^2}$ as ϵ, we recognize Dirichlet's Theorem on rational approximation (see [18] page 60, for example). We can then say that there exists two integers k and r satisfying the inequality and such that $1 \le r \le \frac{1}{\epsilon} = \frac{p^2}{MNT}$. We also know that k and r can be found using the continued fraction method (see D page 237, for a brief description).

Since the upper bound on r from Dirichlet's Theorem is tight, and since $0 \le \alpha_3 < M$, the best bounds we can give on $\alpha_3 r$ in general are $0 \le \alpha_3 r < \frac{p^2}{NT}$. To ensure that $\alpha_3 r < T$, we must therefore require

$$\frac{p^2}{NT} \le T,$$

or equivalently $T \ge p/\sqrt{N}$. Since we want T as small as possible (we will end up with an interval of size $2T$ for the kangaroo method), we fix $T = p/\sqrt{N}$.

This means that r can be selected such that computing α' with the kangaroo method can be done in a time bounded above by $O(\sqrt{2}p^{1/2}/N^{1/4})$. From $\alpha' = r\alpha_3 + s\alpha_1$, we have to solve an easy diophantine equation to obtain α_1 and α_3 (see Appendix C for details), and Equation (1) gives us α.

We can therefore reduce the search time for the discrete log from $O(\sqrt{p})$ for Pollard rho (or the kangaroo method applied directly on the possible values of α) by a factor of at least $O(N^{1/4})$ in general, and up to $O(\sqrt{N})$ in the best situations, while keeping the memory requirement of $O(\log p)$ of the Kangaroo algorithm.

4   Scenario II – Partial Information About the Square and Multiply Chain Is Revealed

In order to do modular exponentiation efficiently, typically one uses the *square and multiply algorithm*. We will illustrate the working of this algorithm by means of an example here and refer the reader to [10], page 71, for its formal description and analysis.

For example, suppose we want to compute g^{43}. We will first write down 43 in binary as 101011. Starting after the leading 1, replace each 0 by S (Square) and each 1 by SM (Square and Multiply) to get the string $SSMSSMSM$. Such a string goes by the name *square and multiply chain*. We start with $h = g$ (i.e. with g^1, corresponding to the leading bit) and do the operations (Squaring h and Multiplying h by g) specified by a scan of the above string from left to right, storing the result back in h each time. At the end, h will have the desired result. In this particular example, the successive values assumed by h would be $g \rightarrow g^2 \rightarrow g^4 \rightarrow g^5 \rightarrow g^{10} \rightarrow g^{20} \rightarrow g^{21} \rightarrow g^{42} \rightarrow g^{43}$. Note that, the final value of h is g^{43} as desired.

In this section, we assume that exponentiation is done using the square and multiply algorithm. As the power consumption profile of squaring operation is often considerably different from that of multiplication operation, one could figure out which one of the two operations is being performed using side channel information (unless sufficient countermeasures are used). In the following, we assume that some partial information about the square and multiply chain is revealed by side channel attacks.

Specifically, we assume that a side channel attack revealed the position of some of the multiplications (M) and squares (S) of the square and multiply chain. Note that once a multiplication is identified, the operations next to it are known to be squares (i.e. we have the substring SMS) since there are no consecutive multiplications in the square and multiply chain. We will assume that n elements of the square and multiply chain have not been identified, of which i are multiplications (M). The problem that we address is how to exploit the partial information effectively to figure out the entire square and multiply chain.

As the chain is made up of only S's and M's, if we can figure out the positions of all the M's, the string is completely determined, so we need to figure out the exact positions of the remaining i M's. A naive approach to solving the problem consists in guessing the positions of the i remaining M's. This will determine the chain completely and hence the key. We can then verify whether our guess is correct by checking if g^α equals β. If we guessed correctly we can stop, otherwise we can make new guesses until we eventually succeed. In this approach, we are essentially doing an exhaustive search, so the complexity would be $O(n^i \log p)$ as there are $\binom{n}{i}$ possible guesses for the missing M's, each of which requires $O(\log p)$ time to test.

Also note that even though we may know the relative position of some of the multiplications in the square and multiply chain, this does not readily translate into information on the bitstring as the number of the remaining M's after and between two known M's will change the position of the corresponding nonzero bits in the binary representation of α (in particular, this is why the algorithms of Stinson [22] and Cheng [3] cannot easily be adapted to work in this situation).

We could use the fact that no two M's are next to one another in the square and multiply chain to reduce the number of possible guesses to test. However, the overall effect will usually be small, and the worst case complexity will continue to be essentially $O(n^i \log p)$. To solve this, we develop a more sophisticated approach of exploiting the partial information that will make an impact on the worst case complexity.

First, we assume that we can somehow split the chain into two parts, left and right, such that $\frac{i}{2}$ of the remaining M's are on the left part and the other $\frac{i}{2}$ are on the right part. We can now make use of the time-memory trade off technique and determine the entire chain in time $O(n^{\frac{i}{2}} \log p)$. The details are explained below.

Suppose we make a specific guess for the $\frac{i}{2}$ M's on the left part and another guess for the $\frac{i}{2}$ M's on the right part. This determines the square and multiply chain completely and hence the bit string representation of the key α. Let a be the number represented by the bit string corresponding to the left part and let b be the number represented by the bit string corresponding to the right part. Let x be the length of the bitstring corresponding to the right part. Note that, we do know x as we are assuming, for the moment, that the position of the split is given to us. Then, clearly

$$\beta = g^\alpha$$
$$= g^{a2^x + b}$$
$$= \left(g^{2^x}\right)^a g^b \ . \tag{5}$$

If we denote g^{-2^x} by h, then the above equation reduces to

$$g^b = h^a \beta \ . \tag{6}$$

We can use Equation (6) to check whether our guess is correct. However, even if we use Equation (6) to verify a guess, the worst case complexity will still be $O(n^i \log p)$. This is because there will be $O(n^{\frac{i}{2}})$ guesses for a, $O(n^{\frac{i}{2}})$ guesses for b and any guess for a can be paired with any guess for b to make up a complete guess.

Instead, we shall use a time-memory trade off technique. Consider all different possible guesses for the left part. This will yield all different possible guesses for a. For each such guess we compute h^a and we record the pairs (a, h^a). We then sort all the pairs into an ordered table based on the second column, viz. the value of h^a.

Next we make a guess for the right part. This will yield a guess for b. We can now compute $y = \beta^{-1} \times g^b$. If our guess for b is indeed correct, then y will be present in the second column of some row in the table we built. The first column of that row will produce the matching guess for a. So, all that we have to do is search for the presence of y in the second column of the table. This can be done using the *binary search* algorithm as the table is already sorted as per the second column. If y is present, we are done and have determined the key. If y is not present, we make a different guess for the right part and continue until we eventually succeed. Since there are $n^{\frac{i}{2}}$ guesses for the right part, the time complexity of this algorithm will be $O(n^{\frac{i}{2}} \log p)$. As there are $n^{\frac{i}{2}}$ guesses for the left part, the table that we are building will have that many entries and so the space complexity of our algorithm will be $O(n^{\frac{i}{2}} \log p)$.

Hence, both the time and space complexity of our algorithm will be $O(n^{\frac{i}{2}})$ (ignoring the $\log p$ term). In contrast, the naive approach would have a time complexity of $O(n^i)$ and space complexity of $O(\log p)$ as only constant amount of storage space is needed. This is why this is called a time-memory trade off technique.

In the analysis above, we ignored the costs associated with handling the table (sorting and searching). Let $m = n^{\frac{i}{2}}$. Whereas it takes time $O(m \log p)$ (oracle operations) to compute the elements of the table, sorting it will require a time of $O(m \log m)$ bit operations even with an optimal sorting algorithm (such as merge sort). Similarly, to search in a sorted table of size $O(m)$ even with an optimal searching algorithm (such as binary search) will take $O(\log m)$ bit operations. Note that in practice it is common to use a hash table when m is large, but this does not change the form of the asymptotic cost. As we have $O(m)$ guesses for the right part and a search is needed for each guess, the total time spent after the table is built would be $O(m \log m)$. So, technically speaking, the true complexity of our algorithm is $O(m \log m) = O(n^{i/2} \log p \log \log p)$ bit operations (since both n and i are $O(\log p)$). However, in practice our "unit time" of oracle (group) operation is more expensive than a bit operation (requiring at least $\log p$ bit operations), whereas the $\log \log p$ terms grows extremely slowly, and we can safely assume that the main cost (in oracle time) also covers the table costs.

Recall that we assumed that we can somehow split the chain into two parts, left and right, such that $\frac{i}{2}$ of the remaining M's are in each of the two parts. This is, of course, an unjustified assumption. Although we made this assumption for ease of exposition, this assumption is not really needed. If we consider the set of remaining M's as ordered, then we can easily define one of them as the "middle one". We will use the position of the "middle" M as our splitting position. There are $n - i = O(n)$ possible positions for the "middle" M (of which only one is the true position).

For each of the possible positions, we try to obtain a match by assuming the "middle" M is in this position and placing the $i-1$ others. If i is odd, then we have $\frac{i-1}{2} = \lfloor \frac{i}{2} \rfloor$ of the remaining M's on each side, and if i is even then we have $\frac{i}{2} - 1$ of the remaining M's on one side (say, on the left) and $\frac{i}{2}$ on the other (with $\frac{i}{2} = \lfloor \frac{i}{2} \rfloor$). Using the time-memory trade off technique, we have a complexity of $O(n^{\lfloor \frac{i}{2} \rfloor} \log p)$ for each of the possible positions for the "middle" M, and we obtain a total complexity of $O(n^{\lfloor \frac{i}{2} \rfloor + 1} \log p)$.

With our algorithm, we are able to cut down the complexity from $O(n^i)$ for the naive approach to $O(n^{\lfloor \frac{i}{2} \rfloor + 1})$ (ignoring logarithmic terms). This is a significant improvement considering that the exponent of n has been reduced to about half the original value and n will typically be very large (but still $O(\log p)$) in practical implementations of elliptic curve based cryptosystems.

5   Concluding Remarks

To summarize, in this article we considered the problem of determining the key used in discrete logarithm based systems when partial knowledge of the key is obtained by side channel attacks. We considered two different scenarios of partial information viz. knowing a sequence of contiguous bits in the key and knowing some part of the square and multiply chain. In both scenarios, we were able to develop better algorithms in comparison to both using a square-root algorithm (ignoring the partial information available to us) and doing an exhaustive search using the extra information available. In particular, in the second scenario, our algorithm is almost asymptotically optimal considering that its complexity is very close to the square root of the order of the remaining key space.

Although we have made some progress, many more situations could be considered. We give the following as examples:

1. Consider the first scenario where we assume that a sequence of contiguous bits in the middle of the key corresponding to a set of size N have been revealed by side channel attacks (Section 3.1). Although we were able to reduce the search time, only some situations will match the optimal search time of $O(\sqrt{p/N})$. For a general combination of p, M and N, we would still have to reduce the search time by a factor of $O(N^{1/4})$ to obtain an asymptotically optimal algorithm.
2. In the first scenario, we assumed that the known bits are contiguous bits. It is possible that in some circumstances, we may get to know some bits of the key, but the known bits may not be contiguous.
3. Finally, the Non-adjacent Form Representation (NAF) of the key is sometimes used to do the exponentiation operation more efficiently (in the average case) [5]. If a sequence of bits is known, the situation is very similar to that of Section 3. When partial information about the square and multiply chain is obtained, the situation changes significantly compared to the binary square and multiply, since it is usually assumed that multiplications by g and g^{-1} are indistinguishable. Although it is easy to adapt the algorithm presented in

Section 4 to locate the position of the multiplications coming from nonzero bits, a factor of $O(2^m)$ (where m is the number of nonzero bits in the NAF representation of the key) would be included in the complexity to deal with the signs of the bits, which often cancels any gains we obtained.

In these situations, we leave the development of optimal algorithms, whose complexity would be the square root of the order of the remaining key space (or close to it), as an open problem.

Acknowledgments. This paper is an outcome of a research project proposed at the RMMC Summer School in Computational Number Theory and Cryptography which was held at the University of Wyoming in 2006. We would like to thank the sponsors for their support.

References

1. Brier, É., Déchène, I., Joye, M.: Unified point addition formulæ for elliptic curve cryptosystems. In: Embedded Cryptographic Hardware: Methodologies and Architectures, pp. 247–256. Nova Science Publishers (2004)
2. Brier, É., Joye, M.: Weierstraß elliptic curves and side-channel attacks. In: Naccache, D., Paillier, P. (eds.) PKC 2002. LNCS, vol. 2274, pp. 335–345. Springer, Heidelberg (2002)
3. Cheng, Q.: On the bounded sum-of-digits discrete logarithm problem in finite fields. SIAM J. Comput. 34(6), 1432–1442 (2005)
4. Gandolfi, K., Mourtel, C., Olivier, F.: Electromagnetic analysis: Concrete results. In: Koç, Ç.K., Naccache, D., Paar, C. (eds.) CHES 2001. LNCS, vol. 2162, pp. 251–261. Springer, Heidelberg (2001)
5. Gordon, D.M.: A survey of fast exponentiation methods. Journal of Algorithms 27, 129–146 (1998)
6. Howgrave-Graham, N., Smart, N.P.: Lattice attacks on digital signature schemes. Des. Codes Cryptogr. 23(3), 283–290 (2001)
7. Kocher, P.C.: Timing attacks on implementations of Diffie-Hellman, RSA, DSS and other systems. In: Koblitz, N. (ed.) CRYPTO 1996. LNCS, vol. 1109, pp. 104–113. Springer, Heidelberg (1996)
8. Kocher, P.C., Jaffe, J., Jun, B.: Differential power analysis. In: Wiener, M.J. (ed.) CRYPTO 1999. LNCS, vol. 1666, pp. 388–397. Springer, Heidelberg (1999)
9. Leadbitter, P.J., Page, D., Smart, N.P.: Attacking DSA under a repeated bits assumption. In: Joye, M., Quisquater, J.-J. (eds.) CHES 2004. LNCS, vol. 3156, pp. 428–440. Springer, Heidelberg (2004)
10. Menezes, A.J., van Oorschot, P.C., Vanstone, S.A.: Handbook of Applied Cryptography. CRC Press, Inc., Boca Raton (1996)
11. Nguyen, P.Q., Shparlinski, I.E.: The insecurity of the digital signature algorithm with partially known nonces. J. Cryptology 15(3), 151–176 (2002)
12. Nguyen, P.Q., Shparlinski, I.E.: The insecurity of the elliptic curve digital signature algorithm with partially known nonces. Des. Codes Cryptogr. 30(2), 201–217 (2003)
13. Odlyzko, A.M.: Discrete logarithms: The past and the future. Designs, Codes and Cryptography 19, 129–145 (2000)

14. Pohlig, S.C., Hellman, M.E.: An improved algorithm for computing logarithms over $GF(p)$ and its cryptographic significance. IEEE Transactions on Information Theory 24, 106–110 (1978)
15. Pollard, J.M.: Monte Carlo methods for index computation (mod p). Mathematics of Computation 32(143), 918–924 (1978)
16. Pollard, J.M.: Kangaroos, Monopoly and discrete logarithms. Journal of Cryptology 13(4), 437–447 (2000)
17. Quisquater, J.-J., Samyde, D.: Electromagnetic analysis (EMA): Measures and counter-measures for smart cards. In: Attali, I., Jensen, T. (eds.) E-smart 2001. LNCS, vol. 2140, pp. 200–210. Springer, Heidelberg (2001)
18. Schrijver, A.: Theory of Linear and Integer Programming. In: Wiley-Interscience Series in Discrete Mathematics, John Wiley & Sons, Chichester (1986)
19. Shanks, D.: Class number, a theory of factorization and genera. In: Proc. Symp. Pure Math., vol. 20, pp. 415 440 (1971)
20. Shoup, V.: Lower bounds for discrete logarithms and related problems. In: Fumy, W. (ed.) EUROCRYPT 1997. LNCS, vol. 1233, pp. 256–266. Springer, Heidelberg (1997)
21. Stebila, D., Thériault, N.: Unified point addition formulae and side-channel attacks. In: Goubin, L., Matsui, M. (eds.) CHES 2006. LNCS, vol. 4249, pp. 354–368. Springer, Heidelberg (2006)
22. Stinson, D.: Some baby-step giant-step algorithms for the low hamming weight discrete logarithm problem. Math. Comp. 71(237), 379–391 (2002)
23. Teske, E.: Square-root algorithms for the discrete logarithm problem (a survey). In: Public-Key Cryptography and Computational Number Theory, pp. 283–301. Walter de Gruyter, Berlin (2001)
24. Walter, C.D.: Simple power analysis of unified code for ECC double and add. In: Joye, M., Quisquater, J.-J. (eds.) CHES 2004. LNCS, vol. 3156, pp. 191–204. Springer, Heidelberg (2004)

A   Case I – Left Part

Here we assume that contiguous most significant bits of the key are known. Let z denote the bit string formed by the sequence of known bits. Suppose that l is the length of the key. Suppose that l_1 is the length of the known sequence of contiguous bits and l_2 is the length of the remaining bits so that $l = l_1 + l_2$.

Then the smallest possible value for α is the number a represented (in unsigned binary notation) by z concatenated with a sequence of l_2 zeroes. The largest possible value for α is the number b represented by z concatenated with a sequence of l_2 ones. We do know that $a \leq \alpha \leq b$.

Since α could take all the values in the interval $[a, b]$, we are in the ideal situation for Pollard's *Kangaroo Algorithm*. This probabilistic algorithm [15,16] was developed to compute the discrete logarithm when it is known to lie in an interval $[a, b]$. It also can be implemented in a space efficient manner and has expected running time of $O(\sqrt{b - a})$ under some heuristic assumptions.

Since we know both a and b, we can make use of Pollard's Kangaroo algorithm in this case. Note that the binary representation of $b - a$ in our case is simply a sequence of l_2 1's and so $b - a$ is $2^{l_2} - 1$. The expected running time to determine α using the Kangaroo algorithm will then be $O(2^{l_2/2})$.

Observe that if we ignored the partial information available to us and solved the DLP by using the rho method or the baby-step giant-step method, the running time would be $O(2^{l/2})$ which is much higher. Also, observe that if we had exhaustively searched for a key consistent with our partial knowledge, the running time would be $O(2^{l_2})$. So, we are able to do better than these two obvious ways.

B   Case II – Right Part

Here we assume that contiguous least significant bits of the key are known. Let z denote the bit string formed by the sequence of known bits and let α_2 denote the number represented (in unsigned binary notation) by z. Suppose that l is the length of the key. Suppose that l_2 is the length of the known sequence of contiguous bits and l_1 is the length of the remaining bits so that $l = l_1 + l_2$.

Observe that if we ignored the partial information available to us and solved the DLP by using the rho method or the baby-step giant-step method, the running time would be $O(2^{l/2})$. Also, observe that if we had exhaustively searched for a key consistent with our partial knowledge, the running time would be $O(2^{l_1})$.

We know that $\alpha \equiv \alpha_2 \bmod 2^{l_2}$ (since we know the l_2 right-most bits of α), and we let α_1 be the integer corresponding to the l_1 left-most bits of α, i.e.

$$\alpha = \alpha_1 \times 2^{l_2} + \alpha_2 . \tag{7}$$

Let $M = \left\lfloor \frac{p - \alpha_2 - 1}{2^{l_2}} \right\rfloor$, then we know that $0 \le \alpha_1 \le M$ since $0 \le \alpha \le p-1$. Raising g to both sides of Equation (7), we can write

$$\begin{aligned}
\beta = g^\alpha &= g^{\alpha_1 \times 2^{l_2} + \alpha_2} \\
&= g^{\alpha_1 \times 2^{l_2}} g^{\alpha_2} \\
&= \left(g^{2^{l_2}} \right)^{\alpha_1} g^{\alpha_2} .
\end{aligned} \tag{8}$$

If we denote $g^{2^{l_2}}$ by g' and $\beta \times g^{-\alpha_2}$ by β', then Equation (8) reduces to $\beta' = (g')^{\alpha_1}$. As Teske [23] observed, we can then solve this DLP by using Pollard's Kangaroo Algorithm on g' and β' in $O(\sqrt{M})$ time as $0 \le \alpha_1 \le M$. Once α_1 is known, Equation (7) gives the value of α. As $M = \left\lfloor \frac{p - \alpha_2 - 1}{2^{l_2}} \right\rfloor$, the complexity is easily seen to be $O(\sqrt{\frac{p}{2^{l_2}}})$, which is same as $O(2^{l_1/2})$. Thus, we are able to cut down the complexity to square root of the size of the remaining key space.

C   Solving the Diophantine Equation

In Section 3.1, we use Pollard's kangaroo algorithm to compute $\alpha' = \alpha_3 r + \alpha_1 s$, for some carefully chosen r and s. However, this raises the question of how to extract α_1 and α_3 from α', after which we can use Equation (1) to obtain α.

To do this, we first show that r and s may be assumed to be coprime. Since $s = rMN - kp$ with

$$\left| \frac{MN}{p} - \frac{k}{r} \right| < \frac{\left(\frac{MNT}{p^2} \right)}{r} ,$$

we can assume that k and r are coprime: if $\gcd(k,r) = d > 1$, we can replace r and k with $\tilde{r} = r/d$ and $\tilde{k} = k/d$, which gives us $\tilde{s} = \tilde{r}MN - \tilde{k}p = s/d$. The inequality clearly still holds for $\tilde{r}$ and $\tilde{k}$ since $\frac{\tilde{k}}{\tilde{r}} = \frac{k}{r}$ and $\frac{\left(\frac{MNT}{p^2} \right)}{r} < \frac{\left(\frac{MNT}{p^2} \right)}{\tilde{r}}$, but we have smaller values for the interval, which is clearly more advantageous for the search. Furthermore, r is also coprime to p (since p is a prime and $0 < r < p$), so we have

$$\gcd(r,s) = \gcd(r, rMN - kp) = \gcd(r, kp) = 1 .$$

Once we have that $\gcd(r,s) = 1$, finding all integer solutions of $\alpha' = s\alpha_1 + r\alpha_3$ is straightforward. Well known number theoretic techniques give us that all solutions are of the form

$$\alpha_1 = b + ir$$
$$\alpha_3 = \frac{\alpha' - sb}{r} - is$$

where $b \equiv \alpha' s^{-1} \bmod r$. The problem is then to restrict the number of possible solutions, and the choice of r and s helps us once again. Recall that we started with the condition $|s| < p/2$, which forces $r > \frac{p}{2MN}$. Since $0 \leq \alpha_1 < \frac{p}{MN}$, there are at most two possible (and easily determined) values of i, and we can verify each one in time $O(\log p)$.

D   Computing r and k

We now give a brief description of how to compute r and k using the continued fraction method. For the theoretical background, the reader can refer to [18]. In our context, we are trying to approximate the number $\gamma = \frac{MN}{p}$ with a rational $\frac{k}{r}$ such that the approximation error is less than $\frac{\epsilon}{r}$ with $\frac{1}{\epsilon} = \frac{p^2}{MNT} = \frac{p}{M\sqrt{N}}$.

The continued fraction method works iteratively, giving approximations k_i/r_i of γ ($i \geq 0$). Since $0 < \gamma < 1$, we can initialize the process with $\gamma_0 = \gamma$, $r_0 = 1$, $k_0 = 0$, $r_{-1} = 0$ and $k_{-1} = 1$. At each iterative step, we let $a_i = \lfloor \gamma_{i-1}^{-1} \rfloor$ and we compute $r_i = a_i r_{i-1} + r_{i-2}$, $k_i = a_i k_{i-1} + k_{i-2}$, and $\gamma_i = \gamma_{i-1}^{-1} - a_i$. The continued fraction expansion of γ will be $[0; a_1, a_2, a_3, \ldots]$.

To find an optimal pair (r,s), i.e. a pair that minimizes $Mr + \frac{p}{MN}|s|$ (the size of the interval that will be searched with the kangaroo method), we proceed as follows. Once we have a first approximation $\frac{k_i}{r_i}$ of γ such that $r_{i+1} < \frac{1}{\epsilon}$, we evaluate $L_i = Mr_i + \frac{p}{MN}|s_i|$ (with $s_i = r_i MN - k_i p$). We then continue the iterations, keeping track of the best pair r, s found so far (in the form of a triple (r_j, s_j, L_j)). Once $Mr_{i+1} > L_j$, we are done and we can set $r = r_j$, $s = s_j$. To see that we find the optimal pair, observe that the value of r_i never decreases, so once $Mr_{i+1} > L_j$ all further iterations will produce an L_i greater than L_j. This will take no more than $O(\log p)$ steps.

New Description of SMS4 by an Embedding over $GF(2^8)$

Wen Ji and Lei Hu

State Key Laboratory of Information Security,
Graduate School of Chinese Academy of Sciences,
Beijing 100049, China

Abstract. SMS4 is a 128-bit block cipher which is used in the WAPI standard in China for protecting wireless transmission data. Due to the nature that the functions deployed in the round transformations of SMS4 operate on two different fields $GF(2^8)$ and $GF(2)$, it is difficult to analyze this cipher algebraically. In this paper we describe a new block cipher called ESMS4, which uses only algebraic operations over $GF(2^8)$. The new cipher is an extension of SMS4 in the sense that SMS4 can be embedded into ESMS4 with restricted plaintext space and key spaces. Thus, the SMS4 cipher can be investigated through this embedding over $GF(2^8)$. Based on this new cipher, we represent the SMS4 cipher with an overdetermined, sparse multivariate quadratic equation system over $GF(2^8)$. Furthermore, we estimate the computational complexity of the XSL algorithm for solving the equation system and find that the complexity is 2^{77} when solving the whole system of equations.

Keywords: block cipher, SMS4, ESMS4, algebraic equation, XSL algorithm.

1 Introduction

SMS4 is a block cipher published in 2006, which was designed to be used in the WAPI (Wired Authentication and Privacy Infrastructure) standard [1], and WAPI is officially mandated for securing wireless networks within China. The SMS4 cipher was carefully designed to resist standard block cipher attacks such as linear and differential attacks. As far as we know, since its publication, there have been only two papers analyzing this cipher [10,13], from the aspects of differential power attack and integral attack, respectively. In this paper we analyze the structure of SMS4 from a viewpoint of algebra.

In the last few years, algebraic attack on block ciphers has received much attention. This mainly comes from the fact that a block cipher can be described as a system of quadratic equations, and that solving this system can break the cipher [4].

For solving a large nonlinear equation system some linearization methods were presented, such as relinearization [8], XL [5], variants of XL [6] and XSL [3,4]. Among these methods which all have large complexity, XSL is claimed to

K. Srinathan, C. Pandu Rangan, M. Yung (Eds.): Indocrypt 2007, LNCS 4859, pp. 238–251, 2007.

be more efficient than other methods for solving some special equation systems derived from ciphers like AES [3,4].

In Crypto'02, Murphy and Robshow introduced a new cipher BES which uses simple algebraic operations over GF(2^8). With restricted plaintext and key spaces, AES can be regarded as being identical to the BES cipher. Therefore, the AES cipher can be described by using only algebraic operations over the field GF(2^8) [11]. In the same paper they described the AES cipher with an overdetermined multivariate system over GF(2^8), and indicated that this equation system is more sparse than the equation system on GF(2). Additionally, the XSL algorithm is presented and it can solve this large equation system over GF(2^8) far faster than the system over GF(2) in [4].

As in AES, the critical structure of the round transformations in SMS4 consists of an S-box substitution and linear diffusion transformations, and they operate over two different fields, GF(2^8) and GF(2). For example, the inversion transformation of the S-box operates on GF(2^8), while the left cyclicly shifts in linear diffusion transformations operate on GF(2). Thus, cryptanalysis on SMS4 becomes complicated due to the nature of transformations operating two different fields. To avoid this drawback for the investigation of SMS4, we introduce a new cipher called ESMS4, which is an extension of SMS4 by using the conjugate transformation of bytes. All the transformations in ESMS4 operate on GF(2^8) and SMS4 can be embedded in ESMS4 with restricted plaintext and key spaces. Thus, we can discuss the SMS4 cipher over a unique field GF(2^8), and the algebraic cryptanalysis to SMS4 can be then simplified. With this new construction in mind, we describe the SMS4 cipher with a system of multivariate quadratic equations over GF(2^8) and find that the system of equations is very sparse. Along this line, we discuss the computational complexity of XSL algorithm for solving the equation system of the SMS4 cipher over GF(2^8), and we get that the complexity is 2^{77}, which is much smaller than the resulting complexity 2^{401} of BES-128 [12]. We also find that with the increasing of the number of rounds, the computational complexity increases slowly.

This paper is organized as follows. The SMS4 cipher is briefly reviewed in Section 2, and its extension, ESMS4, is presented in Section 3. In Section 4, we present a multivariate equation system over GF(2^8), which describes the SMS4 cipher. In Section 5, the computational complexity of XSL algorithm for solving the system is investigated. Finally, the paper is concluded in Section 6.

1.1   Notation

In this paper we use the following notations. An 8-bit value is simply called a byte, a 32-bit value a word, and a 128-bit value is called a block. In SMS4 the input of each round is a 128-bit block, but the round transformations operate on words. For convenience, the values over which the functions in round transformations operate are considered as an element of GF(2^8) or a column vector in GF(2)^8. When a column vector is multiplied by a matrix, the matrix will occur on the left. Let $\mathbb{F}$ denote the field GF(2^8), then the SMS4 cipher is investigated over the vector space $\mathbb{F}^4$ which is denoted with $\mathcal{A}$, and the new cipher ESMS4 is described over the vector space $\mathbb{F}^{32}$ which is denoted with $\mathcal{E}$.

2   A Brief Description of SMS4

In this section we briefly review the SMS4 block cipher. SMS4 encrypts a 128-bit plaintext with a 128-bit key and outputs 128-bit ciphertext after 32 rounds of nonlinear iterations.

Let $(X_i, X_{i+1}, X_{i+2}, X_{i+3}) \in (\mathrm{GF}(2)^{32})^4$ be the input block of the i-th round, rk_i be the corresponding round key, $X_{i+j}, rk_i \in \mathrm{GF}(2)^{32}$, $i = 0, \cdots, 31$, $j = 0, \cdots, 3$. The i-th round transformation of SMS4 is defined as:

$$
\begin{aligned}
T : \mathrm{GF}(2)^{128} \times \mathrm{GF}(2)^{32} &\to \mathrm{GF}(2)^{128} \\
(X_i, X_{i+1}, X_{i+2}, X_{i+3}, rk_i) &\mapsto (X_{i+1}, X_{i+2}, X_{i+3}, \\
&\quad X_i \oplus L(S^*(X_{i+1} \oplus X_{i+2} \oplus X_{i+3} \oplus rk_i))),
\end{aligned}
$$

where L is a linear diffusion function and S^* is a brick-layer function applying an 8-bit S-box to the input 4 times in parallel. According to [10], the S-box used in S^* is a composition of three transformations, that is,

$$
S(\boldsymbol{x}) = L_1 \circ I \circ L_1(\boldsymbol{x}), \qquad \forall \boldsymbol{x} \in \mathrm{GF}(2)^8,
$$

where L_1 is an affine transformation defined by

$$
L_1(\boldsymbol{x}) = A_1 \boldsymbol{x} \oplus \boldsymbol{c}, \qquad \forall \boldsymbol{x} \in \mathrm{GF}(2)^8,
$$

here

$$
A_1 = \begin{pmatrix}
1 & 1 & 1 & 0 & 0 & 1 & 0 & 1 \\
1 & 1 & 1 & 1 & 0 & 0 & 1 & 0 \\
0 & 1 & 1 & 1 & 1 & 0 & 0 & 1 \\
1 & 0 & 1 & 1 & 1 & 1 & 0 & 0 \\
0 & 1 & 0 & 1 & 1 & 1 & 1 & 0 \\
0 & 0 & 1 & 0 & 1 & 1 & 1 & 1 \\
1 & 0 & 0 & 1 & 0 & 1 & 1 & 1 \\
1 & 1 & 0 & 0 & 1 & 0 & 1 & 1
\end{pmatrix}
$$

is an 8×8 matrix over $\mathrm{GF}(2)$ and $\boldsymbol{c} = (1, 1, 0, 0, 1, 0, 1, 1)^T$ is a constant column vector. I is a nonlinear transformation on $\mathrm{GF}(2)^8$ derived from the multiplicative inversion on the field $\mathrm{GF}(2^8) \cong \mathrm{GF}(2)[x]/(f(x))$, where

$$
f(x) = x^8 + x^7 + x^6 + x^5 + x^4 + x^2 + 1
$$

is a binary irreducible polynomial and I maps the zero vector into itself. Finally, the linear diffusion function L is defined as follows:

$$
\begin{aligned}
L : \mathrm{GF}(2)^{32} &\to \mathrm{GF}(2)^{32} \\
x &\mapsto x \oplus (x <<< 2) \oplus (x <<< 10) \oplus (x <<< 18) \oplus (x <<< 24).
\end{aligned}
$$

The key scheduling algorithm of SMS4 operates in a similar manner as its encryption. It generates 32 round keys rk_i of 32-bit length from a 128-bit cipher key with 32 rounds of iteration. The round transformation of key schedule is almost identical to that of the encryption algorithm, and a unique difference is the linear diffusion function. In the key schedule the linear diffusion function L' is replaced by

$$L' : GF(2)^{32} \to GF(2)^{32}$$
$$x \mapsto x \oplus (x <<< 13) \oplus (x <<< 23).$$

To obtain the round keys, the cipher key $K = (K_0, K_1, K_2, K_3)$ is first masked with a system parameter $FK = (FK_0, FK_1, FK_2, FK_3)$, where

$$FK = 0\text{xa3b1bac656aa3350677d9197b27022dc (hexadecimal)}$$

and then one sets

$$rk_{-4} = K_0 \oplus FK_0, \ rk_{-3} = K_1 \oplus FK_1,$$
$$rk_{-2} = K_2 \oplus FK_2, \ rk_{-1} = K_3 \oplus FK_3.$$

The round key for the i-th round is computed as follows:

$$rk_i = rk_{i-4} \oplus L'(S^*(rk_{i-3} \oplus rk_{i-2} \oplus rk_{i-1} \oplus ck_i)),$$

where the ck_i $(0 \le i \le 31)$ are fixed constant vectors.

3   An Extended Cipher of SMS4

In this section we introduce a new block cipher named ESMS4, which is an Extension of **SMS4**.

3.1   Extension Maps

Inversion. In the SMS4 cipher, the inversion transformation is identical to the standard field inversion with $0^{(-1)} = 0$.

For any n-dimensional vector

$$a = (a_0, \cdots, a_{n-1}) \in \mathbb{F}^n,$$

the inverse transformation on a is viewed as a componentwise operation

$$a^{(-1)} = (a_0^{(-1)}, \cdots, a_{n-1}^{(-1)}).$$

Vector Conjugate. For any $a \in \mathbb{F}$, the vector conjugate of a is defined as

$$\phi : \mathbb{F} \to \mathbb{F}^8$$
$$a \mapsto \widehat{a} = (a^{2^0}, a^{2^1}, \cdots, a^{2^7}).$$

This map can be extended in an obvious way to the n-dimensional vector conjugate map

$$\phi' : \mathbb{F}^n \to \mathbb{F}^{8n},$$

namely, for any $a = (a_0, \cdots, a_{n-1}) \in \mathbb{F}^n$,

$$\widehat{a} = \phi'(a) = (\phi(a_0), \cdots, \phi(a_{n-1})).$$

It is obvious that the vector conjugate transformation is additive and preserves the inversion, i.e.,

$$\phi'(a + a') = \phi'(a) + \phi'(a'),$$
$$\phi'(a^{(-1)}) = \phi'(a)^{(-1)}.$$

Thus, any state vector of SMS4 in vector space $\mathcal{A}$ can be embedded into the vector space $\mathcal{E}$ of ESMS4 with the vector conjugate map ϕ'.

In the following subsection we will describe the ESMS4 cipher in detail.

3.2   The ESMS4 Cipher

According to Section 2, the input block of each round in SMS4 is 128 bits, i.e., 4 words, and the last three words first exclusive or to get a 32-bit word that is a basis operated by the remaining functions in the round transformation. Therefore, in the following we only consider the state vectors in the vector space $\mathcal{A}$.

For any $a = (a_0, a_1, a_2, a_3)^T \in \mathcal{A}$, the corresponding column vector in the vector space $\mathcal{E}$ of ESMS4 is represented as:

$$\begin{aligned} e &= (\phi(a_0), \phi(a_1), \phi(a_2), \phi(a_3)) \\ &= (e_{00}, \cdots, e_{07}, e_{10}, \cdots, e_{17}, e_{20}, \cdots, e_{27}, e_{30}, \cdots, e_{37}). \end{aligned}$$

This extension ensures that each function of the round transformation in SMS4 can be expressed by algebraic operations over $\mathbb{F}$. So, the ESMS4 cipher can be obtained from the SMS4 cipher by this corresponding relationship.

The basic operations in the round transformation of SMS4 include the exclusive or operation, the inversion transformation, a GF(2)-affine transformation in S-box, and a linear diffusion function. In the following we will get the ESMS4 cipher by extending these basic operations of SMS4 from GF(2) to $\mathbb{F} = \mathrm{GF}(2^8)$.

First, the operations of exclusive or and inversion can be extended to $\mathbb{F}$ in an obvious manner. This is due to that the two operations operate on bytes componentwisely, the corresponding operations in ESMS4 only need preserving this transformation by conjugate mapping ϕ. Next, we discuss the extension of the other two operations detailedly.

The GF(2)-affine transformation in the S-box. We know that the GF(2)-affine transformation is the composition of a GF(2)-linear transformation and an exclusive or with a constant vector. The exclusive or operation can be extended in the previous manner, that is, replicating the conjugate vector of the constant vector 4 times, we will then get the corresponding constant vector $C \in \mathbb{F}^{32}$.

As for a GF(2)-linear transformation l defined by

$$l(x) = A_1 x, \quad \forall x \in \mathrm{GF}(2)^8,$$

we have the following extension.

According to [7,9], l can be represented with a linear combination of conjugates, i.e.,

$$l(a) = \sum_{k=0}^{7} \lambda_k a^{2^k}, \qquad \forall a \in \mathbb{F},$$

where $\lambda_i \in \mathbb{F}, i = 0, \cdots, 7$. With this equation in mind, we extend l over $\mathbb{F}$ to the following 8×8 matrix L'_E:

$$L'_E = \begin{pmatrix} (\lambda_0)^{2^0} & (\lambda_1)^{2^0} & \cdots & (\lambda_7)^{2^0} \\ (\lambda_7)^{2^1} & (\lambda_0)^{2^1} & \cdots & (\lambda_6)^{2^1} \\ \vdots & \vdots & \cdots & \vdots \\ (\lambda_1)^{2^7} & (\lambda_2)^{2^7} & \cdots & (\lambda_0)^{2^7} \end{pmatrix}.$$

This matrix replicates the action of l on the first byte of a vector conjugate set and ensures the vector conjugate property is preserved on the remaining bytes. As for the entire GF(2)-linear operation l, it can be extended to $\mathbb{F}$ with a 32×32 matrix L_E over $\mathbb{F}$. And, L_E is a block diagonal matrix with 4 identical blocks L'_E.

Linear transformation. The linear diffusion function L of SMS4 includes 4 left cyclicly shift transformations, i.e., they left cyclicly shift 2, 10, 18, and 24 bits, respectively. Since these transformations only operate on bits instead of on bytes, therefore, in order to extend these operations to $\mathbb{F}$, we would transform them to operate on bytes. In the following, we just discuss the left cyclicly 2-bit shift and the other three cases can be discussed in a similar way.

For $a = (a_0, a_1 \cdots, a_{31}) \in \mathrm{GF}(2)^{32}$, after the transformation of left cyclicly shifting 2 bits, a is transformed to $b = (a_2, \cdots, a_{31}, a_0, a_1)$. Suppose the first two bytes of a are denoted with $a_1 = (a_0, \cdots, a_7)^T$ and $a_2 = (a_8, \cdots, a_{15})^T$, respectively. The first byte in b denoted $b_1 = (a_2, \cdots, a_8, a_9)^T$ can be obtained from a_1 and a_2 with the following equation:

$$b_1 = A_1 a_1 + B_1 a_2,$$

where

$$A_1 = \begin{pmatrix} 0 & 0 & 1 & 0 & 0 & 0 & 0 & 0 \\ 0 & 0 & 0 & 1 & 0 & 0 & 0 & 0 \\ 0 & 0 & 0 & 0 & 1 & 0 & 0 & 0 \\ 0 & 0 & 0 & 0 & 0 & 1 & 0 & 0 \\ 0 & 0 & 0 & 0 & 0 & 0 & 1 & 0 \\ 0 & 0 & 0 & 0 & 0 & 0 & 0 & 1 \\ 0 & 0 & 0 & 0 & 0 & 0 & 0 & 0 \\ 0 & 0 & 0 & 0 & 0 & 0 & 0 & 0 \end{pmatrix}, B_1 = \begin{pmatrix} 0 & 0 & 0 & 0 & 0 & 0 & 0 & 0 \\ 0 & 0 & 0 & 0 & 0 & 0 & 0 & 0 \\ 0 & 0 & 0 & 0 & 0 & 0 & 0 & 0 \\ 0 & 0 & 0 & 0 & 0 & 0 & 0 & 0 \\ 0 & 0 & 0 & 0 & 0 & 0 & 0 & 0 \\ 0 & 0 & 0 & 0 & 0 & 0 & 0 & 0 \\ 1 & 0 & 0 & 0 & 0 & 0 & 0 & 0 \\ 0 & 1 & 0 & 0 & 0 & 0 & 0 & 0 \end{pmatrix}.$$

Let A be a block diagonal 32×32 matrix over GF(2) consisting of 4 identical blocks A_1, and B a block diagonal 32×32 matrix with 4 identical blocks B_1. Thus, for any column vector $a \in \mathrm{GF}(2)^{32}$, the left cyclicly 2-bit shift transformation can be described with the following equation:

$$b = Aa + B \cdot Ca,$$

where

$$C = \begin{pmatrix} \mathbf{0_8} & \mathbf{I_8} & \mathbf{0_8} & \mathbf{0_8} \\ \mathbf{0_8} & \mathbf{0_8} & \mathbf{I_8} & \mathbf{0_8} \\ \mathbf{0_8} & \mathbf{0_8} & \mathbf{0_8} & \mathbf{I_8} \\ \mathbf{I_8} & \mathbf{0_8} & \mathbf{0_8} & \mathbf{0_8} \end{pmatrix}$$

is a 32×32 matrix over $\mathrm{GF}(2)$ consisting of four 8×8 identity matrices $\mathbf{I_8}$ and twelve 8×8 zero matrices $\mathbf{0_8}$. Thus, the operation of left cyclicly shifting 2 bits is transformed to operate on bytes.

In the following we extend this left cyclicly 2-bit shift transformation to $\mathbb{F}$. For any column vector $a \in \mathrm{GF}(2)^8$, set $f_1(a) = A_1 a$ and $g_1(a) = B_1 a$. Again according to [7,9], these two transformations can be represented as the following form:

$$f_1(a) = \sum_{k=0}^{7} \lambda_{1k} a^{2^k}, \quad g_1(a) = \sum_{k=0}^{7} \lambda_{2k} a^{2^k},$$

where, $a, \lambda_{ij} \in \mathbb{F}$, $i = 0, 1, j = 0, \cdots, 7$. In ESMS4 the corresponding operations of f_1 and g_1 can be represented by the following two 8×8 matrices over $\mathbb{F}$

$$E_{A_1} = \begin{pmatrix} (\lambda_{20})^{2^0} & (\lambda_{21})^{2^0} & \cdots & (\lambda_{27})^{2^0} \\ (\lambda_{27})^{2^1} & (\lambda_{20})^{2^1} & \cdots & (\lambda_{26})^{2^1} \\ \vdots & \vdots & \cdots & \vdots \\ (\lambda_{21})^{2^7} & (\lambda_{22})^{2^7} & \cdots & (\lambda_{20})^{2^7} \end{pmatrix}$$

and

$$E_{B_1} = \begin{pmatrix} (\lambda_{30})^{2^0} & (\lambda_{31})^{2^0} & \cdots & (\lambda_{37})^{2^0} \\ (\lambda_{37})^{2^1} & (\lambda_{30})^{2^1} & \cdots & (\lambda_{36})^{2^1} \\ \vdots & \vdots & \cdots & \vdots \\ (\lambda_{31})^{2^7} & (\lambda_{32})^{2^7} & \cdots & (\lambda_{30})^{2^7} \end{pmatrix},$$

respectively. These two matrices replicate the action of the transformation f_1 and g_1 on the first byte of a vector conjugate set and ensure the vector conjugate property is preserved on the remaining bytes.

Therefore, for any $b \in \mathcal{E}$, the operation of left cyclicly shifting 2 bits can be extended to ESMS4 with the following way:

$$E_A b + C \cdot E_B b,$$

where, E_A is a block diagonal matrix with 4 identical blocks E_{A_1}, and E_B also is a block diagonal matrix with 4 identical blocks E_{B_1}. Let L_{σ_1} denote $E_A + C \cdot E_B$, thus this transformation can be written as $L_{\sigma_1} b$.

With a similar method, the operations of left cyclicly shifting 10, 18, and 24 bits can all be extended to ESMS4, and let L_{σ_2}, L_{σ_3}, and L_{σ_4} denote corresponding 32×32 matrices of them over $\mathbb{F}$, respectively.

Now we can extend the linear diffusion function L from GF(2) to $\mathbb{F}$, and the corresponding transformation is still denoted with L, which can be written as:

$$L(\boldsymbol{b}) = I\boldsymbol{b} + L_{\sigma_1}\boldsymbol{b} + L_{\sigma_2}\boldsymbol{b} + L_{\sigma_3}\boldsymbol{b} + L_{\sigma_4}\boldsymbol{b}, \tag{1}$$

where, $\boldsymbol{b} \in \mathcal{E}$, I is a 32×32 identity matrix over GF(2).

Let L_σ denote $I + L_{\sigma_1} + L_{\sigma_2} + L_{\sigma_3} + L_{\sigma_4}$. Thus, the equation (1) can be simplified and represented as $L(\boldsymbol{b}) = L_\sigma \boldsymbol{b}$.

By far we have completed the extension of basic operations in SMS4 to ESMS4. From Section 2, the basic operations in the key schedule of SMS4 are same to those in encryption algorithm. Therefore, similarly to the encryption algorithm the key schedule of SMS4 can be extended to ESMS4. Thus, the ESMS4 cipher is obtained completely.

We define $\mathcal{E}_\mathcal{A}$ to be the embedded image of the SMS4 state space in the ESMS4 state space. We have the following commuting diagram shown in Fig. 1.

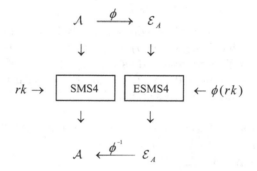

Fig. 1. Relationship between SMS4 and ESMS4

4   Multivariate Quadratic Equations

In this section we show that the SMS4 cipher can be expressed with a system of sparse multivariate quadratic equations over $\mathbb{F}$. To do this we need to describe the ESMS4 cipher in such an equation system.

From the previous section we know that the ESMS4 cipher is an extension of SMS4 by the conjugate transformation of bytes, and completely describes the encryption of SMS4. Before describing the ESMS4 cipher with the system of equations over $\mathbb{F}$, we first give some notations.

Suppose $(X_0, X_1, X_2, X_3) \in (\mathbb{F}^{32})^4$ is the input block of ESMS4 and the corresponding ciphertext is $(X_{32}, X_{33}, X_{34}, X_{35}) \in (\mathbb{F}^{32})^4$. And for the i-th round $(0 \le i \le 31)$, the input of the affine transformation in S-box is denoted with W_i; the input and output variables of inversion transformation are denoted by V_i and

Y_i, respectively; the input and output variants of linear diffuse transformation are denoted with U_i and Z_i, respectively, and the round key is denoted with RK_i.

Then, the encryption algorithm of ESMS4 can be described with the following equation system:

$$
\begin{aligned}
W_i &= X_i + X_{i+1} + X_{i+2} + RK_i \\
V_i &= L_E W_i + C \\
Y_i &= V_i^{(-1)} \\
U_i &= L_E Y_i + C \\
Z_i &= L_\sigma U_i \\
X_{i+3} &= X_{i-1} + Z_i, \text{ all for } 0 \le i \le 31.
\end{aligned}
$$

Let L_B, D denote $L_\sigma L_E$ and $L_\sigma C$, respectively, the system of equations can be simply written as:

$$
\begin{aligned}
W_i &= X_i + X_{i+1} + X_{i+2} + RK_i \\
V_i &= L_E W_i + C \\
Y_i &= V_i^{(-1)} \\
Z_i &= L_B Y_i + D \\
X_{i+3} &= X_{i-1} + Z_i.
\end{aligned} \tag{2}
$$

Now we consider these equations componentwisely. For convenience, the matrix L_E is denoted with (α), L_B with (β), respectively; the $(8j+m)$-th component of W_i, V_i, Y_i, Z_i, and X_i is represented as $W_{i,(j,m)}$, $V_{i,(j,m)}$, $Y_{i,(j,m)}$, $Z_{i,(j,m)}$, and $X_{i,(j,m)}$, respectively for $m = 0, \cdots, 7$, $j = 0, \cdots, 3$. With these notations the equation system (2) can be expressed as :

$$
\begin{aligned}
W_{i,(j,m)} &= X_{i,(j,m)} + X_{i+1,(j,m)} + X_{i+2,(j,m)} + K_{i,(j,m)} \\
V_{i,(j,m)} &= (L_E W_i)_{(j,m)} + C_{(j,m)} \\
Y_{i,(j,m)} &= V_{i,(j,m)}^{(-1)} \\
Z_{i,(j,m)} &= (L_B Y_i)_{(j,m)} + D_{(j,m)} \\
X_{i+3,(j,m)} &= X_{i-1,(j,m)} + Z_{i,(j,m)},
\end{aligned}
$$

for $m = 0, \cdots, 7$, $j = 0, \cdots, 3$. By investigating the S-box in SMS4, we found that the probability of 0-inverse occurring is only $O(2^{-10})$. In this case, we could assume there is no 0-inverse in the equations discussed. Based on this assumption, the previous system of equations has the following form:

$$
\begin{aligned}
0 &= X_{i,(j,m)} + X_{i+1,(j,m)} + X_{i+2,(j,m)} + K_{i,(j,m)} + W_{i,(j,m)} \\
0 &= (L_E W_i)_{(j,m)} + C_{(j,m)} + V_{i,(j,m)} \\
0 &= V_{i,(j,m)} Y_{i,(j,m)} + 1 \\
0 &= (L_B Y_i)_{(j,m)} + D_{(j,m)} + Z_{i,(j,m)} \\
0 &= X_{i-1,(j,m)} + Z_{i,(j,m)} + X_{i+3,(j,m)}.
\end{aligned} \tag{3}
$$

It is obvious that the equation system (3) fully describes the encryption of ESMS4. Using the notations (α) and (β), this system can be further represented as:

$$0 = X_{i,(j,m)} + X_{i+1,(j,m)} + X_{i+2,(j,m)} + K_{i,(j,m)} + W_{i,(j,m)}$$
$$0 = C_{(j,m)} + V_{i,(j,m)} + \sum_{(j',m')} \alpha_{(j,m),(j',m')} W_{i,(j',m')}$$
$$0 = V_{i,(j,m)} Y_{i,(j,m)} + 1$$
$$0 = D_{(j,m)} + Z_{i,(j,m)} + \sum_{(j',m')} \beta_{(j,m),(j',m')} Y_{i,(j',m')}$$
$$0 = X_{i-1,(j,m)} + Z_{i,(j,m)} + X_{i+3,(j,m)}.$$

If considering the conjugate property of variables, we could get more quadratic equations. Adding these equations to the equation system, we get the resulting system which describes the SMS4 encryption over $\mathbb{F}$,

$$0 = X_{i,(j,m)} + X_{i+1,(j,m)} + X_{i+2,(j,m)} + K_{i,(j,m)} + W_{i,(j,m)}$$
$$0 = C_{(j,m)} + V_{i,(j,m)} + \sum_{(j',m')} \alpha_{(j,m),(j',m')} W_{i,(j',m')}$$
$$0 = D_{(j,m)} + Z_{i,(j,m)} + \sum_{(j',m')} \beta_{(j,m),(j',m')} Y_{i,(j',m')}$$
$$0 = X_{i-1,(j,m)} + Z_{i,(j,m)} + X_{i+3,(j,m)}$$
$$0 = V_{i,(j,m)} Y_{i,(j,m)} + 1$$
$$0 = V^2_{i,(j,m)} + V_{i,(j,m+1)}$$
$$0 = Y^2_{i,(j,m)} + Y_{i,(j,m+1)},$$

where, $i = 0, \cdots, 31$, $m = 0, \cdots, 7$, and $j = 0, \cdots, 3$.

In this equation system, there are totally $32 \times 5 \times 32 + 32 \times 2 \times 32 = 7168$ equations over $\mathbb{F}$, of which $32 \times 3 \times 32 = 3027$ quadratic equations and 4096 linear equations. In addition,

$$36 \times 32 + 32 \times 2 \times 32 + 32 \times 32 \times 2 + 32 \times 32 + 32 \times 32 \times 2 = 8320$$

terms appear in the system in total, and they comprise $36 \times 32 + 32 \times 32 \times 4 = 1152 + 4096 = 5248$ variables, and among them, there are 1152 key variables and 4096 intermediate variables.

With a similar method, the key schedule of SMS4 can also be described with a multivariate quadratic system over $\mathbb{F}$. After computation, this system comprises 7168 equations, of which 4096 linear equations and 3027 quadratic equations; all these equations consist of 6144 variables.

Thus, for the SMS4 cipher we can totally obtain $7168 \times 2 = 14336$ equations over $\mathbb{F}$, and these equations are composed of $4096 \times 2 = 8192$ linear equations and $3072 \times 2 = 6144$ quadratic equations. The whole system includes only $8320 + 9216 = 17536$ terms in total, which shows this multivariate quadratic system is very sparse. In the next section we discuss the computational complexity of XSL algorithm for solving the equation system.

5   Solving the Equation System of SMS4 with the XSL Algorithm

XSL algorithm has two versions. The first one is introduced in [3], and the second one is called "compact XSL" and proposed in [4]. In the first version, XSL algorithm presents two different attacks, one eliminates the key schedule equations but requires a number of pairs of plaintexts and ciphertexts, while the other uses the key schedule equations and only works with a single plaintext-ciphertext pair.

According to [3] the second attack is more specific for the structure of the cipher itself, in this paper we base our discussion of XSL technique on this attack.

According to [2,3], there are four main steps in XSL algorithm, and the following is a brief description.

Step 1. Process the existing set of equations, by choosing certain sets of monomials and equations that will be used during the later steps of the algorithm.

Step 2. Select the value of the parameter P, and multiply the chosen equations by the product of $P - 1$ selected monomials. This is the "core" of XSL and should generate a large number of equations whose terms are the product of the monomials chosen earlier.

Step 3. Perform the "T' method", in which some selected equations are multiplied by a single variable. The goal is to generate new equations without creating any new monomials. Iterate with as many variables as necessary until the system has enough linearly independent equations to apply linearization.

Step 4. Apply linearization, by considering each monomial as a new variable and performing Gaussian elimination. This should yield a solution for the system.

The Step 3 is usually called "T' method", which is a critical technique of the XSL algorithm. All the new equations are generated in this step. These new equations can be divided into two sets; one is obtained by exploiting S-boxes and the other generated by the linear diffusion layers.

Before investigating the computational complexity of XSL algorithm for solving the equation system which describes the SMS4 cipher over $\mathbb{F}$, we give some notations from [3].

B: the number of S-boxes in each round;
N_r: the number of encryption rounds;
$\mathcal{R}$: the set of all equations generated by S-boxes of the cipher and R denotes the cardinality of $\mathcal{R}$;
$\mathcal{R}'$: the set of all equations generated by the equations of the linear diffusion layer and R' denotes the cardinality of $\mathcal{R}'$;
$\mathcal{T}$: the set of all monomials generated by S-boxes of the cipher and T denotes the cardinality of $\mathcal{T}$;
$\mathcal{T}_i'$: the set of monomials in the system such that $x_i \cdot \mathcal{T}_i' \subseteq \mathcal{T}$ and T' denotes the cardinality of $\mathcal{T}_i'$;
t: the number of monomials in the S-box equation;
t': the number of monomials in the S-box equations to be used in the T' method;
s: the number of bits operated by the S-box;
r: the number of equations in an S-box;
S: the total number of S-boxes in the cipher and the key schedule;
P: the critical parameter used in XSL.

In the following discussion, we consider the encryption algorithm and the key schedule simultaneously. We first present some formulas for computing the computational complexity of XSL algorithm without interpretation. For the details, see [3].

The total number of equations generated by the S-boxes is about:

$$R \approx r * S * t^{(P-1)} * \binom{S-1}{P-1}.$$

The total number of terms in these equations is about:

$$T \approx t^P * \binom{S}{P}. \tag{4}$$

After eliminating obvious linear dependencies, the number of equations could be computed by the following formula:

$$R \approx \sum_{i=1,\cdots,P} \binom{S}{i} r^i * \binom{S-i}{P-i} (t-r)^{(p-i)} = \binom{S}{P} * (t^P - (t-r)^P).$$

The total number of terms in these equations still can be computed with formula (4).

The number of equations generated by linear diffusion layer (including the equations generated by the linear diffusion layer of the key schedule) can be computed with the following formula:

$$R' \approx 2 * 2 * s * B * Nr * (t-r)^{(P-1)} * \binom{S}{P-1}.$$

At last, the number of terms generated by "T' method" could be represented as:

$$T' \approx t' * t^{(P-1)} * \binom{S-1}{P-1}.$$

According to [3], the working condition of the "T' method" is:

$$\frac{R+R'}{T-T'} > 1,$$

that is,

$$T - R - R' < T'. \tag{5}$$

With this formula, we could compute the least value P which makes the XSL technique work. Suppose ω is the constant used for computing the complexity of the Gaussian elimination, then the complexity of the XSL attack is about

$$T^\omega \approx t^{\omega P} * \binom{S}{P}^\omega.$$

These equations all take on their general forms in the XSL algorithm. However, when XSL algorithm applied to different ciphers, these formulas need to be changed slightly.

In the following, we present formulas to compute the complexity of the XSL algorithm for solving the equation system describing SMS4 over $\mathbb{F}$ by investigating the ESMS4 cipher.

Based on the previous formulas, the number of equations generated by S-boxes in ESMS4 can be computed with the following formula:

$$R \approx \binom{S}{P} * (t^P - (t-r)^P),$$

the number of terms in these equations is described as:

$$T \approx t^p * \binom{S}{P},$$

and the total number of equations generated through the linear diffusion layers is about:

$$R' \approx 2 * 2 * s * B * Nr * (t-r)^{(P-1)} * \binom{S}{P-1}.$$

The number of terms generated by "T' method" could be estimated by the formula:

$$T' \approx t' * t^{(P-1)} * \binom{S-1}{P-1}.$$

In the case of ESMS4, we have the following values: B=4, N_r=32, S=256, t'=3, $t = 5 \times 8 + 1 = 41$, $r = 3 \times 8 = 24$, s=8.

After computation with the formula (5), we get the least value P=2 which makes the XSL algorithm work. In general we consider ω=3, then the computational complexity of XSL algorithm attacking to ESMS4 is

$$T^3 = 41^6 \cdot \binom{256}{2}^3 = 2^{77}.$$

In other words, when solving the multivariate quadratic system of SMS4 over $\mathbb{F}$ by XSL algorithm, the computational complexity is only 2^{77}. Table 1 lists the computational complexity of the XSL algorithm on the different number of rounds under the condition P=2.

Table 1. The complexity of different rounds

Nr	4	5	6	7	8	9	10	11	12	13
complexity	2^{59}	2^{61}	2^{62}	2^{64}	2^{65}	2^{66}	2^{67}	2^{68}	2^{68}	2^{69}
Nr	14	15	16	17	18	19	20	21	22	23
complexity	2^{70}	2^{70}	2^{71}	2^{71}	2^{72}	2^{72}	2^{73}	2^{73}	2^{74}	2^{74}
Nr	24	25	26	27	28	29	30	31	32	
complexity	2^{74}	2^{75}	2^{75}	2^{75}	2^{76}	2^{76}	2^{76}	2^{77}	2^{77}	

From this table we can see that the computational complexity increases slowly with the increasing of the number of rounds.

6   Conclusion

In this paper we extended the SMS4 cipher to a new cipher ESMS4 over GF(2^8) by the conjugate transformation. With restricted plaintext and key spaces SMS4 is identical to ESMS4. In this case, all operations in the round transformation of SMS4 can be considered operating on GF(2^8). Thus, the difficulty that investigating the SMS4 cipher operating over two different fields GF(2^8) and GF(2) is avoided, which makes algebraic cryptanalysis for SMS4 easier. We described SMS4 with an extremely sparse overdetermined multivariate quadratic system over GF(2^8), whose solution will recover a cipher key of SMS4. Based on this fact, we used the XSL algorithm to solve the equation system describing the SMS4 cipher over GF(2^8), and estimated the computational complexity. As a result, we found that the complexity of solving the whole system with XSL algorithm is 2^{77}, which is much smaller than the resulting complexity for the embedding cipher of AES.

References

1. Beijing Data Security Company, The SMS4 Block Cipher (in Chinese), Beijing (2006), available at http://www.oscca.gov.cn/UpFile/200621016423197990.pdf
2. Cid, C., Leurent, G.: An Analysis of the XSL Algorithm. In: Roy, B. (ed.) ASIACRYPT 2005. LNCS, vol. 3788, pp. 333–352. Springer, Heidelberg (2005)
3. Courtois, N., Pieprzyk, J.: Cryptanalysis of Block Ciphers with Overdefined Systems of Equations, Cryptology ePrint Archive, Report, /044, 2002 (2002), available at http://eprint.iacr.org/2002/044
4. Courtois, N., Pieprzyk, J.: Cryptanalysis of Block Ciphers with Overdefined Systems of Equations. In: Zheng, Y. (ed.) ASIACRYPT 2002. LNCS, vol. 2501, pp. 267–287. Springer, Heidelberg (2002)
5. Courtois, N., Klimov, A., Patarin, J., Shamir, A.: Efficient Algorithms for Solving Overdefined Systems of Multivariate Polynomial Equations. In: Preneel, B. (ed.) EUROCRYPT 2000. LNCS, vol. 1807, pp. 392–407. Springer, Heidelberg (2000)
6. Courtois, N., Patarin, J.: About the XL Algorithm over GF(2). In: Joye, M. (ed.) CT-RSA 2003. LNCS, vol. 2612, pp. 141–157. Springer, Heidelberg (2003)
7. Daemen, J., Rijmen, V.: AES proposal: The Rijndael block cipher. Springer, Heidelberg (1999)
8. Kipnis, A., Shamir, A.: Cryptanalysis of the HFE Public Key Cryptosystem by Relinearization. In: Wiener, M.J. (ed.) CRYPTO 1999. LNCS, vol. 1666, pp. 19–30. Springer, Heidelberg (1999)
9. Lidl, R., Niederreiter, H.: Introduction to Finite Fields and Their Applications. Cambridge University Press, Cambridge (1984)
10. Liu, F., Ji, W., Hu, L., Ding, J., Lv, S., Pyshkin, A., Weinmann, R.: Analysis of the SMS4 Block Cipher. In: ACISP 2007. LNCS, vol. 4586, pp. 158–170. Springer, Heidelberg (2007)
11. Murphy, S., Robshaw, M.: Essential Algebraic Structure within the AES. In: Yung, M. (ed.) CRYPTO 2002. LNCS, vol. 2442, pp. 1–16. Springer, Heidelberg (2002)
12. Lim, C., Khoo, K.: An Analysis of XSL Applied to BES. In: FSE 2007. LNCS, vol. 4593, pp. 242–253. Springer, Heidelberg (2007)
13. Zhang, L., Wu, W.: Difference Fault Attack on the SMS4 Encryption Algorithm (in Chinese). Chinese Journal of Computers 29(9), 1596–1602 (2006)

Tweakable Enciphering Schemes from Hash-Sum-Expansion

Kazuhiko Minematsu[1,2] and Toshiyasu Matsushima[2]

[1] NEC Corporation, 1753 Shimonumabe, Nakahara-Ku, Kawasaki, Japan
k-minematsu@ah.jp.nec.com
[2] Waseda University, 3-4-1 Okubo Shinjuku-ku Tokyo, Japan

Abstract. We study a tweakable blockcipher for arbitrarily long message (also called a tweakable enciphering scheme) that consists of a universal hash function and an expansion, a keyed function with short input and long output. Such schemes, called HCTR and HCH, have been recently proposed. They used (a variant of) the counter mode of a blockcipher for the expansion. We provide a security proof of a structure that underlies HCTR and HCH. We prove that the expansion can be instantiated with any function secure against Known-plaintext attacks (KPAs), which is called a weak pseudorandom function (WPRF). As an application of our proof, we provide efficient blockcipher-based schemes comparable to HCH and HCTR. For the double-block-length case, our result is an interesting extension of previous attempts to build a double-block-length cryptographic permutation using WPRF.

Keywords: Mode of operation, HCTR, HCH, Weak Pseudorandom Function.

1 Introduction

A tweakable blockcipher, introduced by Liskov, Rivest, and Wagner [8], is a blockcipher that accepts an additional input called tweak. Formally, a ciphertext of a tweakable blockcipher, $\widetilde{E}$, is $C = \widetilde{E}_K(M, T)$, where M is a plaintext, K is the key, and T is the tweak. It is length-preserving, i.e., $|C| = |M|$ always holds. In this paper, we study the tweakable blockcipher with arbitrarily long message, which is also called the tweakable enciphering scheme (TES)[20]. A typical application of TES is the disk sector encryption, where a plaintext is a content of a sector, and a tweak is used to specify the sector number. The first approach to TES is Naor and Reingold [18]. They proposed to use the ECB mode with lightweight mixing layers applied to the top and bottom. Although they did not consider the tweak, it is easy to make their proposal tweakable. Their scheme was proved to be secure against any combination of Chosen-plaintext attack (CPA) and Chosen-ciphertext attack (CCA). Since then, many TESs have been proposed (e.g., [6][5][14]).

Our target is based on the Naor and Reingold's approach. This uses the Sum-Expansion instead of ECB mode. The Sum-Expansion for ℓ-bit message consists of an n-bit blockcipher, E_K, and a keyed function with n-bit input and

K. Srinathan, C. Pandu Rangan, M. Yung (Eds.): Indocrypt 2007, LNCS 4859, pp. 252–267, 2007.

$(\ell - n)$-bit output, denoted by $F_{K'}$. Let Σ^n denote $\{0,1\}^n$. Then, for message $x = (x_L, x_R)$, $x_L \in \Sigma^n$, $x_R \in \Sigma^{\ell-n}$, the Sum-Expansion is defined as $\varphi[E_K, F_{K'}](x_L, x_R) = (E_K(x_L), F_{K'}(x_L \oplus E_K(x_L)) \oplus x_R)$, where $x_L \oplus E_K(x_L)$ is called the I/O sum. An unbalanced Feistel permutation using a universal hash function is applied to the top and bottom of Sum-Expansion. This structure, which we call Hash-Sum-Expansion (HSE), has been employed by previous proposals. HCTR [20] uses the Sum-Expansion $\varphi[E_K, \text{CTR}[E_K]]$, where $\text{CTR}[E_K]$ is the counter mode of the blockcipher E_K that uses I/O sum as a counter (See Sect. 4). HCH [2] is similar, but the I/O sum is encrypted by E_K before it is given to $\text{CTR}[E_K]$. Our purpose is to prove the security of HSE based on the abstract properties of its components, in particular the property of the expansion. That is, we want to identify the sufficient security conditions for each component. Our main contribution is to prove that, if the expansion is an independently-keyed function secure against any practical *known-plaintext attack* (KPA), the resulting HSE is provably secure. Such a KPA-secure function is called a weak pseudorandom function (WPRF). Its construction and application have been studied in many papers [17][3][12][13]. Our result demonstrates the soundness of using the counter mode, as we prove that the counter mode of a CPA-secure blockcipher is a (length-expanding) WPRF. However our proof can not be obtained by extending the proofs of previous modes. Moreover, our proof enables us to use any WPRF other than the counter mode, in order to improve the efficiency or security. Although we do not have a practical proposal at this moment, we briefly discuss this issue in Sect. 4.2.

Our result is also useful in designing secure and efficient TESs. Combined with the idea of *tagged* tweakable blockcipher (See Sect. 4.1), we provide an implementation of HSE based on a blockcipher. The resulting mode is similar to HCH and HCTR. However, it has a better security bound than that of HCTR and is comparable to HCH (or, slightly faster if preprocessing is allowed). We also describe how to improve the security of HCTR with small changes.

A more theoretical (but still interesting) application of our result is building a double-block-length (i.e., $\ell = 2n$, DBL for short) permutation. Recent studies [12][16] showed that a CPA-secure DBL permutation can be built using one CPA-secure and KPA-secure functions. However, it was not clear if a similar construction is possible for building a CCA-secure DBL permutation: previous approaches needed at least two CPA-secure functions (or CCA-secure permutations, see Sect. 4.3). In contrast, HSE for $\ell = 2n$ is a CCA-secure DBL permutation using one CCA-secure permutation and one KPA-secure function.

2   Preliminaries

Let Σ^n denote $\{0,1\}^n$ and Σ^* denote the set of all finite-length binary sequences. The bit length of x is denoted by $|x|$. A uniform random function (URF) with n-bit input and ℓ-bit output, denoted by $\mathsf{R}_{n,\ell}$, is a random variable uniformly distributed over $\{f : \Sigma^n \to \Sigma^\ell\}$. Similarly, a random variable uniformly distributed over all n-bit permutations is an n-bit block uniform random permutation (URP) and is denoted by P_n. A tweakable n-bit URP with n'-bit tweak is defined by the set of $2^{n'}$ independent URPs (i.e., an independent n-bit URP is used for each tweak) and is denoted by $\widetilde{\mathsf{P}}_{n,n'}$. We write $\widetilde{\mathsf{P}}_n$ if n' is clear from the context. If $F_K : \mathcal{X} \to \mathcal{Y}$ is a keyed function, then F_K is a random variable (not necessarily uniformly) distributed over $\{f : \mathcal{X} \to \mathcal{Y}\}$. If F_K is a keyed permutation, F_K^{-1} will denote its inversion. If its key, K, is uniform over $\mathcal{K}$, we have $\Pr(F_K(x) = y) = \{k \in \mathcal{K} : f(k,x) = y\}/|\mathcal{K}|$ for some function $f : \mathcal{K} \times \mathcal{X} \to \mathcal{Y}$. We will omit the subscript K and write $F : \mathcal{X} \to \mathcal{Y}$, when K is clear from the context.

Elements of GF(2^n). Following [19], we express the elements of field GF(2^n) by the n-bit coefficient vectors of the polynomials in the field. We alternatively represent n-bit coefficient vectors by integers $0, 1, \ldots, 2^n - 1$. For example, 5 corresponds to the coefficient vector $(00\ldots0101)$ (which corresponds to the polynomial $\mathsf{x}^2 + 1$) and 1 corresponds to $(00\ldots01)$, i.e., the identity element. For $x, y \in \Sigma^n$, we define xy as the field multiplication of corresponding elements.

Security notions. Consider the game in which we want to distinguish two keyed functions, G and G', using a black-box access to them. We define classes of attacks: Chosen-plaintext attack (CPA), Known-plaintext attack (KPA), where plaintexts are independent and uniformly random, and Chosen-ciphertext attack (CCA), and CCA with tweak, i.e., a tweak and a plaintext (or ciphertext) can be arbitrarily chosen by the adversary. Obviously, CCA (with tweak) can be defined when G and G' are keyed (tweakable) permutation. Let $\mathtt{atk} \in \{\mathtt{cpa}, \mathtt{kpa}, \mathtt{cca}, \widetilde{\mathtt{cca}}\}$, where $\widetilde{\mathtt{cca}}$ denotes CCA with tweak. The maximum advantage of adversary using $\mathtt{atk}$ in distinguishing G and G' is:

$$\mathrm{Adv}_{G,G'}^{\mathtt{atk}}(\theta) \stackrel{\text{def}}{=} \max_{A:\theta-\mathtt{atk}} \left| \Pr[A^G = 1] - \Pr[A^{G'} = 1] \right|, \tag{1}$$

where $A^G = 1$ denotes that A's guess is 1, which indicates one of G or G'. The parameter θ denotes the attack resource, such as the number of queries, q, and time complexity, τ. If θ does not contain τ, the adversary has no computational restriction. The maximum is taken for all $\mathtt{atk}$-adversaries having θ. For KPA, we assume a generation of q random n-bit plaintexts requires $O(q)$ time. For $G : \Sigma^n \to \Sigma^m$, we have

$$\mathrm{Adv}_G^{\mathtt{prf}}(\theta) \stackrel{\text{def}}{=} \mathrm{Adv}_{G,\mathsf{R}_{n,m}}^{\mathtt{cpa}}(\theta), \text{ and } \mathrm{Adv}_G^{\mathtt{wprf}}(\theta) \stackrel{\text{def}}{=} \mathrm{Adv}_{G,\mathsf{R}_{n,m}}^{\mathtt{kpa}}(\theta),$$

and if G is an n-bit keyed permutation, $\mathrm{Adv}_G^{\mathtt{sprp}}(\theta) \stackrel{\text{def}}{=} \mathrm{Adv}_{G,\mathsf{P}_n}^{\mathtt{cca}}(\theta)$. Moreover, if G is an n-bit keyed tweakable permutation, we define $\mathrm{Adv}_G^{\widetilde{\mathtt{sprp}}}(\theta) \stackrel{\text{def}}{=} \mathrm{Adv}_{G,\widetilde{\mathsf{P}}_n}^{\widetilde{\mathtt{cca}}}(\theta)$. If $\mathrm{Adv}_G^{\mathtt{prf}}(\theta)$ is negligibly small for any practical θ including τ, G is a pseudorandom function (PRF)[4]. If G is invertible, it is also called a pseudorandom permutation (PRP). Similarly, if $\mathrm{Adv}_G^{\mathtt{wprf}}(\theta)$ is negligibly small for any practical θ, G is a weak pseudorandom function (WPRF), and if G is a keyed permutation with negligibly small $\mathrm{Adv}_G^{\mathtt{sprp}}(\theta)$, G is a strong pseudorandom permutation (SPRP). Finally, if G is a tweakable keyed permutation with negligibly small $\mathrm{Adv}_G^{\widetilde{\mathtt{sprp}}}(\theta)$, G is called a tweakable SPRP.

For keyed functions $E : \Sigma^s \to \Sigma^n$ and $F : \Sigma^n \to \Sigma^\ell$, let $E \unlhd F : \Sigma^s \to \Sigma^{n+\ell}$ be the composition such that $E \unlhd F(x) = (E(x), F(E(x)))$. For F and $G : \Sigma^n \to \Sigma^\ell$, we have the following relationships between cpa and kpa-advantages. For the proofs, see [15] (the second claim is Lemma 2.1 of [15]).

$$\mathrm{Adv}_{F,G}^{\mathrm{kpa}}(q, \tau) = \mathrm{Adv}_{\mathrm{R}_{m,n} \unlhd F, \mathrm{R}_{m,n} \unlhd G}^{\mathrm{cpa}}(q, \tau') \text{ for any } m \geq q, \text{ and} \tag{2}$$

$$\mathrm{Adv}_{E \unlhd F, E \unlhd G}^{\mathrm{cpa}}(q, \tau) \leq 2\mathrm{Adv}_{E, \mathrm{R}_{s,n}}^{\mathrm{cpa}}(q, \tau) + \mathrm{Adv}_{F,G}^{\mathrm{kpa}}(q, \tau'), \text{ where } \tau' = \tau + O(q). \tag{3}$$

These definitions are for the fixed input length (FIL) setting. However, they can be naturally extend for the variable input length (VIL) setting, where an adversary can change the input length, by considering VIL versions of URF and URP. Every security notion will denote the security under FIL (VIL) model if the target functions are FIL (VIL).

3   The Security of Hash-Sum-Expansion

3.1   Main Theorem

We prove the $\widetilde{\mathrm{cca}}$-security of HSE. The basic HSE described in Introduction has no tweak, however it can be tweakable with a slight modification. We assume the unit block length is n-bit, and the message length is ℓ-bit. Here, ℓ is not necessarily a multiple of n. For simplicity, we assume $\ell > n$. All proposals of this paper can be easily extended to the case $\ell = n$. We also assume the tweak length is n-bit. The tweakable HSE uses three keyed functions. The first is $\widetilde{E}$, an n-bit block tweakable blockcipher with n-bit tweak. If tweak is not needed, it is an n-bit blockcipher, E. The second and third are keyed functions $F : \Sigma^n \to \Sigma^{\ell-n}$ and $H : \Sigma^{\ell-n} \to \Sigma^n$. We first consider the FIL scheme. The VIL scheme will be described later. Let $\mathrm{HSE}[H, \widetilde{E}, F]$ denote HSE using H, $\widetilde{E}$, and F. The encryption of $\mathrm{HSE}[H, \widetilde{E}, F]$ is as follows. For plaintext $x \in \Sigma^\ell$ and tweak $t \in \Sigma^n$, we first partition x into $x_L \in \Sigma^n$ and $x_R \in \Sigma^{\ell-n}$. Then, we compute $S = x_L \oplus H(x_R)$, $U = \widetilde{E}(S, t)$, and $V = S \oplus U$. The ciphertext $y = (y_L, y_R)$ is computed as $y_R = F(V) \oplus x_R$ and $y_L = U \oplus H(y_R)$. The Sum-Expansion, $\varphi[\widetilde{E}, F]$, is a tweakable permutation defined as $\varphi[\widetilde{E}, F](x, t) = (\widetilde{E}(x_L, t), F(x_L \oplus \widetilde{E}(x_L, t)) \oplus x_R)$. See Fig. 1. The decryption procedure is clear, thus omitted here.

If $\Pr[H(x) \oplus H(x') = c] \leq \epsilon$ for any $c \in \Sigma^n$ and any distinct inputs x and x', H is called an ϵ-almost XOR universal (ϵ-AXU) hash function [21]. We prove that $\mathrm{HSE}[H, \widetilde{E}, F]$ is $\widetilde{\mathrm{cca}}$-secure, if $\widetilde{E}$ is a tweakable SPRP, and H is an ϵ-AXU hash function, and F is a WPRF.

Theorem 1. *If H, and $\widetilde{E}$, and F are independent and H is ϵ-AXU, we have*

$$\mathrm{Adv}_{\mathrm{HSE}[H,\widetilde{E},F]}^{\widetilde{\mathrm{sprp}}}(\ell, q, \tau) \leq \mathrm{Adv}_{\widetilde{E}}^{\widetilde{\mathrm{sprp}}}(q, \tau') + \mathrm{Adv}_F^{\mathrm{wprf}}(q, \tau') + q^2 \epsilon + \frac{2q^2}{2^n},$$

where the message is ℓ-bit, and $\tau' = \tau + O(q)$.

Algorithm 3.1: $\mathrm{HSE}[H, \widetilde{E}, F](x, t)$

Partition x into $(x_L \in \Sigma^n, x_R \in \Sigma^{\ell-n})$
$S \leftarrow x_L \oplus H(x_R)$
$U \leftarrow \widetilde{E}(S, t)$
$V \leftarrow S \oplus U$
$y_R \leftarrow F(V) \oplus x_R$
$y_L \leftarrow U \oplus H(y_R)$
$y \leftarrow (y_L, y_R), \mathbf{return}\ (y)$

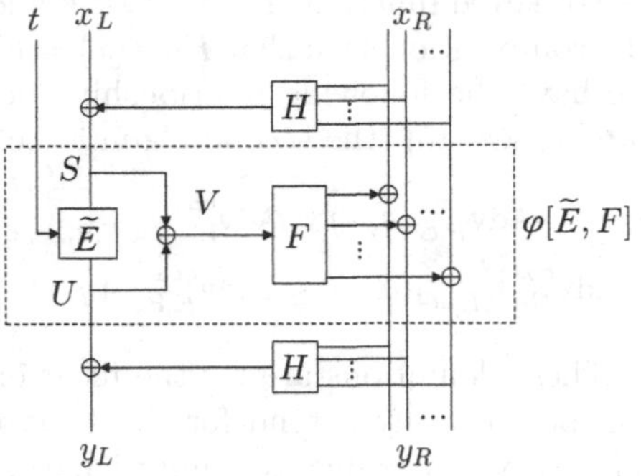

Fig. 1. Encryption procedure of HSE (left) and the structure of HSE (right)

Proof. Let Q be $\mathrm{HSE}[H, \widetilde{E}, F]$, and let Q_{pf} be $\mathrm{HSE}[H, \widetilde{\mathsf{P}}_n, F]$, and let Q_{pr} be $\mathrm{HSE}[H, \widetilde{\mathsf{P}}_n, \mathsf{R}_{n,\ell-n}]$. Using triangle inequality, we have

$$\mathbf{Adv}_Q^{\widetilde{\mathrm{sprp}}}(q, \tau) = \mathbf{Adv}_{Q, \widetilde{\mathsf{P}}_\ell}^{\widetilde{\mathrm{cca}}}(q, \tau) \le \mathbf{Adv}_{Q, Q_{\mathrm{pf}}}^{\widetilde{\mathrm{cca}}}(q, \tau) + \mathbf{Adv}_{Q_{\mathrm{pf}}, Q_{\mathrm{pr}}}^{\widetilde{\mathrm{cca}}}(q, \tau) + \mathbf{Adv}_{Q_{\mathrm{pr}}, \widetilde{\mathsf{P}}_\ell}^{\widetilde{\mathrm{cca}}}(q, \tau). \tag{4}$$

First, we have

$$\mathbf{Adv}_{Q, Q_{\mathrm{pf}}}^{\widetilde{\mathrm{cca}}}(q, \tau) \le \mathbf{Adv}_{\widetilde{E}, \widetilde{\mathsf{P}}_n}^{\widetilde{\mathrm{cca}}}(q, \tau') = \mathbf{Adv}_{\widetilde{E}}^{\widetilde{\mathrm{sprp}}}(q, \tau'), \tag{5}$$

which follows from the standard arguments (e.g., see [1]). Next, we focus on $\mathbf{Adv}_{Q_{\mathrm{pf}}, Q_{\mathrm{pr}}}^{\widetilde{\mathrm{cca}}}(q, \tau)$. Observe that any CCA can be described as a CPA having an additional 1-bit input, w, as oracle indicator. Here, $w = 0$ ($w = 1$) means the access to encryption (decryption) oracle. Since H is common to both Q_{pf} and Q_{pr}, we have

$$\mathbf{Adv}_{Q_{\mathrm{pf}}, Q_{\mathrm{pr}}}^{\widetilde{\mathrm{cca}}}(q, \tau) \le \mathbf{Adv}_{\varphi[\widetilde{\mathsf{P}}_n, F], \varphi[\widetilde{\mathsf{P}}_n, \mathsf{R}_{n,\ell-n}]}^{\widetilde{\mathrm{cca}}}(q, \tau') \le \mathbf{Adv}_{\mathsf{P}_n^\oplus \trianglelefteq F, \mathsf{P}_n^\oplus \trianglelefteq \mathsf{R}_{n,\ell-n}}^{\mathrm{cpa}}(q, \tau'), \tag{6}$$

where $\mathsf{P}_n^\oplus$ is a keyed function: $\Sigma^n \times \Sigma^n \times \Sigma \to \Sigma^n$ that uses $\widetilde{\mathsf{P}}_n$ such as

$$\mathsf{P}_n^\oplus(v_{[1]}, v_{[2]}, w) = \begin{cases} \widetilde{\mathsf{P}}_n(v_{[1]}, v_{[2]}) \oplus v_{[1]} & \text{if } w = 0, \\ \widetilde{\mathsf{P}}_n^{-1}(v_{[1]}, v_{[2]}) \oplus v_{[1]} & \text{if } w = 1. \end{cases}$$

Here, $(v_{[1]}, v_{[2]})$ corresponds to (input, tweak) for $\widetilde{\mathsf{P}}_n$. Note that, a pair of queries that is adaptively chosen can cause a $\mathsf{P}_n^\oplus$-output collision with probability one. Since such a pair is pointless, we only have to consider cpa without such queries, which is denoted by cpa'. Then we have

$$\mathbf{Adv}_{\mathsf{P}_n^\oplus \trianglelefteq F, \mathsf{P}_n^\oplus \trianglelefteq \mathsf{R}_{n,\ell-n}}^{\mathrm{cpa}}(q, \tau') = \mathbf{Adv}_{\mathsf{P}_n^\oplus \trianglelefteq F, \mathsf{P}_n^\oplus \trianglelefteq \mathsf{R}_{n,\ell-n}}^{\mathrm{cpa}'}(q, \tau')$$

$$\le 2\mathbf{Adv}_{\mathsf{P}_n^\oplus, \mathsf{R}_{2n+1,n}}^{\mathrm{cpa}'}(q, \tau') + \mathbf{Adv}_{F, \mathsf{R}_{n,\ell-n}}^{\mathrm{kpa}}(q, \tau'). \tag{7}$$

The inequality of Eq. (7) follows from Eq. (3), as Eq. (3) holds for a constrained CPA such as cpa′. A simple collision analysis provides

$$\text{Adv}^{\text{cpa}'}_{\text{P}^{\oplus}_n,\text{R}_{2n+1,n}}(q,\tau') \leq \text{Adv}^{\text{cpa}'}_{\text{P}^{\oplus}_n,\text{R}_{2n+1,n}}(q) \leq \binom{q}{2}\frac{1}{2^n}. \tag{8}$$

Thus, combining Eqs. (6)(7)(8), we obtain

$$\text{Adv}^{\widetilde{\text{cca}}}_{Q_{\text{pf}},Q_{\text{pr}}}(q,\tau) \leq \text{Adv}^{\text{kpa}}_{F,\text{R}_{n,\ell-n}}(q,\tau') + \binom{q}{2}\frac{2}{2^n} \leq \text{Adv}^{\text{wprf}}_F(q,\tau') + \frac{q^2}{2^n}. \tag{9}$$

For the last term of Eq. (4), we have the following information-theoretic bound.

Lemma 1. $\text{Adv}^{\widetilde{\text{cca}}}_{Q_{\text{pr}},\widetilde{\text{P}}_\ell}(q,\tau) \leq \text{Adv}^{\widetilde{\text{cca}}}_{Q_{\text{pr}},\widetilde{\text{P}}_\ell}(q) \leq q^2\epsilon + q^2/2^n.$

The proof of Lemma 1 is based on Maurer's method (see Appendix. A) and written in Appendix B. The proof of Theorem 1 is completed by combining Eqs. (4)(5)(9) and Lemma 1.

3.2 Variants of HSE

Different tweak processing. We can think of another tweak processing, where tweak t is a part of H's input. With this tweak processing, we use $H(x_R\|t)$ and $H(y_R\|t)$ instead of $H(x_R)$ and $H(y_R)$, and U is defined as $E(S)$ for a (non-tweakable) blockcipher E. This variant is called HSE type 2. It has almost the same security proof as that of HSE defined in Sect. 3.1 (we call it HSE type 1 if it is confusing). The following is the security proof of HSE type 2.

Theorem 2. Let $\text{HSE2}[H, E, F]$ be the HSE type 2. If H, E, and F are independent and H is ϵ-AXU for ℓ-bit inputs (as $|x_R\|t| = \ell$), $\text{Adv}^{\widetilde{\text{sprp}}}_{\text{HSE2}[H,E,F]}(q,\tau)$ is at most $\text{Adv}^{\text{sprp}}_E(q,\tau') + \text{Adv}^{\text{wprf}}_F(q,\tau') + 2q^2\epsilon + 2q^2/2^n$, where $\tau' = \tau + O(q)$.

The proof of Theorem 2 will be given in the full paper[1]. Type 2 tweak processing has been employed by HCTR. In practice, the difference in the tweak processing type may slightly affect the performance characteristic (see Sect. 4.1).

VIL schemes. Converting HSE (of both types) into a VIL scheme requires H to be a VIL-AXU hash function: it must assure a small differential probability of outputs for any pair of finite-length inputs. Typically, such hash functions can be obtained from FIL-AXU hash functions with some input encodings. For example, the polynomial evaluation hash over $\text{GF}(2^n)$, defined as $\text{poly}_K(x) \overset{\text{def}}{=} K^m \cdot x_{[1]} + \cdots + K \cdot x_{[m]}$, where $x = (x_{[1]}\|\ldots\|x_{[m]})$, $x_{[i]} \in \Sigma^n$, $|x| = nm$, is $m/2^n$-AXU if $K \in \Sigma^n$ is uniform (note that we interchangeably use Σ^n and $\text{GF}(2^n)$ to denote n-bit variable) for any m-block inputs. Let $\text{len}(x) \in \Sigma^n$ denote the

---

[1] The proof structure is the same as Theorem 1. However, the proof will require few more steps for bounding the $\widetilde{\text{cca}}$-advantage between Q_{pr} and $\widetilde{\text{P}}_\ell$. This results in the additional constant ϵq^2.

n-bit representation of $|x|$. We define $\texttt{pad}(x) = x$ if $\gamma = |x| \bmod n = 0$, and $\texttt{pad}(x) = x\|0^*$, where 0^* denotes the $(n - \gamma)$-bit sequence of all zeros. Then,

$$H(x) = \text{poly}_K(\texttt{pad}(x)\|\texttt{len}(x)) \tag{10}$$

accepts any finite-length input. It is $(m + 1)/2^n$-AXU if $|x| \le nm$. Almost the same hash function was employed in HCTR with tweak processing type 2. For completeness, the encryption procedure of HCTR is in the right of Fig. 2. For both of tweak processing types, the security proof of the VIL scheme is almost the same as Theorems 1 and 2. Thus we omit it here.

4   Applications

4.1   Implementation Based on a Blockcipher

As mentioned, our primal purpose is proving the security of general HSE, rather than specifying a concrete implementation. At the same time, our result can be useful in building concrete TESs. In this section, we describe an implementation of HSE type 1 using a blockcipher. As well as HCTR, we use counter mode and field multiplication. The resulting scheme is similar to HCTR and HCH. However, it has a better security bound than that of HCTR and is comparable to HCH. We first describe a FIL scheme using one blockcipher key. To start with, we prove that the counter mode of E is a WPRF if E is a PRP.

Lemma 2. *Let* $\text{CTR}_m[E] : \Sigma^n \to \Sigma^m$ *be the counter mode of E, such as*

$$\text{CTR}_m[E](x) \stackrel{\text{def}}{=} (E(x), E(x \oplus 1), \dots, \text{cut}(E(x \oplus \lceil m/n \rceil - 1))),$$

where $\text{cut}(v)$ *is the first* $(m \bmod n)$-*bit of v. Then* $\text{Adv}^{\text{wprf}}_{\text{CTR}_m[E]}(q, \tau)$ *is at most* $\text{Adv}^{\text{prp}}_E(\omega q, \tau') + \text{Adv}^{\text{wprf}}_{\text{CTR}_m[P_n]}(q)$ *and* $\text{Adv}^{\text{wprf}}_{\text{CTR}_m[P_n]}(q)$ *is at most* $\binom{\omega q}{2}\frac{2}{2^n}$, *where* $\omega = \lceil m/n \rceil$ *and* $\tau' = \tau + O(\omega q)$.

Proof. The first claim follows from the triangle inequality and the fact that $\text{Adv}^{\text{kpa}}_{\text{CTR}_m[E], \text{CTR}_m[P_n]}(q, \tau)$ is no larger than $\text{Adv}^{\text{prp}}_E(\omega q, \tau')$. The second follows from a simple collision analysis similar to the analysis of the XOR mode (randomized counter mode) provided by Bellare et al. [1].

The components $(H, \widetilde{E}, F)$ are defined as follows. First, we define $\mathtt{u}_0$, $\mathtt{u}_1$, $\mathtt{u}_2$, and $\mathtt{u}_3$ as distinct constants of $\text{GF}(2^n) \setminus \{0\}$. For $\widetilde{E}$, we use a scheme based on LRW [8] and XEX [19], which was provided by Minematsu [16]. Let $\text{TW}[E] : \Sigma^n \times \Sigma^n \to \Sigma^n$ be the tweakable blockcipher defined as

$$\text{TW}[E](x, t) \stackrel{\text{def}}{=} E(x \oplus \text{mul}'(L, t)) \oplus \text{mul}'(L, t), \text{ where } L = E(0) \text{ and}$$

$$\text{mul}'(L, t) = \begin{cases} tL & \text{if } t \in \Sigma^n \setminus \{0, 1\}, \\ \mathtt{u}_1 L^2 & \text{if } t = 1, \\ \mathtt{u}_0 L^2 & \text{if } t = 0. \end{cases}$$

Algorithm 4.1: $\mathrm{HSE}_f[E](x,t)$

Setup :
$L \leftarrow E(0), L_{sq} \leftarrow L^2, K \leftarrow E(\mathsf{u}_2 L_{sq})$
Partition x into $(x_L \in \Sigma^n, x_R \in \Sigma^{\ell-n})$
$S \leftarrow x_L \oplus \mathrm{poly}_K(\mathbf{pad}(x_R))$
$U \leftarrow E(S \oplus \mathrm{mul}'(L,t))) \oplus \mathrm{mul}'(L,t)$
$V \leftarrow S \oplus U$
$y_R \leftarrow \mathrm{CTR}_{\ell-n}[E](V \oplus \mathsf{u}_3 L_{sq}) \oplus x_R$
$y_L \leftarrow U \oplus \mathrm{poly}_K(\mathbf{pad}(y_R))$
$y \leftarrow (y_L, y_R), \mathbf{return}\ (y)$

Algorithm 4.2: $\mathrm{HCTR}[E](x,t)$

Partition x into $(x_L \in \Sigma^n, x_R \in \Sigma^{\ell-n})$
$S \leftarrow x_L \oplus \mathrm{poly}_K(\mathbf{pad}(x_R\|t)\|\mathbf{len}(x_R\|t))$
$U \leftarrow E(S)$ (1)
$V \leftarrow S \oplus U$ (2)
$y_R \leftarrow \mathrm{CTR}_{\ell-n}[E](V) \oplus x_R$ (3)
$y_L \leftarrow U \oplus \mathrm{poly}_K(\mathbf{pad}(y_R\|t)\|\mathbf{len}(x_R\|t))$
$y \leftarrow (y_L, y_R), \mathbf{return}\ (y)$

Fig. 2. Encryptions of HSE_f (Left) and HCTR (Right), where the hash key, K, is independent. In the original HCTR, the counter mode is slightly different from this.

Recall that $t = 0$ and $t = 1$ represent $(0\ldots0)$ and $(0\ldots01)$ in the n-bit representation, which correspond to the zero and identity elements in $\mathrm{GF}(2^n)$. We use L^2 since using tL for all $t \in \Sigma^n$ leads to an attack [19][16]. We use $\mathrm{TW}[E]$ as $\widetilde{E}$ of HSE. Moreover, H and F are defined as: $H(x) = \mathrm{poly}_K(\mathbf{pad}(x))$, where $K = E(\mathsf{u}_2 L^2)$, and $F(x) = \mathrm{CTR}_{\ell-n}[E](x \oplus \mathsf{u}_3 L^2)$. Note that $F(x)$ is equivalent to $\mathrm{CTR}_{\ell-n}[E^{\oplus \mathsf{u}_3 L^2}](x)$, where $E^{\oplus c}$ denotes a cipher with input mask: $E^{\oplus c}(x) = E(x \oplus c)$. Our scheme is $\mathrm{HSE}[H, \mathrm{TW}[E], \mathrm{CTR}_{\ell-n}[E^{\oplus \mathsf{u}_3 L^2}]]$, which will be denoted by $\mathrm{HSE}_f[E]$ (f for FIL). Its encryption procedure is shown in Fig. 2. This scheme requires several constant-variable multiplications. Generally such multiplications can be done with few simple operations. Typically, we define the field $\mathrm{GF}(2^n)$ with a primitive polynomial and set $\mathsf{u}_i = 2^i$. This allows very efficient and simple implementations, as described by [19].

Security. To prove the security of HSE_f, we need the idea of *tagged* tweakable blockcipher introduced by Rogaway [19], where a tweak is always specified with 1-bit tag. When the tag is 1, the adversary can access to either encryption or decryption oracle, and when the tag is 0, it can only access to the encryption oracle. Once a tag has been set for a tweak, it can not be changed. Tags can be adaptively determined by the adversary, or fixed in advance to the attack. Here, we only need to consider the latter case, which we call a (static) tagged CCA, denoted by $\widetilde{\mathrm{t\text{-}cca}}$. The maximum $\widetilde{\mathrm{t\text{-}cca}}$-advantage for a tagged tweakable n-bit blockcipher, $\widetilde{E}$, can be naturally defined as $\mathbf{Adv}_{\widetilde{E}}^{\mathrm{t\text{-}sprp}}(\theta) \overset{\mathrm{def}}{=} \mathbf{Adv}_{\widetilde{E},\widetilde{\mathsf{P}}_n}^{\mathrm{t\text{-}cca}}(\theta)$. We need the following theorem to prove the security of HSE_f.

Theorem 3. *Let* $\mathrm{TW}^*[E]$ *be a tagged tweakable blockcipher induced by* $\mathrm{TW}[E]$. *That is, it has tweak space* $\mathcal{T} = \Sigma^n \cup \{\alpha, \beta\}$, *and tag* $\mathbf{tag} \in \Sigma$, *defined as*[2] $\mathrm{TW}^*[E](x, \mathbf{tag}, t) = E(x \oplus \mathrm{mul}''(L, t)) \oplus \mathrm{sel}(\mathbf{tag}, L, t)$, *where* $L = E(0)$ *and*

---

[2] The inverse is $\mathrm{TW}^*[E]^{-1}(x, \mathbf{tag}, t) = E^{-1}(x \oplus \mathrm{sel}(\mathbf{tag}, L, t)) \oplus \mathrm{mul}''(L, t)$. Note that the inverse can not be accessed when $\mathbf{tag} = 0$.

$$\mathrm{mul}''(L,t) = \begin{cases} \mathrm{mul}'(L,t) & \text{if } t \in \Sigma^n, \\ u_2 L^2 & \text{if } t = \alpha, \\ u_3 L^2 & \text{if } t = \beta, \end{cases}$$

and $\mathrm{sel}(\mathtt{tag}, L, t) = \mathrm{mul}''(L, t)$ if $\mathtt{tag} = 1$, 0 if $\mathtt{tag} = 0$. For any tagged CCA with $\mathtt{tag} = 1$ for all $t \in \Sigma^n$ and $\mathtt{tag} = 0$ for $t \in \{\alpha, \beta\}$, we have $\mathbf{Adv}_{\mathrm{TW}^*[\mathrm{P}_n]}^{\widetilde{\mathrm{t\text{-}sprp}}}(q) \leq 4q^2/2^n$.

TW^* is based on the idea of Rogaway's tagged tweakable cipher, XEX^* (also called XEorXEX) [19]. The proof of Theorem 3 is obtained by extending Theorem 4 of [16], which will be given in the full paper.

Using Theorem 3, the proof of HSE_f is as follows.

Corollary 1. *Let the message length of* $\mathrm{HSE}_f[E]$ *be* ℓ-*bit and* $\omega = \lceil \ell/n \rceil$. *Then*

$$\mathbf{Adv}_{\mathrm{HSE}_f[E]}^{\widetilde{\mathrm{sprp}}}(q, \tau) \leq \mathbf{Adv}_E^{\mathrm{sprp}}(\omega q + 2, \tau') + \frac{5(\omega q + 1)^2}{2^n}, \text{ where } \tau' = \tau + O(\omega q).$$

Proof. First, note that $\mathrm{HSE}_f[\mathrm{P}_n]$ can be obtained as a function of $\mathrm{TW}^*[\mathrm{P}_n]$. Let $\widetilde{\mathrm{P}}_n^*$ be the n-bit tagged tweakable URP with tweak space $\Sigma^n \cup \{\alpha, \beta\}$ (note that a tag can be attached to any tweakable URP). Then, let HSE_f^* use $\widetilde{\mathrm{P}}_n^*$ instead of $\mathrm{TW}^*[\mathrm{P}_n]$. From Theorem 3, we have

$$\mathbf{Adv}_{\mathrm{HSE}_f[\mathrm{P}_n], \mathrm{HSE}_f^*}^{\widetilde{\mathrm{cca}}}(q) \leq \mathbf{Adv}_{\mathrm{TW}^*[\mathrm{P}_n], \widetilde{\mathrm{P}}_n^*}^{\widetilde{\mathrm{t\text{-}cca}}}(\omega q + 1) \leq \frac{4(\omega q + 1)^2}{2^n}. \tag{11}$$

Clearly, the components (H, $\widetilde{E}$, and F) of HSE_f^* are independent of each other. Since H's input is $(\ell - n)$-bit, H is $(\omega - 1)/2^n$-AXU (see Sect. 3.2). Thus we have

$$\mathbf{Adv}_{\mathrm{HSE}_f^*, \widetilde{\mathrm{P}}_\ell}^{\widetilde{\mathrm{cca}}}(q) \leq \mathbf{Adv}_{\widetilde{\mathrm{P}}_n}^{\widetilde{\mathrm{sprp}}}(q) + \mathbf{Adv}_{\mathrm{CTR}_{\ell-n}[\mathrm{P}_n]}^{\mathrm{wprf}}(q) + q^2 \cdot \frac{\omega - 1}{2^n} + \frac{2q^2}{2^n}, \tag{12}$$

$$\leq \binom{(\omega - 1)q}{2} \frac{2}{2^n} + \frac{(\omega + 1)q^2}{2^n}. \tag{13}$$

where the first inequality follows from Theorem 1, and the second follows from the second claim of Lemma 2. Combining Eqs. (11)(13) and triangle inequality, we have

$$\mathbf{Adv}_{\mathrm{HSE}_f[\mathrm{P}_n]}^{\widetilde{\mathrm{sprp}}}(q) \leq \frac{4(\omega q + 1)^2}{2^n} + \binom{(\omega - 1)q}{2} \frac{2}{2^n} + \frac{(\omega + 1)q^2}{2^n},$$

$$\leq \frac{4(\omega q + 1)^2 + (\omega - 1)^2 q^2 + (\omega + 1)q^2}{2^n} \leq \frac{5(\omega q + 1)^2}{2^n}, \tag{14}$$

using $(\omega^2 - \omega + 2) \leq \omega^2$ when $\omega \geq 2$ (as $\ell > n$). We can easily convert Eq. (14) into a computational counterpart. This concludes the proof.

VIL variant. We can also build a VIL variant called HSE_v using the hash function H defined as Eq. (10). This will require one more multiplication using Horner's rule. As the hash function is called twice, HSE_v needs two additional multiplications compared to HSE_f. As mentioned in Sect. 3.2, the security proof HSE_v is almost the same as that of HSE_f, therefore it is omitted.

Comparison. Table 1 shows a comparison between HCTR, HCH, and HSE. HCTR's security bound is cubic. HSE and HCH have quadratic bounds, and their total (i.e., Setup + Encryption) complexities are generally the same. However, in the Encryption phase, HSE requires fewer blockcipher calls (ω for VIL and FIL) than that of HCH ($\omega + 3$ for VIL, $\omega + 2$ for FIL). Since HSE requires a larger number of multiplication than HCH, the gain of HSE will depend on the ratio of speed between the blockcipher and multiplication (and hence it is not always faster than HCH). Note that, as HSE uses a variant of XEX* to implement independent permutations, a naive implementation requires one or two XOR operations of masking values for every blockcipher call, which can be a drawback. However, the number of these XORs can be reduced to three, as shown by Fig. 2. The number of multiplication in Encryption of HSE_v is one less than that of HCTR, as HSE_v needs 2ω multiplications for the hash computations, and (at most) one multiplication for computing $\text{mul}'(L, t)$, while HCTR always needs $2(\omega + 1)$ multiplications for the hash computations. Generally, the difference would be greater if the tweak gets longer. This demonstrates the difference in performance characteristics between HSE types 1 and 2.

4.2   Other Implementations

Improving HCTR with small changes. We describe how to improve the security of HCTR with small changes. This fix is also an implementation of HSE type 2. As shown by Fig. 2, HCTR uses $\varphi[E, \text{CTR}[E]]$. Our proof technique can not be directly applied to this Sum-Expansion, as the same blockcipher keys are used for the I/O sum generation and the expansion. Now, we use $\varphi[E^{\oplus \oplus 2L}, \text{CTR}[E^{\oplus 2^2 L}]]$ instead of $\varphi[E, \text{CTR}[E]]$, where $E^{\oplus \oplus 2L}(x) = E(x \oplus 2L) \oplus 2L$ with $L = E(0)$ (recall $E^{\oplus c}(x) = E(x \oplus c)$). That is, we changes the lines (from (1) to (3)) of Fig. 2 such as

$$U \leftarrow E(S \oplus 2L) \oplus 2L$$
$$V \leftarrow S \oplus U$$
$$y_R \leftarrow \text{CTR}_{\ell-n}[E](V \oplus 2^2 L) \oplus x_R$$

and keep other lines unchanged. This fix requires only one additional blockcipher call in the setup. Since $(E^{\oplus \oplus 2L}, E^{\oplus 2^2 L})$ is an instance of $\text{TW}^*[E]$, we can easily prove the security of the modified HCTR by combining[3] Theorems 2, 3, and Lemma 2, in the same manner as Corollary 1. Although we omit the concrete

---

[3] As HCTR is a VIL scheme, we also need to extend Theorem 2 to the VIL version. This is quite straightforward.

Table 1. Comparison of modes. The blockcipher's key bit length is k, and $\omega = \lceil \ell/n \rceil$ where the message length is ℓ-bit. i/j denotes i blockcipher calls and j multiplications.

Mode	Type	Key	Setup	Encryption	Security
HCTR[20]	VIL	$k+n$	0/0	$\omega/2\omega + 2$	$(\omega q)^3/2^n$
HCH[2]	VIL	k	0/0	$\omega + 3/2\omega - 2$	$(\omega q)^2/2^n$
HCHfp[2]	FIL	$k+n$	0/0	$\omega + 2/2\omega - 2$	$(\omega q)^2/2^n$
HSE$_v$	VIL	k	2/1	$\omega/2\omega + 1$	$(\omega q)^2/2^n$
HSE$_f$	FIL	k	2/1	$\omega/2\omega - 1$	$(\omega q)^2/2^n$

proof due to the space limitation, the constant of its security bound will be improved to quadratic: $O(\omega^2 q^2/2^n)$.

Other expansions. Since Theorem 1 does not rely on a particular expansion, various expansions, e.g., a stream cipher with n-bit initial vectors, can be used. We can also consider using a blockcipher-based expansion with better kpa-advantage than that of the counter mode. A naive approach is to combine multiple independent expansions. Let $m = \lambda \cdot \mu \cdot n$, where λ and μ are positive integers. We use μ independently-keyed blockciphers, $E^{(1)}, \ldots, E^{(\mu)}$, and define the extended counter mode, $\mathrm{eCTR}_m[E^{(1)}, \ldots, E^{(\mu)}](x)$, as the concatenation of $\mathrm{CTR}_{\lambda n}[E^{(i)}](x)$ for $i = 1, \ldots, \mu$. Using Lemma 2 and Lemma 1 of [3], it is easy to prove that the kpa-advantage of $\mathrm{eCTR}_m[E^{(1)}, \ldots, E^{(\mu)}](x)$ is at most $\omega^2 q^2/\mu 2^n$, where $\omega = m/n$. I.e., the bound linearly decreases w.r.t. the number of keys. In the extreme setting $(\lambda, \mu) = (1, \omega)$, the bound is $\omega q^2/2^n$. Of course, this is quite impractical and thus searching for a better expansion would be an interesting future research.

4.3 Construction of a Double-Block-Length SPRP

The construction of double-block-length (DBL) PRP or SPRP, i.e., $2n$-bit blockcipher using n-bit functions, has been an active research topic. Classical proposals [9][10][18] use two or more CPA-secure functions (i.e., PRF or PRP or SPRP) and some universal hash. However, recent studies prove that two CPA-secure functions are not needed to build a DBL PRP. Maurer et al.[12] proved that the last round function of the 3-round Feistel cipher can be any WPRF, if the first is an AXU hash, and the second is a PRF. A variant of this is also seen in [15]. These results demonstrated that an invocation of a CPA-secure function can be substituted with that of a KPA-secure one. Since WPRF is naturally expected to be faster than PRF (e.g., see [13][17]), such combined constructions are *theoretically* faster than the previous ones[4]. However, we still need two CPA-secure functions to build a DBL *SPRP* in the previous proposals. Naor and Reingold's 4-round Feistel cipher [18] required two PRFs and two AXU hashs. They also

---

[4] Obviously, they would require more memory for implementation than the constructions using single primitive.

proposed a construction using two SPRPs and two universal hash-based mixing layers. In contrast, the non-tweakable HSE for $\ell = 2n$ requires one SPRP and one WPRF, and two AXU hashs. To our knowledge, this is the first non-trivial DBL SPRP construction that does not require two CPA-secure functions.

5 Conclusion

We have studied the tweakable enciphering scheme using universal hash functions and a length-expanding function called the expansion, and proved that the expansion can be any weak pseudorandom function. While this structure has been employed by the previous proposals using the counter mode-based expansion, ours is the first to point out this fact. As applications of this result, we have provided some efficient blockcipher-based TESs. We expect that this structure can offer more applications, e.g., TESs beyond the birthday-bound security.

Acknowledgments

We would like to thank the anonymous referees for very useful comments.

References

1. Bellare, M., Desai, A., Jokipii, E., Rogaway, P.: A Concrete Security Treatment of Symmetric Encryption. In: FOCS 1997. Proceedings of the 38th Annual Symposium on Foundations of Computer Science, pp. 394–403 (1997)
2. Chakraborty, D., Sarkar, P.: HCH: A New Tweakable Enciphering Scheme Using the Hash-Encrypt-Hash Approach. In: Barua, R., Lange, T. (eds.) INDOCRYPT 2006. LNCS, vol. 4329, pp. 287–302. Springer, Heidelberg. The full version is available from IACR ePrint 2007/028 (2006)
3. Damgård, I., Nielsen, J.: Expanding Pseudorandom Functions; or: From Known-Plaintext Security to Chosen-Plaintext Security. In: Yung, M. (ed.) CRYPTO 2002. LNCS, vol. 2442, pp. 449–464. Springer, Heidelberg (2002)
4. Goldreich, O.: Modern Cryptography, Probabilistic Proofs and Pseudorandomness. Springer, Heidelberg
5. Halevi, S.: EME*: Extending EME to Handle Arbitrary-Length Messages with Associated Data. In: Canteaut, A., Viswanathan, K. (eds.) INDOCRYPT 2004. LNCS, vol. 3348, pp. 315–327. Springer, Heidelberg (2004)
6. Halevi, S., Rogaway, P.: A Tweakable Enciphering Mode. In: Boneh, D. (ed.) CRYPTO 2003. LNCS, vol. 2729, pp. 482–499. Springer, Heidelberg (2003)
7. Halevi, S., Rogaway, P.: A Parallelizable Enciphering Mode. In: Okamoto, T. (ed.) CT-RSA 2004. LNCS, vol. 2964, pp. 292–304. Springer, Heidelberg (2004)
8. Liskov, M., Rivest, R., Wagner, D.: Tweakable Block Ciphers. In: Yung, M. (ed.) CRYPTO 2002. LNCS, vol. 2442, pp. 31–46. Springer, Heidelberg (2002)
9. Luby, M., Rackoff, C.: How to Construct Pseudo-random Permutations from Pseudo-random functions. SIAM J. Computing 17(2), 373–386 (1988)
10. Lucks, S.: Faster Luby-Rackoff Ciphers. In: Gollmann, D. (ed.) Fast Software Encryption. LNCS, vol. 1039, pp. 189–203. Springer, Heidelberg (1996)

11. Maurer, U.: Indistinguishability of Random Systems. In: Knudsen, L.R. (ed.) EU-ROCRYPT 2002. LNCS, vol. 2332, pp. 110–132. Springer, Heidelberg (2002)
12. Maurer, U., Oswald, Y.A., Pietrzak, K., Sjoedin, J.: Luby-Rackoff Ciphers from Weak Round Functions. In: Vaudenay, S. (ed.) EUROCRYPT 2006. LNCS, vol. 4004, pp. 391–408. Springer, Heidelberg (2006)
13. Maurer, U., Sjoedin, J.: A Fast and Key-Efficient Reduction of Chosen-Ciphertext to Known-Plaintext Security. In: EUROCRYPT 2007. LNCS, vol. 4515, pp. 498–516. Springer, Heidelberg (2007)
14. McGrew, D., Fluhrer, S.: The Extended Codebook (XCB) Mode of Operation. IACR ePrint archive (2004), http://eprint.iacr.org/2004/278
15. Minematsu, K., Tsunoo, Y.: Hybrid Symmetric Encryption Using Known-Plaintext Attack-Secure Components. In: Won, D.H., Kim, S. (eds.) ICISC 2005. LNCS, vol. 3935, pp. 242–260. Springer, Heidelberg (2006)
16. Minematsu, K.: Improved Security Analysis of XEX and LRW Modes. In: SAC 2006 Selected Areas in Cryptography. LNCS, vol. 4356, pp. 96–113 (2007)
17. Naor, M., Reingold, O.: Number-theoretic Constructions of Efficient Pseudo-random Functions. In: 38th Annual Symposium on Foundations of Computer Science, FOCS 1997, pp. 458–467 (1997)
18. Naor, M., Reingold, O.: On the Construction of Pseudorandom Permutations: Luby-Rackoff Revisited. Journal of Cryptology 12(1), 29–66 (1999)
19. Rogaway, P.: Efficient Instantiations of Tweakable Blockciphers and Refinements to Modes OCB and PMAC. In: Lee, P.J. (ed.) ASIACRYPT 2004. LNCS, vol. 3329, pp. 16–31. Springer, Heidelberg (2004)
20. Wang, P., Feng, D., Wu, W.: HCTR: A Variable-Input-Length Enciphering Mode. In: Feng, D., Lin, D., Yung, M. (eds.) CISC 2005. LNCS, vol. 3822, pp. 175–188. Springer, Heidelberg (2005)
21. Wegman, M., Carter, L.: New Hash Functions and Their Use in Authentication and Set Equality. Journal of Computer and System Sciences 22, 265–279 (1981)

A   Maurer's Methodology

We briefly describe the methodology developed by Maurer [11], which will be used in Appendix B. Consider event a_i defined for i input/output pairs (and possibly internal variables) of a keyed function. We assume a_i is monotone, i.e., a_i never occurs if $\overline{a_{i-1}}$, negation of a_{i-1}, occurs. An infinite sequence $\mathcal{A} = a_0 a_1 \ldots$ is called a *monotone event sequence* (MES). Here, a_0 is some tautological event. A random variable and its value are written in capital and small letters, respectively. For random variables $X_1, X_2, \ldots$, let X^i denote $(X_1, \ldots, X_i)$. After this, $\mathrm{dist}(X^i)$ will denote an event where $X_1, X_2, \ldots, X_i$ are distinct. If $\mathrm{dist}(X^i, Y^j)$ holds true, then we have no collision among $\{X_1, \ldots, X_i, Y_1, \ldots, Y_j\}$. Let MESs $\mathcal{A}$ and $\mathcal{B}$ be defined for two keyed functions, $F : \mathcal{X} \to \mathcal{Y}$ and $G : \mathcal{X} \to \mathcal{Y}$, respectively. Let $X_i \in \mathcal{X}$ and $Y_i \in \mathcal{Y}$ be the i-th input and output. Let P^F be the probability space defined by F. For example, $P^F_{Y_i|X^iY^{i-1}}(y^i, x^i)$ means $\Pr[Y_i = y_i|X^i = x^i, Y^{i-1} = y^{i-1}]$ where $Y_j = F(X_j)$ for $j \geq 1$. If $P^F_{Y_i|X^iY^{i-1}}(y^i, x^i) = P^G_{Y_i|X^iY^{i-1}}(y^i, x^i)$ for all possible (y^i, x^i), then we write $P^F_{Y_i|X^iY^{i-1}} = P^G_{Y_i|X^iY^{i-1}}$. This is also represented as $F \equiv G$, and we say they are equivalent. Inequalities such as $P^F_{Y_i|X^iY^{i-1}} \leq P^G_{Y_i|X^iY^{i-1}}$ are similarly defined.

Definition 1. *We write $F^{\mathcal{A}} \equiv G^{\mathcal{B}}$ if $P^F_{Y_i a_i | X^i Y^{i-1} a_{i-1}} = P^G_{Y_i b_i | X^i Y^{i-1} b_{i-1}}$ holds for all possible (y^i, x^i) and $i \geq 1$. We also write $F|\mathcal{A} \equiv G|\mathcal{B}$ if $P^F_{Y_i | X^i Y^{i-1} a_i}(y^i, x^i) = P^G_{Y_i | X^i Y^{i-1} b_i}(y^i, x^i)$ holds for all possible (y^i, x^i) and $i \geq 1$.*

Definition 2. *We define $\nu(F, \overline{a_q})$ as the maximal probability of $\overline{a_q}$ for any CPA with q queries and infinite computational power that interacts with F.*

If F is a (tweakable) keyed permutation, we can think of the maximal probability of $\overline{a_q}$ for any cca ($\widetilde{\text{cca}}$) adversary. For simplicity, it is also written as $\nu(F, \overline{a_q})$. The following theorems and lemmas will be used in Appendix B.

Theorem 4. *(Theorem 1 (i) of [11]) If $F^{\mathcal{A}} \equiv G^{\mathcal{B}}$ or $F|\mathcal{A} \equiv G$ holds true, we have $\text{Adv}^{\text{cpa}}_{F,G}(q) \leq \nu(F, \overline{a_q})$.*

Theorem 5. *(Lemma 1 (iv) and Theorem 1 (iii) of [11]) If $F|\mathcal{A} \equiv G|\mathcal{B}$ and $P^F_{a_i | X^i Y^{i-1} a_{i-1}} \leq P^G_{b_i | X^i Y^{i-1} b_{i-1}}$ for $i \geq 1$, then $F^{\mathcal{A}} \equiv G^{\mathcal{B} \wedge \mathcal{C}}$ for some MES $\mathcal{C}$ defined for G, and $\text{Adv}^{\text{cpa}}_{F,G}(q) \leq \nu(F, \overline{a_q})$.*

Lemma 3. *(Lemma 6 (ii) of [11]) If $F^{\mathcal{A}} \equiv G^{\mathcal{B}}$, then $\nu(F, \overline{a_q}) = \nu(G, \overline{b_q})$.*

Lemma 4. *(Lemma 4 (ii) of [11]) Let $\mathbb{F}$ be the function of F and G (i.e., $\mathbb{F}[F]$ is a function that internally invokes F to process its inputs). If $F^{\mathcal{A}} \equiv G^{\mathcal{B}}$ holds, $\mathbb{F}[F]^{\mathcal{A}'} \equiv \mathbb{F}[G]^{\mathcal{B}'}$ also holds. Here, MES $\mathcal{A}' = a'_0 a'_1 \ldots$ is defined such that a'_i denotes $\mathcal{A}$-event is satisfied for the time period i. $\mathcal{B}'$ is defined in the same way.*

Lemma 5. *(Lemma 6 (iii) of [11]) $\nu(F, \overline{a_q \wedge b_q}) \leq \nu(F, \overline{a_q}) + \nu(F, \overline{b_q})$.*

Lemma 6. *(An extension of Lemma 2 (ii) of [11]) If $F^{\mathcal{A}} \equiv G^{\mathcal{B}}$, then $F^{\mathcal{A} \wedge \mathcal{D}} \equiv G^{\mathcal{B} \wedge \mathcal{D}}$ holds for any MES $\mathcal{D}$ defined on the inputs and/or outputs.*

B Proof of Lemma 1

We use the notations from Appendix A. The first inequality is trivial. For the second, consider the game of distinguishing $\varphi_{\text{pr}} \overset{\text{def}}{=} \varphi[\widetilde{\mathsf{P}}_n, \mathsf{R}_{n,\ell-n}]$ from $\widetilde{\mathsf{P}}_\ell$ (note that this game itself is quite easy to win). We convert CCA into CPA using the oracle indicator. Let $X_i = (M_i, T_i, W_i)$ be the i-th query, where message $M_i \in \Sigma^\ell$, tweak $T_i \in \Sigma^n$, and oracle indicator $W_i \in \Sigma$. Let $Y_i \in \Sigma^\ell$ be the i-th answer. We assume that $M_i \neq M_j$ whenever $(T_i, W_i) = (T_j, W_j)$ with $i < j$, and $Y_i \neq M_j$ whenever $T_i = T_j$ and $W_i \neq W_j$ with $i < j$, since such queries are pointless for a keyed permutation. We call this the invertibility assumption (IA). A CPA with IA is denoted by cpa''. For a tweakable permutation $\widetilde{E}$ with $M \in \Sigma^\ell$ and $T \in \Sigma^n$, we write $\langle \widetilde{E} \rangle$ to denote the corresponding keyed function : $\Sigma^\ell \times \Sigma^n \times \Sigma \to \Sigma^\ell$ such that $\langle \widetilde{E} \rangle(M, T, W)$ equals $\widetilde{E}(M, T)$ if $W = 0$, and $\widetilde{E}^{-1}(M, T)$ if $W = 1$. We define (S_i, U_i) as the first n bits of (M_i, Y_i) when $W_i = 0$, and the first n bits of (Y_i, M_i) when $W_i = 1$. We also define $V_i = S_i \oplus U_i$. In $\langle \varphi_{\text{pr}} \rangle$, S_i (U_i) corresponds to the i-th input (output) of $\widetilde{\mathsf{P}}_n$. Let a_q (b_q) be the event where

$S_i \neq S_j$ ($U_i \neq U_j$) for any $1 \leq i < j \leq q$ such that $T_i \neq T_j$. In addition, d_q denotes the event $\mathrm{dist}(V^q)$. MESs are defined as $\mathcal{A} = a_0 a_1 \ldots$, $\mathcal{B} = b_0 b_1 \ldots$, and $\mathcal{D} = d_0 d_1 \ldots$. Note that $\mathcal{A} \equiv \mathcal{B}$ in $\langle \varphi_{\mathrm{pr}} \rangle$, but not in $\langle \widetilde{\mathsf{P}}_\ell \rangle$. From the properties of φ_{pr} and $\widetilde{\mathsf{P}}_n$, it is easy to check that $\langle \varphi_{\mathrm{pr}} \rangle | \mathcal{ABD} \equiv \langle \varphi_{\mathrm{pr}} \rangle | \mathcal{AD} \equiv \langle \widetilde{\mathsf{P}}_\ell \rangle | \mathcal{ABD}$ holds true. Also, a simple probability analysis provides

$$P^{\langle \widetilde{\mathsf{P}}_\ell \rangle}_{a_i b_i d_i | X^i Y^{i-1} a_{i-1} b_{i-1} d_{i-1}} \leq P^{\langle \varphi_{\mathrm{pr}} \rangle}_{a_i b_i d_i | X^i Y^{i-1} a_{i-1} b_{i-1} d_{i-1}} = P^{\langle \varphi_{\mathrm{pr}} \rangle}_{a_i d_i | X^i Y^{i-1} a_{i-1} d_{i-1}}.$$

From these observations and the second claim of Theorem 5, $\langle \varphi_{\mathrm{pr}} \rangle^{\mathcal{A} \wedge \mathcal{D} \wedge \mathcal{G}} \equiv \langle \widetilde{\mathsf{P}}_\ell \rangle^{\mathcal{A} \wedge \mathcal{B} \wedge \mathcal{D}}$ holds for some MES $\mathcal{G}$. Let $\psi[H]$ be the ℓ-bit unbalanced Feistel permutation using the round function H. We define $\mathsf{HPH} : \Sigma^\ell \times \Sigma^n \times \Sigma \to \Sigma^\ell$ as $\langle \widetilde{\mathsf{P}}_\ell \rangle$ sandwiched between $\psi[H]$: when the query to HPH is (M, T, W), the output is $Y = \psi[H](\widehat{Y})$, where $\widehat{Y} = \langle \widetilde{\mathsf{P}}_\ell \rangle(\psi[H](M), T, W)$. Note that if we use $\langle \varphi_{\mathrm{pr}} \rangle$ instead of $\langle \widetilde{\mathsf{P}}_\ell \rangle$, we obtain $\langle Q_{\mathrm{pr}} \rangle$. From this observation, and $\langle \varphi_{\mathrm{pr}} \rangle^{\mathcal{A} \wedge \mathcal{D} \wedge \mathcal{G}} \equiv \langle \widetilde{\mathsf{P}}_\ell \rangle^{\mathcal{A} \wedge \mathcal{B} \wedge \mathcal{D}}$, and Lemma 4, we have

$$\langle Q_{\mathrm{pr}} \rangle^{\mathcal{A} \wedge \mathcal{D} \wedge \mathcal{G}} \equiv \mathsf{HPH}^{\mathcal{A} \wedge \mathcal{B} \wedge \mathcal{D}}. \tag{15}$$

Then we obtain

$$\mathrm{Adv}^{\widetilde{\mathrm{cca}}}_{Q_{\mathrm{pr}}, \widetilde{\mathsf{P}}_\ell}(q) = \mathrm{Adv}^{\mathrm{cpa}''}_{\langle Q_{\mathrm{pr}} \rangle, \langle \widetilde{\mathsf{P}}_\ell \rangle}(q) = \mathrm{Adv}^{\mathrm{cpa}''}_{\langle Q_{\mathrm{pr}} \rangle, \mathsf{HPH}}(q) \leq \nu(\mathsf{HPH}, \overline{a_q \wedge b_q \wedge d_q}), \tag{16}$$

where the first equality is trivial, the second follows from $\mathsf{HPH} \equiv \langle \widetilde{\mathsf{P}}_\ell \rangle$ (as $\psi[H]$ is invertible) and the inequality follows from Eq. (15) and Theorem 4.

For $\langle \widetilde{\mathsf{P}}_\ell \rangle$ and $\mathsf{R}_{\ell+n+1,\ell}$, let e_q be the event such as $Y_i \neq Y_j$ when $M_i \neq M_j$ and $(T_i, W_i) = (T_j, W_j)$, and $Y_i \neq X_j$ when $T_i = T_j$ and $W_i \neq W_j$, for any $i \neq j \leq q$. Note that the MES $\mathcal{E} = e_0 e_1 \ldots$ always holds true for $\langle \widetilde{\mathsf{P}}_\ell \rangle$ (but not for $\mathsf{R}_{\ell+n+1,\ell}$) under cpa''. From this observation, we have

$$\langle \widetilde{\mathsf{P}}_\ell \rangle \equiv \langle \widetilde{\mathsf{P}}_\ell \rangle | \mathcal{E} \equiv \mathsf{R}_{\ell+n+1,\ell} | \mathcal{E}, \text{ and } \langle \widetilde{\mathsf{P}}_\ell \rangle^{\mathcal{R}} \equiv \mathsf{R}^{\mathcal{E}}_{\ell+n+1,\ell}, \tag{17}$$

for some MES $\mathcal{R} = r_0 r_1 \ldots$. Now we obtain

$$\langle \widetilde{\mathsf{P}}_\ell \rangle^{\mathcal{A} \wedge \mathcal{B} \wedge \mathcal{D} \wedge \mathcal{R}} \equiv \mathsf{R}^{\mathcal{A} \wedge \mathcal{B} \wedge \mathcal{D} \wedge \mathcal{E}}_{\ell+n+1,\ell}, \text{ and } \mathsf{HPH}^{\mathcal{A} \wedge \mathcal{B} \wedge \mathcal{D} \wedge \mathcal{R}} \equiv \mathsf{HRH}^{\mathcal{A} \wedge \mathcal{B} \wedge \mathcal{D} \wedge \mathcal{E}}, \tag{18}$$

where HRH is defined as $\mathsf{R}_{\ell+n+1,\ell}$ sandwiched between $\psi[H]$. The first equivalence of Eq. (18) follows from Eq. (17) and Lemma 6, and the second follows from the first and Lemma 4. From Eq. (18) and Lemma 3, we obtain

$$\nu(\mathsf{HPH}, \overline{a_q \wedge b_q \wedge d_q}) \leq \nu(\mathsf{HPH}, \overline{a_q \wedge b_q \wedge d_q \wedge r_q}) = \nu(\mathsf{HRH}, \overline{a_q \wedge b_q \wedge d_q \wedge e_q}), \tag{19}$$

where the underlying attack is cpa''. We focus on the last term. In HRH, let $M_{i,L}$ and $M_{i,R}$ ($Y_{i,L}$ and $Y_{i,R}$) be the first n bits and remaining $\ell - n$ bits of M_i (Y_i). Note that (S_i, U_i, V_i) is defined on the input/output of $\mathsf{R}_{\ell+n+1,\ell}$ used by HRH. Thus, $\overline{a_q \wedge b_q}$ consists of the following collision events.

Type 1: $i \neq j$, $T_i = T_j$, $W_i = W_j$, $H(M_{i,R}) \oplus H(M_{j,R}) = M_{i,L} \oplus M_{j,L}$

Type 2: $i \neq j$, $T_i = T_j$, $W_i = W_j$, $H(Y_{i,R}) \oplus H(Y_{j,R}) = Y_{i,L} \oplus Y_{j,L}$
Type 3: $i \neq j$, $T_i = T_j$, $W_i \neq W_j$, $H(M_{i,R}) \oplus H(Y_{j,R}) = M_{i,L} \oplus Y_{j,L}$
Type 4: $i \neq j$, $T_i = T_j$, $W_i \neq W_j$, $H(Y_{i,R}) \oplus H(M_{j,R}) = Y_{i,L} \oplus M_{j,L}$

Let $p(h)$ denote the maximum collision probability of type $h = 1, \ldots, 4$. As $\psi[H]$ is invertible, HRH is equivalent to $R_{\ell+n+1,\ell}$ and thus any adversary can not obtain any information of K, the key of H. In addition, the output Y_i is completely independent and random. From this observation and the fact that H is ϵ-AXU, we have $p(h) \leq \epsilon$ for all $h = 1, \ldots, 4$. For each $1 \leq i < j \leq q$, we can think of collision types 1 and 2 or types 3 and 4. Thus we have $\nu(\mathsf{HRH}, \overline{a_q \wedge b_q}) \leq \binom{q}{2} 2\epsilon$. Next, we focus on $\overline{d_q}$. The collision event belonging to $\overline{d_q}$ is

$$H(M_{i,R}) \oplus H(M_{j,R}) \oplus H(Y_{i,R}) \oplus H(Y_{j,R}) = M_{i,L} \oplus M_{j,L} \oplus Y_{i,L} \oplus Y_{j,L}, \text{ for } i \neq j.$$

W.l.o.g., we assume $i < j$. Then, Y_j is independent and uniformly random, even if the adversary is adaptive. Thus the collision probability is $1/2^n$ and we have $\nu(\mathsf{HRH}, \overline{d_q}) \leq \binom{q}{2}/2^n$. Also, it is easy to obtain $\nu(\mathsf{HRH}, \overline{e_q}) \leq \binom{q}{2}/2^\ell$. Combining these bounds and Lemma 5, we obtain $\nu(\mathsf{HRH}, \overline{a_q \wedge b_q \wedge d_q \wedge e_q}) \leq \binom{q}{2}\left(2\epsilon + 1/2^n + 1/2^\ell\right)$. From this and Eqs. (16)(19), the proof is completed.

A Framework for Chosen IV Statistical Analysis of Stream Ciphers

Håkan Englund[1], Thomas Johansson[1], and Meltem Sönmez Turan[2]

[1] Dept. of Electrical and Information Technology, Lund University, Sweden
[2] Institute of Applied Mathematics, METU, Turkey

Abstract. Saarinen recently proposed a chosen IV statistical attack, called the d-monomial test, and used it to find weaknesses in several proposed stream ciphers. In this paper we generalize this idea and propose a framework for chosen IV statistical attacks using a polynomial description. We propose a few new statistical attacks, apply them on some existing stream cipher proposals, and give some conclusions regarding the strength of their IV initialization. In particular, we experimentally detected statistical weaknesses in some state bits of Grain-128 with full IV initialization as well as in the keystream of Trivium using an initialization reduced to 736 rounds from 1152 rounds. We also propose some stronger alternative initialization schemes with respect to these statistical attacks.

1 Introduction

Synchronous stream ciphers are an important part of symmetric cryptosystems and they are suitable for applications where high speed and low delay are required. As examples, the stream cipher family A5 is used in the GSM standard and the cipher E0 is used to supply privacy in Bluetooth applications. In most applications, the transmission of ciphertext is assumed to be done over a noisy channel where the synchronization between sender and receiver can be lost and resynchronization is necessary.

Depending on the protocol, different resynchronization mechanisms can be used. In most applications the message is divided into frames and each frame is encrypted using different publicly known Initialization Vectors (IVs) and the same secret key. In such systems, the ciphers should be designed to resist attacks that use many short keystreams generated by random or chosen IVs.

In [1], an attack on nonlinear filter generators with linear resynchronization and filter function with few inputs is presented and this attack is extended to the case where the filter function is unknown in [2]. More extensions of the resynchronization attack is available in [3].

To avoid such attacks, the initialization of stream ciphers in which the internal state variables are determined using the secret key and the public IV should be designed carefully. In most ciphers, firstly the key and IV are loaded into the state variables, then a next state function is applied to the internal state iteratively for a number of times without producing any output. The number of iterations

K. Srinathan, C. Pandu Rangan, M. Yung (Eds.): Indocrypt 2007, LNCS 4859, pp. 268–281, 2007.

play an important role on both security and the efficiency of the cipher. It should be chosen so that each key and IV bit affect each initial state bit in a complex way. On the other hand, using a large number of iterations is inefficient and may hinder the speed for applications requiring frequent resynchronizations.

In [4], tests were introduced to evaluate the statistical properties of symmetric ciphers using the number of the monomials in the Boolean functions that simulate the action of a given cipher. In [5], Saarinen recently proposed to extend these ideas to a chosen IV statistical attack, called the d-monomial test, and used it to find weaknesses in several proposed stream ciphers.

In this paper we generalize this idea and propose a framework for chosen IV statistical attacks using a polynomial description. The basic idea is to select a subset of IV bits as variables. Assuming all other IV values as well as the key being fixed, we can write a keystream symbol as a Boolean function of the selected IV variables. By running through all possible values of these bits and creating a keystream output for each of them, we create the truth table of this Boolean function. We now hope that this Boolean function has some statistical weaknesses that can be detected. We describe the d-monomial test in this framework, and then we propose two new tests, called the monomial distribution test and the maximal degree monomial test.

We then apply them on some existing stream cipher proposals, and give some conclusions regarding the strength of their IV initialization. In particular, we experimentally detected statistical weaknesses in the keystream of Grain-128 with IV initialization reduced to 192 rounds as well as in the keystream of Trivium using an initialization reduced to 736 rounds. Furthermore, we repeat our experiments to study the statistical properties of internal state bits. Here we could detected statistical weaknesses in some state bits of Grain-128 with full IV initialization. In the context, we also propose alternative initial loadings for some of the ciphers so that the diffusion is satisfied in fewer rounds.

The paper is organized as follows. In the next section, some background information about hypothesis testing and Boolean functions are given. In Section 3, the suggested framework for chosen IV statistical attacks is presented, and in Section 4 some results are presented for reduced round initializations of the ciphers Grain [6], Trivium [7] and Decim [8]. Finally we conclude the paper in Section 5.

2 Preliminaries

2.1 Hypothesis Testing

Assume we have independently and identically distributed (i.i.d.) random variables X_i, the sum is a new random variable, denoted by Y, i.e., $Y = \sum_{i=0}^{n} X_i$. According to the central limit theorem Y is approximately normally distributed if n is large. Let y denote an observation from Y, and assume that we have r observations of random variables Y, i.e., $y_0, \ldots, y_{r-1}$, then the chi-square statistic is

$$\chi^2 = \sum_{k=0}^{r-1} \frac{\left(y_k - E(Y)\right)^2}{E(Y)} \xrightarrow{d} \chi_r^2$$

where $\xrightarrow{d}$ means convergence in distribution, and r is called the degrees of freedom (i.e., number of independent pieces of information).

Our two hypothesis are

- H_0 : $z = 0$, $y_0, \ldots, y_{r-1}$ are samples from Y,
- H_1 : $z \neq 0$, $y_0, \ldots, y_{r-1}$ are not samples from Y.

For a one-sided χ^2-Goodness of fit test, the hypothesis is rejected if the test statistics χ^2 is greater than the tabulated $\chi^2(1-\alpha; r)$ value, for some significance level α with r degrees of freedom.

2.2   Algebraic Normal Form of a Boolean Function

Let $f : \mathbb{F}_2^n \mapsto \mathbb{F}_2$ be a mapping from n binary input bits into one output bit, then f is called a Boolean function. There are many representations of a Boolean function, but in this paper we are mainly interested in the so called Algebraic Normal Form (ANF). The *ANF* is the polynomial

$$f(x_1, x_2, \ldots, x_n) = a_0 \oplus a_1 x_1 \oplus \ldots \oplus a_n x_n \oplus a_{n+1} x_1 x_2 \oplus \ldots \oplus a_{2^n-1} x_1 x_2 \ldots x_n$$

with unique a_i's in $\mathbb{F}_2$.

2.3   Computation of Algebraic Normal Form

Assume the truth table of an n-variable Boolean function is represented in a vector v of size 2^n and the ANF of the Boolean function can be calculated with complexity $O(n2^n)$ using the algorithm presented in Figure 1, which uses two auxiliary vectors t and u, both of size 2^{n-1}.

```
COMPUTE ANF(v)

    for i = 1, ..., n
        for j = 1, ..., 2^{n-1}
            t_j = v_{2j-1}
            u_j = v_{2j-1} ⊕ v_{2j}
        end for
        v = t||u
    end for
```

Fig. 1. Algorithm to compute the ANF in vector v from the truth table in v

2.4 Properties of a Random Boolean Function

Let $f : \mathbb{F}_2^n \mapsto \mathbb{F}_2$ be a Boolean function, and let the number of monomials in the ANF of f be denoted by M. If f is randomly chosen, each monomial is included with probability one half, i.e., a Bernoulli distribution. The sum of Bernoulli distributed random variables is Binomially distributed, hence $M \in \mathrm{Bin}(2^n, \frac{1}{2})$, with expected value $E(M) = 2^{n-1}$. Let's denote the number of monomials of degree k by M_k, i.e., $M = \sum_{k=0}^{n} M_k$. The distribution of M_k is $\mathrm{Bin}(\binom{n}{k}, \frac{1}{2})$ with $E(M_k) = \frac{1}{2}\binom{n}{k}$.

Let m_k be an observation from M_k, then

$$\chi^2 = \sum_{k=0}^{n} \frac{\left(m_k - \frac{1}{2}\binom{n}{k}\right)^2}{\frac{1}{2}\binom{n}{k}} \xrightarrow{d} \chi^2_{n+1}, \quad \text{when} \quad \binom{n}{k} \to \infty.$$

If $\binom{n}{k}$ is large enough, methods described in Section 2.1 can be used to perform a hypothesis test to decide if the function in question has a deviant number of monomials of degree k.

3 A Framework for Chosen IV Statistical Attacks

For an additive synchronous stream cipher, let $K = (k_0, \ldots, k_{N-1})$ denote the secret key. Furthermore, let $IV = (iv_0, \ldots, iv_{M-1})$ denote the public IV value used, and, finally, let $Z = z_0, z_1, \ldots$ denote the keystream sequence. We assume that the attacker has received a number of different keystream sequences generated using different (possibly chosen) IV values.

Different tests have been introduced to evaluate statistical properties of sequences from symmetric ciphers and hash functions. The tests are usually based on taking one long keystream sequence and then applying different statistical tests, like the NIST statistical test suit used in the AES evaluation [9].

However, recently several researchers have noted the possibility to instead generate a lot of short keystream sequences, from different chosen IV values and look at the statistical properties of, say, only the first output symbol of each keystream. One such example is the observation by Shamir and Mantin that the second byte in RC4 is strongly biased [10].

Based on work in [4], Saarinen [5] recently proposed the d-monomial IV distinguisher. The behavior of the keystream is analyzed using a function of n IV bits, i.e., $z = f(iv_0, \ldots, iv_{n-1})$. All other IV and key bits are considered to be constants. In [5] among a few other tests, Saarinen suggested the d-monomial test. For a chosen parameter d (set to be a small value), the test counts the number of monomials of weight d in the ANF of f and compares it to its expected value $\frac{1}{2}\binom{n}{d}$, using the χ^2-Goodness of Fit test with one degree of freedom.

In this paper, we will instead sum the test statistics for each d and evaluate the result using $n+1$ degrees of freedom. The algorithm for the d-Monomial test is summarized in Figure 2.

The complexity of this attack is $O(n2^n)$ operations and it needs memory $O(n2^n)$. The downside of this method is that statistical deviations for lower

d-MONOMIAL TEST

for $iv = 1, \ldots, 2^n - 1$
    Initialize cipher with iv
    $v[iv]$=first keystream bit after initialization
end for
Compute ANF of vector v and store result in v.
for $i = 1, \ldots, 2^n - 1$
    if $v[i] = 1$
        $weight$= weight of monomial i
        distr$[weight] + +$
end for
for $d = 0, \ldots, n$
$$\chi^2 + = \frac{(\text{distr}[d] - \frac{1}{2}\binom{n}{d})^2}{\frac{1}{2}\binom{n}{d}}.$$
if $\chi^2 > \chi^2(1 - \alpha; n + 1)$
    return *cipher*
else
    return *random*

Fig. 2. Summary of the d-monomial test, complexity $O(n2^n)$

and higher degree monomials are hard to detect since their numbers are few. So even if the maximal degree monomial never occurs, the test does not detect this anomaly. In the next section we will present alternative attacks that solves this problem.

3.1 A Generalized Approach

We suggest to use a generalized approach. Instead of analyzing just one function in ANF form, we can study the behavior of more polynomials so that monomials that are more (or less) probable than others can be detected.

Let us select n IV values, denoted $iv_0, \ldots, iv_{n-1}$, as our *variables*. The remaining IV values as well as key bits are kept constant. Using the first output symbol, $z_0 = f_1(iv_0, \ldots, iv_{n-1})$, for each choice of $iv_0, \ldots, iv_{n-1}$, the ANF of f_1 can be constructed.

The new approach is now to do the same again, but using some other choice on IV values outside the IV variables. Running through each choice of $iv_0, \ldots, iv_{n-1}$ in this case gives us a new function f_2. Continuing in this way, we derive P different Boolean functions $f_1, f_2, \ldots, f_P$ in ANF form. In some situations, it might also be possible to obtain polynomials from different keys, where the same IVs have been used.

Having P different polynomials in our possession we can now design any test that looks promising, taken over all polynomials. The d-monomial test would appear for the special case $P = 1$, and the test being counting the number of weight d monomials. We now propose in detail two different tests.

3.2 The Monomial Distribution Test

The attack scenario is similar to the d-monomial test, but instead of counting the number of monomials of a certain degree, we generate P polynomials and calculate in how many of the polynomials each monomial is present. That is, we generate P polynomials of the form (1) and count the number of occurrences of $a_i = 1$, $0 \leq i \leq 2^n - 1$

$$f = a_0 + a_1 x_1 + \ldots + a_{n+1} x_1 x_2 + \ldots + a_{2^n - 1} x_1 x_2 \ldots x_{n-1} x_n \qquad (1)$$

Denote the number of occurrences of coefficient a_i by M_{a_i}, since each monomial should be included in a function with probability $1/2$, i.e., $P(a_i = 1) = 0.5$, $0 \leq i \leq 2^n - 1$, the number of occurrences is binomially distributed with expected value $E(M_{a_i}) = P/2$ for each monomial. We will as previously perform a χ^2-Goodness of fit test with 2^n degrees of freedom, as described by Equation (2).

$$\chi^2 = \sum_{i=0}^{2^n - 1} \frac{(M_{a_i} - \frac{P}{2})^2}{\frac{P}{2}} \qquad (2)$$

If the observed amount is larger than some tabulated limit $\chi^2(1 - \alpha; 2^n)$, for some significance level α, we can distinguish the cipher from a random one. The pseudo-code of the monomial distribution test is given in Figure 3.

MONOMIAL DISTRIBUTION TEST

> **for** $j = 1, \ldots, P$
> > **for** $iv = 1, \ldots, 2^n - 1$
> > > Initialize cipher with iv
> > > $v[iv]$=first keystream bit after initialization
> >
> > **end for**
> > Compute ANF of vector v and store result in v.
> > **for** $i = 1, \ldots, 2^n - 1$
> > > **if** $v[i] = 1$
> > > > $M_{a_i} + +$
> >
> > **end for**
>
> **end for**
> **for** $d = 0, \ldots, 2^n - 1$
> > $\chi^2 + = \frac{(M_{a_d} - \frac{P}{2})^2}{\frac{P}{2}}.$
>
> **end for**
> **if** $\chi^2 > \chi^2(1 - \alpha; 2^n)$
> > **return** *cipher*
>
> **else**
> > **return** *random*

Fig. 3. Summary of Monomial distribution test, complexity $O(Pn2^n)$

This algorithm has a higher computational complexity than the d-Monomial attack, $O(Pn2^n)$, and needs the same amount of memory, $O(n2^n)$. On the other hand, if for a cipher some certain monomials are highly non-randomly distributed, the attack may be successful with less number of IV bits, i.e., smaller n, compared to the d-monomial test. Additionally, although this attack is originally proposed for the chosen IV scenario of a fixed unknown key, it is also possible to apply the test for different key values, if the same IV bits are considered.

3.3   The Maximal Degree Monomial

A completely different and very simple test is to see if the maximal degree monomial can be produced by the keystream generator. The maximal degree monomial is the product of all IV bits and can hence only occur if all the IV bits have been properly mixed. In hardware oriented stream ciphers the IV loading is usually as simple as possible to save gates, e.g., the IV bits are loaded into different memory cells. The update function is then performed a number of steps to produce proper diffusion of the bits, intuitively it will take many clockings before all IV bits meet in the same memory cell and even more clocking before they spread to all the memory cells and are mixed nonlinearly. The aim of the Maximal Degree Monomial is to check in a simple way whether the number of initial clockings are sufficient. Since the maximal degree monomial is unlikely to exist if lower degree monomials do not exist, this is our best candidate to study. Hence, the existence of the maximal degree term in ANFs is a good indication to the satisfaction of diffusion criteria, especially completeness.

According to the Reed-Muller transform the maximal degree monomial can be calculated as the XOR of all entries in the truth table. So the test is similar to the previous tests performed by initializing the cipher with all possible combinations for n IV bits, $z^{iv_0,...,iv_{n-1}} = f(iv_0,...,iv_{n-1})$, all other bits are considered to be constants. The existence of the maximal degree monomial can be checked by XORing the first keystream bit from each initialization, following the notation from Section 2.2, this is equivalent to determining a_{2^n-1}.

$$a_{2^n-1} = \bigoplus_{iv_0,...,iv_{n-1}} z^{iv_0,...,iv_{n-1}}.$$

By for example changing some other IV bit we receive a new polynomial and perform the same procedure again, this is repeated for P polynomials, if the maximal degree polynomial never occurs in any of the polynomials or if it occurs in all of the polynomials we successfully distinguish the cipher. Hence we can, with low complexity, and more importantly, almost no memory, check whether the maximal degree monomial can exist in the output from the cipher. It is possible, with the same complexity, to consider other weak monomials, the coefficient can be calculated according to the Reed-Muller transform. The complexity of the Maximal Degree Attack is $O(P2^n)$ and it only requires $O(1)$ memory. The description of the test is given in Figure 4.

```
MAXIMAL DEGREE MONOMIAL TEST

for j = 1, ..., P
    a_{2^n-1} = 0
    for iv = 1, ..., 2^n - 1
        Initialize cipher with iv
        z = first keystream bit after initialization
        a_{2^n-1} = a_{2^n-1} ⊕ z
    end for
    if a_{2^n-1} = 1
        ones++
end for
if  ones=0 or ones=P
    return cipher
else
    return random
```

Fig. 4. Summary of Maximal Degree Test, complexity $O(P2^n)$

3.4 Other Possible Tests

We have proposed two specific tests that we will use in the sequel to analyze different stream ciphers. Our framework gives us the possibility to design many other interesting tests. As an example, a monomial distribution test restricted to only monomials with very high weight could be an interesting test. Another possibility would be to examine properties of the Walsh transform of each polynomial. These tests have not been experimentally examined in this work.

4 Experimental Results

We applied the proposed tests described above on some of the Phase III eSTREAM candidates to evaluate their efficiency of initializations. We evaluated their security margin by testing reduced round versions of the ciphers. We also presented some results on the statistical properties of the internal state variables.

The significance level of the hypothesis tests is chosen to be approximately $1 - \alpha = 1 - 2^{-10}$. The tabulated results have a success rate of at least 90%. The required number of IVs, polynomials and the amount of memory needed to attack the ciphers are given in tables. Also, the results for initial state variables are presented with the percentage of weak initial state variables.

Hardware oriented stream ciphers use simple initial key and IV loading compared to software oriented ciphers. Generally, key and IV bits affect one initial state variable. Therefore, they require a large number of clockings to satisfy the diffusion of each input bit on each state bit. We repeated some of our simulations using alternative key/IV loadings in which each IV bit is assigned to more than one internal state bit and compared the results to the original settings. In the alternative loadings the hardware complexity is slightly higher, however on the other hand the cipher has more resistance to chosen IV attacks.

4.1   Grain-128

Grain-128 [6] is a hardware oriented stream cipher using a LFSR and a NFSR together with a nonlinear filter function. In the initialization of Grain, a 128 bit key is loaded into the NFSR and a 96 bit IV is loaded into the first 96 positions of the LFSR, the rest of the LFSR is filled with ones. The cipher is then clocked 256 times and for each clock the output bit is fed back into both the LFSR and the NFSR.

In Table 1, the results obtained for reduced version of Grain are given. The highest number of rounds, we succeeded to break is 192 out of the original 256 which corresponds to the 75% of the initialization phase.

Table 1. Number of IV bits needed to attack the first keystream bit of Grain-128 for different number of rounds in the initialization (out of 256 rounds)

Rounds	d-Monomial test			Monomial distr. test			Max. degree monomial		
	P	IVs	Memory	P	IVs	Memory	P	IVs	Memory
160	1	14	2^{14}	2^6	7	2^7	2^5	11	1
192	1	25	2^{25}	2^6	22	2^{22}	2^5	22	1

In Table 2, the results of the experiments for initial state variables are presented. The number of weak initial state variables are three times better in the maximum degree test compared to the d-monomial test. The statistical deviations in state bits remain even after full initialization. These weak state bits are located in the left most positions of the feedback shift registers. To remove the statistical deviations in state variables, at least 320 initial clockings are needed. It is possible that if we use larger number of IV bits, the weaknesses in state variables may also be observed from the keystream bits (See Appendix A).

Table 2. Number of IV bits needed to attack the initial state variables Grain-128 for different number of rounds in the initialization (out of 256 rounds)

Rounds	d-Monomial test				Monomial distr. test				Max. degree monomial			
	P	IVs	Memory	Fraction	P	IVs	Memory	Fraction	P	IVs	Memory	Fraction
256	1	14	2^{14}	33/256	2^6	8	2^8	20/256	2^5	14	1	108/256
256	1	16	2^{16}	40/256	2^6	10	2^{10}	35/256	2^5	16	1	120/256
256	1	20	2^{20}	56/256	2^6	15	2^{15}	44/256	2^5	20	1	138/256
288	1	20	2^{20}	0/256	2^6	20	2^{20}	0/256	2^5	20	1	73/256

Table 3. Number of IV bits needed to attack the first keystream bit of Grain-128 with alternative Key/IV loading for different number of rounds in the initialization (out of 256 rounds)

Rounds	d-Monomial test			Monomial distr. test			Max. degree monomial		
	P	IVs	Memory	P	IVs	Memory	P	IVs	Memory
160	1	19	2^{19}	2^6	20	2^{20}	2^5	21	1

Alternative Key/IV Loading for Grain-128. Here we propose an alternative Key/IV loading in which only the loading of the first 96 bits of the NFSR is different from the original. Instead of directly assigning the key, we assign the modulo 2 summation of IV and the first 96 bits of the key. The proposed loading is very similar to the original and the increase in number of gates required is approximately 10-15%. In an environment where many resynchronizations are expected, one can reduce the number of initial clockings by using some more gates in the hardware implementation. In the new loading, each IV bit affects two internal state variables. We repeated our experiments using the new loading and the results are given in Table 3 and Table 4. Using alternative loading, Grain shows more resistance to the presented attacks, but still the statistical deviations in the state bits remain after full initialization.

Table 4. Number of IV bits needed to attack the initial state variables of Grain-128 with alternative Key/IV loading for different number of rounds in the initialization (out of 256 rounds)

Rounds	d-Monomial test				Monomial distr. test				Max. degree monomial			
	P	IVs	Memory	Fraction	P	IVs	Memory	Fraction	P	IVs	Memory	Fraction
256	1	14	2^{14}	1/256	2^6	8	2^8	4/256	2^5	14	1	100/256
256	1	16	2^{16}	5/256	2^6	10	2^{10}	10/256	2^5	20	1	108/256
288	1	20	2^{20}	0/256	2^6	20	2^{20}	0/256	2^5	20	1	47/256

4.2 Trivium

Trivium [7] is another hardware oriented stream cipher based on NFSRs. The state is divided into three registers which in total stores 288 bits. During the initialization the 80-bit key is inserted into the first register while an 80-bit IV is inserted into the second register. The cipher is clocked 4 full cycles before producing any keystream, i.e., 1152 clockings.

The results for Trivium are given in Table 5 and Table 6. The attacks on 736 and more rounds, the d-Monomial and the Monomial distribution attacks suffer from too large memory requirements. The maximal degree monomial test can be used to attack even 736 rounds (approximately 64% of initialization) using 33 IV bits, the attack on 736 rounds has only been performed a handful of times so the success rate is still an open issue in this case. The percentage of weak initial state

Table 5. Number of IV bits needed to attack the first keystream bit of Trivium for different number of rounds in the initialization (out of 1152 rounds)

Rounds	d-Monomial test			Monomial distr. test			Max. degree monomial		
	P	IVs	Memory	P	IVs	Memory	P	IVs	Memory
608	1	12	2^{12}	2^5	9	2^9	2^5	9	1
640	1	15	2^{15}	2^6	13	2^{13}	2^5	13	1
672	1	20	2^{20}	2^8	18	2^{17}	2^5	18	1
704	1	27	2^{27}	2^6	23	2^{23}	2^5	24	1
736	–	–	–	–	–	–	–	33 *	1

Table 6. Number of IV bits needed to attack the initial state variables of Trivium for different number of rounds in the initialization (out of 1152 rounds)

Rounds	d-Monomial test				Monomial distr. test				Max. degree monomial			
	P	IVs	Memory	Fraction	P	IVs	Memory	Fraction	P	IVs	Memory	Fraction
608	1	12	2^{12}	144/288	2^5	12	2^{12}	105/288	2^5	12	1	169/288
640	1	12	2^{12}	57/288	2^5	12	2^{12}	29/288	2^5	12	1	86/288
672	1	15	2^{15}	87/288	2^5	15	2^{15}	0/288	2^5	15	1	108/288
704	1	20	2^{20}	74/288	2^5	20	2^{20}	12/288	2^5	20	1	76/288

variables for Trivium are approximately same using d-monomial and maximal degree tests.

Alternative Key/IV Loading for Trivium. In the original key/IV loading, 128 bits of the initial state are assigned to constants and the key and IV bits affect only one state bit. Here, we propose an alternative initial Key/IV loading in which the first register is filled with the modulo 2 summation of key and IV, the second register is filled with IV and the last register is filled with the complement of key plus IV. In this setting, each IV bit affects 3 internal state bits, therefore the diffusion of IV bits to the state bits is satisfied in less number of clockings. We repeated the tests using the alternative loading and obtained the results given in Table 7 and Table 8. In the alternative loading, the required number of IV bits and memory needed to attack Trivium are approximately 50 percent more compared to the original loading.

4.3 Decim

Decim-v2 [8] is also a hardware oriented stream cipher based on a nonlinearly filtered LFSR and the irregularly decimation mechanism, ABSG. The internal state size of Decim-v2 is 192 bit and it is loaded with 80 bit Key and 64 bit IV. The first 80 bits of the LFSR are filled with the key, the bits between 81 and 160 are filled with linear functions of key and IV and the last 32 bits are filled with a linear function of IV bits.

Table 7. Number of IV bits needed to attack the first keystream bit of Trivium with alternative Key/IV loading for different number of rounds in the initialization (out of 1152 rounds)

Rounds	d-Monomial test			Monomial distr. test			Max. degree monomial		
	P	IVs	Memory	P	IVs	Memory	P	IVs	Memory
608	1	18	2^{18}	2^5	22	2^{22}	2^5	17	1

Table 8. Number of IV bits needed to attack the initial state variables of Trivium with alternative Key/IV loading for different number of rounds in the initialization (out of 1152 rounds)

Rounds	d-Monomial test				Monomial distr. test				Max. degree monomial			
	P	IVs	Memory	Fraction	P	IVs	Memory	Fraction	P	IVs	Memory	Fraction
608	1	12	2^{12}	4/288	2^5	12	2^{12}	2/288	2^5	12	1	21/288
640	1	18	2^{18}	17/288	2^5	18	2^{18}	19/288	2^5	18	1	24/288
672	1	20	2^{20}	0/288	2^5	20	2^{20}	0/288	2^5	20	1	0/288

The results we obtained for Decim-v2 are given in Table 9 and Table 10. The security margin for Decim against chosen IV attacks is very large, the cipher can only be broken when not more than about 3% of the initialization is used. This is mainly because of the initial loading of key and IV in which each IV bits affect 3 state variables and the high number of quadratic terms in the filter function. The weakness in initial state variables can be observed for higher number of clockings. The number of weak initial state variables are approximately same for all attacks.

Table 9. Number of IV bits needed to attack the first keystream bit of Decim-v2 for different number of rounds in the initialization (out of 768 rounds)

Rounds	d-Monomial test			Monomial distr. test			Max. degree monomial		
	P	IVs	Memory	P	IVs	Memory	P	IVs	Memory
20	1	16	2^{16}	2^5	13	2^{13}	2^5	19	1

Table 10. Number of IV bits needed to attack the initial state variables of Decim-v2 for different number of rounds in the initialization (out of 768 rounds)

Rounds	d-Monomial test				Monomial distr. test				Max. degree monomial			
	P	IVs	Memory	Fraction	P	IVs	Memory	Fraction	P	IVs	Memory	Fraction
160	1	12	2^{12}	47/192	2^5	17	2^{17}	47/192	2^5	12	1	44/192
192	1	20	2^{20}	18/192	2^5	20	2^{20}	13/192	2^5	20	1	17/192

5   Conclusions

In this study, we generalize the idea of d-monomial attacks and propose a framework for chosen IV statistical analysis. The proposed framework can be used as an instrument for designing good initialization procedures. It can be used to verify the effectiveness of the initialization, but also to help designing a well-balanced initialization, e.g., prevent an unnecessary large number of initial clockings or even reduce the number of gates used in an hardware implementation by being able to use a simpler loading procedure.

Also, we propose a few new statistical attacks, apply them on some existing stream cipher proposals, and give some conclusions regarding the strength of their IV initialization. In particular, we experimentally detected statistical weaknesses in the keystream of Trivium using an initialization reduced to 736 rounds as well as in some state bits of Grain-128 with full IV initialization. It is an open question how to utilize these weaknesses of state bits to attack the cipher.

For ciphers Grain and Trivium, we also propose alternative initialization schemes with slightly higher hardware complexity. In the proposed loadings, each IV and key bit affects more than one state bit and the resistance of the ciphers to the proposed attacks increases about 50%. Decim seems to have a high security margin and it is an interesting question whether a simpler loading procedure could be used in Decim which could mean a smaller footprint in hardware, fewer intial clockings could also be used for a faster intialization procedure.

References

1. Daemen, J., Govaerts, R., Vandewalle, J.: Resynchronization weaknesses in synchronous stream ciphers. In: Helleseth, T. (ed.) EUROCRYPT 1993. LNCS, vol. 765, pp. 159–167. Springer, Heidelberg (1994)
2. Golic, J.D., Morgari, G.: On the resynchronization attack. In: Johansson, T. (ed.) FSE 2003. LNCS, vol. 2887, pp. 100–110. Springer, Heidelberg (2003)
3. Armknecht, F., Lano, J., Preneel, B.: Extending the resynchronization attack. In: Handschuh, H., Hasan, M.A. (eds.) SAC 2004. LNCS, vol. 3357, pp. 19–38. Springer, Heidelberg (2004)
4. Filiol, E.: A new statistical testing for symmetric ciphers and hash functions. In: Varadharajan, V., Mu, Y. (eds.) ACISP 2001. LNCS, vol. 2119, pp. 21–35. Springer, Heidelberg (2001)
5. Saarinen, M.J.O.: Chosen-iv statistical attacks on estream stream ciphers. eSTREAM, ECRYPT Stream Cipher Project, Report 2006/013 (2006), http://www.ecrypt.eu.org/stream
6. Hell, M., Johansson, T., Maximov, A., Meier, W.: A stream cipher proposal: Grain-128. ISIT, Seattle, USA (2006), available at http://www.ecrypt.eu.org/stream
7. De Cannière, C., Preneel, B.: Trivium - specifications. eSTREAM, ECRYPT Stream Cipher Project, Report 2005/030 (2005), available at http://www.ecrypt.eu.org/stream

8. Berbain, C., Billet, O., Canteaut, A., Courtois, N., Debraize, B., Gilbert, H., Goubin, L., Gouget, A., Granboulan, L., Lauradoux, C., Minier, M., Pornin, T., Sibert, H.: Decim v2. eSTREAM, ECRYPT Stream Cipher Project, Report 2006/004 (2006), http://www.ecrypt.eu.org/stream
9. Rukhin, A., Soto, J., Nechvatal, J., Smid, M., Barker, E., Leigh, S., Levenson, M., Vangel, M., Banks, D., Heckert, A., Dray, J., Vo, S.: A statistical test suite for random and pseudorandom number generators for cryptographic applications (2001), http://www.nist.gov
10. Fluhrer, S., Mantin, I., Shamir, A.: Weaknesses in the key scheduling algorithm of RC4. In: Vaudenay, S., Youssef, A.M. (eds.) SAC 2001. LNCS, vol. 2259, pp. 1–24. Springer, Heidelberg (2001)

A   Linear Regression Model for d-Monomial Test of Grain

In this part, we model the relationship between the number of IVs and number of rounds using linear regression. We fit a linear equation to the observed data of d-monomial test of Grain as given in Figure 5.

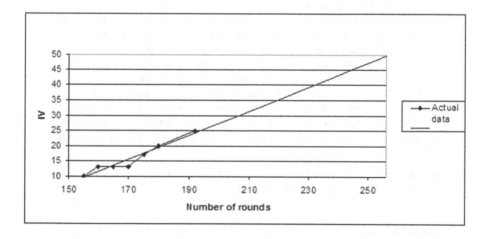

Fig. 5. The linear regression model for d-monomial test of Grain

The trend equation is obtained as $y = 0.3981092437x - 52.21953782$ where y represents the number of IVs and x represents the number of rounds in initialization. The correlation coefficient of the model is 0.96850. Using this model, the prediction of required number of IVs to attack Grain with d-monomial test is $y = 0.3981092437(256) - 52.21953782 = 49.69643 \approx 50$. However, note that this model is just a guess, more points needs to be calculated before a conclusion about the security can be drawn.

Public Key Encryption with Searchable Keywords Based on Jacobi Symbols

Giovanni Di Crescenzo[1] and Vishal Saraswat[2]

[1] Telcordia Technologies, Piscataway-NJ, USA
giovanni@research.telcordia.com
[2] University of Minnesota, Minneapolis-MN, USA
vishal@math.umn.edu

Abstract. Public-key encryption schemes with searchable keywords are useful to delegate searching capabilities on encrypted data to a third party, who does not hold the entire secret key, but only an appropriate token which allows searching operations but preserves data privacy. Such notion was previously proved to imply identity-based public-key encryption [5] and to be equivalent to anonymous (or key-private) identity-based encryption which are useful for fully-private communication.

So far all presented public-key encryption with keyword search (PEKS) schemes were based on bilinear forms and finding a PEKS that is not based on bilinear forms has been an open problem since the notion of PEKS was first introduced in [5]. We construct a public-key encryption scheme with keyword search based on a variant of the *quadratic residuosity problem*. We obtain our scheme using a non-trivial transformation of Cocks' identity-based encryption scheme [9]. Thus we show that the primitive of PEKS can be based on additional intractability assumptions which is a conventional desiderata about all cryptographic primitives.

Keywords: Public-Key Encryption, Searchable Public-Key Encryption, Quadratic Residuosity, Jacobi Symbol.

1 Introduction

A classical research area in Cryptography is that of designing candidates for cryptographic primitives under different intractability assumptions, so to guarantee that the cryptographic primitive does not depend on the supposed hardness of a single computational problem and its fortune against cryptanalytic research. In this paper we concentrate on a recently introduced primitive, *public-key encryption with keyword search* (PEKS) [5], for which all constructions in the literature were based on assumptions related to bilinear forms. We present a PEKS scheme based on a new assumption that can be seen as a variant of the classical assumption on the hardness of deciding quadratic residuosity modulo composite integers.

Motivation. PEKS allows a sender to compute an encrypted message, so that the receiver can allow a third party to search keywords in the encrypted message

K. Srinathan, C. Pandu Rangan, M. Yung (Eds.): Indocrypt 2007, LNCS 4859, pp. 282–296, 2007.

without (additional) loss of privacy on the content of the message. The following motivating example for PEKS is taken almost verbatim from [5]. Suppose user Alice wishes to read her email on a number of devices: laptop, desktop, pager, etc. Alice's mail gateway is supposed to route email to the appropriate device based on the keywords in the email. For example, when Bob sends email with the keyword "urgent" the mail is routed to Alice's pager. When Bob sends email with the keyword "lunch" the mail is routed to Alice's desktop for reading later. One expects each email to contain a small number of keywords. For example, all words on the subject line as well as the sender's email address could be used as keywords. Now, suppose Bob sends encrypted email to Alice using Alice's public key. Both the contents of the email and the keywords are encrypted. In this case the mail gateway cannot see the keywords and hence cannot make routing decisions. With public-key encryption with keyword search one can enable Alice to give the gateway the ability to test whether "urgent" is a keyword in the email, but the gateway should learn nothing else about the email. More generally, Alice should be able to specify a few keywords that the mail gateway can search for, but learn nothing else about incoming mail.

Previous work. In its non-interactive variant, constructions for this primitive were showed to be at least as hard to obtain as constructions for identity-based encryption (as proved in [5]). Moreover, the existence of PEKS was proved to follow from the existence of "anonymous" or "key-private" identity-based encryption (this was noted in [5] and formally proved in [1]); namely, encryption where the identity of the recipient remains unknown. Anonymous encryption is well-known to be an attractive solution to the problem of fully-private communication (i.e., sender-anonymous and receiver-anonymous ciphertexts, as well as protection against traffic analysis, by using bulletin boards); see, e.g., discussions in [2,8]). It is a natural goal then to try to convert the existing identity-based public-key cryptosystems into their anonymous variant, so that a PEKS is automatically obtained. In fact, the anonymity or key-privacy property for a public-key encryption scheme (whether it is identity-based or not), is itself a property of independent interest, as already discussed in [2], where this property was defined and investigated for conventional (i.e., not identity-based) public-key encryption schemes. So far, however, all presented public-key encryption schemes with keyword search were transformations of identity-based cryptosystems based on bilinear forms. Even the authors of [5] noted the difficulty of coming up with other examples of public-key encryption schemes with keyword search, by observing that the only identity-based cryptosystem not based on bilinear forms (namely, Cocks' scheme [9]) does not seem to have a direct transformation into an anonymous variant and thus into a public-key encryption scheme with keyword search. Further work on PEKS (e.g., [17,15,13,1,8]) did not contribute towards this goal, but further studied schemes and variations based on bilinear forms.

Our results. In this paper we construct the first public-key encryption scheme with keyword search which is not based on bilinear forms but is based on a new assumption that can be seen as a variant of the well-known hardness of deciding quadratic residues modulo a large composite integer. Our scheme is obtained as a non-trivial transformation of Cocks' identity-based encryption scheme [9]. By the known equivalence of public-key encryption scheme with keyword search and anonymous identity-based encryption, our scheme immediately gives the first anonymous identity-based encryption scheme which is not based on bilinear forms, a problem left open in [5]. Our scheme essentially preserves the time efficiency of the (not anonymous) identity-based encryption of Cocks' scheme, which was claimed in the original paper [9] to be satisfactory in a hybrid encryption mode (that is, when used to encrypt first a short session key and then using this key to produce a symmetric encryption of a large message). We do note however that the decryption time of Cocks' scheme (and thus of our scheme too) is less efficient than the known schemes based on bilinear forms.

The construction of a new identity-based encryption scheme based on quadratic residuosity [7] and having short ciphertexts was claimed very recently, and after seeing our present work [3]. This scheme is also anonymous, like ours, but is based on very different techniques. Although their scheme is quite elegant, encryption and decryption operations are estimated [4] to be significantly less efficient than in Cocks' scheme. Instead, when used as an anonymous identity-based encryption scheme, our scheme is only less efficient than the original (and not anonymous) Cocks' scheme [9] by a small constant factor.

Organization of the paper. In what follows, we start by reviewing in Section 2 the formal definitions related to the notion of interest in this paper: public-key encryption with keyword search. In Section 3 we present our public-key cryptosystem with keyword search and in Section 4 we prove its properties.

2   Definitions and Preliminaries

We recall the known notion and formal definition of public-key encryption with keyword search (as defined in [5,1]). We assume familiarity with the notion of identity-based public-key cryptosystems (as defined, for instance, in [6,9]).

An identity-based public-key cryptosystem can be defined as a 4-tuple of algorithms (Setup, KeyGen, Encrypt, Decrypt), with the following semantics: Setup is used by the trusted authority TA to generate public parameters PK and a master secret key SK; KeyGen is used by the trusted authority TA to generate a trapdoor key t_{ID} given a party's ID; Encrypt is used by a sender who wants to encrypt a message to a receiving party and only uses the receiver's ID and the public parameters PK; Decrypt is used by a receiver to decrypt a ciphertext and only uses the trapdoor t_{ID} and the public parameters PK. We denote the identity-based cryptosystem in [9] as CC-IBE = (CC-Setup, CC-KeyGen, CC-Encrypt, CC-Decrypt):

2.1   Public-Key Encryption with Keyword Search

Informally speaking, in a public-key encryption scheme with keyword search, a sender would like to send a message in encrypted form to a receiver, so that the receiver can allow a third party to search keywords in the encrypted message without losing (additional) privacy on the message's content. According to [5], a non-interactive implementation of this task can be performed as follows. The sender encrypts her message using a conventional public key cryptosystem, and then appends to the resulting ciphertext a *Public-key Encryption with Keyword Search* (PEKS) of each keyword. Specifically, to encrypt a message M with searchable keywords $W_1, \ldots, W_m$, the sender computes and sends to the receiver

$$E_{A_{pub}}(M) \; \| \; \mathsf{ksEnc}(A_{pub}, W_1) \; \| \; \cdots \; \| \; \mathsf{ksEnc}(A_{pub}, W_m), \qquad (1)$$

where A_{pub} is the receiver's public key and E is the encryption algorithm of the conventional public-key cryptosystem. Based on this encryption, the receiver can give the third party a certain trapdoor T_W which enables the third party to test whether one of the keywords associated with the message is equal to the word W of the receiver's choice. Specifically, given $\mathsf{ksEnc}(A_{pub}, W')$ and T_W, the third party can test whether $W = W'$; if $W \neq W'$ the third party learns nothing more about W'. Note that sender and receiver do not communicate in this entire process, as the sender generates the searchable ciphertext for W' just given the receiver's public key (thus, the term *"public-key* encryption with keyword search" is used here).

More formally, we consider a setting with three parties: a sender, a receiver, and a third party (representing the e-mail gateway in the application example given in the introduction). In this setting, a public-key encryption with keyword search is defined as follows.

Definition 1. A (non-interactive) public-key encryption scheme with keyword search (PEKS) consists of the following polynomial time randomized algorithms:

1. $\mathsf{KeyGen}(1^m)$: on input security parameter 1^m in unary, it returns a pair (A_{pub}, A_{priv}) of public and private keys.
2. $\mathsf{ksEnc}(A_{pub}, W)$: on input a public key A_{pub} and a keyword W, it returns a ciphertext, also called the *searchable encryption* of W.
3. $\mathsf{Trapdoor}(A_{priv}, W)$: on input Alice's private key and a keyword W, it returns a trapdoor T_W.
4. $\mathsf{Test}(A_{pub}, S, T_W)$: on input Alice's public key, a searchable encryption $S = \mathsf{ksEnc}(A_{pub}, W')$, and a trapdoor $T_W = \mathsf{Trapdoor}(A_{priv}, W)$, it returns 'yes' or 'no'.

Given the above definition, an execution of a public-key encryption scheme with keyword search goes as follows. First, the receiver runs the KeyGen algorithm to generate her public/private key pair. Then, she uses the $\mathsf{Trapdoor}$ algorithm to generate trapdoors T_W for any keywords W which she wants the third party to search for. The third party uses the given trapdoors as input to the Test algorithm to determine whether a given message encrypted by any sender using algorithm ksEnc contains one of the keywords W specified by the receiver.

We now define three main properties which public-key encryption schemes with keyword search may satisfy: (two variants of) consistency and security. The following basic definition will be useful towards that: we say that a given function $f : N \rightarrow [0,1]$ is *negligible in* n if $f(n) < 1/p(n)$ for any polynomial p and sufficiently large n.

CONSISTENCY. Next, we consider definitions of consistency for a PEKS (following definitions in [5,1]). We consider two variants: right-keyword consistency and adversary-based consistency for a PEKS in the random oracle model.

Informally, in right-keyword consistency, we require the success of the search of any word W for which the encryption algorithm had computed a searchable encryption.

Definition 2. We say that a PEKS is *right-keyword consistent* if it holds that for any word W, the probability that $\mathsf{Test}(A_{pub}, C, T_W) \neq$ 'yes' is negligible in m, where A_{pub} was generated using the KeyGen algorithm and input 1^m, C was computed as $\mathsf{ksEnc}(A_{pub}, W)$ and T_W was computed as $\mathsf{Trapdoor}(A_{priv}, W)$.

Informally, in adversary-based consistency, one would like to ensure that even an adversary that has access to the public parameters PK and to a (uniformly distributed) random oracle cannot come up with two different keywords such that the testing algorithm returns 'yes' on input a trapdoor for one word and a public-key encryption with keyword search of the other. Formally, we define consistency against an attacker $\mathcal{A}$ using the following game between a challenger and an attacker. Here, we denote by m the security parameter, given in unary as input to both players, and by k the length of the keywords, where we assume that $k = \Theta(m^c)$, for some constant $c > 0$ (this assumption is seen to be wlog using simple padding).

PEKS Adversary-Based Consistency Game

1. The challenger runs the $\mathsf{KeyGen}(1^m)$ algorithm to generate A_{pub} and A_{priv}. It gives A_{pub} to the attacker.
2. The attacker returns two keywords $W_0, W_1 \in \{0,1\}^k$.
3. Encryption $C = \mathsf{ksEnc}(A_{pub}, W_0)$ and trapdoor $T_{W_1} = \mathsf{Trapdoor}(A_{priv}, W_1)$ are computed.
4. The attacker wins the game if $W_0 \neq W_1$ and $\mathsf{Test}(A_{pub}, C, T_{W_1})$ returns 'yes'.

We define $\mathcal{A}$'s advantage $\mathrm{Adv}_{\mathcal{A}}(m, k)$ in breaking the consistency of PEKS as the probability that the attacker wins the above game.

Definition 3. We say that a PEKS satisfies (computational) *adversary-based consistency* if for any attacker $\mathcal{A}$ running in time polynomial in m, we have that the function $\mathrm{Adv}_{\mathcal{A}}(m)$ is negligible in m.

SECURITY. Finally, we recall the definition of security for a PEKS (in the sense of semantic-security). Here, one would like to ensure that an $\mathsf{ksEnc}(A_{pub}, W)$ does not reveal any information about W unless T_w is available. This is done by considering an attacker who is able to obtain trapdoors T_W for any W of his choice, and require that, even under such attack, the attacker should not be able to distinguish an encryption of a keyword W_0 from an encryption of a keyword W_1 for which he did not obtain the trapdoor. Formally, we define security against an active attacker $\mathcal{A}$ using the following game between a challenger and the attacker. Here, we denote by m the security parameter, given in unary as input to both players, and by k the length of the keywords, where we assume that $k = \Theta(m^c)$, for some constant $c > 0$ (this assumption is seen to be wlog using simple padding).

PEKS Security Game

1. The challenger runs the KeyGen(1^m) algorithm to generate A_{pub} and A_{priv}. It gives A_{pub} to the attacker.
2. The attacker can adaptively ask the challenger for the trapdoor T_W for any keyword $W \in \{0, 1\}^k$ of his choice.
3. At some point, the attacker $\mathcal{A}$ sends the challenger two keywords W_0, W_1 on which it wishes to be challenged. The only restriction is that the attacker did not previously ask for the trapdoors T_{W_0} or T_{W_1}. The challenger picks a random $b \in \{0, 1\}$ and gives the attacker $C = \mathsf{ksEnc}(A_{pub}, W_b)$. We refer to C as the challenge ciphertext.
4. The attacker can continue to ask for trapdoors T_W for any keyword W of his choice as long as $W \neq W_0, W_1$.
5. Eventually, the attacker $\mathcal{A}$ outputs $d \in \{0, 1\}$.

Here, the attacker wins the game if its output differs significantly depending on whether he was given the challenge ciphertext corresponding to W_0 or W_1. This is formalized as follows. First, for $b = 0, 1$, let $\mathcal{A}^b = 1$ denote the event that $\mathcal{A}$ returns 1 given that $C = \mathsf{ksEnc}(A_{pub}, W_b)$. Then, define $\mathcal{A}$'s advantage in breaking the PEKS scheme as

$$\mathrm{Adv}_{\mathcal{A}}(m) = \big| \mathrm{Prob}[\mathcal{A}^0 = 1] - \mathrm{Prob}[\mathcal{A}^1 = 1] \big|$$

Definition 4. We say that a PEKS is semantically secure against an adaptive chosen-keyword attack if for any attacker $\mathcal{A}$ running in time polynomial in m, we have that the function $\mathrm{Adv}_{\mathcal{A}}(m)$ is negligible in m.

Remarks. We defined right-keyword consistency as done in [5,1] (although the name was first used in [1]). The (computational version of the) adversary-based consistency was defined as a relaxed version of what is called just consistency in [1]; the relaxation consisting in only restricting the adversary to return keywords which have a known upper-bounded length. Although guaranteeing a slightly weaker property, this is essentially not a limitation in practical scenarios where a (small) upper bound on the length of keywords is known to all parties. We also note that such a relaxation is always done, for instance, in the definition of conventional public-key cryptosystems.

3   Our Construction

In this section we present our main construction: a public-key encryption with keyword search under an intractability assumption related to quadratic residuosity modulo Blum-Williams integers. We first define the intractability assumption which we use and formally state our main result; then, in Subsection 3.1 we give an informal discussion where we sketch a preliminary (but flawed) construction, explain why it does not work, and how we fix it; finally, in Subsection 3.2, we formally describe our public-key cryptosystem with keyword search.

An intractability assumption. Our cryptosystem is based on the following assumption which is a variation of the well-known quadratic residuosity problem.

Quadratic Indistinguishability Problem (QIP). Let m be a security parameter. Let $(p, q) \leftarrow \mathrm{BW}(1^m)$ denote the random process of uniformly and independently choosing two m-bit primes p, q such that $p = q = 3 \bmod 4$. Let $QR(n)$ denote the set of quadratic residues modulo n and let $\mathbb{Z}_n^{+1}$ (resp. $\mathbb{Z}_n^{-1}$) be the set of positive integers which are $< n$, coprime with n and have Jacobi symbol equal to $+1$ (resp. -1). Also, let $s \leftarrow \mathrm{CS}(n, \alpha)$ denote the random process of randomly and independently choosing an integer s in $\mathbb{Z}_n^{+1}$ such that the condition α holds. The QIP *problem* consists of efficiently distinguishing the following two distributions:

$$D_0(1^m) = \{(p, q) \leftarrow \mathrm{BW}(1^m); n \leftarrow p \cdot q; h \leftarrow \mathbb{Z}_n^{+1};$$
$$s \leftarrow \mathrm{CS}(n, s^2 - 4h \in \mathbb{Z}_n^{-1} \cup QR(n)) : (n, h, s)\}$$
$$D_1(1^m) = \{(p, q) \leftarrow \mathrm{BW}(1^m); n \leftarrow p \cdot q; h \leftarrow \mathbb{Z}_n^{+1}; s \leftarrow \mathbb{Z}_n^* : (n, h, s)\}.$$

We say that algorithm $\mathcal{A}$ has *advantage* ϵ in solving QIP if we have that:

$$\left| \Pr[(n, h, s) \leftarrow D_0(1^m) : \mathcal{A}(n, h, s) = 1] \right.$$
$$\left. - \Pr[(n, h, s) \leftarrow D_1(1^m) : \mathcal{A}(n, h, s) = 1] \right| = \epsilon. \quad (2)$$

We say that QIP is *intractable* if all polynomial time (in m) algorithms have a negligible (in m) advantage in solving QIP.

Our result. In the rest of the paper we prove the following.

Theorem 1. Assume that the QIP problem is intractable. Then there exists (constructively) a public-key encryption scheme with keyword search.

3.1   An Informal Discussion

A first (not yet anonymous) construction. The first approach is a very natural one — we simply apply the Cocks' IBE scheme in place of Boneh-Franklin scheme as done in the public-key encryption scheme with keyword search presented in [5]. Given a security parameter 1^m, for $m \in \mathbb{Z}^+$, the user Alice uses algorithm CC-Setup to generate two sufficiently large primes p, q such that both p and q are congruent to 3 mod 4 and a cryptographic hash function H (assumed to behave like a random oracle in the analysis); then she outputs the public parameters $A_{pub} = (n, H)$ and keeps secret the master secret key $A_{priv} = (p, q)$. Alice treats each keyword W as an identity and, using algorithm CC-KeyGen, computes a square root g of $h = H(W)$ or $-h$ depending on which one is a square modulo n, and supplies the mail server with $T_W = g$ as the trapdoor for W. A user Bob wishing to send an encrypted email to Alice with the keyword W uses algorithm CC-Encrypt to encrypt the string 1^k, where $k = |W|$, using W as the necessary input identity, thus obtaining $\mathsf{ksEnc}(A_{pub}, W) = s = (s_1, \ldots, s_k, s_{k+1}, \ldots, s_{2k})$. The server then decrypts s as in algorithm CC-Decrypt and outputs 'yes' if the decryption returns 1^k, or outputs 'no' otherwise.

The problem with the first approach. The above scheme does not satisfy anonymity because Cocks' IBE scheme is not *public-key private* or *anonymous*, in the sense of Bellare et al. [2], and thus the ciphertext returned by algorithm PEKS may reveal more information than desired about the searchable keyword. This fact was already briefly mentioned in [5], but it is useful to analyze it in greater detail here to understand how we will obtain our main construction. Specifically, to clarify this fact, we note that for $i = 1, \ldots, k$, it holds that

$$s_i^2 - 4h = (t_i + h/t_i)^2 - 4h = (t_i - h/t_i)^2 \mod n$$

and therefore, except with negligible probability, $\left(\frac{s_i^2 - 4h}{n}\right) = +1$ and, analogously, for $i = k + 1, \ldots, 2k$, it holds that

$$s_i^2 + 4h = (t_i - h/t_i)^2 + 4h = (t_i + h/t_i)^2 \mod n$$

and therefore, except with negligible probability, $\left(\frac{s_i^2 + 4h}{n}\right) = +1$.

On the other hand, for any other keyword $W' \neq W$, if $h' = H(W')$, the quantities $s_i^2 - 4h'$ and $s_i^2 + 4h'$ are not necessarily squares and their Jacobi symbols $\left(\frac{s_i^2 \pm 4h'}{n}\right)$ may be -1. In fact, when h' is randomly chosen in $\mathbb{Z}_n^*$, for each $i \in \{1, \ldots, 2k\}$, it holds that $\left(\frac{s_i^2 \pm 4h'}{n}\right) = -1$ exactly half the time $s_i^2 \pm 4h'$ is in $\mathbb{Z}_n^*$. Thus, an outsider can easily find out whether a keyword W is in the message or not with some non-negligible probability, which is not desirable.

Fixing the problems and ideas behind our construction. At a very high level, we still would like to use the approach in [5]; which, very roughly speaking, might be abstracted as follows: a searchable ciphertext is 'carefully computed' as the output of an identity-based encryption algorithm on input a plaintext sent in the clear and a function of the keyword W as the identity; the computation of the searchable ciphertext is such that (with high probability) the plaintext sent in the clear is the actual decryption of this ciphertext if and only if the trapdoor associated to the same keyword W is used to decrypt. However, the main difficulty in implementing this approach with (a modification of) the Cocks' scheme CC-IBE is in obtaining a modification which additionally satisfies the 'public-key privacy' or 'anonymity' property. We solve this problem by modifying the distribution of the ciphertext in CC-IBE, so that its modified distribution is 'properly randomized', and, when used in the context of a ciphertext associated with our public-key encryption scheme with keyword search, does not reveal which keyword is being used. The randomization of the ciphertext has to guarantee not only that the ciphertext does not reveal the identity used (or, in other words, the integer $h = H(ID)$), but also has to guarantee that the distribution remains the same when it is matched with another identity (e.g., another integer $h' = H(ID')$). In this randomization process, we have to take care of two main technical obstacles, one related to the distribution of the integers s_i with respect to the Jacobi symbols $\left(\frac{s_i^2 \pm 4h}{n}\right)$; and another one, related to efficiently guaranteeing that all values s_i are constructed using uniformly and independently distributed hashes (or, functions of the keyword) playing as the identity. We achieve this through two levels of randomization. First, the ciphertext contains $4k$ integers s_i in $\mathbb{Z}_n^*$ such that the Jacobi symbols of the related expressions $s_i^2 - 4h$ are uniformly distributed in $\{-1, +1\}$ whenever $s_i^2 \pm 4h$ is in $\mathbb{Z}_n^*$. Second, to make sure that these Jacobi symbols are also independently distributed, we do not use a single value h, but use a uniformly and independently distributed h_i for each index i.

3.2   Formal Description

We denote our public-key encryption with keyword search scheme as MainScheme = (M-KeyGen, M-ksEnc, M-Trapdoor, M-Test). MainScheme uses a cryptographic hash function $H : \{0,1\}^k \rightarrow \mathbb{Z}_n^{+1}$ (which is assumed in the analysis to behave as a random oracle). We denote by m the security parameter and by k the length of keywords. We assume wlog that $k = \Theta(m^c)$ for some constant $c > 0$ (concrete values for m, k can be $m = 1024$ and $k = 160$). MainScheme can be described as follows:

M-KeyGen(1^m):   On input security parameter 1^m in unary, for $m \in \mathbb{Z}^+$, do the following:
    1. randomly choose two primes p, q of length $m/2$, and such that both p and q are congruent to $3 \mod 4$ and set $n = pq$;
    2. Set $A_{pub} = (n, 1^k)$ and $A_{priv} = (p, q)$, and output: (A_{pub}, A_{priv}).
M-ksEnc(A_{pub}, W):   Let $A_{pub} = (n, 1^k)$, $W \in \{0,1\}^k$, and do the following:
    1. For each $i = 1, \ldots, 4k$,
        compute $h_i = H(W|i)$;
        randomly and independently choose $u_i \in \mathbb{Z}_n^*$;
        if $\left(\frac{u_i^2 - 4h_i}{n}\right) = +1$ then
            randomly and independently choose $t_i \in \mathbb{Z}_n^{+1}$;
            set $s_i = (t_i + h_i/t_i) \mod n$.
        if $\left(\frac{u_i^2 - 4h_i}{n}\right) \in \{-1, 0\}$ then set $s_i = u_i$.
    2. Output $s = (s_1, \ldots, s_{4k})$.
M-Trapdoor(A_{priv}, W):   Let $A_{priv} = (p, q)$, $W \in \{0,1\}^k$, and do the following:
    1. For $i = 1, \ldots, 4k$;
        compute $h_i = H(W|i)$;
        use p, q to randomly choose $g_i \in \mathbb{Z}_n^*$ (if any) such that $g_i^2 = h_i \mod n$;
        if h_i has no square root modulo n, then set $g_i = \perp$;
    2. return: $(g_1, \ldots, g_{4k})$.

M-Test(A_{pub}, s, T_W): Let $T_W = (g_1, \ldots, g_{4k})$ and $s = (s_1, \ldots, s_{4k})$, and do the
following:
1. For $i = 1, \ldots, 4k$,
   if $g_i = \perp$ then set $\bar{t}_i = \perp$;
   if $g_i^2 = h_i \mod n$ then
      if $\left(\frac{s_i^2 - 4h_i}{n} \right) = +1$ then set $\bar{t}_i = \left(\frac{s_i + 2g_i}{n} \right)$;
      otherwise set $\bar{t}_i = \perp$;
2. output 'yes' if $\bar{t}_i \in \{+1, \perp\}$ for all $i = 1, \ldots, 4k$; otherwise output 'no'.

Remarks: ciphertext distribution and scheme parameters. We note the
distribution of the ciphertext $s = (s_1, \ldots, s_{4k})$ returned by algorithm M-ksEnc
has only negligible statistical distance from the distribution where each element
s_i is uniformly distributed among the integers such that $s_i^2 - 4h_i \in QR(n)$ with
probability $1/2$, or $s_i^2 - 4h_i \in \mathbb{Z}_n^{-1}$ with probability $1/2$.

We note that it is essential to choose our scheme's parameter $k = \Theta(m^c)$,
for some $c > 0$, to guarantee that the consistency properties of MainScheme
are satisfied in an asymptotic sense. Good practical choices for parameters m, k
include setting $m = 1024$ and $k = 160$.

4   Properties of Our Construction

In Subsections 4.1 and 4.2 we prove the consistency and security properties of
our public-key encryption scheme with keyword search.

4.1   Proof of Consistency

We prove the right-keyword consistency here and omit the proof of adversary-
based consistency of MainScheme to meet space constraints.

Right-keyword consistency. For $i = 1, \ldots, 4k$, whenever $\left(\frac{s_i^2 - 4h_i}{n} \right) = +1$, it
always holds that $h_i = g_i^2 \mod n$ and it never holds that $\bar{t}_i = \left(\frac{s_i + 2g_i}{n} \right) = -1$.
The latter fact is proved by observing that, for $i = 1, \ldots, 4k$, it holds that

$$s_i + 2g_i = t_i + h_i/t_i + 2g_i = t_i \cdot (1 + g_i^2/t_i^2 + 2g_i/t_i) = t_i \cdot (1 + g_i/t_i)^2 \mod n,$$

and thus, except with negligible probability,

$$\bar{t}_i = \left(\frac{s_i + 2g_i}{n} \right) = \left(\frac{t_i}{n} \right) = +1.$$

Now, the above equalities do not hold only when $s_i + 2g_i \mod n$ is not in $\mathbb{Z}_n^*$,
in which case it still holds that $\left(\frac{s_i + 2g_i}{n} \right) = 0 \neq -1$. As a consequence of these
two facts, the right-keyword consistency property holds with probability 1.

4.2   Proof of Security

Let $\mathcal{A}$ be a polynomial-time algorithm that attacks MainScheme and succeeds in breaking with advantage ϵ, and while doing that, it makes at most $q_H > 0$ queries to the random oracle H and at most $q_T > 0$ trapdoor queries. We would like to show that ϵ is negligible in m or otherwise $\mathcal{A}$ can be used to construct an algorithm $\mathcal{B}$ that violates the intractability of the QIP problem. More precisely, we will attempt to violate the intractability of one among two problems that we call QIP_1 and QIP_2, and that are easily seen to be computationally equivalent to QIP.

We prove this by defining a sequence of games, which we call 'MainScheme Security Game t', for $t = 0, \ldots, 4k$, which are all variations of the PEKS Security Game defined in Section 2.

MainScheme Security Game t

1. Algorithm $\mathcal{B}$ takes as input (n, h_0, h_1, s), where n is a Blum-Williams integer, and $h_0, h_1 \in \mathbb{Z}_n^{+1}$ and $s \in \mathbb{Z}_n^*$.
2. First of all $\mathcal{B}$ runs the M-KeyGen(1^m) algorithm to generate $A_{pub} = (n, 1^k)$ and $A_{priv} = (p, q)$; afterwards, it gives A_{pub} to the attacker $\mathcal{A}$.
3. $\mathcal{A}$ can adaptively ask for outputs from the random oracle H to any inputs of its choice. To respond to H-queries, algorithm $\mathcal{B}$ maintains a list of tuples $\langle W_i, j, h_{i,j}, g_{i,j}, d(i,j), c(i,j) \rangle$ called the H-list. The list is initially empty. When $\mathcal{A}$ queries the random oracle H at a point $(W_i|j)$, for $W_i \in \{0,1\}^k$ and $j \in \{1, \ldots, 4k\}$, algorithm $\mathcal{B}$ responds as follows.

   If tuple $\langle W_i, j, h_{i,j}, g_{i,j}, d(i,j), c(i,j) \rangle$ appears on the H-list then algorithm $\mathcal{B}$ responds with $H(W_i|j) = h_{i,j} \in \mathbb{Z}_n^{+1}$.

   Otherwise, $\mathcal{B}$ uniformly chooses $d(i,j) \in \{0,1\}$, $r_{i,j} \in \mathbb{Z}_n^*$, and randomly choose $c(i,j) \in \{0,1\}$ such that $c(i,j) = 0$ with probability $1/(q_T + 1)$ and $c(i,j) = 1$ with probability $1 - 1/(q_T + 1)$.

   If $c(i,j) = 1$ then $\mathcal{B}$ computes $h_{i,j} = (-1)^{d(i,j)} \cdot r_{i,j}^2 \mod n$; sets $g_{i,j} = \perp$ if $d(i,j) = 1$, or $g_{i,j} = r_{i,j}$ if $d(i,j) = 0$; adds $\langle W_i, j, h_{i,j}, g_{i,j}, d(i,j), c(i,j) \rangle$ to the H-list and responds with $h_{i,j}$ to the H-query $(W_i|j)$.

   If $c(i,j) = 0$, then $\mathcal{B}$ sets $d = d(i,j)$, computes $h_{i,j} = h_d \cdot r_{i,j}^2 \mod n$, sets $g_{i,j} = r_{i,j}$, adds $\langle W_i, j, h_{i,j}, g_{i,j}, d(i,j), c(i,j) \rangle$ to the H-list and responds with $h_{i,j}$ to the H-query $(W_i|j)$.
4. $\mathcal{A}$ can adaptively ask for the trapdoor T_W for any keyword $W \in \{0,1\}^k$ of his choice, to which $\mathcal{B}$ responds as follows.

   If tuple $\langle W_i, j, h_{i,j}, g_{i,j}, d(i,j), c(i,j) \rangle$ already appears on the H-list, for some $j \in \{1, \ldots, 4k\}$ and $W_i = W$, then $\mathcal{B}$ responds with $(g_{i,1}, \ldots, g_{i,4k})$ to the trapdoor query W if $c(i,j) = 1$ or reports failures and halts if $c(i,j) = 0$.

   Otherwise $\mathcal{B}$ randomly chooses $d(i,j) \in \{0,1\}$ and $r_{i,j} \in \mathbb{Z}_n^*$; computes $h_{i,j} = (-1)^{d(i,j)} \cdot r_{i,j}^2 \mod n$; sets $g_{i,j} = \perp$ if $d(i,j) = 1$ or $g_{i,j} = r_{i,j}$ otherwise, responds with $(g_{i,1}, \ldots, g_{i,4k})$ to the trapdoor query W and inserts $\langle W_i, j, h_{i,j}, g_{i,j}, d(i,j), c(i,j) \rangle$ in H-list.
5. The attacker $\mathcal{A}$ sends the two keywords W_0, W_1 on which it wishes to be challenged (for which it did not previously ask for trapdoors T_{W_0} or T_{W_1}).

If the two tuples

$$\langle W_u, j, h_{u,j}, g_{u,j}, d(u,j), c(u,j) \rangle \text{ and } \langle W_v, j, h_{v,j}, g_{v,j}, d(v,j), c(v,j) \rangle$$

satisfying $W_u = W_0$, $W_v = W_1$, $j = t$, and $((c(u,j) = 0) \vee (c(v,j) = 0))$, are not in H-list, then $\mathcal{B}$ reports failures and halts.

Otherwise $\mathcal{B}$ computes $(s_1, \ldots, s_{4k})$ as follows:
 - $s_1, \ldots, s_{t-1}$ are computed as from algorithm M-ksEnc on input A_{pub}, W_0;
 - s_t is set equal to $s \cdot r_{i,t} \mod n$;
 - $s_{t+1}, \ldots, s_{4k}$ are computed as from algorithm M-ksEnc on input A_{pub}, W_1.
6. Given challenge $(s_1, \ldots, s_{4k})$, $\mathcal{A}$ can continue to ask for random oracle H's outputs for any input of its choice, and for trapdoors T_W for any keyword W of his choice as long as $W \neq W_0, W_1$; these are answered as in items 3 and 4, respectively.
7. $\mathcal{A}$ outputs $out \in \{0, 1\}$.

By using a standard hybrid argument on our assumption that $\mathcal{A}$ breaks the security of MainScheme with probability ϵ, we obtain that there exists $t \in \{1, \ldots, 4k\}$ such that

$$|\operatorname{Prob}[\mathcal{A}_t = 1] - \operatorname{Prob}[\mathcal{A}_{t+1} = 1]| \geq \epsilon/4k, \tag{3}$$

where by $\mathcal{A}_t = 1$ we denote the event that $\mathcal{A}$ returns 1 in the real attack game given that the challenge ciphertext s had been computed as follows: $s_1, \ldots, s_t$ are computed as in algorithm M-ksEnc on input A_{pub}, W_0; and $s_{t+1}, \ldots, s_{4k}$ are computed as in algorithm M-ksEnc on input A_{pub}, W_1. Similarly, as in [5], we can obtain that with probability at least $\epsilon/4k$, $\mathcal{A}$ queries at least one of the two H-queries $(W_0|t)$, $(W_1|t)$. (The proof is omitted for lack of space.)

The proof continues by considering two cases according to whether only one of the two queries are made or both of them are made. In the first case, we show that $\mathcal{B}$ violates the intractability of the QIP_1 problem, and in the second case we show that it violates the intractability of the QIP_2 problem. Both the QIP_1 and QIP_2 are minor variants of the QIP problem and easily seen to be computationally equivalent to it.

Case (a). We now consider the case when only one of the two H-queries; say, $(W_0|t)$, is made by algorithm $\mathcal{A}$. We define the QIP_1 *problem* as the problem of efficiently distinguishing the following two distributions:

$$D_{1,0}(1^m) = \{(p,q) \leftarrow \text{BW}(1^m); n \leftarrow p \cdot q; d \leftarrow \{0,1\}; h_0, h_1 \leftarrow \mathbb{Z}_n^{+1};$$
$$s \leftarrow \text{CS}(n, s^2 - 4h_d \in \mathbb{Z}_n^{-1} \cup QR(n)) : (n, h_0, h_1, s)\}$$

$$D_{1,1}(1^m) = \{(p,q) \leftarrow \text{BW}(1^m); n \leftarrow p \cdot q; h_0, h_1 \leftarrow \mathbb{Z}_n^{+1}; s \leftarrow \mathbb{Z}_n^* : (n, h_0, h_1, s)\}$$

We say that algorithm $\mathcal{A}$ has *advantage* ϵ in solving QIP_1 if we have that:

$$\big| \operatorname{Pr}[(n, h_0, h_1, s) \leftarrow D_{1,0}(1^m) : \mathcal{A}(n, h_0, h_1, s) = 1]$$
$$- \operatorname{Pr}[(n, h_0, h_1, s) \leftarrow D_{1,1}(1^m) : \mathcal{A}(n, h_0, h_1, s) = 1] \big| = \epsilon. \tag{4}$$

We say that QIP_1 is *intractable* if all polynomial time (in m) algorithms have a negligible (in m) advantage in solving QIP_1.

By a simple simulation argument, we can prove the following theorem:

Theorem 2. The QIP_1 problem is intractable if and only if the QIP problem is so.

We continue the proof by noting that bit $c(i,t)$ associated to the query $(W_0|t)$, where the i-th queried keyword is W_0, satisfies $c(i,t) = 0$ with probability $1/(q_T + 1)$. Assuming that $c(i,t) = 0$, we evaluate the distribution of ciphertext s in MainScheme Security Game t, for $t = 1, \ldots, 4k$.

First, we let $d = d(i,t)$ and observe that when $(n, h_0, h_1, s) \in D_{1,0}(1^m)$, the ciphertext s in MainScheme Security Game t appears to $\mathcal{A}$ to be distributed exactly as if $s_1, \ldots, s_t$ were computed as in algorithm M-ksEnc on input A_{pub}, W_0, and $s_{t+1}, \ldots, s_{4k}$ were computed as in algorithm M-ksEnc on input A_{pub}, W_1. This can be seen by observing that we assumed that $c(i,t) = 0$ and thus $H(W_0|t) = h_d \cdot r_{i,t}^2$; then, it holds that s_t is randomly distributed among the integers such that $s_t^2 - 4H(W_0|t) \in \mathbb{Z}_n^{-1} \cup QR(n))$ as it satisfies $s_t^2 - 4H(W_0|t) = (s r_{i,t})^2 - 4h_d r_{i,t}^2 = r_{i,t}^2(s^2 - 4h_d)$, where $s^2 - 4h_d$ is also randomly distributed among the integers in $\mathbb{Z}_n^{-1} \cup QR(n))$ as $(n, h_0, h_1, s) \in D_{1,0}(1^m)$. Therefore, the probability that $\mathcal{A}$ returns 1 in MainScheme Security Game t when $(n, h_0, h_1, s) \in D_{1,0}(1^m)$ is the same as the probability that $\mathcal{A}_t = 1$.

We now consider the case when $(n, h_0, h_1, s) \in D_{1,1}(1^m)$, the ciphertext s in MainScheme Security Game t appears to $\mathcal{A}$ to be distributed exactly as if $s_1, \ldots, s_{t-1}$ were computed as in algorithm M-ksEnc on input A_{pub}, W_0, and $s_t, \ldots, s_{4k}$ were computed as in algorithm M-ksEnc on input A_{pub}, W_1. This can be seen by observing that s_t is uniformly distributed in $\mathbb{Z}_n^*$ by definition of $D_{1,1}$, and that if s_t were computed as in algorithm M-ksEnc on input A_{pub}, W_1, it would appear to $\mathcal{A}$ to have the same distribution, as we assumed that $(W_1|t)$ was not queried by $\mathcal{A}$. Therefore, the probability that $\mathcal{A}$ returns 1 in MainScheme Security Game t when $(n, h_0, h_1, s) \in D_{1,1}(1^m)$ is the same as the probability that $\mathcal{A}_{t-1} = 1$.

This implies that the probability that $\mathcal{B}$ distinguishes $D_{1,0}(1^m)$ from $D_{1,1}(1^m)$ is the probability $1/(e \cdot q_T)$ that $\mathcal{B}$ does not halt in MainScheme Security Game t, times the probability $\epsilon/(4k \cdot (q_T + 1))$ that $\mathcal{A}$ makes only one H-queries among $(H_0|t), (H_1|t)$ and it holds that the associated bit $c_{\cdot,t} = 0$.

Since ϵ is assumed to be not negligible, then so is the quantity $\epsilon/(e \cdot 4k \cdot (q_T + 1))$, and therefore $\mathcal{B}$ violates the intractability of the QIP_1 problem.

Case (b). We now consider the case when both H-queries $(W_0|t), (W_1|t)$ are made by algorithm $\mathcal{A}$. We define the QIP_2 *problem* as the problem of efficiently distinguishing the following two distributions:

$$D_{2,0}(1^m) = \{(p,q) \leftarrow BW(1^m); n \leftarrow p \cdot q; h_0, h_1 \leftarrow \mathbb{Z}_n^{+1};$$
$$s \leftarrow CS(n, s^2 - 4h_0 \in \mathbb{Z}_n^{-1} \cup QR(n)) : (n, h_0, h_1, s)\}$$
$$D_{2,1}(1^m) = \{(p,q) \leftarrow BW(1^m); n \leftarrow p \cdot q; h_0, h_1 \leftarrow \mathbb{Z}_n^{+1};$$
$$s \leftarrow CS(n, s^2 - 4h_1 \in \mathbb{Z}_n^{-1} \cup QR(n)) : (n, h_0, h_1, s)\}$$

We say that algorithm $\mathcal{A}$ has *advantage* ϵ in solving QIP_2 if we have that:

$$\left| \Pr[(n, h_0, h_1, s) \leftarrow D_{2,0}(1^m) : \mathcal{A}(n, h_0, h_1, s) = 1] \right.$$
$$\left. - \Pr[(n, h_0, h_1, s) \leftarrow D_{2,1}(1^m) : \mathcal{A}(n, h_0, h_1, s) = 1] \right| = \epsilon. \quad (5)$$

We say that QIP_2 is *intractable* if all polynomial time (in m) algorithms have a negligible (in m) advantage in solving QIP_2.

By a simple hybrid argument, we can prove the following theorem:

Theorem 3. The QIP_2 problem is intractable if and only if the QIP problem is so.

We continue the proof by noting that bits $c(i, t), c(j, t)$ associated to the two queries, where the i-th queried keyword is W_i and the j-th queried keyword is W_1, satisfy $c(i, t) = c(j, t) = 0$ with probability at least $1/(q_T + 1)^2$. Under this setting, we evaluate the distribution of ciphertext s in MainScheme Security Game t, for $t = 1, \ldots, 4k$.

First, we observe that when $(n, h_0, h_1, s) \in D_{2,0}(1^m)$, the ciphertext s in MainScheme Security Game t appears to $\mathcal{A}$ to be distributed exactly as if $s_1, \ldots, s_t$ were computed as in algorithm M-ksEnc on input A_{pub}, W_0, and $s_{t+1}, \ldots, s_{4k}$ were computed as in algorithm M-ksEnc on input A_{pub}, W_1. This can be seen by observing that we assumed that $c(i, t) = 0$ and thus $H(W_0|t) = h_0 \cdot r_{i,t}^2$; then, it holds that s_t is randomly distributed among the integers such that $s_t^2 - 4H(W_0|t) \in \mathbb{Z}_n^{-1} \cup QR(n))$ as it satisfies $s_t^2 - 4H(W_0|t) = (sr_{i,t})^2 - 4h_0 r_{i,t}^2 = r_{i,t}^2(s^2 - 4h_0)$, where $s^2 - 4h_0$ is also randomly distributed among the integers in $\mathbb{Z}_n^{-1} \cup QR(n))$ as $s \in D_{2,0}(1^m)$. Therefore, the probability that $\mathcal{A}$ returns 1 in MainScheme Security Game t when $(n, h_0, h_1, s) \in D_{2,0}(1^m)$ is the same as the probability that $\mathcal{A}_t = 1$.

Analogously, when $(n, h_0, h_1, s) \in D_{2,1}(1^m)$, the ciphertext s in MainScheme Security Game t appears to $\mathcal{A}$ to be distributed exactly as if $s_1, \ldots, s_{t-1}$ were computed as in algorithm M-ksEnc on input A_{pub}, W_0, and $s_t, \ldots, s_{4k}$ were computed as in algorithm M-ksEnc on input A_{pub}, W_1. This can be seen as before by again observing that we assumed that $c(j, t) = 0$ and thus $H(W_1|t) = h_1 \cdot r_{j,t}^2$; then, it holds that s_t is randomly distributed among the integers such that $s_t^2 - 4H(W_1|t) \in \mathbb{Z}_n^{-1} \cup QR(n))$ as it satisfies $s_t^2 - 4H(W_1|t) = (sr_{j,t})^2 - 4h_1 r_{j,t}^2 = r_{j,t}^2(s^2 - 4h_1)$, where $s^2 - 4h_1$ is also randomly distributed among the integers in $\mathbb{Z}_n^{-1} \cup QR(n))$ as $s \in D_{2,1}(1^m)$. Therefore, the probability that $\mathcal{A}$ returns 1 in MainScheme Security Game t when $(n, h_0, h_1, s) \in D_{2,1}(1^m)$ is the same as the probability that $\mathcal{A}_{t-1} = 1$.

This implies that $\mathcal{B}$ distinguishes $D_{2,0}(1^m)$ from $D_{2,1}(1^m)$ is the probability $1/(e \cdot q_T)$ that $\mathcal{B}$ does not halt in MainScheme Security Game t, times the probability $\epsilon/(4k \cdot (q_T + 1)^2)$ that $\mathcal{A}$ makes both H-queries $(H_0|t), (H_1|t)$ and it holds that $c_{i,t} = c_{j,t} = 0$.

Since ϵ is assumed to be not negligible, then so is the quantity $\epsilon/(e \cdot 4k \cdot q_T (q_T + 1)^2)$, and therefore $\mathcal{B}$ violates the intractability of the QIP_2 problem.

Acknowledgements

We thank Fadil Santosa, Minnesota Center for Industrial Mathematics, University of Minnesota, and Andrew Odlyzko, Digital Technology Center, University of Minnesota, for support and interesting discussions.

References

1. Abdalla, M., Bellare, M., Catalano, D., Kiltz, E., Kohno, T., Lange, T., Malone-Lee, J., Neven, G., Paillier, P., Shi, H.: Searchable Encryption Revisited: Consistency Properties, Relation to Anonymous IBE, and Extensions. In: Shoup, V. (ed.) CRYPTO 2005. LNCS, vol. 3621, Springer, Heidelberg (2005)
2. Bellare, M., Boldyreva, A., Desai, A., Pointcheval, D.: Key-Privacy in Public-Key Encryption. In: Boyd, C. (ed.) ASIACRYPT 2001. LNCS, vol. 2248, Springer, Heidelberg (2001)
3. Boneh, D.: Private communication (February 2007)
4. Boneh, D.: Private communication (August 2007)
5. Boneh, D., Di Crescenzo, G., Ostrovsky, R., Persiano, G.: Public Key Encryption with Keyword Search. In: Cachin, C., Camenisch, J.L. (eds.) EUROCRYPT 2004. LNCS, vol. 3027, pp. 506–522. Springer, Heidelberg (2004)
6. Boneh, D., Franklin, M.: Identity-based Encryption from the Weil Pairing. SIAM J. of Computing 32(3), 586–615 (2003) (Extended abstract in Crypto 2001)
7. Boneh, D., Gentry, C., Hamburg, M.: Space-Efficient Identity Based Encryption Without Pairings (in submission)
8. Boyen, X., Waters, B.: Anonymous Hierarchical Identity-Based Encryption (without Random Oracles). In: Dwork, C. (ed.) CRYPTO 2006. LNCS, vol. 4117, Springer, Heidelberg (2006)
9. Cocks, C.: An Identity Based Encryption Echeme based on Quadratic Residues. In: Eighth IMA International Conference on Cryptography and Coding, Royal Agricultural College, Cirencester, UK (December 2001)
10. Cohen, H.: A Course in Computational Algebraic Number Theory. In: Graduate Texts in Mathematics, vol. 138, Springer, Heidelberg (1993)
11. Coron, J.: On the Exact Security of Full-Domain-Hash. In: Bellare, M. (ed.) CRYPTO 2000. LNCS, vol. 1880, pp. 229–235. Springer, Heidelberg (2000)
12. Dolev, D., Dwork, C., Naor, M.: Non-Malleable Cryptography. SIAM Journal on Computing (2000) Early version in Proc. of STOC 1991
13. Golle, P., Staddon, J., Waters, B.R.: Secure Conjunctive Keyword Search over Encrypted Data. In: Jakobsson, M., Yung, M., Zhou, J. (eds.) ACNS 2004. LNCS, vol. 3089, Springer, Heidelberg (2004)
14. Maniatis, P., Roussopoulos, M., Swierk, E., Lai, K., Appenzeller, G., Zhao, X., Bake, M.: The Mobile People Architecture. ACM Mobile Computing and Communications Review (MC2R) 3(3) (July 1999)
15. Park, D.J., Kim, K., Lee, P.J.: Public Key Encryption with Conjunctive Keyword Search. In: Lim, C.H., Yung, M. (eds.) WISA 2004. LNCS, vol. 3325, Springer, Heidelberg (2005)
16. Shamir, A.: Identity-based Cryptosystems and Signature Schemes. In: Blakely, G.R., Chaum, D. (eds.) CRYPTO 1984. LNCS, vol. 196, Springer, Heidelberg (1985)
17. Waters, B., Balfanz, D., Durfee, G., Smetters, D.: Building an Encrypted and Searchable Audit Log. In: Proc. of NDSS 2004 (2004)

A Certificate-Based Proxy Cryptosystem with Revocable Proxy Decryption Power

Lihua Wang[1], Jun Shao[2], Zhenfu Cao[2],
Masahiro Mambo[3], and Akihiro Yamamura[1]

[1] Information Security Research Center, National Institute of Information and
Communications Technology, Tokyo 184-8795, Japan
{wlh,aki}@nict.go.jp
[2] Department of Computer Science and Engineering, Shanghai Jiao Tong University,
Shanghai 200240, P.R. China
chn.junshao@gmail.com, zfcao@sjtu.edu.cn
[3] Graduate School of Systems and Information Engineering, University of Tsukuba,
Tsukuba 305-8573, Japan
mambo@cs.tsukuba.ac.jp

Abstract. We present a proxy cryptosystem based on a certificate-based encryption scheme. The proposed scheme inherits the merits of certificate-based encryption systems: no-key-escrow and implicit certification. In addition, the proposed scheme allows the proxy's decryption power to be revoked even during the valid period of the proxy key without changing the original decryptor's public information. Few proxy schemes have this property, and ours is more efficient than the existing ones. We show that our proposal is IND-CBPd-Rev-CCA secure under the bilinear Diffie-Hellman assumption in the random oracle model.

Keywords: proxy cryptosystem, pairing, certificate-based encryption (CBE).

1 Introduction

Proxy cryptosystems were invented by Mambo and Okamoto [7] for the delegation of the power to decrypt ciphertexts. In a proxy cryptosystem, there are three roles of participants involved: encryptor Alice, original decryptor Bob, and Bob's proxy decryptors Charlie, Clara and so on.

Let us consider the following scenario. A busy corporate manager, Bob, receives a great number of e-mails encrypted using his public key every day. To reduce the burden of decrypting all of the ciphertexts, Bob partly delegates his decryption power to his secretaries by subject assigned to them. The subjects may be project names, trade names, or simply topics written in the subject lines of e-mails. For example, Bob delegates the decryption power corresponding to a certain subject job_c (say it is related to a project during the fiscal year 2007) to Charlie. Charlie is then given proxy decryption power for the period from the first of January to the thirty-first of December, 2007. As long as the proxy behaves well, the proxy decryption power will be valid until the end of 2007. However,

K. Srinathan, C. Pandu Rangan, M. Yung (Eds.): Indocrypt 2007, LNCS 4859, pp. 297–311, 2007.

the following unexpected cases may occur before the proxy decryption power expires: Charlie's occupation changes, Charlie's proxy key has disappeared, or Charlie becomes corruptible so that Bob dose not trust him anymore.

In such cases, it is desirable for Bob to be able to revoke Charlie's proxy decryption power even if his proxy key has not expired. We name this ability *revocability*. If a proxy cryptosystem has revocability, we call it a proxy cryptosystem with *revocable proxy decryption power*. To avoid unnecessary tasks on encryptors, Bob's public information, public key, and subjects assigned to proxy decryptors should remain unchanged when a proxy's decryption power is revoked.

In the Mambo-Okamoto proxy cryptosystem [7], Bob controls his proxy decryption power by transforming an original ciphertext into another ciphertext for the proxy. The proxy cannot decrypt a ciphertext at all before Bob transfers the ciphertext for him. Therefore, the Mambo-Okamoto proxy cryptosystem can revoke decryption power without changing Bob's public information. However, the ciphertext transformation needs to be executed for every ciphertext, so the Mambo-Okamoto proxy cryptosystem is not efficient enough for practical use.

To make the cryptosystem more efficient, many ciphertext Transformation-Free Proxy cryptosystems (TFP systems) were studied [8,10,11,12,14]. In these schemes, each proxy can decrypt a ciphertext directly without ciphertext transformation. However, none of them is a proxy cryptosystem with revocable proxy decryption power.

Furthermore, all proxy cryptosystems (including those with ciphertext transformation and the above TFP systems) are constructed based either on directory-based public key cryptography (PKC) or identity-based cryptography (IBC). It is well known that directory-based PKC suffers from the certificate revocation problem and IBC suffers from the key escrow problem. To overcome these disadvantages in directory-based PKC and IBC, Gentry [6] introduced the certificate-based encryption (CBE) scheme, which combines the best aspects of IBCs (implicit certification) and of directory-based PKCs (no-key-escrow). To the best of our knowledge, there is no proxy cryptosystem constructed based on CBE.

In this paper, we give the first construction of a certificate-based proxy cryptosystem (CBPd)[1]. In addition to the inherent merits of the CBE scheme, our CBPd scheme can revoke proxy decryption power without changing Bob's public information even during the valid period of the proxy key. Our scheme is also efficient because it is a ciphertext transformation-free proxy cryptosystem. The proposed CBPd scheme with revocable proxy decryption power is semantically secure against adaptive chosen ciphertext attack (IND-CBPd-Rev-CCA secure) under the random oracle model. Therefore, our scheme is the first certificate-based proxy cryptosystem that has revocability, transformation-freeness, and IND-CBPd-Rev-CCA security.

The rest of this paper is organized as follows. In Section 2, we recall the definition of pairings and computational complexity assumptions. In Section 3, we first formally define a certificate-based proxy cryptosystem and then propose

---

[1] "d" is used here to distinguish proxy decryption from proxy signature.

a concrete scheme with revocable proxy decryption power. In Section 4, we prove that our scheme is IND-CBPd-Rev-CCA secure in the random oracle model. We discuss several issues in Section 5 and make brief concluding remarks in Section 6.

2   Bilinear Pairings and Complexity Assumption

Our scheme is based on an admissible pairing that was first used to construct cryptosystems independently by Sakai et al. [9] and Boneh et al. [2]. The modified Weil pairing and the Tate pairing associated with supersingular elliptic curves are examples of such admissible pairings. However, we describe pairings and the related mathematics in a more general format here.

Let $\mathbb{G}_1$, $\mathbb{G}_2$ be two multiplicative groups with prime order p and g be a generator of $\mathbb{G}_1$. $\mathbb{G}_1$ has an admissible bilinear map into $\mathbb{G}_2$, $\hat{e} : \mathbb{G}_1 \times \mathbb{G}_1 \longrightarrow \mathbb{G}_2$, if the following three conditions hold:

(1) *Bilinear.* $\hat{e}(g^a, g^b) = \hat{e}(g, g)^{ab}$ for all $a, b \in \mathbb{Z}_p^*$.
(2) *Non-degenerate.* $\hat{e}(g, g) \neq 1_{\mathbb{G}_2}$.
(3) *Computable.* There is an efficient algorithm to compute $\hat{e}(f, h)$ for any $f, h \in \mathbb{G}_1$.

Bilinear Diffie-Hellman (BDH) Parameter Generator [2,6]: A randomized algorithm $\mathcal{IG}$ is a BDH parameter generator if $\mathcal{IG}$ takes a security parameter $k > 0$, runs in time polynomial in k, and outputs the description of two groups $\mathbb{G}_1$ and $\mathbb{G}_2$ of the same prime order p and the description of an admissible pairing $\hat{e} : \mathbb{G}_1 \times \mathbb{G}_1 \rightarrow \mathbb{G}_2$.

The following three problems are assumed to be intractable for any polynomial time algorithm.

Discrete Logarithm Problem: Given $g, g^a \in \mathbb{G}_1$, or $\mu, \mu^a \in \mathbb{G}_2$, find $a \in \mathbb{Z}_q^*$.

Computational Diffie-Hellman (CDH) Problem [2]: Given $g, g^a, g^b \in \mathbb{G}_1$, find $g^{ab} \in \mathbb{G}_1$.

Bilinear Diffie-Hellman (BDH) Problem [2]: Given $g, g^a, g^b, g^c \in \mathbb{G}_1$, find $\hat{e}(g, g)^{abc} \in \mathbb{G}_2$.

3   A CBPd Scheme with Revocable Proxy Decryption Power

3.1   Definitions

Definition 1 (CBPd system). *A certificate-based proxy cryptosystem (CBPd system) consists of the following five algorithms:*

Setup: This algorithm takes as input a security parameter 1^k and returns the certifier's master secret key msk and master public key mpk. It also outputs a public parameter params, which is shared in the system.

Original Decryption Key Generation

- **PartialKeyGen:** *This algorithm takes as input system parameter params. It outputs a user U's partial secret/public key pair (sk_U, pk_U).*
- **Certification:** *This algorithm takes as input master secret/public key pair (msk, mpk), system parameter params, user U's information which includes his or her identity ID_U and public key pk_U. It outputs a certificate $Cert_U$.*
- **DecKeyGen:** *This algorithm takes as input certificate $Cert_U$ and user U's partial secret/public key pair (sk_U, pk_U). It outputs user U's original decryption key $sk_U^{(0)}$.*

Proxy Decryption Key Distribution ($\mathcal{PDK}$): *This algorithm takes as input master public key mpk, system parameter params, user U's original decryption key $sk_U^{(0)}$, and delegation subject job. It outputs a proxy key $sk_U^{(job)}$.*

Encryption ($\mathcal{E}$): *The encryption algorithm, $\mathcal{E}$, is a probabilistic algorithm that takes as input a plaintext m, U's public key pk_U and identity ID_U, and the subject line, job. It outputs ciphertext C_{job} of m.*

Decryption: *The decryption algorithm, $(\mathcal{D}_U, \mathcal{D}_{job})$, is a deterministic algorithm that consists of an original/basic decryption algorithm $\mathcal{D}_U$ and a proxy decryption algorithm $\mathcal{D}_{job}$.*
- *The original/basic decryption algorithm, $\mathcal{D}_U$, takes as input a ciphertext C_{job}, and U's original decryption key $sk_U^{(0)}$. It outputs plaintext m of C_{job}.*
- *The proxy decryption algorithm, $\mathcal{D}_{job}$, takes as input a ciphertext C_{job}, proxy key $sk_U^{(job)}$, and user U's public information. It outputs plaintext m of C_{job}.*

In the above CBPd system, the delegation subject, *job*, acts as the "identity" of the proxy. Therefore, the encryptor does not need to know who the proxy of Bob is.

Definition 2 (Revocability). *A CBPd system with revocable proxy decryption power satisfies the following basic security requirements:*

- *(General requirements) On receipt of a ciphertext to Bob, no one can do basic decryption operation without Bob's partial secret key sk_B or decryption key S_B; The proxy for the subject job cannot decrypt the ciphertext corresponding to subject $job' \neq job$.*
- *(Requirements in view of revocability) When the proxy key is marked void by Bob forcibly on time τ, Charlie can no longer decrypt ciphertexts under subject job_c on time $\tau' > \tau$ even if the valid period of the proxy key has not expired.*

3.2   Our Scheme

In our scheme, proxy decryption key distribution algorithm $\mathcal{PDK}$ and proxy decryption algorithm $\mathcal{D}_{job}$ are devised as follows to meet revocability requirements:

- Bob distributes proxy keys with a valid period to proxies by subject, e.g., $sk_B^{(job_c)}$ denotes the proxy key that Bob distributes to Charlie, who is delegated proxy decryption power for subject job_c.
- Bob distributes short-term common data and often (e.g., once a day) renews the data for all of his proxies. Let $Y(\tau_i)$ denote the data on day τ_i where $i = 1, 2, ...$
- On receipt of the ciphertext C in terms of the subject job_c and time τ, Charlie decrypts the ciphertext C, using both $sk_B^{(job_c)}$ and $Y(\tau)$.

Assume that there are n assistants who are delegated to do proxy decryption for Bob. Their proxy authorities are for different subjects $job_1, ..., job_n$. Using ideas from the CBE of Gentry [6], the IBE of Waters [13], and Fujisaki and Okamoto's scheme [5], we construct the following concrete scheme.

Setup: The CA[2] does as follows:

- Run the BDH parameter generator $\mathcal{IG}$ on input 1^k to generate a prime p, two multiplicative groups $\mathbb{G}_1, \mathbb{G}_2$ of order p, and an admissible bilinear map $\hat{e} : \mathbb{G}_1 \times \mathbb{G}_1 \to \mathbb{G}_2$. Choose an arbitrary generator $g \in \mathbb{G}_1$.
- Pick a random $s_C \in \mathbb{Z}_p^*$, and set $g_{pub} = g^{s_C}$.
- Choose six cryptographic hash functions $H_0 : \mathbb{G}_1 \times \{0,1\}^{l_{id}} \times \mathbb{G}_1 \to \mathbb{G}_1$, $H_1 : \{0,1\}^{l_{id}} \times \mathbb{G}_1 \to \mathbb{G}_1$, $H_2 : \{0,1\}^{l_{job}} \to \mathbb{G}_1$, $H_3 : \{0,1\}^{l_{time}} \to \mathbb{G}_1$, $H_4 : \{0,1\}^{2l_\mathcal{M}} \to \mathbb{Z}_p^*$, and $H_5 : \{0,1\}^{l_\mathcal{M}} \to \{0,1\}^{l_\mathcal{M}}$, where $l_{id}, l_{job}, l_{time}$ and $l_\mathcal{M}$ denote the lengths of user identity, subject job, time τ, and plaintext $m \in \mathcal{M}$, respectively. Choose a key derivation function $F : \mathbb{G}_2 \to \{0,1\}^{l_\mathcal{M}}$ for $l_\mathcal{M}$, such as KDF1 defined in IEEE Standard 1363-2000. The security analysis will view F, $H_i(i \in \{0,1,2,3,4,5\})$ as random oracles.

The system parameters are

$$params = (\mathbb{G}_1, \mathbb{G}_2, p, \hat{e}, g, g_{pub}, F, H_i(i \in \{0,1,2,3,4,5\})).$$

The message space is $\mathcal{M} = \{0,1\}^{l_\mathcal{M}}$. The CA's secret is $s_C \in \mathbb{Z}_p^*$.

Original Decryption Key Generation

- Bob generates his partial secret/public key pair (s_B, pk_B), where $pk_B = g^{s_B}$ is computed according to the parameters issued by the CA.
- Bob sends his information to the CA, which includes his public key pk_B and any necessary additional identifying information, such as his name and mail address. The CA verifies Bob's information, then computes

$$Cert_B = H_0(g_{pub}, ID_B, pk_B)^{s_C} \in \mathbb{G}_1.$$

Certificate $Cert_B$ is, in fact, a signature (the signature in [4]) by the CA on Bob's information.

---

[2] The main difference from the CA in PKC is that the CA in CBE uses an IBE scheme to generate the certificate ([6]).

- Before doing decryptions, Bob also signs on his own information by producing $Sign_B = H_1(ID_B, pk_B)^{s_B}$. Then Bob will use this two person aggregate signature [3]

$$S_B = Cert_B \cdot Sign_B = H_0(g_{pub}, ID_B, pk_B)^{s_C} H_1(ID_B, pk_B)^{s_B}$$

  as his decryption key.

Proxy Decryption Key Distribution ($\mathcal{PDK}$): Bob selects $A \in_R \mathbb{G}_1$, before generating a long-term personal proxy key for each proxy and short-term common data for all of his proxies.

- Long-term personal proxy key: Charlie, one of Bob's proxies, is delegated the decryption power for subject, job_c. Bob selects $d_{job_c} \in_R \mathbb{Z}_p^*$ then computes proxy key $sk_B^{(job_c)} = (sk_1^{(job_c)}, sk_2^{(job_c)})$ for Charlie as follows:

$$sk_1^{(job_c)} = S_B H_2(job_c)^{d_{job_c}} A, \quad sk_2^{(job_c)} = g^{d_{job_c}}.$$

  Bob sends $sk_B^{(job_c)}$ to Charlie via a secure channel. Let job_c denote a project during the fiscal year 2007, then the proxy key's valid period should be until the end of 2007. Note that, $(sk_1^{(job_c)}/A, sk_2^{(job_c)})$ is the signature in [13] with a hash function.
- Short-term common data: Bob selects $d_\tau \in_R \mathbb{Z}_p^*$, $\tau = \{\tau_1, \tau_2, ...\}$, then computes common data $Y(\tau) = (y_1(\tau), y_2(\tau))$ on τ as follows:

$$y_1(\tau) = A^{-1} H_3(\tau)^{d_\tau}, \quad y_2(\tau) = g^{d_\tau}.$$

  Bob renews $Y(\tau)$ once a day by changing random number d_τ, in which no proxies' information is included, so $Y(\tau)$ is the common data for all of his proxies.
- Verification of common data of time τ: Charlie computes $\eta = \hat{e}(H_0(g_{pub}, ID_B, pk_B), g_{pub}) \cdot \hat{e}(H_1(ID_B, pk_B), pk_B)$ and then checks whether

$$\hat{e}(sk_1^{(job_c)} y_1(\tau), g) \stackrel{?}{=} \eta \cdot \hat{e}(H_2(job_c), sk_2^{(job_c)}) \cdot \hat{e}(H_3(\tau), y_2(\tau)).$$

  If it is correct, together with proxy key $sk_B^{(job_c)}$, common data $Y(\tau)$ is used to do proxy decryption operation.

Note that parameters A, d_{job_c} for $job_c \in \{job_1, ..., job_n\}$, and d_τ for $\tau = \tau_1, \tau_2, ...$ must be kept secretly by Bob himself.

Encryption ($\mathcal{E}$): To encrypt message $m \in \mathcal{M}$ under subject job_c on day τ, encryptor Alice

- computes $\eta = \hat{e}(H_0(g_{pub}, ID_B, pk_B), g_{pub}) \cdot \hat{e}(H_1(ID_B, pk_B), pk_B)$;
- selects $\sigma \in_R \{0,1\}^{l_\mathcal{M}}$, let $r = H_4(m||\sigma)$, and computes ciphertext $C = (C_1, C_2, C_3, C_4, C_5)$, where

$$C_1 = m \oplus H_5(\sigma), C_2 = \sigma \oplus F(\eta^r), C_3 = g^r, C_4 = H_2(job_c)^r, C_5 = H_3(\tau)^r;$$

- sends $\langle C, job_c, \tau \rangle$ to Bob.

Decryption

– $\mathcal{D}_B$: Bob can decrypt the ciphertext under any subject by computing

$$\sigma' = C_2 \oplus F(\hat{e}(S_B, C_3)),$$

and

$$m' = C_1 \oplus H_5(\sigma').$$

Then he can check whether

$$C_3 \overset{?}{=} g^{H_4(m'||\sigma')}.$$

If this is correct, then $m = m'$ is accepted as the plaintext. Otherwise abort.

Accordingly, this proposed CBPd scheme will degenerate into Gentry's CBE scheme if Alice sends $C = (C_1, C_2, C_3)$ to Bob as a ciphertext. This implies that encryptor Alice can decide whether to let the proxy read m or not.

– $\mathcal{D}_{job_c}$: Charlie performs the proxy decryption operation for the ciphertext for his own subject, job_c, by computing

$$\sigma' = C_2 \oplus F(\frac{\hat{e}(sk_1^{(job_c)}y_1(\tau), C_3)}{\hat{e}(C_4, sk_2^{(job_c)}) \cdot \hat{e}(C_5, y_2(\tau))}),$$

and

$$m' = C_1 \oplus H_5(\sigma').$$

Then he checks whether

$$C_3 \overset{?}{=} g^{H_4(m'||\sigma')}, \quad C_4 \overset{?}{=} H_2(job_c)^{H_4(m'||\sigma')}, \quad C_5 \overset{?}{=} H_3(\tau)^{H_4(m'||\sigma')}.$$

If they are correct, then $m = m'$ is accepted as the plaintext. Otherwise abort.

3.3 Correctness

Correctness can be proved as follows:

For $\mathcal{D}_B$, we have

$$\begin{aligned}
\hat{e}(S_B, C_3) &= \hat{e}(H_0(g_{pub}, ID_B, pk_B)^{s_C} H_1(ID_B, pk_B)^{x_B}, g^r) \\
&= [\hat{e}(H_0(g_{pub}, ID_B, pk_B), g_{pub}) \cdot \hat{e}(H_1(ID_B, pk_B), pk_B)]^r \\
&= \eta^r.
\end{aligned}$$

For $\mathcal{D}_{job_c}$, we have

$$\begin{aligned}
&\hat{e}(sk_1^{(job_c)}y_1(\tau), C_3) \\
&= \hat{e}(S_B H_2(job_c)^{d_{job_c}} H_3(\tau)^{d_\tau}, g^r) \\
&= \hat{e}(S_B, g^r) \cdot \hat{e}(H_2(job_c)^{d_{job_c}}, g^r) \cdot \hat{e}(H_3(\tau)^{d_\tau}, g^r) \\
&= \hat{e}(S_B, C_3) \cdot \hat{e}(H_2(job_c)^r, g^{d_{job_c}}) \cdot \hat{e}(H_3(\tau)^r, g^{d_\tau}) \\
&= \eta^r \cdot \hat{e}(C_4, sk_2^{(job_c)}) \cdot \hat{e}(C_5, y_2(\tau)).
\end{aligned}$$

Therefore,

$$\frac{\hat{e}(sk_1^{(job_c)}y_1(\tau), C_3)}{\hat{e}(C_4, sk_2^{(job_c)}) \cdot \hat{e}(C_5, y_2(\tau))} = \eta^r.$$

4 Security

4.1 How to Revoke Proxy Decryption Power

According to the description of revocable proxy decryption power in Section 1, *Bob should be able to revoke Charlie's proxy decryption power even if his proxy key has not expired.* He should have this ability in order to deal with the cases such as Charlie's occupation changing, Charlie's proxy key disappearing, or Bob losing trust in Charlie.

For the communication for subject job_c to continue securely and, at the same time, avoid unnecessarily burdening encryptors, Bob's public information, including his public key and subjects assigned to Charlie, should remain unchanged when Charlie's decryption power is revoked. Bob can cope with this situation by changing original parameters A, d_{job} into new parameters $A', d'_{job} \in \mathbb{Z}_p^*$, where $A \neq A'$, and $d_{job} \neq d'_{job}$ for each $job \in \{job_1, ..., job_n\}$. The process in detail is:

- Renew the data $Y(\tau) = (y_1(\tau), y_2(\tau))$ using the new parameter A'. Then

$$y_1(\tau) = A'^{-1}H_3(\tau)^{d_\tau}, \quad y_2(\tau) = g^{d_\tau}.$$

- Renew the proxy keys for other proxies in terms of the new parameter A'. Then the proxy keys

$$sk_1^{(job)} = S_B H_2(job)^{d'_{job}} A', \quad sk_2^{(job)} = g^{d'_{job}},$$

where $job \in \{job_1, ..., job_n\} \setminus \{job_c\}$.
- Execute the original decryption, $\mathcal{D}_B$, before delegating the decryption power in terms of the above new parameters A' and d'_{job_c}, to another assistant.

The above new common data, $Y(\tau)$, corresponding to A' cannot pass verification using the revoked proxy key $(sk_1^{(job_c)}, sk_2^{(job_c)}) = (S_B H_2(job_c)^{d_{job_c}} A, g^{d_{job_c}})$, to say nothing of proxy decryption operation. In fact, on receipt of the ciphertext of m' on time $\tau' > \tau$, using the revoked proxy key $(sk_1^{(job_c)}, sk_2^{(job_c)})$ and new common data $Y(\tau')$, Charlie can only obtain $\eta^r \hat{e}(AA'^{-1}, g)^r \neq \eta^r$. So, $m'' = C_1 \oplus F(\eta^r \hat{e}(AA'^{-1}, g)^r) = m' \oplus F(\eta^r) \oplus F(\eta^r \hat{e}(AA'^{-1}, g)^r) \neq m'$. Accordingly, Charlie's proxy decryption power is revoked.

4.2 IND-CBPd-Rev-CCA

For a secure proxy cryptosystem with revocable proxy decryption power, the following requirements should be considered: the proxy whose decryption power has been revoked cannot decrypt any new ciphertext for his or her subject using the revoked proxy key, even if he or she can obtain the new common data, $Y(\tau)$, and/or collude with other proxies for different subjects. Therefore, we say that a certificate-based proxy cryptosystem scheme with revocable proxy decryption power is semantically secure against an adaptive chosen ciphertext attack (IND-CBPd-Rev-CCA) if no polynomially bounded adversary $\mathcal{A}$ has a non-negligible advantage against the challenger in the following game.

Setup. The challenger generates system parameters and gives the parameters to the adversary. Furthermore, after the challenger generates master key, the private key of original decryptor Bob's and the corresponding public keys, he or she gives the public keys to the adversary.

Phase 1. The adversary is permitted to make the following queries:

- (*Certification query*) On input (ID, pk) by the adversary, the challenger returns the corresponding certification.
- (*Proxy secret key query*) On input (τ, job) by the adversary, the challenger returns the corresponding long-term proxy key.
- (*Common data query*) On input τ by the adversary, the challenger returns the corresponding common data.
- (*Basic decryption oracle query*) On input the basic ciphertext[3] by the adversary, the challenger returns the corresponding plaintext.
- (*Proxy decryption oracle query*) On input the proxy ciphertext[4] for (τ, job) by the adversary, the challenger returns the corresponding plaintext.
- (*Revocation oracle query*) On input (τ, job) by the adversary, the challenger returns the corresponding long-term proxy key and common data. In fact, this query is the combination of *proxy secret key query* and *common data query*. For soundness, we define this oracle here.

Challenge. The adversary submits a target subject, job^*, and two messages (m_0, m_1) on time τ^*. The adversary's choice of job^* is restricted to the subjects that he did not request a proxy key for on time τ^* in Phase 1. The challenger flips a fair binary coin, γ, and returns an encryption c^* of m_γ for subject job^* on τ^*.

Phase 2. Phase 1 is repeated with the following constraints:

- the adversary cannot request the proxy secret key oracle query on (τ^*, job^*);
- the adversary cannot request the basic decryption oracle query on c^*;
- the adversary cannot request the proxy decryption oracle query on (c^*, τ^*, job^*);
- the adversary cannot request the revocation oracle query on (τ^*, job^*).

Guess. The adversary submits a guess, γ', of γ.

Then the advantage of $\mathcal{A}$ is

$$Adv_{\mathcal{A}}^{CBPd} = |Pr[\gamma' = \gamma'] - \frac{1}{2}|.$$

---

[3] In our proposal, the basic ciphertext is (C_1, C_2, C_3).
[4] In our proposal, the proxy ciphertext is $(C_1, C_2, C_3, C_4, C_5)$.

Definition 3 (IND-CBPd-Rev-CCA). *A CBPd scheme with revocable proxy decryption power is semantically secure against an adaptive chosen ciphertext attack (IND-CBPd-Rev-CCA Secure) if $Adv_{\mathcal{A}}^{CBPd}$ is negligible.*

Theorem 1. *The proposed scheme is an IND-CBPd-Rev-CCA secure certificate-based proxy cryptosystem under the assumption that there is no polynomial algorithm to solve the BDH problem with non-negligible probability.*

Proof Sketch. We show that if an adversary $\mathcal{A}$ can break our CBPd scheme with non-negligible probability, then we have another algorithm $\mathcal{B}$ that, using $\mathcal{A}$, can solve the BDH problem with non-negligible probability (refer to Appendix A for more details).

5   Discussion

Advantages of Our Proposed CBPd Scheme
- Our scheme is the first certificate-based proxy cryptosystem, so it inherits the merits of certificate-based encryption systems: no-key-escrow and implicit certification.
- Our scheme is a TFP system and has the property of revocability.

The Efficiency. The Mambo-Okamoto scheme [7] supplies a revocable proxy decryption power, but they are inefficient, because a proxy cannot decrypt a ciphertext at all before Bob transfers the ciphertext for him. In our scheme, Bob controls proxy decryption power by renewing the common data for all of his proxies once a day.

Let Exp denote exponential computation, $N(\tau)$ denote the mail number that Bob received on day τ, $N_i(\tau)$ denote the mail number for subject job_i, $i = 1, ..., n$, and $Cost_{Bob}^{\tau}(\cdot)$ denote the computation cost that Bob incurs on day τ. Then Bob's computation in our scheme is $2 \cdot Exp$, while in the Mambo-Okamoto scheme, it is $1 \cdot Exp \times N(\tau)$, where $N(\tau) = N_1(\tau) + ... + N_i(\tau) + ... + N_n(\tau)$. Accordingly,

$$Cost_{Bob}^{\tau}(Ours) \approx \frac{2}{N(\tau)} Cost_{Bob}^{\tau}(MO).$$

That is to say, on day τ, the computation cost that Bob incurs in our scheme is nearly $\frac{2}{N(\tau)}$ of that in the Mambo-Okamoto scheme. Therefore, our new CBPd scheme with revocable proxy decryption power is much more flexible and efficient than the Mambo-Okamoto scheme.

Problems. In our scheme, Bob controls proxy decryption power by renewing the short-term common data, which are generated for all of the proxies. Accordingly, Bob can use one data $Y(\tau)$ to control all of his proxies when every thing is normal. However, Bob has to reissue all of the proxy keys when he wants to revoke one of the proxy keys.

Table 1. Related work

	CBPd	TFP	Revocability	Bob's computation cost on day τ
Mambo-Okamoto [7]	$\times$	$\times$	$\checkmark$	$Cost^\tau_{Bob}(MO) = 1 \cdot Exp \times N(\tau)$
MOV [8]	$\times$	$\checkmark$	$\times$	-
HEAD [10]	$\times$	$\checkmark$	$\times$	-
AL-TFP [11,12]	$\times$	$\checkmark$	$\times$	-
ZCC [14]	$\times$	$\checkmark$	$\times$	-
Proposed CBPd	$\checkmark$	$\checkmark$	$\checkmark$	$Cost^\tau_{Bob}(Ours) = 2 \cdot Exp$

6 Concluding Remarks

We presented a certificate-based proxy cryptosystem that is ciphertext transformation-free and has revocability, which means that proxy decryption power
can be revoked even if the validity of the proxy key has not expired. Therefore,
our scheme is more practical than the existing schemes in protecting messages
of the original decryptor from his proxies who have become untrustworthy. The
security of our scheme is based on the well-known BDH assumption.

References

1. Boneh, D., Boyen, X.: Secure identity based encryption without random oracles.
   In: Franklin, M. (ed.) CRYPTO 2004. LNCS, vol. 3152, pp. 443–459. Springer,
   Heidelberg (2004)
2. Boneh, D., Franklin, M.: Identity-based encryption from the Weil pairing. In: Kilian, J. (ed.) CRYPTO 2001. LNCS, vol. 2139, pp. 213–229. Springer, Heidelberg
   (2001)
3. Boneh, D., Gentry, C., Lynn, B., Shacham, H.: Aggregate and verifiably encrypted
   signatures from bilinear maps. In: Biham, E. (ed.) EUROCRPYT 2003. LNCS,
   vol. 2656, pp. 416–432. Springer, Heidelberg (2003)
4. Boneh, D., Lynn, B., Shacham, H.: Short signatures from the Weil pairing. In:
   Boyd, C. (ed.) ASIACRYPT 2001. LNCS, vol. 2248, pp. 514–532. Springer, Heidelberg (2001)
5. Fujisaki, E., Okamoto, T.: Secure integration of asymmetric and symmetric encryption schemes. In: Wiener, M. (ed.) CRYPTO 1999. LNCS, vol. 1666, pp. 537–554.
   Springer, Heidelberg (1999)
6. Gentry, C.: Certificate-based encryption and the certificate revocation problem.
   In: Biham, E. (ed.) EUROCRPYT 2003. LNCS, vol. 2656, pp. 272–293. Springer,
   Heidelberg (2003)
7. Mambo, M., Okamoto, E.: Proxy cryptosystem: delegation of the power to decrypt
   ciphertexts. IEICE Trans. Fundamentals E80-A(1), 54–63 (1997)
8. Mu, Y., Varadharajan, V., Nguyen, K.Q.: Delegation decryption. In: Walker, M.
   (ed.) IMA - Crypto & Coding 1999. LNCS, vol. 1746, pp. 258–269. Springer, Heidelberg (1999)
9. Sakai, R., Ohgishi, K., Kasahara, M.: Cryptosystems based on pairing, SCIS2000-
   C20 (2000)
10. Sarkar, P.: HEAD: hybrid encryption with delegated decryption capbility. In: Can-
    teaut, A., Viswanathan, K. (eds.) INDOCRYPT 2004. LNCS, vol. 3348, pp. 230–
    244. Springer, Heidelberg (2004)

11. Wang, L., Cao, Z., Okamoto, E., Miao, Y., Okamoto, T.: Transformation-free proxy cryptosystems and their applications to electronic commerce. In: Proceeding of International Conference on Information Security (InfoSecu 2004), pp. 92–98. ACM Press, New York (2004)
12. Wang, L., Cao, Z., Okamoto, T., Miao, Y., Okamoto, E.: Authorization-limited transformation-free proxy cryptosystems and their security analyses. IEICE Trans. Fundamentals E89-A(1), 106–114 (2006)
13. Waters, B.: Efficient identity-based encryption without random oracles. In: Cramer, R. (ed.) EUROCRYPT 2005. LNCS, vol. 3494, pp. 114–127. Springer, Heidelberg (2005)
14. Zhou, Y., Cao, Z., Chai, Z.: Constructing secure proxy cryptosystem. In: Feng, D., Lin, D., Yung, M. (eds.) CISC 2005. LNCS, vol. 3822, pp. 150–161. Springer, Heidelberg (2005)

A   Proof of Theorem 1

We show that if an adversary $\mathcal{A}$ can break our CBPd scheme with non-negligible probability, then we have another algorithm $\mathcal{B}$ that, using $\mathcal{A}$, can solve the BDH problem with non-negligible probability. That is, given $\langle \mathbb{G}_1, \mathbb{G}_2, p, \hat{e}; g, g^a, g^b, g^c \in \mathbb{G}_1 \rangle$, $\mathcal{B}$ aims to output $\hat{e}(g,g)^{abc} \in \mathbb{G}_2$. We denote the challenge ciphertext as $(C_1^*, C_2^*, C_3^*, C_4^*, C_5^*)$ for (τ^*, job^*).

Setup. $\mathcal{B}$ generates master public parameters $\langle \mathbb{G}_1, \mathbb{G}_2, p, \hat{e}, g, g_{pub}, F, H_i(i \in \{0,1,2,3,4,5\}) \rangle$ where $g_{pub} = g^{sc}$. And set $pk_B = g^a$, $H_1(ID_B, pk_B) = g^b$, $C_3^* = g^c$. Then the game between $\mathcal{A}$ and $\mathcal{B}$ is as follows.

Phase 1. $\mathcal{B}$ builds the following oracles.

$\mathcal{O}_F$: On input R_i, if there is a tuple (R_i, R_i') in table T_F, then return R_i'; otherwise, choose a random number R_i', return it and record (R_i, R_i') in table T_F.

$\mathcal{O}_{H_0}$: On input (g_{pub}, ID_i, pk_i), if there is a tuple $(g_{pub}, ID_i, pk_i, r_i^{(0)})$ in table T_{H_0}, then return $r_i^{(0)}$; otherwise, choose a random number $r_i^{(0)} \in \mathbb{G}_1$, return it and record $(g_{pub}, ID_U, pk_U, r_i^{(0)})$ in table T_{H_0}.

$\mathcal{O}_{H_1}$: On input (ID_i, pk_i), if there is a tuple $(ID_i, pk_i, r_i^{(1)})$ in table T_{H_1}, then return $r_i^{(1)}$; otherwise, choose a random number $r_i^{(1)} \in \mathbb{G}_1$, return it and record $(ID_i, pk_i, r_i^{(1)})$ in table T_{H_1}. Note that, at the beginning of game, $\mathcal{B}$ should set $H_1(ID_B, pk_B) = g^a$, and record (ID_B, pk_B, g^a) into table T_{H_1}.

$\mathcal{O}_{H_2}$: On input job_i, if there is a tuple $(job_i, r_i^{(2)})$ in table T_{H_2}, then return $H_1(ID_B, pk_B)^{r_i^{(2)}}$; otherwise, choose a random number $r_i^{(2)} \in \mathbb{Z}_p^*$, return $H_1(ID_B, pk_B)^{r_i^{(2)}}$ and record $(job_i, r_i^{(2)})$ in table T_{H_2}. Note that, at the beginning of game, $\mathcal{B}$ should guess which query will be job^* (the probability of correctness is $1/q_{H_2}$, where q_{H_2} is the total number of queries to H_3), and set $H_2(job^*) = g^{a^*}$, and record (job^*, a^*) into table T_{H_2}.

$\mathcal{O}_{H_3}$: On input τ_i, if there is a tuple $(\tau_i, r_i^{(3)})$ in table T_{H_3}, then return $H_1(ID_B, pk_B)^{r_i^{(3)}}$; otherwise, choose a random number $r_i^{(3)} \in \mathbb{Z}_p^*$, return $H_1(ID_B,$

$pk_B)^{r_i^{(3)}}$, and record $(\tau_i, r_i^{(3)})$ in table T_{H_3}. Note that, at the beginning of the game, $\mathcal{B}$ should guess which query will be τ^* (the probability of correctness is $1/q_{H_3}$, where q_{H_3} is the total number of queries to H_3), set $H_3(\tau^*) = g^{\beta^*}$, and record (τ^*, β^*) into table T_{H_3}.

$\mathcal{O}_{H_4}$: On input (m_i, σ_i), if there is a tuple $(m_i, \sigma_i, r_i^{(4)})$ in table T_{H_4}, then return $r_i^{(4)}$; otherwise, choose a random number $r_i^{(4)} \in \mathbb{Z}_p^*$, return it, and record $(m_i, \sigma_i, r_i^{(4)})$ into table T_{H_4}.

$\mathcal{O}_{H_5}$: On input σ_i, if there is a tuple $(\sigma_i, r_i^{(5)})$ in table T_{H_5}, then return $r_i^{(5)}$; otherwise, choose a random number $r_i^{(5)} \in \{0, 1\}$, return it, and record $(\sigma_i, r_i^{(5)})$ into table T_{H_5}.

Certificate query: This is the same as real execution because $\mathcal{B}$ knows the value of s_C.

Proxy decryption key query: On input (τ, job_c), $\mathcal{B}$ searches whether $(\tau, job_c, \bigstar_1, \bigstar_2)$ exists in table T_{key}. If it does, return $(\bigstar_1, \bigstar_2)$; otherwise, $\mathcal{B}$ does the following.

Case 1. $\tau \neq \tau^*$: Search whether $(\tau, \bigstar_1, \bigstar_2, \bigstar_3)$ exits in table T_{com} (see the following common data query).

- If it is there, set $sk_1^{job_c} = H_0(g_{pub}, ID_B, pk_B)^{s_C} H_3(\tau)^{\bigstar_1} \cdot H_2(job_c)^{d_{job_c}} / \bigstar_2$, $sk_2^{job_c} = g^{d_{job_c}}$, where d_{job_c} is a random number from Z_q^*. Finally, record $(\tau, job_c, sk_1^{job_c}, sk_2^{job_c})$ into table T_{key}.

- If it does not exist, choose two random numbers $d \in Z_q^*$, $y \in \mathbb{G}_1$. Set $sk_1^{job_c} = H_0(g_{pub}, ID_B, pk_B)^{s_C} H_3(\tau)^d \cdot H_2(job_c)^{d_{job_c}} / y$, $sk_2^{job_c} = g^{d_{job_c}}$, where d_{job} is a random number from Z_q^*. Finally, record $(\tau, job_c, sk_1^{job_c}, sk_2^{job_c})$ into table T_{key}, and $(\tau, d, y, pk_B^{-1/r^{(3)}} g^d)$ into table T_{com}, where $H_3(\tau) = H_1(ID_B, pk_B)^{r^{(3)}}$.

Case 2. $\tau = \tau^*$ and $job_c \neq job^*$: Search whether $(\tau, \bigstar_1, \bigstar_2, \bigstar_3)$ exists in table T_{com}.

- If it exists, set $sk_1^{job_c} = H_0(g_{pub}, ID_B, pk_B)^{s_C} H_3(\tau)^{\bigstar_1} \cdot H_2(job_c)^d / \bigstar_2$, $sk_2^{job_c} = pk_B^{-1/r^{(2)}} g^d$, where d is a random number from Z_q^* and $H_2(job_c) = H_1(ID_B, pk_B)^{r^{(2)}}$. Finally, record $(\tau, job_c, sk_1^{job_c}, sk_2^{job_c})$ into table T_{key}.

- If it does not exist, choose two random numbers $d \in Z_q^*$, $y \in \mathbb{G}_1$. Set $sk_1^{job_c} = H_0(g_{pub}, ID_B, pk_B)^{s_C} H_3(\tau)^d \cdot H_2(job_c)^{d'} / y$, $sk_2^{job_c} = pk_B^{-1/r^{(2)}} g^d$, where d' is a random number from Z_q^* and $H_2(job_c) = H_1(ID_B, pk_B)^{r^{(2)}}$. Finally, record $(\tau, job_c, sk_1^{job_c}, sk_2^{job_c})$ into table T_{key}, and (τ, d, y, g^d) into table T_{com}.

Note that,

$$H_0(g_{pub}, ID_B, pk_B)^{s_C} (H_1(ID_B, pk_B)^r)^d$$
$$= H_0(g_{pub}, ID_B, pk_B)^{s_C}$$

$$\cdot H_1(ID_B, pk_B)^{s_B}(H_1(ID_B, pk_B)^r)^{\frac{-s_B}{r}} \cdot (H_1(ID_B, pk_B)^r)^d$$
$$= H_0(g_{pub}, ID_B, pk_B)^{s_C} H_1(ID_B, pk_B)^{s_B} \cdot (H_1(ID_B, pk_B)^r)^{d-\frac{s_B}{r}}$$
$$= S_B H_2(job_c)^{d-\frac{s_B}{r}},$$

and

$$pk_B^{-1/r} g^d = (g^{s_B})^{\frac{-1}{r}} g^d = g^{d-\frac{s_B}{r}}.$$

Hence, we have the following equations for item $(\tau, job_c, \square_1, \square_2)$ in table T_{key} and item $(\tau, \blacksquare_1, \blacksquare_2, \blacksquare_3)$ in table T_{com}.

$$\square_1 \cdot \blacksquare_2 = S_B H_2(job_c)^{d_{job_c}} H_3(\tau)^{d_\tau},$$
$$\square_2 = g^{d_{job_c}}, \quad \blacksquare_3 = g^{d_\tau}.$$

Common data query: On input τ, $\mathcal{B}$ searches whether $(\tau, \star_1, \star_2, \star_3)$ exists in table T_{com}. If it does, return $(\star_2, \star_3)$, Otherwise, $\mathcal{B}$ chooses two random numbers $d \in Z_q^*$ and $y \in \mathbb{G}_1$.

- If $\tau = \tau^*$, return (y, g^d) and record (τ, d, y, g^d) into table T_{com}.
- If $\tau \neq \tau^*$, return $(y, pk_B^{-1/r^{(3)}} g^d)$ and record $(\tau, d, y, pk_B^{-1/r^{(3)}} g^d)$ into table T_{com}, where $H_3(\tau) = H_1(ID_B, pk_B)^{r^{(3)}}$.

Basic decryption oracle query: On input (C_1, C_2, C_3), $\mathcal{B}$ does the following.

1. Set two empty sets S_1 and S_2.
2. Find $(m_i, \sigma_i, r_i^{(4)})$ in table T_{H_4}, such that $C_3 = g^{r_i^{(4)}}$ and put the triples into set S_1. If S_1 is empty, abort. This step makes this oracle distinguishable in real execution when the adversary can guess the correct value of $H_4(m_i \| \sigma_i)$ without querying $\mathcal{O}_{H_4}$. The probability of this event is q_{H_4}/p, where q_{H_4} is the number of queries to $\mathcal{O}_{H_4}$.
3. Find $(\sigma_j, r_j^{(5)})$ in table T_{H_5}, such that $\sigma_j = \sigma_i$, where $(m_i, \sigma_i, r_i^{(4)})$ is the item in set S_1. Put $((m_i, \sigma_i, r_i^{(4)}), (\sigma_j, r_j^{(5)}))$ into set S_2. If S_2 is empty, abort. This step makes this oracle distinguishable in real execution when the adversary can guess the correct value of $H_5(\sigma_j)$ without querying $\mathcal{O}_{H_5}$. The probability of this event is $q_{H_5}/2^{l_\mathcal{M}}$, where q_{H_5} is the number of queries to $\mathcal{O}_{H_5}$.
4. Find the tuple $((m_i, \sigma_i, r_i^{(4)}), (\sigma_j, r_j^{(5)}))$ in set S_2, such that $C_1 = m_i \oplus r_j^{(5)}$, and $C_2 = \sigma_i \oplus F(\eta^{r_i^{(4)}})$. If the ciphertext is valid, then there is only one such tuple. This step makes this oracle distinguishable in real execution when the adversary can guess the correct value of $F(\eta^{r_i^{(4)}})$ without querying $\mathcal{O}_F$. The probability of this event is $q_F/2^{l_\mathcal{M}}$, where q_F is the number of queries to $\mathcal{O}_F$.
5. Output m_i.

As a result, this oracle is indistinguishable in real execution with $(1 - q_{H_4}/p)(1 - q_{H_5}/2^{l_\mathcal{M}})(1 - q_F/2^{l_\mathcal{M}})$.

Proxy decryption oracle query: On input $(C_1, C_2, C_3, C_4, C_5, job, \tau)$, $\mathcal{B}$ does the following

1. Set two empty sets S_1 and S_2.
2. Find $(m_i, \sigma_i, r_i^{(4)})$ in table T_{H_4}, such that $C_3 = g^{r_i^{(4)}}$, $C_4 = H_2(job)^{r_i^{(4)}}$, $C_5 = H_3(\tau)^{r_i^{(4)}}$, and put the triples into set S_1. If S_1 is empty, abort.
3. Find $(\sigma_j, r_j^{(5)})$ in table T_{H_5}, such that $\sigma_j = \sigma_i$, where $(m_i, \sigma_i, r_i^{(4)})$ is the item in set S_1, and put $((m_i, \sigma_i, r_i^{(4)}), (\sigma_j, r_j^{(5)}))$ into set S_2. If S_2 is empty, abort.
4. Find the tuple $((m_i, \sigma_i, r_i^{(4)}), (\sigma_j, r_j^{(5)}))$ in set S_2, such that $C_1 = m_i \oplus r_j^{(5)}$, and $C_2 = \sigma_i \oplus F(\eta^{r_i^{(4)}})$.
5. Output m_i.

The analysis of probability for indistinguishability between this oracle and real execution is the same as for the above oracle and the probability is also $(1 - q_{H_4}/p)(1 - q_{H_5}/2^{l_M})(1 - q_F/2^{l_M})$.

Challenge. The adversary submits a target subject, job^*, and two messages (m_0, m_1) in time τ^*. The adversary's choice of job^* is restricted to the subjects that he did not request a proxy key for in time τ^* in Phase 1. The challenger flips a fair binary coin, γ, and returns an encryption of m_γ for subject job^* on τ^*. $C_1^* = m_\gamma \oplus H_5(\sigma^*)$, C_2^*, $C_3^* = g^c$, $C_4^* = H_2(job^*)^c = (g^c)^{\alpha^*}$, $C_5^* = H_3(\tau^*)^c = (y^c)^{\beta^*}$, where o^* and C^* are two random numbers from $\{0,1\}^{l_M}$.

Phase 2. Phase 1 is repeated with the restriction that
- the adversary cannot request the proxy secret key oracle query on (τ^*, job^*);
- the adversary cannot request the basic decryption oracle query on c^*;
- the adversary cannot request the proxy decryption oracle query on (c^*, τ^*, job^*);
- the adversary cannot request the revocation oracle query on (τ^*, job^*).

Guess. The adversary submits a guess, γ', of γ. $\mathcal{B}$ outputs $\frac{R_i}{\hat{e}(H_0(g_{pub}, ID_B, pk_B)^{sC}, g^c)}$ $= e(g, g)^{abc}$ as the BDH solution, where R_i is a random number chosen from table T_F.

If the probability of $\gamma' = \gamma$ is $1/2 + \epsilon$ and the game is completed, then $\frac{R_i}{\hat{e}(H_0(g_{pub}, ID_B, pk_B)^{sC}, g^c)}$ is the correct BDH solution with probability ϵ/q_F, where q_F is the total number of queries to F. However,
- $\mathcal{B}$ guesses the correct target τ^* and job^* with probability $1/q_{H_2}$ and $1/q_{H_3}$, respectively.
- certificate query oracle, proxy decryption key query oracle, and common data query oracle can always succeed.
- basic decryption query oracle and proxy decryption query oracle can both succeed with probability $(1 - q_{H_4}/p)(1 - q_{H_5}/2^{l_M})(1 - q_F/2^{l_M})$.

Hence, if an adversary $\mathcal{A}$ can break our scheme with advantage ϵ, then we have another algorithm $\mathcal{B}$ solves BDH problem with probability $\frac{\epsilon}{q_{H_2} q_{H_3} q_F}(1 - q_{H_4}/p)^2(1 - q_{H_5}/2^{l_M})^2(1 - q_F/2^{l_M})^2$, which is non-negligible. $\qquad \square$

Computationally-Efficient Password Authenticated Key Exchange Based on Quadratic Residues

Muxiang Zhang

Verizon Communications Inc.
40 Sylvan Road, Waltham, MA 02451, USA
muxiang.zhang@verizon.com

Abstract. In this paper, we present a computationally efficient password authenticated key exchange protocol based on quadratic residues. The protocol, called $QR\text{-}CEKE$, is derived from the protocol $QR\text{-}EKE$, a previously published password authenticated key exchange protocol based on quadratic residues. The computational time for the client, however, is significant reduced in the protocol $QR\text{-}CEKE$. In comparison with $QR\text{-}EKE$, the protocol $QR\text{-}CEKE$ is more suitable to an imbalanced computing environment where a low-end client device communicates with a powerful server over a broadband network. Based on number-theoretic techniques, we show that the computationally efficient password authenticated key exchange protocol is secure against *residue attacks*, a special type of off-line dictionary attack against password-authenticated key exchange protocols based on factorization. We also provide a formal security analysis of $QR\text{-}CEKE$ under the factoring assumption and the random oracle model.

1 Introduction

Password-authenticated key exchange protocols allow two entities who share a small password to authenticate each other and agree on a large session key between them. Such protocols are attractive for their simplicity and convenience and have received much interest in the research community. A major challenge in designing password-authenticated key exchange protocols is to deal with the so-called exhaustive guessing or off-line dictionary attack, as passwords are generally drawn from a small space enumerable, *off-line*, by an adversary. In 1992, Bellovin and Merritt [3] presented a family of protocols, known as *Encrypted Key exchange* (EKE), which was shown to be secure against off-line dictionary attack. Following EKE, a number of protocols for password-based authentication and key exchange have been proposed; a comprehensive list of such protocols can be found in Jablon's research link [6]. Over the last decade, many researchers have investigated the feasibility of implementing EKE using different types of public-key cryptosystems such as RSA, ElGamal, and Diffie-Hellman key exchange. Nonetheless, most of the well-known and secure variants of EKE

K. Srinathan, C. Pandu Rangan, M. Yung (Eds.): Indocrypt 2007, LNCS 4859, pp. 312–321, 2007.
© Springer-Verlag Berlin Heidelberg 2007

are based on Diffie-Hellman key exchange. It seems that EKE works well with Diffie-Hellman key exchange, but presents subtleties one way or the other when implemented with RSA and other public-key cryptographic systems. In their original paper [3], Bellovin and Merritt pointed out that the RSA-based EKE variant is subject to a special type of dictionary attack, called residue attack. In 1997, Lucks [7] proposed an RSA-based password-authenticated key exchange protocol (called OKE) which was claimed to be secure against residue attacks. Later, Mackenzie et al. [8] found that the OKE protocol is still subject to residue attacks. In [8], Mackenzie et al. proposed an RSA-based EKE variant (called SNAPI) and provided a formal security proof in the random oracle model. Unfortunately, the SNAPI protocol has to use a prime public exponent e which is larger than the RSA modulus n. This renders the SNAPI protocol impractical in resource-limited platforms. To avoid using large public exponents, Zhu et al. [28] proposed an "interactive" protocol which is revised from an idea of [3]. The interactive protocol requires a large communication overhead in order to verify the RSA public key. Bao [1] and Zhang [14] also pointed out some weaknesses of Zhu et al.'s password-authenticated key exchange protocol. In 2004, Zhang [12] presented an RSA-based password authenticated key exchange protocol (called $PEKEP$) which can use both large and small primes as RSA public exponents, but without inducing large communication overhead on communication entities. Alternatively, Zhang [13] also presented a password authenticated key exchange protocol (called $QR\text{-}EKE$) based on quadratic resides.

In comparison with the RSA based password authenticated key exchange protocol $PEKEP$, the quadratic residue based password authenticated key exchange protocol $QR\text{-}EKE$ has a merit that one of the entities (i.e., client) does not need to perform primality test. In the protocol $QR\text{-}EKE$, the computational time for both entities, i.e., client and server, is $\mathcal{O}(\log_2 n)^3$. In many applications, however, it is highly desirable that the computational time for a low-end client device be much less than $\mathcal{O}(\log_2 n)^3$, in order to support resource-limited computing platforms, such as mobile phones and personal digital assistants. To reduce the computational burden on client devices, we present a computationally efficient password authenticated key exchange protocol in this paper. The protocol, called $QR\text{-}CEKE$, is derived from $QR\text{-}EKE$ by adding two additional flows between the client and the server. The two additional flows increase the communication overhead by $\log_2 n + 2k$ bits, where k is the security parameter (e.g., $k = 160$). With the two additional flows, we show that the probability for an adversary to launch a successful residue attack against $QR\text{-}CEKE$ is less than or equal to 2ε, where ε is a small number (e.g., $0 < \varepsilon \leq 2^{-80}$) selected by the client. In the protocol $QR\text{-}CEKE$, the computational time for the client is $\mathcal{O}(\log_2 \varepsilon^{-1}(\log_2 n)^2)$, which is much less than $\mathcal{O}((\log_2 n)^3)$. When $\varepsilon = 2^{-80}$, $\log_2 n = 1024$, $k = 160$, for example, the computational time for the client in the protocol $QR\text{-}CEKE$ is about 13.5 times less than that in the protocol $QR\text{-}EKE$, while the communication overhead in $QR\text{-}CEKE$ is just about 1 mini-second (1 ms) more than that in $QR\text{-}EKE$ when both protocols are running in a communication network of 1 megabits per second (1 mbps) bandwidth. Hence, the protocol $QR\text{-}CEKE$ is more suitable to an imbalanced computing environment where a low-end client device communicates with a powerful server over a broadband network. Under the factoring assumption and the random oracle model, we also provide a formal security analysis of the $QR\text{-}CEKE$ protocol.

2   Security Model

Our formal model of security for password-authenticated key exchange protocols is based on that of [2].

INITIALIZATION. Let I denote the identities of the protocol participants. Elements of I will often be denoted A and B (Alice and Bob). Each pair of entities, $A, B \in I$, are assigned a password w which is randomly selected from the password space $\mathcal{D}$. The initialization process may also specify a set of cryptographic functions (e.g., hash functions) and establish a number of cryptographic parameters.

RUNNING THE PROTOCOL. Mathematically, a protocol Π is a probabilistic polynomial-time algorithm which determines how entities behave in response to received message. For each entity, there may be multiple instances running the protocol in parallel. We denote the i-th instance of entity A as Π_A^i. The adversary $\mathcal{A}$ can make queries to any instance; she has an endless supply of Π_A^i oracles ($A \in I$ and $i \in \mathbb{N}$). In response to each query, an instance updates its internal state and gives its output to the adversary. At any point in time, the instance may accept and possesses a session key sk, a session id sid, and a partner id pid. The query types, as defined in [2], include:

- Send(A, i, M): This sends message M to instance Π_A^i. The instance executes as specified by the protocol and sends back its response to the adversary.
- Execute(A, i, B, j): This call carries out an honest execution between two instances Π_A^i and Π_B^j, where $A, B \in I, A \neq B$ and instances Π_A^i and Π_B^j were not used before. At the end of the execution, a transcript is given to the adversary, which logs everything an adversary could see during the execution (for details, see [2]).
- Reveal(A, i): The session key sk_A^i of Π_A^i is given to the adversary.
- Test(A, i): The instance Π_A^i generates a random bit b and outputs its session key sk_A^i to the adversary if $b = 1$, or else a random session key if $b = 0$. This query is allowed only once, at any time during the adversary's execution.
- Oracle(M): This gives the adversary oracle access to a function h, which is selected at random from some probability space Ω. The choice of Ω determines whether we are working in the standard model, or in the random-oracle model (see [2] for further explanations).

Let Π_A^i and Π_B^i, $A \neq B$, be a pair of instances. We say that Π_A^i and Π_B^i are *partnered* if both instances have accepted and hold the same session id sid and the same session key sk. Here, we define the sid of Π_A^i (or Π_B^i) as the concatenation of all the messages sent and received by Π_A^i (or Π_B^i). We say that

Π_A^i is *fresh* if: i) it has accepted; and ii) a Reveal query has not been called either on Π_A^i or on its partner (if there is one). With these notions, we now define the advantage of the adversary $\mathcal{A}$ in attacking the protocol. Let Succ denote the event that $\mathcal{A}$ asks a single Test query on a fresh instance, outputs a bit b', and $b' = b$, where b is the bit selected during the Test query. The advantage of the adversary $\mathcal{A}$ is defined as $\mathsf{Adv}_{\mathcal{A}}^{ake} = 2Pr(\mathsf{Succ}) - 1$.

As suggested in [5], we use the Send query type to count the number of on-line guesses performed by the adversary. We only count *one* Send query for each entity instance, that is, if the adversary sends two Send queries to an entity instance, it should still count as a single password guess. Based on this idea, we have the following definition of secure password-authenticated key exchange protocol, which is the same as in [5].

Definition 1. *A protocol Π is called a secure password-authenticated key exchange protocol if for every polynomial-time adversary $\mathcal{A}$ that makes at most Q_{send} ($Q_{send} \leq |\mathcal{D}|$) queries of Send type to different instances, the following two conditions are satisfied:*

(1) *Except with negligible probability, each oracle call* Execute(A, i, B, j) *produces a pair of partnered instances Π_A^i and Π_B^j.*
(2) $\mathsf{Adv}_{\mathcal{A}}^{ake} \leq Q_{send}/|\mathcal{D}| + \epsilon$, *where $|\mathcal{D}|$ denotes the size of the password space and ϵ is a negligible function of security parameters.*

3   Computationally-Efficient Password Authenticated Key Exchange Based on Quadratic Residues

Define hash functions $H_1, H_2, H_3 : \{0,1\}^* \rightarrow \{0,1\}^k$ and $H : \{0,1\}^* \rightarrow \mathbb{Z}_n$, where k is a security parameter, e.g., $k = 160$. The protocol $QR\text{-}CEKE$, which is described in Fig. 1, is based on $QR\text{-}EKE$, but the number of squaring operations performed by Bob is much less than $\lfloor \log_2 n \rfloor$. In the protocol $QR\text{-}CEKE$, Bob selects a small number ε, $0 < \varepsilon \leq 2^{-80}$, which determines the probability of a successful residue attack against the protocol $QR\text{-}CEKE$. Alice starts the protocol $QR\text{-}CEKE$ by sending a Blum integer n and two random numbers $\rho, r_A \in_R \{0,1\}^k$ to Bob. Bob verifies if n is an odd integer. If n is not odd, Bob rejects. Else, Bob computers $m = \lceil \log_2 \varepsilon^{-1} \rceil$. Then Bob selects a random number $\varrho \in_R \{0,1\}^k$ such that $\gamma = H(n, e, \rho, \varrho, A, B, m)$ satisfying $\gcd(\gamma, n) = 1$ and $\left(\frac{\gamma}{n}\right) = 1$. Bob sends ϱ and m to Alice. After receiving ϱ and m, Alice checks if $\gamma = H(n, e, \rho, \varrho, A, B, m)$ is a quadratic residue. If γ is not a quadratic residue, then $-\gamma$ must be a quadratic residue since n is a Blum integer. Next, Alice computes an integer u satisfying $u^{2^m} = \pm\gamma \mod n$, and sends u back to Bob. Subsequently, Bob verifies if Alice has made the right computation, i.e., $u^{2^m} = \pm\gamma \mod n$. If not, Bob rejects. Else, Alice and Bob executes the rest of the protocol as in $QR\text{-}EKE$.

Note that in the protocol $QR\text{-}CEKE$, Bob only verifies that the integer n received from Alice is an odd number; he does not verify that n is the product of two distinct primes p and q and $p \equiv q \equiv 3 \pmod 4$. This may foster the

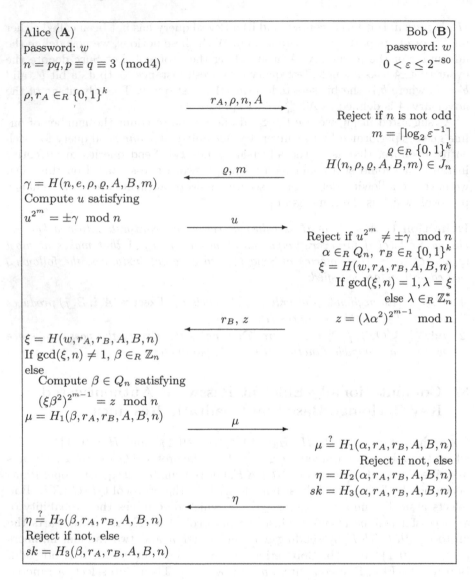

Fig. 1. The Protocol QR-CEKE

so-called residue attack as described in [3]. In such an attack, an adversary, say, *Eva*, selects a password π_0 at random from $\mathcal{D}$ and an odd integer n which may not necessarily be a Blum integer. Then Eva impersonates as Alice and starts the protocol by sending r_E, n, A to Bob. After receiving r_B and z from Bob, Eva Computes μ and sends it back to Bob. If Bob accepts, then Eva has a successful guess of Alice's password. If Bob rejects, on the other hand, Eva excludes her guess (i.e., π_0) from the password space $\mathcal{D}$. Furthermore, Eva may exclude more passwords by repeating, *off-line*, the following three steps:

1) Eva selects a password π from $\mathcal{D}$.
2) Eva computes $\gamma = H(\pi, r_E, r_B, A, B, n)$.
3) Eva tests if $\gcd(\gamma, n) = 1$. If not, Eva returns to step 1; otherwise, Bob verifies if the congruence $(\gamma x^2)^{2^t} \equiv z \pmod{n}$ has a solution in Q_n. If the congruence has a solution, Eva returns to step 1. If the congruence has no solution in Q_n, then Eva is ensured that π is not the password of Alice. Next Eva excludes π from $\mathcal{D}$ and returns to step 1.

Theorem 1. *Let n, $n > 1$, be an odd integer with prime-power factorization $n = p_1^{a_1} p_2^{a_2} \dots p_r^{a_r}$. Let m be a positive integer. If there exists a prime power, say $p_i^{a_i}$, of the factorization of n such that $2^m \mid \phi(p_i^{a_i})$, then for an integer γ randomly selected from J_n, the probability that γ is an 2^m-th power residue of n is less than or equal to 2^{-m+1}.*

Proof. Let $n_i = p_i^{a_i}$ be a prime power of the factorization of n such that $2^m \mid \phi(n_i)$. Since n is odd, n_i possesses a primitive root. Let g be a primitive root of n_i. For an integer γ randomly selected from J_n, let $\text{ind}_g \gamma$ denote the index of γ to the base g modulo n_i. Then γ is an 2^m-th power residue of n_i if and only if the congruence $x^{2^m} \equiv \gamma \pmod{n_i}$ has a solution, or equivalently, if and only if

$$g^{2^m \text{ind}_g x - \text{ind}_g \gamma} \equiv 1 \pmod{n_i},$$

which is equivalent to

$$2^m \text{ind}_g x \equiv \text{ind}_g \gamma \pmod{\phi(n_i)}.$$

Since $2^m \mid \phi(n_i)$, γ is an 2^m-th power residue of n_i if and only if $2^m \mid \text{ind}_g \gamma$.

Let $n_i' = n/n_i$, then n_i and n_i' are relatively prime. For any integer $\beta \in \mathbb{Z}_n^*$, it is clear that $\beta \bmod n_i$ and $\beta \bmod n_i'$ are integers of $\mathbb{Z}_{n_i}^*$ and $\mathbb{Z}_{n_i'}^*$, respectively. On the other hand, for two integers $\alpha_1 \in \mathbb{Z}_{n_i}^*$ and $\alpha_2 \in \mathbb{Z}_{n_i'}^*$, by the Chinese Remainder Theorem, there is an unique integer $\alpha \in \mathbb{Z}_n^*$, such that $\alpha \equiv \alpha_1 \pmod{n_i}$, and $\alpha \equiv \alpha_2 \pmod{n_i'}$. So, the number of integers $\alpha \in \mathbb{Z}_n^*$ which satisfy the congruence $\alpha \equiv \alpha_1 \pmod{n_i}$ is $\phi(n_i')$. If γ is randomly selected from J_n, then for any integer s, $0 \leq s \leq \phi(n_i) - 1$, we have

$$Pr(g^s = \gamma \bmod n_i) \leq \frac{\phi(n_i')}{|J_n|} \leq \frac{2}{\phi(n_i)}.$$

Note that in last inequality described above, we make use of the fact $|J_n| \geq \phi(n)/2$. Thus, we have $Pr(\text{ind}_g \gamma = s) \leq 2/\phi(n_i)$. Therefore,

$$Pr(2^m \mid \text{ind}_g \gamma) = \sum_{2^m \mid s,\, 0 \leq s < \phi(n_i)} Pr(\text{ind}_g \gamma = s)$$

$$\leq 2\phi(n_i) 2^{-m} / \phi(n_i)$$

$$= 2^{-m+1}$$

which indicates that, for an integer γ randomly selected from J_n, the probability that γ is an 2^m-th power residue of n_i is less than or equal to 2^{-m+1}. So, the probability that γ is an 2^m-th power residue of n does not exceed 2^{-m+1}. $\square$

Theorem 1 demonstrates that, if there exits a prime-power $p_i^{a_i}$ of the factorization of n such that $2^m \mid \phi(p_i^{a_i})$, then for a random number $\gamma \in J_n$, the probability that Alice can take square roots of γ or $-\gamma$ repetitively m times is less than or equal to 2^{-m+1}.

Theorem 2. *Let n, $n > 1$, be an odd integer with prime-power factorization $n = p_1^{a_1} p_2^{a_2} \ldots p_r^{a_r}$. Let m be a positive integer such that for any prime-power $p_i^{a_i}$ of the factorization of n, $2^{m+1} \nmid \phi(p_i^{a_i}), 1 \leq i \leq r$. If z is an 2^m-th power residue of n, then for any $\lambda \in \mathbb{Z}_n^*$, the congruence $(\lambda x^2)^{2^m} \equiv z \pmod{n}$ has a solution in Q_n.*

Proof. To prove that $(\lambda x^2)^{2^m} \equiv z \pmod{n}$ has a solution in Q_n, we only need to prove that, for each prime power $p_i^{a_i}$ of the factorization of n, the following congruence

$$(\lambda x^2)^{2^m} \equiv z \pmod{p_i^{a_i}} \tag{1}$$

has a solution in $Q_{p_i^{a_i}}$.

Let $n_i = p_i^{a_i}, 1 \leq i \leq r$. Then $\phi(n_i) = p_i^{a_i-1}(p_i - 1)$. Since n is odd, p_i is an odd prime. Thus, the integer n_i possesses a primitive root. Let g be a primitive root of n_i, that is, $g^{\phi(n_i)} = 1 \bmod n_i$, and for any $0 \leq i, j \leq \phi(n_i) - 1, i \neq j$, $g^i \neq g^j \bmod n_i$. Let $\gcd(2^m, \phi(n_i)) = 2^c, 1 \leq c \leq m$. We consider the following two cases:

(1) If $c = 1$, then $d = \phi(n_i)/2$ must be an odd integer. For any integer $a \in \mathbb{Z}_{n_i}^*$, $a^{2d} \equiv 1 \pmod{n_i}$, which implies that $a^d \equiv 1$ or $-1 \pmod{n_i}$. We claim that $a^d \equiv -1 \pmod{n_i}$ if and only if a is a quadratic non-residue of n_i. If $a^d \equiv -1 \pmod{n_i}$, it is obvious that $a \in \bar{Q}_{n_i}$. On the other hand, if $a \in \bar{Q}_{n_i}$, then there exists an odd integer s such that $a = g^s \bmod n_i$ since g is the primitive root of n_i. As the order of g is $2d$, not d, we have $a^d = (g^d)^s = -1 \bmod n_i$. Similarly, we can also prove that, if $a \in Q_{n_i}$, then the congruence $x^2 \equiv a \pmod{n_i}$ has two solutions, with one solution in Q_{n_i} and another in $\bar{Q}_{n_i}$. Hence, for any $\gamma \in \mathbb{Z}_n^*$, there exists a solution x_j, $0 \leq j \leq 1$, such that $x_j\gamma \in Q_{n_i}$, that is, $(x_j\gamma)^d = 1 \pmod{n_i}$. Hence, congruence (3) has a solution in Q_{n_i}.

(2) Next, we consider the case that $2 \leq c \leq m$. Since z is a 2^m-th power residue modulo n, the congruence $x^{2^m} \equiv z \pmod{n}$ has solutions in $\mathbb{Z}_n^*$. By the Chinese Remainder Theorem, the following congruence

$$y^{2^m} \equiv z \pmod{n_i} \tag{2}$$

has solutions in $\mathbb{Z}_{n_i}^*$. Let $\text{ind}_g z$ denote the index of z to the base g modulo n_i and let $y \in \mathbb{Z}_{n_i}^*$ be a solution of (4). Then, $g^{2^m \text{ind}_g y - \text{ind}_g z} \equiv 1 \pmod{n_i}$. Since the order of g modulo n_i is $\phi(n_i)$, it follows that

$$2^m \text{ind}_g y \equiv \text{ind}_g z \pmod{\phi(n_i)} \tag{3}$$

Also since $\gcd(2^m, \phi(n_i)) = 2^c$, equation (5) has exactly 2^c incongruent solutions modulo $\phi(n_i)$ when taking $\text{ind}_g y$ as variable. This indicates that equation (4) has exactly 2^c incongruent solutions modulo n_i. Let y_0 be one of the solutions of equation (4), then the 2^c incongruent solutions of (5) are given by

$$\text{ind}_g y = \text{ind}_g y_0 + j\phi(n_i)/2^c \mod \phi(n_i), \quad 0 \le j \le 2^c - 1.$$

For any $\gamma \in \mathbb{Z}_n^*$, we have

$$\text{ind}_g y - \text{ind}_g \gamma = \text{ind}_g y_0 - \text{ind}_g \gamma + j\phi(n_i)/2^c \mod \phi(n_i), \quad 0 \le j \le 2^c - 1.$$

Without loss of generality, let's assume that $\text{ind}_g y_0 - \text{ind}_g \gamma \ge 0$; otherwise we consider $\text{ind}_g \gamma - \text{ind}_g y$. Under the condition that $2^{m+1} \nmid \phi(n_i)$ it is clear that $\phi(n_i)/2^c$ is an odd integer. Hence, there exist an integer $j, 0 \le j \le 3 \le 2^c - 1$, such that

$$\text{ind}_g y_0 - \text{ind}_g \gamma + j\phi(n_i)/2^c \equiv 0 \ (\text{mod } 4),$$

which implies that there exists an integer $y \in \mathbb{Z}_{n_i}^*$ such that $y^{2^m} \equiv z \ (\text{mod } n_i)$ and $y\gamma^{-1}$ is a 4-th power residue of n_i. Therefore, the congruence (3) has a solution in Q_{n_i}, which proves the theorem. $\qquad\square$

Based on Theorem 1 and Theorem 2, we can conclude that the probability for an adversary to launch a successful residue attack against the protocol QR-$CEKE$ is less than or equal to 2ε.

In the protocol QR-$CEKE$, Alice proves to Bob in an interactive manner (via flow 2 and flow 3) that for every prime-power $p_i^{a_i}$ of the factorization of n, $2^m \nmid \phi(p_i^{a_i})$. The interactive procedure increases the communication overhead on Alice and Bob by $\log_2 n + 2k$ bits. When $\log_2 n = 1024$ and $k = 160$, for example, the communication overhead induced by the interactive procedure is about 1 mini-second (1 ms) over a broadband network of 1 megabits per second (1 mbps) bandwidth. In QR-$CEKE$, the computational burden on Bob includes two modulo exponentiations, i.e., $u^{2^m} \mod n$ and $(\lambda a^2)^{2^{m-1}} \mod n$, where $m = \lceil \log_2 \varepsilon^{-1} \rceil$. The computation time for the two modulo exponentiations is $\mathcal{O}((\log_2 \varepsilon^{-1})(\log_2 n)^2)$. When $\varepsilon^{-1} \ll n$, the computational load on Bob is greatly reduced in QR-$CEKE$ in comparison with that in QR-EKE (or in SNAPI).

4   Formal Security Analysis

In this section, we analyze the security of QR-$CEKE$ within the formal model of security given in Section 2. Our analysis is based on the random-oracle model. In this model, a hash function is modeled as an oracle which returns a random number for each new query. If the same query is asked twice, identical answers are returned by the oracle. In our analysis, we also assume the intractability of the Factoring problem.

Factoring Assumption: Let GE be a probabilistic polynomial-time algorithm that on input 1^ℓ returns a product of two distinct primes p and q of length $\ell/2$

satisfying $p \equiv q \equiv 3 \pmod 4$. For any probabilistic polynomial-time algorithm $\mathcal{C}$ of running time t, the following probability

$$\mathsf{Adv}_{\mathcal{C}}^{fac}(t) = Pr(\mathcal{C}(n) = (p,q), pq = n : n \leftarrow GE(1^\ell))$$

is negligible. In the following, we use $\mathsf{Adv}^{fac}(t)$ to denote $\max_{\mathcal{C}}\{\mathsf{Adv}_{\mathcal{C}}^{fac}(t)\}$, where the maximum is taken over all polynomial-time algorithms of running time t.

Under the above assumptions, we have the following Theorem 4.

Theorem 3. *Let $\mathcal{A}$ be an adversary which runs in time t and makes Q_{send}, $Q_{send} \leq |\mathcal{D}|$, queries of type Send to different instances. Then the adversary's advantage in attacking the protocol QR-CEKE is bounded by*

$$\mathsf{Adv}_{\mathcal{A}}^{ake} \leq \frac{Q_{send}}{|\mathcal{D}|} + 4\varepsilon + (Q_{execute} + 5Q_{send})\mathsf{Adv}^{fac}(\mathcal{O}(t)) + + \frac{Q_{send}}{2^{k-1}}$$
$$+ \frac{(Q_{execute} + 2Q_{send})Q_{oh}}{\phi(n)},$$

where $Q_{execute}$ denotes the number of queries of type Execute and Q_{oh} denotes the number of random oracle calls.

5   Conclusion

In this paper, we present a computationally efficient password authenticated key exchange protocol based on quadratic residues. The protocol QR-$CEKE$ is derived from the protocol QR-EKE, a previously published password authenticated key exchange protocol based on quadratic residues. However, the computational time for the client is significant reduced in the protocol QR-$CEKE$. In comparison with QR-EKE, the protocol QR-$CEKE$ is more suitable to an imbalanced computing environment where a low-end client device communicates with a powerful server over a broadband network. Based on number-theoretic techniques, we show that the computationally efficient password authenticated key exchange protocol is secure against *residue attacks*, a special type of off-line dictionary attack against password-authenticated key exchange protocols based on RSA and quadratic residues. We also provide a formal security analysis of QR-$CEKE$ under the factoring assumption and the random oracle model.

References

1. Bao, F.: Security analysis of a password authenticated key exchange protocol. In: Boyd, C., Mao, W. (eds.) ISC 2003. LNCS, vol. 2851, pp. 208–217. Springer, Heidelberg (2003)
2. Bellare, M., Pointcheval, D., Rogaway, P.: Authenticated key exchange secure against dictionary attack. In: Preneel, B. (ed.) EUROCRYPT 2000. LNCS, vol. 1807, pp. 139–155. Springer, Heidelberg (2000)

3. Bellovin, S.M., Merritt, M.: Encrypted key exchange: Password-based protocols secure against dictionary attacks. In: Bellovin, S.M., Merritt, M. (eds.) Proc. of the IEEE Symposium on Research in Security and Privacy, Oakland, pp. 72–84 (May 1992)
4. Catalano, D., Pointcheval, D., Pornin, T.: IPAKE: Isomorphisms for Password-based Authenticated Key Exchange. In: Franklin, M. (ed.) CRYPTO 2004. LNCS, vol. 3152, Springer, Heidelberg (to appear, 2004)
5. Gennaro, R., Lindell, Y.: A framework for password-based authenticated key exchange. In: Biham, E. (ed.) EUROCRPYT 2003. LNCS, vol. 2656, pp. 524–542. Springer, Heidelberg (2003)
6. Jablon, D.: http://www.integritysciences.com
7. Lucks, S.: Open key exchange: How to defeat dictionary attacks without encrypting public keys. In: Christianson, B., Lomas, M. (eds.) Proc. Security Protocol Workshop. LNCS, vol. 1361, pp. 79–90. Springer, Heidelberg (1997)
8. MacKenzie, P., Patel, S., Swaminathan, R.: Password-authenticated key exchange based on RSA. In: Okamoto, T. (ed.) ASIACRYPT 2000. LNCS, vol. 1976, pp. 599–613. Springer, Heidelberg (2000)
9. Menezes, A., van Oorschot, P.C., Vanstone, S.A.: Handbook of Applied Cryptography. CRC Press, Boca Raton (1997)
10. Patel, S.: Number theoretic attacks on secure password schemes. In: IEEE Symposium on Security and Privacy, Oakland, California (May 5-7, 1997)
11. Zhu, F., Wong, D., Chan, A., Ye, R.: RSA-based password authenticated key exchange for imbalanced wireless networks. In: Chan, A.H., Gligor, V.D. (eds.) ISC 2002. LNCS, vol. 2433, pp. 150–161. Springer, Heidelberg (2002)
12. Zhang, M.: New approaches to password authenticated key exchange based on RSA. In: Lee, P.J. (ed.) ASIACRYPT 2004. LNCS, vol. 3329, pp. 230–244. Springer, Heidelberg (2004)
13. Zhang, M.: Password Authenticated Key exchange using quadratic residues. In: Jakobsson, M., Yung, M., Zhou, J. (eds.) ACNS 2004. LNCS, vol. 3089, pp. 248–262. Springer, Heidelberg (2004)
14. Zhang, M.: Further analysis of password authenticated key exchange protocol based on RSA for imbalanced wireless networks. In: Zhang, K., Zheng, Y. (eds.) ISC 2004. LNCS, vol. 3225, pp. 12–24. Springer, Heidelberg (2004)

On the k-Operation Linear Complexity of Periodic Sequences

(Extended Abstract)

Ramakanth Kavuluru and Andrew Klapper

Department of Computer Science, University of Kentucky
Lexington, KY, 40506, USA
ramakanth.kavuluru@uky.edu, klapper@cs.uky.edu

Abstract. Non-trivial lower bounds on the linear complexity are derived for a sequence obtained by performing k or fewer operations on a single period of a periodic sequence over $\mathbb{F}_q$. An operation is a substitution, an insertion, or a deletion of a symbol. The bounds derived are similar to those previously established for either k substitutions, k insertions, or k deletions within a single period. The bounds are useful when $T/2k < L < T/k$, where L is the linear complexity of the original sequence and T is its period.

Keywords: Periodic sequence, linear complexity, k-error linear complexity, k symbol insertion, k symbol deletion.

1 Introduction

The linear complexity of a sequence is the length of the shortest linear feedback shift register (LFSR) that can generate the sequence and is an important measure of randomness of a sequence. Given at least the first $2L$ symbols of the sequence one can determine the LFSR that generates it using the Berlekamp-Massey algorithm, where L is the linear complexity of the sequence. Hence for cryptographic purposes sequences with high linear complexity are necessary. Otherwise with only a small initial segment an adversary can recover the LFSR and its initial state and hence the sequence.

Linear complexity might decrease drastically by altering a few symbols in the sequence. This instability can be measured using k-error linear complexity which, for a periodic sequence, is the smallest complexity value that can be obtained by changing k or fewer elements in a single period of the sequence. Counting functions and expected values for linear complexity and k-error linear complexity were extensively explored by Meidl and Niederreiter [2, 3].

Linear complexities of periodic sequences obtained by substituting a few symbols, inserting a few symbols, or deleting a few symbols were also determined [4, 5, 6, 7]. It is well accepted that a cryptographically strong sequence should have high linear complexity and that the linear complexity should not decrease considerably with substitutions, insertions, and deletions of a few symbols.

K. Srinathan, C. Pandu Rangan, M. Yung (Eds.): Indocrypt 2007, LNCS 4859, pp. 322–330, 2007.

Let $\{a_i\}$ be a periodic sequence over $\mathbb{F}_q$ with period T and let $LC(\{a_i\})$ denote the linear complexity of $\{a_i\}$. Let $(\{\hat{a}_i\})$ be a sequence obtained from $\{a_i\}$ by either substituting k symbols, inserting k symbols, or deleting k symbols within one period and periodically repeating the modified period.

Jiang, Dai, and Imamura [1] gave a proof that $LC(\{\hat{a}_i\}) \geq T/k - LC(\{a_i\})$ in each of the following three separate cases:

1. at most k substitutions are performed;
2. at most k insertions are performed; or
3. at most k deletions are performed.

Their analysis did not allow any combination of these operations.

Definition 1. The k-operation linear complexity of a periodic sequence $\{a_i\}$ is the smallest linear complexity obtained by performing any combination of up to k substitutions, insertions, and deletions in a single period of $\{a_i\}$ and then repeating the period.

In this paper we prove

1. Jiang, Dai, and Imamura's bound should be $\min(LC(\{a_i\}), T/k - LC(\{a_i\}))$.
2. This bound holds for *any combination* of up to k substitutions, insertions, and deletions. Thus we derive a lower bound on the k-operation linear complexity of a periodic sequence.

2   Preliminaries

Let $\mathbb{F}_q$ denote the finite field with q elements, where $q = p^r, r \geq 1$, and p is prime. Let $\{a_i\} = (a_0, a_1, \cdots)$ be a periodic sequence over $\mathbb{F}_q$ with period T. Let $a(x) = a_0 + a_1 x + \ldots + a_{T-1}x^{T-1}$. The sequence $\{a_i\}$ can be represented as the power series

$$\sum_{i \geq 0} a_i x^i = \frac{a(x)}{1 - x^T} = \frac{g(x)}{f(x)}, \quad \gcd(g(x), f(x)) = 1, \tag{1}$$

with $\deg(g(x)) < \deg(f(x))$. Then the linear complexity of $\{a_i\}$ is

$$LC(\{a_i\}) = \deg\left(\frac{1 - x^T}{\gcd(a(x), 1 - x^T)}\right) = \deg(f(x)). \tag{2}$$

We can see that $LC(\{a_i\}) \leq T$.

We use the following lemma to derive bounds for the linear complexity after k operation modification of a single period in the later sections. The proof is due to Jiang et al. [1].

Lemma 1. *Let* $C(x), D(x) \in \mathbb{F}_q[x]$ *with* $\deg(D(x)) < \deg(C(x))$ *and* $C(x) \neq 0$. *Define a periodic sequence* $\{c_i\}$ *over* $\mathbb{F}_q$ *by*

$$\sum_{i \geq 0} c_i x^i = \frac{D(x)}{C(x)}.$$

Define another sequence $\{\tilde{c}_i\}$ by

$$\sum_{i\geq 0} \tilde{c}_i x^i = \frac{[H(x)D(x)] \mod C(x)}{C(x)},$$

where $H(x) \in \mathbb{F}_q[x]$. Then

$$LC(\{\tilde{c}_i\}) \leq LC(\{c_i\}). \tag{3}$$

If $\gcd(C(x), H(x)) = 1$, then equality holds in equation (3).

We can apply Lemma 1 with $D(x) = a(x), C(x) = 1 - x^T$, and $H(x) = x^s$. Let $\{a_i\}$ be as in equation (1) and define $\{\tilde{a}_i\}$ by

$$\sum_{i\geq 0} \tilde{a}_i x^i = \frac{[x^s a(x)] \mod (1 - x^T)}{1 - x^T}.$$

We have, $\{\tilde{a}_i\} = \{a_{i-s}\}$ for $1 \leq s \leq T-1$. So using the fact that $\gcd(x^s, x^T - 1) = 1$, Lemma 1 implies that

$$LC(\{a_i\}) = LC(\{a_{i-s}\}), \quad 1 \leq s \leq T - 1. \tag{4}$$

3   Notation for k Operation Modification

In this section we describe k operation modification of a sequence and establish the notation we use in the next section to prove the main result of the paper. We ultimately want a bound that applies when up to k modifications are made. We first prove a lower bound assuming exactly k modifications.

Let $\{a_i\}$ be the original sequence and $\{\hat{a}_i\}$ be the sequence obtained after exactly k operations are performed on a single period of $\{a_i\}$. Say there are k_S substitutions, k_D deletions, and k_I insertions. Say the k_I insertions are performed as k_L blocks. So

$$k = k_S + k_D + k_I, \quad k_L \leq k_I. \tag{5}$$

Let $S, D, I \subset \{0, \cdots, T - 1\}$ be sets that denote the positions of substitutions, deletions, and insertions respectively. Substitutions and deletions are performed on the elements at the positions in sets S and D respectively. Insertions occur before the positions in the set I. Note that $|S| = k_S, |D| = k_D$, and $|I| = k_L$. If there are a deletion and substitution at the same place we can remove the substitution and obtain the same modified sequence. Thus we can replace our list of k modifications by a list of $l \leq k$ modifications with no deletions and substitutions at the same place. Similarly we can replace an insertion and a deletion at the same position by a substitution of the element at that position with the element to be inserted. That is, we may assume that

$$D \cap S = D \cap I = \emptyset. \tag{6}$$

However, an insertion and a substitution can occur at the same position. Hence if k' is the cardinality of $S \cup D \cup I$, from equations (5) and (6) we have $k' = |S \cup D \cup I| \leq k_S + k_D + k_L \leq k$. Let $t_1, \cdots, t_{k'}$ be the list of the distinct elements of $S \cup D \cup I$ so that

$$t_1 < t_2 < \cdots < t_{k'}, \quad k' = |S \cup D \cup I|. \tag{7}$$

Let n_i, $i = 1 \cdots k'$, denote the number of symbols to be inserted before a_{t_i}, so that

$$\sum_{t_i \in I} n_i = k_I.$$

From equation (4), by replacing $\{a_i\}$ by a cyclic shift $\{a_{i-s}\}$, $0 \leq s \leq T - 1$, we can make $t_1 = 0$ and

$$T - t_{k'} = \max(t_2, t_3 - t_2, \cdots, T - t_{k'}) \geq \frac{T}{k'}. \tag{8}$$

So from equation (8) we have

$$t_{k'} \leq \frac{(k' - 1)T}{k'} \leq \frac{(k - 1)T}{k} \tag{9}$$

Consider the subsequence of the sequence in equation (7) with all the elements from the set $D \cup I$ and no others,

$$t_{q_1} < t_{q_2} < \cdots < t_{q_{|D \cup I|}}. \tag{10}$$

We write $a(x)$ in equation (1) as $a(x) = A_0(x) + A_1(x) + \ldots + A_{|D \cup I|}(x)$, where

$$A_0(x) = \sum_{i=0}^{t_{q_1} - 1} a_i x^i,$$

$$A_m(x) = \sum_{i=t_{q_m}}^{t_{q_{(m+1)}} - 1} a_i x^i \quad (m = 1, \cdots, |D \cup I| - 1), \text{ and} \tag{11}$$

$$A_{|D \cup I|}(x) = \sum_{i=t_{q_{|D \cup I|}}}^{T-1} a_i x^i.$$

For our analysis we need the differences between numbers of insertions and deletions at each position where an operation is performed. Let $r_m = \sum_{j \leq m, \, t_j \in I} n_j - |\{j \leq m : t_j \in D\}|$, where $m \in \{1, \cdots, k'\}$. Then r_m denotes the net change in the index of a_{t_m}.

4   Main Result

With the notation established in the previous section, we obtain a lower bound on the linear complexity of the modified sequence.

Theorem 1. *Let $\{a_i\}$ be a sequence over $\mathbb{F}_q$ of period T. Let $\{\hat{a}_i\}$ be a sequence obtained after any combination of k substitutions, insertions, and deletions is performed on a single period of $\{a_i\}$ and repeated periodically. Then*

1. *$LC(\{\hat{a}_i\}) \geq \min(LC(\{a_i\}), T/k - LC(\{a_i\}))$ if the number of deletions is greater than or equal to the number of insertions.*
2. *$LC(\{\hat{a}_i\}) \geq \min(LC(\{a_i\}), (T+1)/k - LC(\{a_i\}))$ if the number of deletions is less than the number of insertions.*

Proof. The polynomial $\hat{a}(x)$ corresponding to the new sequence as in equation (1) can be written as

$$
\hat{a}(x) = A_0(x) + \sum_{t_j \in S} (b_{t_j} - a_{t_j}) x^{t_j + r_j} + \sum_{t_j \in I, \; j = q_m} A_m(x) x^{r_j}
$$

$$
+ \sum_{t_j \in I} \sum_{z = t_j + r_j - 1}^{t_j + r_{j-1} + n_j - 1} c_z x^z + \sum_{t_j \in D, \; j = q_m} (A_m(x) - a_{t_j} x^{t_j}) x^{r_j}, \tag{12}
$$

where the b_{t_j}s and c_zs are the new symbols for substitutions and insertions respectively. Now the new sequence can be represented as

$$
\sum_{i \geq 0} \hat{a}_i x^i = \frac{\hat{a}(x)}{1 - x^{T + k_I - k_D}}. \tag{13}
$$

We consider two cases based on whether the number of insertions is greater than the number of deletions.

Case 1: $k_I \leq k_D$
Let

$$
B(x) = x^{k_D - k_I} \hat{a}(x) - a(x) = \sum_{i=1}^{5} T_i(x), \tag{14}
$$

where

$$
\begin{aligned}
T_1(x) &= A_0(x)(x^{k_D - k_I} - 1), \\
T_2(x) &= \sum_{t_j \in S} (b_{t_j} - a_{t_j}) x^{t_j + r_j + k_D - k_I}, \\
T_3(x) &= \sum_{t_j \in I, \; j = q_m} A_m(x)(x^{r_j + k_D - k_I} - 1), \\
T_4(x) &= \sum_{t_j \in I} \sum_{z = t_j + r_{j-1}}^{t_j + r_{j-1} + n_j - 1} c_z x^{z + k_D - k_I}, \\
T_5(x) &= \sum_{t_j \in D, \; j = q_m} (A_m(x)(x^{r_j + k_D - k_I} - 1) - a_{t_j} x^{t_j + r_j + k_D - k_I}).
\end{aligned} \tag{15}
$$

From equation (14), $B(x)$ is a polynomial since $\hat{a}(x)$ and $a(x)$ are polynomials. Now we show that $\deg B(x) \leq t_{k'}$. We make the following observations.

1. We have $r_j = i_j - d_j$ where i_j and d_j are the numbers of insertions and deletions, respectively, up to and including those that are performed at position t_j. For each $t_j \in S \cup D \cup I$ there must be at least $k_D - d_j$ positions after t_j to account for the rest of the deletions. From this observation and equation (7) we have the inequalities

$$t_j \le t_{k'} - (k_D - d_j), t_j \in S \cup D \cup I. \tag{16}$$

2. From equation (15), the term of T_3 or T_5 with $j = q_{|I \cup D|}$ is zero since it has a factor $(x^{k_D - k_I + r_j} - 1)$, and $r_j = k_I - k_D$ for $j = q_{|I \cup D|}$. That is

$$A_{|I \cup D|}(x)(x^{r_{q_{|I \cup D|}} + k_D - k_I} - 1) = 0. \tag{17}$$

From equations (11), (15), (16), and (17) we have

$$\deg T_2(x) \le \max_{t_j \in S}(k_D - k_I + t_j + r_j)$$
$$\le \max_{t_j \in S}(t_{k'} + i_j - k_I) \le t_{k'}, \tag{18}$$

$$\deg T_3(x) \le \max_{t_j \in I}(t_{j+1} - 1 + r_j + k_D - k_I)$$
$$\le \max_{t_j \in I}(t_{k'} + (i_j - k_I) + (d_{j+1} - (d_j + 1))) \le t_{k'}, \tag{19}$$

$$\deg T_1(x) \le (t_{q_1} + k_D) - (k_I + 1) \le t_{k'} - 1 - (k_I + 1) \le t_{k'},$$

and

$$\deg T_4(x) \le \max_{t_j \in I}(k_D - k_I + t_j + r_{j-1} + n_{t_j} - 1)$$
$$\le \max_{t_j \in I}(k_D - k_I + t_j + r_j)$$
$$\le \max_{t_j \in I}(t_{k'} + i_j - k_I) \le t_{k'}.$$

From equations (18) and (19), using a similar derivation we obtain

$$\deg T_5(x) \le t_{k'}.$$

Thus

$$\deg B(x) \le t_{k'}. \tag{20}$$

From equations (1), (13), and (14) we have

$$\sum_{i \ge 0} \hat{a}_i x^i = \frac{\hat{a}(x)}{1 - x^{T + k_I - k_D}}$$
$$= \frac{x^{k_I - k_D}(a(x) + B(x))}{1 - x^{T + k_I - k_D}} \tag{21}$$
$$= \frac{g(x)(1 - x^T) + f(x)B(x)}{f(x)(x^{k_D - k_I} - x^T)}.$$

From equations (1) and (20) we have $\deg(g(x)(1 - x^T) + f(x)B(x)) < \deg$ $(f(x)(x^{k_D - k_I} - x^T))$, and we can apply Lemma 1 with $\{c_i\} = \{\hat{a}_i\}$ and $H(x) = f(x)$. Hence $\{\tilde{c}_i\}$ is the sequence represented by

$$
\begin{aligned}
&\sum_{i \geq 0} \tilde{c}_i x^i \\
&= \frac{[f(x)(g(x)(1 - x^T) + f(x)B(x))] \mod (f(x)(x^{k_D - k_I} - x^T))}{f(x)(x^{k_D - k_I} - x^T)} \\
&= \frac{[g(x)(1 - x^{k_D - k_I}) + f(x)B(x)] \mod (x^{k_D - k_I} - x^T)}{x^{k_D - k_I} - x^T}.
\end{aligned}
\tag{22}
$$

We have the following two subcases based on the degree of the numerator in equation (22).

Case 1a: $[g(x)(1 - x^{k_D - k_I}) + f(x)B(x)] \not\equiv 0 \mod (x^{k_D - k_I} - x^T)$
Since $t_{k'} \geq k_D$, from equations (1), (2), and (20), we have

$$
\begin{aligned}
\deg(g(x)(1 - x^{k_D - k_I}) + f(x)B(x)) &= \deg(f(x)B(x)) \\
&\leq LC(\{a_i\}) + t_{k'}.
\end{aligned}
\tag{23}
$$

From Lemma 1 and equations (22) and (23) we have

$$
\begin{aligned}
LC(\{\hat{a}_i\}) &\geq LC(\{\hat{c}_i\}) \\
&\geq T - \deg(f(x)B(x)) \\
&\geq T - (LC(\{a_i\}) + t_{k'}).
\end{aligned}
$$

Hence from equation (9) we have

$$
LC(\{\hat{a}_i\}) \geq \frac{T}{k} - LC(\{a_i\}).
\tag{24}
$$

Case 1b: $[g(x)(1 - x^{k_D - k_I}) + f(x)B(x)] \equiv 0 \mod (x^{k_D - k_I} - x^T)$
If $LC(\{a_i\}) \geq T/k$, then the right hand side of equation (24) is at most 0 and so the result is trivial. Hence we may assume that

$$
LC(\{a_i\}) < \frac{T}{k}.
\tag{25}
$$

Let

$$
g(x)(1 - x^{k_D - k_I}) + f(x)B(x) = l(x)(x^{k_D - k_I} - x^T)
\tag{26}
$$

for some $l(x) \in \mathbb{F}_q[x]$. From equations (23) and (9) we have

$$
\deg(l(x)(x^{k_D - k_I} - x^T)) \leq LC(\{a_i\}) + \frac{(k-1)T}{k}.
$$

So from equation (25) $\deg l(x) \leq LC(\{a_i\}) - T/k < 0$. From equation (26) this implies that $g(x)(1 - x^{k_D - k_I}) + f(x)B(x) = 0$. Hence we have

$$
B(x) = \frac{g(x)(x^{k_D - k_I} - 1)}{f(x)}.
\tag{27}
$$

From equations (1), (13), (14), and (27) we have

$$\sum_{i \geq 0} \hat{a}_i x^i = \frac{\hat{a}(x)}{1 - x^{T+k_I-k_D}}$$

$$= \frac{B(x) + a(x)}{x^{k_D-k_I} - x^T}$$

$$= \frac{1}{x^{k_D-k_I} - x^T} \left(\frac{g(x)(x^{k_D-k_I} - 1)}{f(x)} + \frac{g(x)(1 - x^T)}{f(x)} \right) \qquad (28)$$

$$= \frac{g(x)}{f(x)} = \sum_{i \geq 0} a_i x^i.$$

From equations (24) and (28) Case 1 is proved.

Case 2: $k_I > k_D$
This can be derived by using the result of Case 1 by switching the roles of $\{a_i\}$ and $\{\hat{a}_i\}$.

Example 1. For a simple example of Case 1b let $T - 10$, sequence $a = (0101010101)^\infty$ and $k = 2$. Hence $T/k - LC(\{a_i\}) = 3$ which is not a lower bound for the linear complexity of the modified sequence because we can delete any two consecutive symbols to have a sequence with linear complexity 2. Similarly we can insert two symbols and use a combination of an insertion and a deletion to obtain the same linear complexity as that of the original sequence. This shows that we must include $LC(\{a_i\})$ in our lower bound. It is this term that was missing from Jiang, et al.'s lower bound [1].

Corollary 1. *Let $\{a_i\}$ be a sequence over $\mathbb{F}_q$ of period T. Let $\{\hat{a}_i\}$ be a sequence obtained after any combination of up to k substitutions, insertions, and deletions is performed on a single period of $\{a_i\}$ and repeated periodically. Then*

1. *$LC(\{\hat{a}_i\}) \geq \min(LC(\{a_i\}), T/k - LC(\{a_i\}))$ if the number of deletions is greater than or equal to the number of insertions.*
2. *$LC(\{\hat{a}_i\}) \geq \min(LC(\{a_i\}), (T+1)/k - LC(\{a_i\}))$ if the number of deletions is less than the number of insertions.*

Proof. The lower bound established in Theorem 1 is monotonically non-increasing in k. Thus if we make up to k modifications, the bound for exactly k modifications still applies.

We note that there are examples of arbitrary large period sequences that meet the bounds.

5   Conclusion

A derivation of non-trivial lower bounds for the linear complexity of a sequence over $\mathbb{F}_q$ obtained by performing k or fewer operations on a single period of a periodic sequence is presented, where an operation is a substitution, insertion, or a deletion of a symbol. The bounds are useful when the linear complexity of the original sequence is less then T/k and greater than $T/2k$ where T is the period. Since the information about the positions and the corresponding values of the new elements to be inserted, deleted, or substituted is not used, the bounds are not always tight. However, it is interesting to see that the bounds using any combination of the three operations are similar to those proved by Jiang et al. when only one type of operation at a time is allowed [1]. In fact the three bounds they derived for k symbol substitution, insertion, and deletion are corollaries of Theorem 1 of this paper.

References

[1] Jiang, S., Dai, Z., Imamura, K.: Linear complexity of a sequence obtained from a periodic sequence by either substituting, insertion or deleting k symbols within one period. IEEE Trans. Inf. Theory 44, 1328–1331 (1998)

[2] Meidl, W., Niederreiter, H.: Counting Functions and Expected Values for the k-error Linear Complexity. Finite Fields Appl. 8, 142–154 (2002)

[3] Meidl, W., Niederreiter, H.: On the Expected Value of Linear Complexity and the k-error Linear Complexity of Periodic Sequences. IEEE Trans. Inf. Theory 48(11), 2817–2825 (2002)

[4] Dai, S., Imamura, K.: Linear complexity for one-symbol substitution of a periodic sequence over GF(q). IEEE Trans. Inf. Theory 44, 1328–1331 (1998)

[5] Uehara, S., Imamura, K.: Linear complexity of periodic sequences obtained from $GF(q)$ sequences with period $q^n - 1$ by one-symbol insertion. IEICE Trans. Fundamentals E79-A, 1739–1740 (1996)

[6] Uehara, S., Imamura, K.: Linear complexity of peridic sequences obtained from a sequence over $GF(p)$ with period $p^n - 1$ by one-symbol deletion. IEICE Trans. Fundamentals E80-A, 1164–1166 (1997)

[7] Uehara, S., Imamura, K.: Linear complexities of periodic sequences obtained from an m-sequence by two-symbol substitution. In: Proc. 1998 Int. Symp. Inf. Theory and Its Appl., Mexico City, Mexico, pp. 690–692 (October 1998)

Trade-Off Traitor Tracing

Kazuto Ogawa[1], Go Ohtake[1], Goichiro Hanaoka[2], and Hideki Imai[2]

[1] Japan Broadcasting Corporation, Japan
{ogawa.k-cm, ohtake.g-fw}@nhk.or.jp
[2] National Institute of Advanced Industrial Science and Technology, Japan
{hanaoka-goichiro,h-imai}@aist.go.jp

Abstract. There has been a wide ranging discussion on the contents copyright protection in digital contents distribution systems. Fiat and Tassa proposed the framework of *dynamic traitor tracing*. Their framework requires dynamic computation transactions according to the real-time responses of the pirate, and it presumes real-time observation of contents redistribution and therefore cannot be simply utilized in an application where such an assumption is not valid. In this paper, we propose a new scheme that not only provides the advantages of dynamic traitor tracing schemes but also overcomes their problems.

1 Introduction

There are a lot of approaches to protect contents copyrights in contents distribution services. Watermarking technology is one of the most important primitives for them and a lot of methods employ the technology. The framework of *dynamic traitor tracing* proposed by Fiat and Tassa is one of them. It assigns user subsets in order to trace illegal redistributors (traitors) in real time dynamically according to the illegally redistributed contents. Dynamic assignment enables tracers to obtain most useful information to trace traitors according to the traitors' strategy. However, it needs a real-time feedback channel and thus it is well known that the delayed attack, which redistributes contents with some delay, is effective. As the above implies, it does not provide a practical protection scheme. Therefore, our goal is to develop a new traitor tracing scheme that has the advantages of dynamic traitor tracing and less of its shortcomings.

1.1 Related Works

Traitor Tracing. Traitor tracing is one of the major schemes for protecting copyrighted works. In a system, a contents provider encrypts contents and distributes them, and each user decrypts them with his/her decryption key, which is distributed prior to the service. Each user's decryption key is unique, so if the user illegally redistributes the decryption key, it is possible to identify the decryption key's owner [4,5,6,10,11]. However, these traitor tracing schemes cannot protect the decrypted contents from illegal copying.

K. Srinathan, C. Pandu Rangan, M. Yung (Eds.): Indocrypt 2007, LNCS 4859, pp. 331–340, 2007.
© Springer-Verlag Berlin Heidelberg 2007

Watermarking. One sort of countermeasure is watermarking. A simple watermarking scheme works as follows [15,3]. The contents provider produces different contents and distributes them to users. These contents are generated from a single original, but their embedded information is different. Effectively then, each user gets contents that are different from any other user's contents. Unfortunately, as part of a broadcasting service, this scheme requires a network capacity in proportion to the number of users and thus is not practical.

Dynamic Traitor Tracing. Fiat and Tassa proposed a scheme such that multiple watermarks are assigned dynamically for users. They assume that contents are redistributed in real time, and that the redistributed contents can be obtained in real time [8,9]. In addition, Berkman, Parnas, and Sgall improved it [1,2]. The system dynamically assigns watermark patterns soon after getting the redistributed contents, and this enables one to identify traitors at a low network cost. Such schemes are called *dynamic traitor tracing* (DTT).

However, it has some shortcomings. One is that a real-time feedback channel to get redistributed contents in real time is required for this system. Another one is that a real-time dynamic watermark assignment is required, which implies that the CPU cost of the watermark assignment server is extremely high. Moreover, there is an effective attack whereby contents are redistributed with some delay and the attack makes it hard for the contents provider to assign watermarks dynamically.

Sequential Traitor Tracing. Safavi-Naini and Wang proposed another approach to solve these problems, called *sequential traitor tracing* (STT) [13,14]. In this scheme, even if there are no traitors, it is necessary to distribute multiple contents, and the number of the contents is in proportion to the number of traitors whom the contents provider assumes to collude and to redistribute contents. Hence, the scheme's network cost is high.

The above discussion illustrates that while DTT and STT are effective ways of tracing traitors, but they do not meet all of the requirements.

1.2   Our Contribution

Our goal is to develop a new scheme that has the advantages of DTT and has less of its shortcomings (we call it *trade-off traitor tracing*). In our scheme, several segments are stored and the watermarks embedded into them are detected. Then, the successive next pattern of watermarks embedded into the next several segments is determined after analyzing the detected watermarks pattern and with one dynamic computation. The determination must be made in a way that the information obtained from the next detected watermarks works most effectively to identify traitors. This method makes it robust against delayed attacks [13,14].

For example, consider a likely scenario in which attackers try to redistribute a serial drama episode the day after it was shown. The conventional scheme would not work in this case, but ours would. In the service, every episode is broadcast every day. The attackers perform delayed attacks and illegally redistribute the jth episode the next day. The tracers, who would like to specify the attackers,

Table 1. Comparison of three schemes: in terms of delayed attack security (DA-security), number of dynamic computations (# DC), and number of variants (# Var)

Scheme	[8,9]	[1,2]	Ours	
			$p < 60$	$p \geq 60$
DA-security	–	–	$\sqrt{}$	
# DC	$p(\log_2 n + 1)$	$p(\log_3 n + 1)$	$p(\log_4(n/p)+4)-3$	$p(\log_4 n + 1)$
# Var	$2p+1$	$3p+1$	$3p+1$	

determine the watermark patterns at the $j+1^{th}$ episode broadcasting after they found illegal redistribution of the jth episode. In the case of DTT, only one new watermark pattern is determined for the next episode. On the other hand in our scheme, one episode is divided into multiple segments (two segments in the following construction), and distinct watermark pattern is assigned for each segment. It means that one episode is considered as *one segment* in DTT, and as *multiple (two) segments* in our scheme. When the tracers find the next illegal redistribution of the $j+1^{th}$ episode, the tracer can decrease the number of users, who include the attackers, to 1/2 in DTT and to 1/4 in our scheme. Hence, our scheme works more effectively against delayed attacks than DTT.

Moreover, the computational cost (the number of dynamic watermark assignment to trace all traitors) of our proposal is less than that of DTT. Table 1 shows a comparison with the schemes of [8,9,1,2], when the number of segments is set to two and the number of watermark variants to specify one traitor is set to three. The comparison shows that our proposal can decrease the number of dynamic computations to about $79\%(\simeq \log_4 3 = (p\log_4 n)/(p\log_3 n))$, where p is the number of traitors and n is the number of total users $(p \ll n)$.

Totally, our scheme has all the advantages and less shortcomings of DTT, and it is more practical than DTT.

2   Model: Trade-Off Traitor Tracing

Our model is similar to that of DTT [8,9], but real-time contents feedback is unnecessary. The next watermark pattern is determined with *adaptive and dynamic* computations, depending on the watermark information detected from illegally redistributed contents. Contents providers distribute contents and then traitors redistribute them illegally. The providers can see the contents redistributed. One content consists of multiple segments and it is possible to generate multiple variants of each segment. Distinct information is embedded in each variant. In addition, users are divided into multiple subsets, and each variant is distributed to each subset. In the process, several current segments are stored and the watermark information is detected. The detected watermark pattern is then used to determine the next pattern, which would be embedded in the next several segments. That is, assignment to *distinct* subsets for *each* segment is determined at an assignment timing dynamically. On the other hand, in DTT, the assignment to the *same* subsets for *all* segments is determined.

Through this modification, the computational cost can decrease and the robustness against delayed redistribution attacks can be achieved. We describe the details of our model below, and in the following we use the notation in Table 2.

Table 2. Notation

m	: the number of total segments
t	: the number of segments which are used at each assignment (that is, the number of segments which are used for one TRC (see the following part))
U	: the set of all users, $\|U\|=n$
n	: the number of all users
T	: a set of users (traitors) who collude and redistribute contents illegally, $T \subset U$ and $\|T\|=p$
p	: the number of traitors
Σ	: the set of alphabets that are used for watermarking, $\Sigma=\{\sigma_1,\cdots,\sigma_r\}$
r	: the number of different variants
v_k^j	: the variant of the jth segment into which σ_k is embedded, $1 \leq j \leq m$, $1 \leq k \leq r$
S_k^j	: the set of the users who received the variant v_k^j
h	: side information to trace traitors. It includes the attributes of each subset of authorized users. The subset is generated in the tracing process.

Content Structure. One content is divided into several sequential segments. In the case of video contents, for example, one content is a program and one content is divided into minute-long segments. We assume that there exists an ideal watermarking scheme. Multiple *variants* of one segment are generated with this watermarking scheme, and the information embedded in each variant is different from the others. The following conditions are required for these variants:

- Fundamentally, all variants carry the same information to the extent that humans cannot easily distinguish between them.
- Given any set of variants of the jth segment $(1 \leq j \leq m)$, $v_1^j,\cdots,v_\lambda^j$, it is impossible to generate another variant that can not be traced back to one of the original variants v_i^j $(1 \leq i \leq \lambda)$.

Clearly, assuming that there exists a watermarking scheme which meets the above requirements, it would be possible to identify at least one variant with illegally redistributed variants, and prove that there is one traitor among the set of users who received the same variant with the identified variant.

Algorithms. Trade-off traitor tracing consists of two algorithms, WMK and TRC. WMK is the algorithm to embed watermarks. TRC is the algorithm to trace traitors who redistribute contents illegally.

WMK: This is an algorithm which takes as inputs U, t consecutive segments (from j^{th} to $j+t-1^{th}$ segments), and h. It generates multiple variants v_k^i $(1 \leq k \leq r)$ for all i^{th} segment $(j \leq i \leq j+t-1)$. For all i and k $(j \leq i \leq j+t-1, 1 \leq k \leq r)$, it determines the sets of users S_k^i and the variant v_k^i distributed to S_k^i, updates side information h, and returns v_k^i, S_k^i and h.

TRC: This is an algorithm which takes as inputs U, the detected watermark information from t consecutive segments (from j^{th} to $j+t-1^{th}$ segments), S_k^i ($j \leq i \leq j+t-1, 1 \leq k \leq r$), and h, and returns the updated h.

These two algorithms are used as follows. When the content is distributed, WMK generates multiple variants v_k^i ($1 \leq k \leq r$) for each segment i ($j \leq i \leq j+t-1$). If the illegally redistributed content is found, the variants detected in the content, the user set information S_k^i ($j \leq i \leq j+t-1, 1 \leq k \leq r$) and information h, which shows the relationship between the sets of users and the distributed variants, are inputted to TRC. TRC analyzes these data and reduces the number of suspicious users. It outputs h, which includes information of new subsets to collect the most meaningful information. After that, WMK takes as inputs these new subsets, and generates new variants v_k^i and sets S_k^i ($1 \leq k \leq r$) for the next t consecutive segments ($j+t \leq i \leq j+2t-1$). This process is repeated until all traitors are identified.

3   Concrete Construction of Trade-Off Traitor Tracing

We show a construction of the trade-off traitor tracing scheme. Although the trade-off traitor tracing scheme is constructed based on [8,9], it is significantly different from them in regard to their strategy to identify traitors. That is, trade-off traitor tracing collects t consecutive segments and analyzes them simultaneously. Different user subsets are created at each segment and this is the trick to get the most meaningful and most effective information.

On the other hand, direct use of the scheme described in [8,9] is not effective, since the traitors can adaptively choose which segments to redistribute and thus the contents provider (tracer) collects less information. That is, once a set of users is divided into subsets, the subsets are not changed until illegal redistribution is found. One subset assignment is used for all segments, regardless of the number of segments. The traitors can, then, take such a strategy that they redistribute only the segments distributed to one subset.

Strategy. The strategy to collect the most meaningful information to identify traitors from the variants of two segments is as follows. In the following, we set t to two and the number of watermark variants to specify one traitor to three to simplify our explanation, even though the larger these numbers are, the more effective our scheme becomes. Let S be a set of users to which an illegal redistributed variant is distributed. Regarding the two segments the contents provider will distribute next, the contents provider makes four subsets of S, $S_\oplus$, $\overline{S_\oplus}$, $S_\otimes$ and $\overline{S_\otimes}$, where $S_\oplus \cup \overline{S_\oplus} = S_\otimes \cup \overline{S_\otimes} = S$, $|S_\oplus| = |\overline{S_\oplus}| = |S_\otimes| = |\overline{S_\otimes}| = \frac{1}{2}|S|$, and $|S_\oplus \cap S_\otimes| = |S_\oplus \cap \overline{S_\otimes}| = |\overline{S_\oplus} \cap S_\otimes| = |\overline{S_\oplus} \cap \overline{S_\otimes}| = \frac{1}{4}|S|$. For the first segment, one variant is distributed to $S_\oplus$ and another variant to $\overline{S_\oplus}$. For the next segment, one is to $S_\otimes$ and another one to $\overline{S_\otimes}$. When, for example, the variants assigned to $S_\oplus$ and $S_\otimes$ are found to be illegally redistributed, the following situations can be imagined:

(i) At least one traitor is in $S_\oplus \cap S_\otimes$.
(ii) At least one of them is in $S_\oplus \cap \overline{S_\otimes}$ and at least one of them is in $\overline{S_\oplus} \cap S_\otimes$.

In particular, when only one traitor exists in S, (i) is true, and hence one dynamic computation enables tracers to decrease the number of suspicious users to $1/4$. In contrast, when the conventional traitor tracing scheme [8,9] is used, two dynamic computations are necessary to decrease the number to $1/4$, but the scheme described above can achieve it through only one dynamic computation.

However, realistically (especially at the beginning of tracing) it is natural to suppose that multiple traitors exist in S. A decision such that (i) or (ii) is true may make it so that traitors can not be identified in the subsequent tracing process. Hence, the decision made from one piece of collected information is not likely to be effective, and we need another strategy to get a correct result for identifying the true traitors. The correct strategy is that the tracers set a high possibility to (i), and that they utilize a scheme to identify traitors in $S_\oplus \cap S_\otimes$. Simultaneously, considering the case that (ii) is true, they utilize another scheme. If (ii) is true and the latter scheme works effectively, the tracers can get more information about traitors than in the first scheme. As a result, traitors cannot help but perform in the way in which (i) is true.

State Transition. In our scheme, there are four states, State0 to State3, and seven transitions, Case1 to Case7. Figure 1 shows the state transition diagram. State$i$-j denotes a State$i$ with an index j, where j is used only to make a distinction with another State$i$. In addition, we use the following variables.

I : The set of users in which traitors have not been found.

$C_{\oplus,l}, \overline{C_{\oplus,l}}, C_{\otimes,l}, \overline{C_{\otimes,l}}$ $(1 \leq l \leq p)$: These are sets of users such that $C_{\oplus,l} \cup \overline{C_{\oplus,l}} = C_{\otimes,l} \cup \overline{C_{\otimes,l}} (:= C_l)$ and there exists at least one traitor in C_l.

$C'_{\oplus,l}, \overline{C'_{\oplus,l}}, C'_{\otimes,l}, \overline{C'_{\otimes,l}}, L'_l, R'_l$ $(1 \leq l \leq p)$: These are sets of users such that $C'_{\oplus,l} \cup \overline{C'_{\oplus,l}} = C'_{\otimes,l} \cup \overline{C'_{\otimes,l}} (:= C'_l)$ and C'_l includes at least one traitor or L'_l and R'_l include at least one traitor.

C''_l, L''_l, R''_l $(1 \leq l \leq p)$: These are sets of users such that two sets among C''_l, L''_l, and R''_l include at least one traitor.

$\emptyset$: This is an empty set of users.

Diagram

- State0 is the state in which the set is I, C_ω or C_l. When illegal redistribution is detected in State0, the state changes into State1. This is Case1 transition.
- When illegal redistribution is detected in State1 and the detected subsets' pair is one (($C_{\oplus,l}, C_{\otimes,l}$) in Fig. 1) of $\{(C_{\oplus,l}, C_{\otimes,l}), (C_{\oplus,l}, \overline{C_{\otimes,l}}), (\overline{C_{\oplus,l}}, C_{\otimes,l}), (\overline{C_{\oplus,l}}, \overline{C_{\otimes,l}})\}$, the state changes into State2. This is Case2 transition.
- When illegal redistribution is detected in State2 and the detected subsets' pair is one (($C'_{\oplus,l}, C'_{\otimes,l}$) in Fig. 1) of $\{(C'_{\oplus,l}, C'_{\otimes,l}), (C'_{\oplus,l}, \overline{C'_{\otimes,l}}), (\overline{C'_{\oplus,l}}, C'_{\otimes,l}), (\overline{C'_{\oplus,l}}, \overline{C'_{\otimes,l}})\}$, the state changes into another State2. This is Case3 transition. This state is different from previous State2, in that the number of $|I|$ increases and the numbers of $|C'_{\oplus,l}|, |\overline{C'_{\oplus,l}}|, |C'_{\otimes,l}|, |\overline{C'_{\otimes,l}}|, |L'_l|$ and $|R'_l|$ decrease to $1/4$.

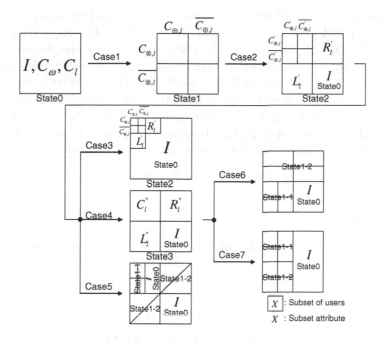

Fig. 1. State Transition Diagram

- When illegal redistribution is detected in State2 and the detected subsets'
  pair is $(L_l' \cup R_l', L_l' \cup R_l')$, the state changes into State3. This is **Case4**
  transition.
- When illegal redistribution is detected in State2 and one of the detected
  subsets is a subset ($C_{\oplus,l}'$ in Fig. 1) among $\{C_{\oplus,l}', \overline{C_{\oplus,l}'}, C_{\otimes,l}', \overline{C_{\otimes,l}'}\}$ and the
  other is $L_l' \cup R_l'$, the state changes into one State0 and two State1 (State1-1
  and State1-2). This is **Case5** transition. In this case, there are at least two
  traitors and at least one of them is in the former subset ($C_{\oplus,l}'$ in Fig. 1), and
  at least one of them is in $L_l' \cup R_l'$. The former subset ($C_{\oplus,l}'$ in Fig. 1) then
  can be treated as C_l in State0, and $L_l' \cup R_l'$ can be treated as C_ω in State0,
  and each of these states changes into State1.
- When illegal redistribution is detected in State3 and the detected subsets'
  pair is one $((L_l'', L_l'')$ in Fig. 1) of $\{(C_l'', C_l''), (L_l'', L_l''), (R_l'', R_l'')\}$, the state
  changes into one State0 and two State1 (State1-1 and State1-2). This is
  Case6 transition. In this case, there are at least two traitors and at least one
  of them is included in a subset (L_l'' in Fig. 1) among $\{C_l'', L_l'', R_l''\}$, and at
  least one of them is included in a combined subset ($C_l'' \cup R_l''$ in Fig. 1), which
  is generated by excluding the detected subset (L_l'' in Fig. 1) from the subset
  $\{C_l'' \cup L_l'' \cup R_l''\}$. The detected subset ($L_l''$ in Fig. 1) can be treated as C_ω in
  State0, another subset ($C_l'' \cup R_l''$ in Fig. 1) can be treated as C_l in State0,
  and each of these states changes into State1.

- When illegal redistribution is detected in State3 and the detected subsets' pair is one $((C_l'', L_l'')$ or (L_l'', C_l'') in Fig. 1) of $\{(C_l'', L_l''), (C_l'', R_l''), (L_l'', C_l''),$ $(L_l'', R_l''), (R_l'', C_l''), (R_l'', L_l'')\}$, the state changes into one State0 and two State1 (State1-1 and State1-2). This is **Case7** transition. In this case, there are at least two traitors and they are in two different subsets $(C_l''$ and L_l'' in Fig. 1) among three possible subset pairs, $(C_l''$ and L_l'', C_l'' and R_l'', or, L_l'' and $R_l'')$. One subset of the two different subsets $(L_l''$ in Fig. 1) can be treated as C_ω in State0, and another one $(C_l''$ in Fig. 1)) can be treated as C_l in State0, and each of these states changes into State1.

Our construction is fully described in the full paper [12].

4   Evaluation

We then address the traceability of the tracing scheme and evaluate our scheme with regard to the number of dynamic computations and compare it with that of DTT [8,9,1,2]. In addition, we evaluate our scheme in terms of delayed attacks.

4.1   Traceability

We show that our tracing algorithm can trace at most p $(1 \le p < \lfloor r/3 \rfloor)$ traitors perfectly, where $\lfloor x \rfloor$ denotes a function which outputs a maximum integer less than or equal to x. Formally, we prove the following theorem.

Theorem 1. *If the number of traitors p is less than $\lfloor r/3 \rfloor$, the tracing algorithm can trace all p traitors.*

For the proof of this theorem, we utilize the following two claims under the condition of $p < \lfloor r/3 \rfloor$. In these claims, which we prove in the full paper [12], the notation Π_l is used, where $\Pi_l \in \{C_l, C_{\oplus,l}, \overline{C_{\oplus,l}}, C_{\otimes,l}, \overline{C_{\otimes,l}}, C'_{\oplus,l}, \overline{C'_{\oplus,l}}, C'_{\otimes,l},$ $\overline{C'_{\otimes,l}}, L'_l \cup R'_l, C''_l, L''_l, R''_l\}$, and its index is l $(1 \le l \le p)$.

Claim 1. *When there are p traitors and they belong to p distinct subsets, which have p distinct indices, $(\Pi_1, \cdots, \Pi_p)$, the tracing algorithm can trace all p traitors.*

Claim 2. *When multiple traitors belong to one subset Π_l and the traitors in Π_l select a variant at every segment for illegal redistribution, such that the traitor whose received variant is used at j^{th} segment is different from the traitor whose received variant is used at $j+1^{th}$ segment, the tracing algorithm can divide the traitors into two subsets Π_l and $\Pi_{l'}$ $(l \ne l')$.*

Proof of Theorem 1 (Sketch). Suppose that there are p $(1 \le p \le \lfloor r/3 \rfloor)$ traitors and that multiple traitors belong to one subset Π_l. The tracing algorithm can lead to the situation, such that the traitors in Π_l are divided into two distinct subset Π_l and $\Pi_{l'}$ from claim 2. By repeating this process, p traitors can be divided into p distinct subsets $\Pi_1, \cdots, \Pi_p$. Moreover, p traitors, who belong to p distinct subsets, can be traced perfectly from claim 1. It shows that the tracing algorithm can trace all p traitors.   □

4.2   Evaluation on the Number of Dynamic Computations

We show that our scheme is better with regard to the number of dynamic computations than [8,9].

Number of Dynamic Computations of Our Scheme. Regarding the number of dynamic computations to trace all p traitors, the following claim holds, and we prove it in the full paper [12].

Claim 3. *The largest number of dynamic computations of our scheme to trace all p traitors is $p \times (\log_4 n - \log_4 p) + 4p - 3$ if $p < 60$, otherwise $p \times (\log_4 n + 1)$.*

Comparison of [8,9] and Our Scheme. Our scheme can decrease the number of suspicious users to $1/4$ with one dynamic computation, whereas the conventional scheme [8,9] can decrease the number only to $1/2$. Thus, our scheme can identify all traitors by using only half the number of dynamic computations. However, while our scheme uses up to three variants to identify one traitor, compared with the conventional scheme's two variants, our scheme does *not always* use three variants.

To evaluate these schemes, a conventional scheme using three variants should be considered, and such an improvement is easy to achieve. It can decrease the number of suspicious users to $1/3$ with one dynamic computation in a way that *always* uses three variants. Such an improvement is described in [1,2]. Table 1 compares the proposed scheme with the conventional schemes.

Generally, we can assume $0 < p \ll n$, and then $p \log_2 n \simeq p(\log_2 n + 1)$, $p \log_3 n \simeq p(\log_3 n + 1)$, and $p \log_4 n \simeq p(\log_4 n + 1) \simeq p(\log_4(n/p) + 4) - 3$. Hence, our scheme can decrease the number of dynamic computations to about $50\% (= \log_4 2 = (p \log_4 n)/(p \log_2 n))$ of the conventional scheme's. Compared with the improved conventional scheme, it can decrease it to about $79\% (\simeq \log_4 3 = (p \log_4 n)/(p \log_3 n))$. This proves that if the contents provider generates enough variants, our scheme is more effective than the conventional scheme. Regarding the number of variants and the number of dynamic computations, our scheme and the conventional scheme have a trade-off relationship; hence, we call our scheme *trade-off traitor tracing*.

4.3   Delayed Attack Resilience

Our scheme is robust against delayed attacks. The subset assignment of each segment is determined before next several segments are distributed, and the assignment is recorded in the side information in order to trace traitors. In the worst case, the traitors wait for the distribution to be completed and then start the redistribution. However, since distinct subsets are assigned for *each* segment, the information to be used for tracing can be obtained even when such an attack is performed and the information can be used for next dynamic computation. This is in contrast to DTT that is completely insecure against a delayed attack.

5 Conclusion

We proposed the *trade-off traitor tracing* scheme. This scheme requires fewer dynamic computations than the conventional scheme does, and it does not need to make a dynamic computation in real time, since the computation is performed after several segments have been stored. Moreover, our scheme is more resilient against delayed attacks than the conventional scheme.

References

1. Berkman, O., Parnas, M., Sgall, J.: Efficient Dynamic Traitor Tracing. In: Proc. of ACM-SODA 2000, pp. 586–595 (2000)
2. Berkman, O., Parnas, M., Sgall, J.: Efficient Dynamic Traitor Tracing. SIAM Journal on Computing 30(6), 1802–1828 (2000) (full version of [1])
3. Blakley, G.R., Meadows, C., Purdy, G.B.: Fingerprinting Long Forgiving Messages. In: Williams, H.C. (ed.) CRYPTO 1985. LNCS, vol. 218, pp. 180–189. Springer, Heidelberg (1986)
4. Boneh, D., Franklin, M.: An Efficient Public Key Traitor Tracing Scheme. In: Wiener, M.J. (ed.) CRYPTO 1999. LNCS, vol. 1666, pp. 338–353. Springer, Heidelberg (1999)
5. Boneh, D., Franklin, M.: An Efficient Public Key Traitor Tracing Scheme, full version of [4], http://crypto.stanford.edu/~dabo/pubs.html
6. Chor, B., Fiat, A., Naor, M.: Tracing Traitors. In: Desmedt, Y.G. (ed.) CRYPTO 1994. LNCS, vol. 839, pp. 252–270. Springer, Heidelberg (1994)
7. Chor, B., Fiat, A., Naor, M., Pinkas, B.: Tracing Traitors. IEEE Trans. on Information Theory 46(3), 893–910 (2000) (full version of [6])
8. Fiat, A., Tassa, T.: Dynamic Traitor Tracing. In: Wiener, M.J. (ed.) CRYPTO 1999. LNCS, vol. 1666, pp. 354–371. Springer, Heidelberg (1999)
9. Fiat, A., Tassa, T.: Dynamic Traitor Tracing. J. of Cryptology 14(3), 211–223 (2001) (full version of [8])
10. Kiayias, A., Yung, M.: Traitor Tracing with Constant Transmission Rate. In: Knudsen, L.R. (ed.) EUROCRYPT 2002. LNCS, vol. 2332, pp. 450–465. Springer, Heidelberg (2002)
11. Kurosawa, K., Desmedt, Y.: Optimum Traitor Tracing and Asymmetric Schemes. In: Nyberg, K. (ed.) EUROCRYPT 1998. LNCS, vol. 1403, pp. 145–157. Springer, Heidelberg (1998)
12. Ogawa, K., Ohtake, G., Hanaoka, G., Imai, H.: Trade-off Traitor Tracing, full version of this paper, available from the first author via e-mail
13. Safavi-Naini, R., Wang, Y.: Sequential Traitor Tracing. In: Bellare, M. (ed.) CRYPTO 2000. LNCS, vol. 1880, pp. 316–332. Springer, Heidelberg (2000)
14. Safavi-Naini, R., Wang, Y.: Sequential Traitor Tracing. IEEE Trans. on Information Theory 49(5), 1319–1326 (2003) (full version of [13])
15. Wagner, N.: Fingerprinting. In: Proc. of IEEE Symposium on S&P 1983, pp. 18–22 (1983)

X-FCSR – A New Software Oriented Stream Cipher Based Upon FCSRs[*]

François Arnault[1], Thierry P. Berger[1], Cédric Lauradoux[2], and Marine Minier[3]

[1] XLIM (UMR CNRS 6172), Université de Limoges
123 avenue Albert Thomas, F-87060 Limoges Cedex - France
arnault@unilim.fr, thierry.berger@unilim.fr
[2] INRIA - projet CODES, B.P. 105, 78153 Le Chesnay Cedex - France
cedric.lauradoux@inria.fr
[3] INSA de Lyon - Laboratoire CITI
21 Avenue Jean Capelle, 69621 Villeurbanne Cedex - France
marine.minier@insa-lyon.fr

Abstract. Feedback with Carry Shift Registers (FCSRs) are a promising alternative to LFSRs in the design of stream ciphers. The previous constructions based on FCSRs were dedicated to hardware applications [3]. In this paper, we will describe X-FCSR a family of software oriented stream ciphers using FCSRs. The core of the system is composed of two 256-bits FCSRs. We propose two versions: X-FCSR-128 and X-FCSR-256 which output respectively 128 and 256 bits at each iteration. We study the resistance of our design against several cryptanalyses. These stream ciphers achieve a high throughput and are suitable for software applications (6.3 cycles/byte).

Keywords: stream cipher, FCSRs, software design, cryptanalysis.

1   Introduction

Following the recent development of algebraic attacks [7,12], it seems difficult to design good stream ciphers using combined or filtered LFSRs. A FCSR is similar to LFSRs, but it performs operations with carries, and so its transition function is not linear. Such an automaton computes the 2-adic expansion of some 2-adic rational number p/q. This can be used to prove several interesting properties of FCSRs: proven period, non-degenerated states, good statistical properties [15,20,21]. The high non-linearity of the FCSR transition function provides an intrinsic resistance to algebraic attacks, and seems also to prevent correlation attacks. There exists a hardware efficient family of stream ciphers based on FCSRs: the filtered FCSR or F-FCSR [1,2,3,5]. In these ciphers, the internal state of the FCSR is filtered by a linear function to provide from 1 to 16 output bits at each iteration. At the present moment, the F-FCSR-H and

---

[*] This work was partially supported by the french National Agency of Research: ANR-06-SETI-013.

K. Srinathan, C. Pandu Rangan, M. Yung (Eds.): Indocrypt 2007, LNCS 4859, pp. 341–350, 2007.
© Springer-Verlag Berlin Heidelberg 2007

F-FCSR-16 are selected for the third and last phase of the European Project eSTREAM for the Profile 2 (i.e. hardware profile) [22].

While F-FCSR stream ciphers have good performances in hardware, they are slow in software since they require many bit manipulation instructions to output only a few bits. In this paper, we propose an efficient way to design a fast stream cipher by generating many keystream bits from the same internal state. Our design is based on two 256-bit FCSRs and on a mechanism of extraction with a 16×256 bits or 16×128 bits memory. The input of the extraction function is the bitwise XOR of the full contents of the two FCSR main registers. This function and the IV setup use some classical and well-known techniques of block cipher design such as the "ShiftRows" and the "MixColumns" operations of the AES.

In this paper, we present two versions of our stream cipher. The first one X-FCSR-256 outputs 256 bits at each iteration. This version is probably riskier but is the most efficient (6.5 cycles/byte). It can be considered as a challenge for cryptanalysts. The second one X-FCSR-128 outputs 128 bits at each iteration. It seems more robust, and have also good performances (8.2 cycles/byte).

Section 2 presents background on FCSRs and their most useful properties for cryptographic use. The stream ciphers X-FCSR-256 and X-FCSR-128 are described in Section 3. We present an analysis of security in Section 4. We give in conclusion detailed results on the performance of our designs.

2   Background on FCSRs and 2-Adic Sequences

The FCSR automaton. The Feedback with Carry Shift Registers (FCSRs) were first introduced by A. Klapper and M. Goresky in [20] (see also [21,15,1,5]). A FCSR automaton with connection integer q and initial state p computes the 2-adic expansion of p/q.

In the sequel, we suppose that q is a negative prime and p satisfies $0 \leq p < |q|$. This ensures that the output sequence is periodic with period T where T is the order of 2 modulo q. This period is maximal if $T = |q| - 1$. In that case the sequence S is called a ℓ-sequence. We define an optimal FCSR as an FCSR generating ℓ-sequences.

We suppose that the size of q is $n + 1$, i.e. $2^n < -q < 2^{n+1}$. Let $d = (1 - q)/2$ and $d = \sum_{i=0}^{n-1} d_i 2^i, d_i \in \{0, 1\}, d_{n-1} = 1$. The FCSR automaton with connection integer q is composed of two registers (sets of cells): a main register M and a carry register C.

The main register M contains n binary cells where each bit is denoted by $m_i(t)$ $(0 \leq i \leq n - 1)$. We call $m(t) = \sum_{i=0}^{n-1} m_i(t) 2^i$ the content of M.

The carry register C contains ℓ cells where $\ell + 1$ is the number of nonzero d_i digits, i.e. the Hamming weight of d. More precisely, the carry register contains one cell for each nonzero d_i with $0 \leq i \leq n - 2$. We denote $c_i(t)$ the binary digit contained in this cell. We put $c_i(t) = 0$ when $d_i = 0$ or when $i = n - 1$. We call the integer $c(t) = \sum_{i=0}^{n-2} c_i(t) 2^i$ the content of C. The Hamming weight of the binary expansion of $c(t)$ is at most ℓ. Note that, if $d_i = 0$, then $c_i(t) = 0$ for all t. At cell level, the transition function of an FCSR is given by:

$$m_i(t+1) = m_{i+1}(t) \oplus d_i c_i(t) \oplus d_i m_0(t)$$
$$c_i(t+1) = d_i \left(m_{i+1}(t)c_i(t) \oplus c_i(t)m_0(t) \oplus m_0(t)m_{i+1}(t) \right).$$

where $\oplus$ denotes bitwise XOR. Note that $m_0(t)$ is the least significant bit of $m(t)$ and represents the feedback bit.

TMD attacks and procedure of initialization of FCSR. The graph of the transition function of a FCSR automaton is well mastered. The following proposition gives its full description [3]:

Proposition 1. *If the order of* $2 \bmod q$ *is* $T = |q| - 1$, *the size of the final cycle of a component of the transition function graph is exactly* T *(except for two degenerated cases:* $m + 2c = 0$ *or* $m + 2c = -q$).

To guarantee a constant entropy equal to $\log(2^n - 1)$ after any number of iterations, we propose to initialize c with 0 and to take a random m from the set $[1, \cdots, 2^n - 1]$. Thus, two distinct initializations cannot converge to the same state after a same number of iterations.

Distinguishing attacks using diffusion of differences. In a FCSR automaton, a small difference between two initial states remains local if the feedback bit is not affected by it. This property has been exploited to mount resynchronization attacks against some stream ciphers with bad IV setup design [18,19].

To avoid this problem, we need to design an IV setup which unables any attacker to master a difference between two internal states by choosing some pairs of IV. Such a procedure can be obtained by essentially two ways: by performing $n + 4$ iterations before to output any value [3] or by digesting the key and the IV using an efficient function such as an hash function or a block cipher. The first solution is very slow for a software purpose, thus we have chosen to initialize the two FCSR main registers with the ciphertext produced by the IV using an AES-like block cipher whereas the round subkeys are derived from the master key.

Algebraic attacks on FCSR automata. For a filtered LFSR of size n, the system of non-linear equations used for algebraic attacks is of the form $f(L^j(X)) = S(t_0 + j)$ where $X = (x_1, \ldots, x_n) = X(t_0)$ is the initial state of the automaton at time t_0, L is the linear transition function of the LFSR, f is the extraction boolean function (the non-linear filter) and $S(t_0 + j)$ is the output at the j-th iteration. The resistance against algebraic attacks depends only on the boolean function f.

For a filtered FCSR, the transition function Q is no more linear but quadratic. The system becomes $f(Q^j(X)) = S(t_0 + j)$ for $0 \le j < r$. In this case, the complexity of the problem depends essentially on the function Q^j and the function f could be linear as done in the F-FCSR stream ciphers. This property ensures a very efficient resistance against algebraic attacks. In fact the difficulty is not only to solve the system, but to compute it.

As soon as the number of required iterations is sufficiently large, the degree of the equations grows up to n and the number of monomials grows exponentially. Following experimental results given in [10], the limit to which it is possible to compute equations seems close to 10 iterations, even for FCSRs of small sizes (typically 64 or 128).

More properties of xored 2-adic sequences. In [15], M. Goresky and A. Klapper studied the general behavior of a sequence of the form: $s = \frac{p}{q} \oplus \frac{p'}{q'}$. The core of our stream cipher is the XOR of two FCSR automata with distinct connection integers q_a and q_b. Set $s = \frac{p_a}{q_a} \oplus \frac{p_b}{q_b}$, $0 < p_a < |q_a|$ and $0 < p_b < |q_b|$ and $S = (s_i)$ with $s = \sum_{i=0}^{\infty} s_i 2^i$. Applying the results given in [15], we have the following properties: the period of S is $T = (|q_a| - 1)(|q_b| - 1)/2$; the sequence S is balanced and the distribution of consecutive pairs in S is uniform.

Until now, there is no known method more efficient than the exhaustive search to recover p/q and p'/q' from the knowledge of s.

3   Design of X-FCSR-128 and X-FCSR-256

X-FCSR is a new family of additive synchronous stream ciphers. The two proposed versions, X-FCSR-128 and X-FCSR-256, essentially differ on the extraction function: for X-FCSR-128, 128 bits of keystream are generated at each iteration, while the extraction of X-FCSR-256 produces 256 bits of keystream. Clearly, the latter version is riskier since it outputs 256 bits - the size of one FCSR main register - and can be viewed as a challenge for cryptanalysts.

Both stream ciphers admit a secret key K of 128-bit length and a public initialization vector IV of bitlength ranging from 64 to 128 as input. These parameters conform to the requirements given in the eSTREAM initial call for Stream Ciphers for software applications [22].

General overview and parameters of the FCSR automata. The core of the design is constituted of two optimal 256 bit FCSRs: The first one with the connection integer q_a is right-clocked whereas the second one is left-clocked with connection integer q_b. q_a and q_b are primes and produces ℓ-sequences (i.e. the corresponding FCSRs are optimal):

$q_a = -23158373676191642998087032666622460867207843241572527691478170790314536991794 7$

$q_b = -1718770051860028145814553936674082372120455831563463236564900047373722326013 07$

At time t, we denote by $M_a(t)$ and $M_b(t)$ the contents of the two main registers. At each iteration, the value $X(t) = M_a(t) \oplus M_b(t)$ feeds an extraction function with memory which computes the output from the current value and from the value obtained at time $t - 16$. The connection integers have been chosen such that, if we denote $X(t)$ by $(x_0(t), \ldots, x_{255}(t))$ at bit level, at least one of the two FCSRs has a feedback between $x_i(t)$ and $x_{i+1}(t)$ for each i, $0 \le i < 255$.

Extraction function of X-FCSR-128. The extraction function is constituted of a function $Round_{128}$ working on 128-bit input/output words, and a memory of 16 128-bit words which stores the output of $Round_{128}$ that will be used 16 iterations later. More formally, the full extraction function works as follows:

- compute the 128-bit word $Y(t) = X^{(0)}(t) \oplus X^{(1)}(t)$,
  with $X(t) = \left(X^{(0)}(t) \| X^{(1)}(t)\right)$, where $\|$ denotes the concatenation.
- Compute $Z(t) = Round_{128}(Y(t))$.
- Store $Z(t)$ in memory (keep it during 16 iterations).
- Output the 128-bit word $Output_{128}(t) = Y(t) \oplus Z(t - 16)$.

$Round_{128}$ is a one-round function from $\{0,1\}^{128}$ into itself: $Round_{128}(a) = Mix_{128}(SR_{128}(SL_{128}(a)))$. If the 128-bit word a is represented at byte level by a 4×4 matrix M where each byte is represented by the word $a_{i,j}$ with $0 \le i, j \le 3$, then the function $Round_{128}$ works as follows:

- $SL_{128}()$ is a S-box layer applied at byte level: each byte $a_{i,j}$ is transformed into an other byte $b_{i,j}$ with $b_{i,j} = S(u_{i,j})$ where S is the S-box given in [4] chosen for its good properties (the differential and linear probabilities are low, the algebraic degree is equal to 7, the nonlinear order is equal to 6, the I/O-degree is equal to 3).
- The $SR_{128}()$ operation corresponds with the ShiftRows() operation of the AES.
- the $Mix_{128}()$ operation is the one used in [16] computed using the operations over $GF(2)$. More precisely for each column of a, we compute $\forall j, 0 \le j \le 3$:

$$
Mix_{128} \begin{pmatrix} a_{0,j} \\ a_{1,j} \\ a_{2,j} \\ a_{3,j} \end{pmatrix} = \begin{pmatrix} a_{3,j} \oplus a_{0,j} \oplus a_{1,j} \\ a_{0,j} \oplus a_{1,j} \oplus a_{2,j} \\ a_{1,j} \oplus a_{2,j} \oplus a_{3,j} \\ a_{2,j} \oplus a_{3,j} \oplus a_{0,j} \end{pmatrix}.
$$

Even if this function is not fully optimal for a diffusion purpose, its branch number is however equal to 4 and its computation is significantly faster than the MixColumns of the AES: Mix_{128} can be computed with only six 32-bit bitwise XORs.

Extraction function of X-FCSR-256. For X-FCSR-256, the extraction function works directly on the 256-bit word $X(t)$ with the function $Round_{256}$ from $\{0,1\}^{256}$ into itself, and a memory of 16×256-bit words which stores the output of $Round_{256}$ that will be used 16 iterations later. More formally, the full extraction function holds as follows:

- Compute the 256-bit word $W(t) = Round_{256}(X(t))$.
- Store $W(t)$ in memory (keep it during 16 iterations).
- Output the 256-bit word $Output_{256}(t) = X(t) \oplus W(t - 16)$.

$Round_{256}$ is a similar round function than the previous one but from $\{0,1\}^{256}$ into itself that can be written as $Round_{256}(a) = Mix_{256}(SR_{256}(SL_{256}(a)))$. If the 256-bit word a is represented at byte level by a 4×8 matrix M where each byte is written $a_{i,j}$ with $0 \le i \le 3$ and $0 \le j \le 7$, then the function $Round_{256}$ can be described as follow:

- the first transformation $SL_{256}()$ consists in the same S-box layer applied at byte level that transforms each byte $a_{i,j}$ into an other byte $b_{i,j}$ such as $b_{i,j} = S(a_{i,j})$ using always the S-box S.

– The $SR_{256}()$ operation corresponds with the ShiftRows() operation of Rijndael described in [13] and consists in Shifting each row of the current matrix on the left at byte level: by 0 for the first row, by one for the second, by three for the third and by four for the fourth one.

– The $Mix_{256}()$ operation is similar to $Mix_{128}()$ but there are here 8 columns to consider.

Key and IV injection. As done in [8], we have split the initialization process into two steps to speed up the IV injection:

-The key schedule, which processes the secret key but does not depend on the IV.

- The IV injection, which uses the output of the key schedule and the IV.

This initializes the stream cipher internal state. Then, the IV setup for a fixed key is less expensive than a complete key setup, improving the common design since changing the IV is more frequent than changing the secret key.

Key schedule. The key setup process used here corresponds to a classical key schedule of a block cipher and is inspired by the one of the DES due to its good resistance against related key attacks [11] and against related key rectangle attacks [17]. The key expansion produces 25×128-bit subkeys denoted $K_0, \cdots, K_{24}$. It works as follow:

– the subkey K_0 is deduced from the master key: $K_0 = (Round_{128}(K))_{<<<23}$ where $_{<<<j}$ denotes a 128-bit left rotation of j positions.

– then K_i is deduced from K_{i-1}: $K_i = Round_{128}((K_{i-1})_{<<<j})$ where $j = 23$ if $i \equiv 3 \mod 4$ and $j = 11$ otherwise.

IV injection. If necessary the IV is extended to a 128-bit word by adding leading zeros. Then, this value is considered as a plaintext that is first enciphered 12 times using the $Round_{128}$ function and then xored with the subkey of the round K_j. More precisely, the process is the following if we denote by V_i the ciphertext after the round i:

$$V_0 = IV \oplus K_0; \qquad \text{for } i \text{ from 1 to 24 do } V_i = Round_{128}(V_{i-1}) \oplus K_i.$$

Then, the values $V_{12}, V_{16}\ V_{20}$ and V_{24} are used to initialize the main registers as follows: $M_a(0) = (V_{12}||V_{20})$ and $M_b(0) = (V_{16}||V_{24})$ whereas the carry registers C_a and C_b are initialized to zero. The two FCSRs are then clocked 16 times to fill the sixteen memory registers of the extraction function. Note that for X-FCSR-256, the registers are directly filled with the 256-bit value $W(t_0) = Round_{256}(M_a(t_0) \oplus M_b(t_0))$ whereas for X-FCSR-128 this value is folded using the 128-bit word $Y(t_0)$.

4   Design Rationale and Security Analysis

Objectives of key and IV injection. In [9], the authors demonstrate that to be secure the key and IV setup (parametrized by the key K) of an IV-

dependent stream cipher must be a pseudo-random function. We have tried to achieve this goal designing our key and IV setup as a block cipher using 12 rounds of $Round_{128}$. Under those conditions, the secret key of the cipher cannot be easily recovered from the initial state of the generator. Once the initial state is recovered, the attacker is only able to generate the output sequence for a particular key and a given IV.

This mechanism already used in [8] also prevents our stream cipher from the distinguishing differential attacks described in Section 2: the values of the two main registers are key dependent and, for a given secret key, no difference could be mastered with a sufficient probability between two distinct IV values and the contents of the main registers. Moreover, and as explained in Section 2, the two carry registers are initialized to 0 to avoid any loss of entropy and prevent TMD attacks.

Role of the core FCSR automata. As noticed in [9], to be secure the keystream generation of an IV-dependent stream cipher must rely on a pseudo-random number generator. Following this requirement and the results of Section 2, we have based our stream cipher on the XOR of two independent 2-adic sequences that provides good pseudo-random sequences. As explained in [5], any cell of the main register of a FCSR automaton provides a 2-adic sequence. Except if there is no feedback bits between two cells, the theoretical dependencies between these sequences cannot be exploited easily. The main idea of X-FCSR is to directly XOR the contents of two distinct FCSRs of size 256 to provide in parallel 256 xored 2-adic sequences, denoted at time t by $X(t) = (x_0(t), \ldots, x_{255}(t))$ in Section 3.

However, the content of $X(t)$ cannot be directly output because in this case we obtain a system of 256 equations with 512 variables at time t that could be easily solved: the condition "q_a and q_b have been chosen such that there is always at least a feedback bit between two consecutive xored cells x_i and x_{i+1}" is necessary but not sufficient.

Moreover, we could also build a guess and determine attack using the knowledge of $X(t)$: first choose an indice i such that there is no carry between $m_{a\,i}$ and $m_{a\,i+1}$, and between $m_{a\,i+1}$ and $m_{a\,i+2}$. So there are carries between $m_{b\,i+1}$ and $m_{b\,i}$ and between $m_{b\,i+2}$ and $m_{b\,i+1}$. Then, guess the contents of the 9 cells $m_{a\,i}$, $m_{a\,i+1}$, $m_{a\,i+2}$, $m_{b\,i}$, $m_{b\,i+1}$, $m_{b\,i+2}$, $c_{b\,i+1}$, $c_{b\,i+2}$ and $m_{b\,255}$ at time t. Using the transition formula, derive the corresponding values at time $t+1$ from the known outputs $x_i(t)$, $x_{i+1}(t)$ and $x_{i+2}(t)$. Repeating this process, we then obtain the consecutive values of the feedback bit $m_{b\,255}$ to deduce the content of the first register and we then could guess the cell values of the second register. In the case of X-FCSR-128 (cf. Section 3), the 128-bit output $Y(t)$ computed from $X(t)$ with the formula $y_i(t) = x_i(t) \oplus x_{i+128}(t)$ prevents this attack from holding even if the information provided by $Y(t)$ seems to be too strong to directly output this value.

An other constraint must be respected: the two registers must be clocked in opposite way because if the two automata are both right-clocked (for example), the values $x_i(t)$ and $x_{i+1}(t+1)$ are correlated according the values of the two feedback bits.

Role of the extraction function. The two round functions $Round_{128}$ and $Round_{256}$ have been chosen for their good diffusion and non-linear properties: they ensure a good resistance against the residual correlations present between the bits of the two main registers of the FCSRs. Their use also prevents attacks that are derived from the ones previously described.

The use of 16 memory registers is a good compromise between a better security and a limited performance cost. First, it increases the number of unknown variables depending on the cells of the main register from 2×256 to 16×256 for X-FCSR-256 (or to 16×128 for X-FCSR-128): solving such a system becomes more expensive than the exhaustive key search. This memory could also be seen as four FCSR automata, since at each operation, the output depends on $X(t)$ and $X(t-16)$. Even if there exists dependencies between $X(t)$ and $X(t-16)$, it is computationally infeasible to determine the values of the main registers at time $t+16$ from the values at time t using the transition functions as noticed in Section 2.

Resistance against known attacks. The good statistical properties (period, balanced sequences and so on) of our constructions are provided by the xored 2-adic properties. We experimentally verified some of them by applying the NIST statistical test suite [23] with success to our two constructions.

As previously mentioned, differential distinguishing attacks, algebraic attacks, Time/Memory/Data trade off attacks and guess and determine attacks are discarded due to the properties described in Section 2 and due to the previous remarks. The fact that 2-adic sequences are xored in the extraction function ensures also a good resistance against 2-adic attacks.

Then, we focus on correlation and fast correlation attacks. In those attacks, the cryptanalyst tries to exploit an existing correlation between some internal bits of the automaton and some output bits using in general linear relations. Since the first FCSR based stream cipher appeared in [5] two years ago, no correlation attack have been exhibited against this construction. This is essentially due to the non-linearity induced by the carries propagation. Even if there exists some 2-adic correlations in a FCSR, those non-linear relations are destroyed by the action of the XOR. Specifically, in the case of X-FCSR, the residual correlations between the neighbor cells of the main registers are stopped by the use of the *Round* functions.

To sum up all the previous analyses, we think that traditional attacks against stream cipher that exploit linear relations built upon the transition function are not realistic in our case. Thus, wanting to cryptanalyse FCSRs leads to create new attacks exploiting other kinds of relations.

5   Conclusion

We have integrated the X-FCSR-128 and X-FCSR-256 stream ciphers to the eSTREAM benchmarking suite [14]. We run the benchmark on an Opteron 250 (1.4Ghz) with GCC 4.1.1 (*-O3-funroll-all-loops -fomit-frame-pointer*) and gather the results in Table 1.

Table 1. X-FCSR performance on an Opteron with the eSTREAM benchmark suite

Algorithm	Keystream speed	cycles/byte			cycles/key	cycles/IV
		40 bytes	576 bytes	1500 bytes	Key setup	IV setup
X-FCSR-256	6.5	50	9.5	7.6	1093	1636
X-FCSR-128	8.21	51	11	9.3	1096	1651
AES-CTR	18.23	22.98	18.3	18.3	172.23	11.74

Our new design X-FCSR-128 and X-FCSR-256 are significantly faster than the AES-CTR except for small plaintexts. The performance of the two proposed stream ciphers are promising. We hope that we have shown how to use FCSRs for software applications. FCSRs have many other advantages such as a simple software implementation or proven properties. In such a design, two important parameters have to be considered: the size of the FCSR automaton, and the size of the output of the extraction function. They have opposite impacts on the throughput and on the security, thus a compromise has to be found. The two stream ciphers presented here correspond to some choice for these parameters. However, the problem of the optimal choice for them remains open.

References

1. Arnault, F., Berger, T.P.: F-FCSR: design of a new class of stream ciphers. In: Gilbert, H., Handschuh, H. (eds.) FSE 2005. LNCS, vol. 3557, pp. 83–97. Springer, Heidelberg (2005)
2. Arnault, F., Berger, T.P., Lauradoux, C.: The FCSR: primitive specification and supporting documentation. ECRYPT - Network of Excellence in Cryptology, Call for stream Cipher Primitives (2005), http://www.ecrypt.eu.org/stream/
3. Arnault, F., Berger, T.P., Lauradoux, C.: Update on F-FCSR stream cipher. ECRYPT - Network of Excellence in Cryptology, Call for stream Cipher Primitives - Phase 2 (2006), http://www.ecrypt.eu.org/stream/
4. Arnault, F., Berger, T.P., Lauradoux, C., Minier, M.: X-FCSR: a new software oriented stream cipher based upon FCSRs (full paper). Cryptology ePrint Archive, Report 2007/380, http://eprint.iacr.org/2007/380
5. Arnault, F., Berger, T.P.: Design and properties of a new pseudorandom generator based on a filtered FCSR automaton. IEEE Trans. Computers 54(11), 1374–1383 (2005)
6. Arnault, F., Berger, T.P., Minier, M.: On the security of FCSR-based pseudorandom generators. In: ECRYPT Network of Excellence - SASC Workshop (2007), Available at: http://sasc.crypto.rub.de/files/sasc2007_179.pdf
7. Ars, G., Faugère, J.-C.: An algebraic cryptanalysis of nonlinear filter generators using Gröbner bases. Research Report INRIA Lorraine, number 4739 (2003)
8. Berbain, C., Billet, O., Canteaut, A., Courtois, N., Gilbert, H., Goubin, L., Gouget, A., Granboulan, L., Lauradoux, C., Minier, M., Pornin, T., Sibert, H.: SOSEMANUK: a fast software-oriented stream cipher. ECRYPT - Network of Excellence in Cryptology, Call for stream Cipher Primitives - Phase 2 (2005), http://www.ecrypt.eu.org/stream/

9. Berbain, C., Gilbert, H.: On the security of IV dependent stream ciphers. In: FSE 2007. LNCS, vol. 4593, Springer, Heidelberg (2007)

10. Berger, T.P., Minier, M.: Two algebraic attacks against the F-FCSRs using the IV mode. In: Maitra, S., Madhavan, C.E.V., Venkatesan, R. (eds.) INDOCRYPT 2005. LNCS, vol. 3797, pp. 143–154. Springer, Heidelberg (2005)

11. Biham, E.: New types of cryptoanalytic attacks using related keys (extended abstract). In: Helleseth, T. (ed.) EUROCRYPT 1993. LNCS, vol. 765, pp. 398–409. Springer, Heidelberg (1994)

12. Courtois, N.: Fast algebraic attacks on stream ciphers with linear feedback. In: Boneh, D. (ed.) CRYPTO 2003. LNCS, vol. 2729, pp. 177–194. Springer, Heidelberg (2003)

13. Daemen, J., Rijmen, V.: AES proposal: Rijndael. In: The Second Advanced Encryption Standard Candidate Conference. N.I.S.T. (1999), available at: http://csrc.nist.gov/encryption/aes/

14. de Cannières, C.: eSTREAM Optimized Code HOWTO (2005), http://www.ecrypt.eu.org/stream/perf

15. Goresky, M., Klapper, A.: Periodicity and distribution properties of combined FCSR sequences. In: Gong, G., Helleseth, T., Song, H.-Y., Yang, K. (eds.) SETA 2006. LNCS, vol. 4086, pp. 334–341. Springer, Heidelberg (2006)

16. Granboulan, L., Levieil, E., Piret, G.: Pseudorandom permutation families over abelian groups. In: Robshaw, M. (ed.) FSE 2006. LNCS, vol. 4047, pp. 57–77. Springer, Heidelberg (2006)

17. Hong, S., Kim, J., Lee, S., Preneel, B.: Related-key rectangle attacks on reduced versions of CHACAL-1 and AES-192. In: Gilbert, H., Handschuh, H. (eds.) FSE 2005. LNCS, vol. 3557, pp. 368–383. Springer, Heidelberg (2005)

18. Jaulmes, E., Muller, F.: Cryptanalysis of ecrypt candidates F-FCSR-8 and F-FCSR-H. ECRYPT Stream Cipher Project Report, 2005/04 (2005), http://www.ecrypt.eu.org/stream

19. Jaulmes, E., Muller, F.: Cryptanalysis of the F-FCSR stream cipher family. In: Preneel, B., Tavares, S. (eds.) SAC 2005. LNCS, vol. 3897, pp. 20–35. Springer, Heidelberg (2005)

20. Klapper, A., Goresky, M.: 2-adic shift registers. In: Anderson, R. (ed.) Fast Software Encryption. LNCS, vol. 809, pp. 174–178. Springer, Heidelberg (1993)

21. Klapper, A., Goresky, M.: Feedback Shift Registers, 2-Adic Span, and Combiners with Memory. J. Cryptol. 10(2), 111–147 (1997)

22. Network of Excellence in Cryptology ECRYPT. Call for stream cipher primitives, http://www.ecrypt.eu.org/stream/

23. National Institute of Standards and Technology. The statistical test suite (v.1.8) (2005), http://csrc.nist.gov/rng/rng2.html

Efficient Window-Based Scalar Multiplication on Elliptic Curves Using Double-Base Number System

Rana Barua, Sumit Kumar Pandey, and Ravi Pankaj

Indian Statistical Institute
205, B.T. Road
Kolkata, India
{rana,mtc0518,mtc0520}@isical.ac.in

Abstract. In a recent paper [10], Mishra and Dimitrov have proposed a window-based Elliptic Curve (EC) scalar multiplication using double-base number representation. Their methods were rather heuristic. In this paper, given the window lengths w_2 and w_3 for the bases 2 and 3, we first show how to fix the number of windows, ρ, and then obtain a Double Base Number System (DBNS) representation of the scalar n suitable for window-based EC scalar multiplication. Using the DBNS representation, we obtain our first algorithm that uses a small table of precomputed EC points. We then modify this algorithm to obtain a faster algorithm by reducing the number of EC additions at the cost of storing a larger number of precomputed points in a table. Explicit constructions of the tables are also given.

1 Introduction

The efficiency of Elliptic Curve Cryptography (ECC) implementation largely depends upon how fast one can compute the point $[n]P = \sum_{i=1}^{n} P$, given a point P on the curve and the integer (scalar) n. Several efficient algorithms for computing $[n]P$ have been proposed. See Avanzi et al [1] and Hankerson et al [8] for detailed discussion on these methods. Several window-based methods have also been proposed, of them w-NAF methods seem to be very efficient.

Recently, Mishra and Dimitrov[10] have proposed a new window-based scalar multiplication algorithm by suitably representing the scalars in DBNS. The DBNS has recently been exploited to compute exponentiation (or scalar multiplication) efficiently[5]. The sparseness of the representation leads to fewer point additions than the usual double-and-add or NAF methods. In fact, one can have a DBNS representation of n having $O(logn/loglogn)$ terms. This together with the fact that $2^a 3^b [P]$, for an EC point P, can be computed efficiently([6]) gives rise to some very efficient algorithms for scalar multiplication. However, the method in [10] is quite heuristic and an explicit method for finding the *partition length* ρ is not given, nor any explicit expression for the cost of scalar multiplication in terms of EC addition, doubling or tripling.

K. Srinathan, C. Pandu Rangan, M. Yung (Eds.): Indocrypt 2007, LNCS 4859, pp. 351–360, 2007.

In this paper, we first show how to fix ρ, the length of the partition i.e. the number of windows, given w_2, w_3, the *lengths of the window corresponding to the bases 2 and 3* respectively. We then obtain a DBNS representation of the scalar n suitable for window-based scalar multiplication. We obtain the DBNS representation more efficiently than in [10] using a much smaller search space. Using our DBNS representation we obtain our scalar multiplication algorithm using a table look-up that stores $(w_2+1)(w_3+1)$ EC points. Explicit construction of the table is also given. Using a larger table that stores $(2^{w_2} \times 3^{w_3})/2$ points, we modify the above algorithm that considerably reduces the number of EC point additions. We also obtain an expression for the average number of EC additions, doubling, tripling required for computing the scalar multiplication.

2   Double-Base Number System

The double base number system (DBNS) [6] is a representation scheme in which every integer n is represented as the sum or difference of numbers of the form $2^a 3^b$(called $\{2,3\}$-integers) i.e.

$$n = \sum_{i=1}^{m} s_i 2^{b_i} 3^{t_i}, \text{ with } s_i \in \{-1,1\} \text{ and } b_i, t_i \geq 0.$$

This representation is very short and the representation scheme is highly redundant. It has been shown that (cf [5], every positive integer n can be represented as the sum of at most $O(logn/loglogn)$ $\{2,3\}$-integers. A simple Greedy algorithm ensures nearly shortest representation for a given integer. A modified Greedy was proposed in [10] suitable for window-based scalar multiplication. Here we propose a more efficient algorithm suitable for window-based method that uses a search space consisting only of $2^{w_2} 3^{w_3}$ integers.

3   Proposed Window-Based Method for Scalar Multiplication

Unlike earlier proposed methods ([10], [6], [11]), by choosing the window sizes, we obtain natural bounds on maximum exponents of bases 2 and 3, and propose a window method so that it reduces the overall storage.

Let n be an r-bit integer. Let w_2, w_3 be the *dimension* of the window. Let max_2, max_3 be integers satisfying $2^{max_2} 3^{max_3} \geq n$ and such that $max_2/w_2 = max_3/w_3 = \rho$, say (as in [10]). Then, we have

$$max_2 + max_3 log_2 3 \geq log_2 n \tag{1}$$

Substituting $max_2 = \rho w_2$, $max_3 = \rho w_3$ in (1), we get

$$\rho \geq \frac{log_2 n}{w_2 + w_3 log_2 3} \tag{2}$$

For fix window lengths, we can obtain the number ρ by choosing ρ to be the least positive integer satisfying inequality (2). If n is r-bit, then we may also choose $\rho = \lceil r/(w_2 + w_3 log_2 3) \rceil$. *Henceforth, we fix such a number ρ.*

3.1   Representation of n

There is no unique representation of n in double base number system. Finding a canonical representation of n, i.e. having least number of terms, is extremely difficult. A short and sparse representation of n results in less computation for scalar multiplication. We propose a representation of n which can be obtained very efficiently and will be suitable for our window-based method. Since we are particularly interested in window method, we will first obtain a DBNS representation of integers m *lying in the window* i.e. $0 \leq m \leq 2^{w_2} 3^{w_3}$ using (distinct) terms in the window.

Proposition 3.1. *Every integer* $0 \leq m \leq 2^{w_2} 3^{w_3}$ *can be represented as* $\sum_j s_j 2^{b_j} 3^{t_j}$, *where* $s_j \in \{-1, 1\}$ *and* $0 \leq b_j \leq w_2, 0 \leq t_j \leq w_3$.

To find a DBNS representation of m lying in a window, we will use a table T such that $T(a, b) = 2^a 3^b$ where $0 \leq a \leq w_2$ and $0 \leq b \leq w_3$. With the help of table T, we can easily find the nearest $\{2, 3\}$-integer lying in a window and hence the double base representation of any integer m, lying in a window.

Algorithm 1 gives the method to find a DBNS representation of n by greedy approach which is almost the same as in [10].

Algorithm 1. Conversion into DBNS

Input : an integer m such that $0 \leq m \leq 2^{w_2} 3^{w_3}$ for a given window
    lengths w_2, w_3 for 2, 3 respectively and table T, where $T(a, b) = 2^a 3^b$
    and $0 \leq a \leq w_2, 0 \leq b \leq w_3$.
Output : The sequence (s_i, b_i, t_i) such that $m = \sum_{i=1}^{l} s_i 2^{b_i} 3^{t_i}$,
    where $s_i \in \{-1, 1\}, 0 \leq b_i \leq w_2, 0 \leq t_i \leq w_3$
1: $i \leftarrow 1$
2: $s_i \leftarrow 1$
3: $A[i] \leftarrow (0, 0, 0)$
4: while $m > 0$ do
5:    define $X = 2^{b_i} 3^{t_i}$, the best approximation of m in T with
       $0 \leq b_i \leq w_2$ and $0 \leq t_i \leq w_3$. If there are two choices,
       choose nearest integer smaller to m.
6:    $A[i] \leftarrow (s_i, b_i, t_i)$
7:    if $m < X$ then
8:       $s_{i+1} \leftarrow -s_i$
9:    $m \leftarrow |m - X|$
10:   $i \leftarrow i + 1$
11:**return** A.

Now, for any integer n, by our choice of ρ we have $0 \leq n \leq (2^{w_2} 3^{w_3})^{\rho}$. The propositon below gives a way to represent n suitable for our purpose.

Proposition 3.2. *Every integer* $0 \leq n \leq (2^{w_2} 3^{w_3})^{\rho}$ *can be represented as* $n = M_{\rho-1}(2^{w_2} 3^{w_3})^{\rho-1} \pm M_{\rho-2}(2^{w_2} 3^{w_3})^{\rho-2} \pm \cdots \pm M_0$ *s.t.* $0 \leq M_{\rho-1} \leq 2^{w_2} 3^{w_3}$ *and* $0 \leq M_j \leq (2^{w_2} 3^{w_3})/2$ *for* $0 \leq j \leq \rho - 1$.

Proof: If $n = (2^{w_2}3^{w_3})^\rho$, then it is obvious. So let $0 \leq n < (2^{w_2}3^{w_3})^\rho$. Then
$n = M'_{\rho-1}(2^{w_2}3^{w_3})^{\rho-1} + R'_{\rho-1}$, where $0 \leq R'_{\rho-1} < (2^{w_2}3^{w_3})^{\rho-1}$. Clearly $M'_{\rho-1} < 2^{w_2}3^{w_3}$, for otherwise $n \geq (2^{w_2}3^{w_3})^\rho$. If $R'_{\rho-1} > (2^{w_2}3^{w_3})^{\rho-1}/2$, take $M_{\rho-1} = M'_{\rho-1} + 1$ and $R_{\rho-1} = R'_{\rho-1} - (2^{w_2}3^{w_3})^{\rho-1}$, else $M_{\rho-1} = M'_{\rho-1}$ and $R_{\rho-1} = R'_{\rho-1}$. So, $n = M_{\rho-1}(2^{w_2}3^{w_3})^{\rho-1} + R_{\rho-1}$, where $0 \leq M_{\rho-1} \leq 2^{w_2}3^{w_3}$ and $-(2^{w_2}3^{w_3})^{\rho-1}/2 < R_{\rho-1} \leq (2^{w_2}3^{w_3})^{\rho-1}/2$.

Now, take $|R_{\rho-1}|$ so that $0 \leq |R_{\rho-1}| \leq (2^{w_2}3^{w_3})^{\rho-1}/2$.
Let $|R_{\rho-1}| = M_{\rho-2}(2^{w_2}3^{w_3})^{\rho-2} + R_{\rho-2}$, where $0 \leq M_{\rho-2} \leq 2^{w_2}3^{w_3}/2$ and $-(2^{w_2}3^{w_3})^{\rho-2}/2 < R_{\rho-2} \leq (2^{w_2}3^{w_3})^{\rho-2}/2$. Observe that $M_{\rho-2} \not> (2^{w_2}3^{w_3})/2$, for otherwise $|R_{\rho-1}| > (2^{w_2}3^{w_3})^{\rho-1}/2$.

Proceeding similarly, we have the result.

Algorithm 2 gives the method to find a representation of n.

Algorithm 2. To find representation of n

Input : an integer n, window dimension w_2, w_3 and ρ.
Output: a seq. of $(s_i, M_i)_{i>0}$ such that $n = \sum_{i=1}^{\rho} s_{\rho-i}M_{\rho-i}(2^{w_2}3^{w_3})^{\rho-i}$,
    where $s_i \in \{-1, 1\}$, $0 \leq M_{\rho-1} \leq 2^{w_2}3^{w_3}$ and
    $0 \leq M_{\rho-i} \leq (2^{w_2}3^{w_3})/2$ for all $2 \leq i \leq \rho$.

```
1:  i ← 1
2:  s_{ρ-1} ← 1
3:  R ← n
4:  X ← (2^{w_2}3^{w_3})^{ρ-1}
5:  while i ≤ ρ do
6:      M_{ρ-i} ← ⌊R/X⌋
7:      R ← R - M_{ρ-i}X
8:      s_{ρ-i-1} ← s_{ρ-i}
9:      if R > X/2 then
10:         M_{ρ-i} ← M_{ρ-i} + 1
11:         R ← X - R
12:         s_{ρ-i-1} ← -s_{ρ-i}
13:     X ← X/2^{w_2}3^{w_3}
14:     i ← i + 1
15:     A[ρ - i] ← (s_{ρ-1}, M_{ρ-i})
16: return A
```

After getting a representation of n, we are in position to find $[n]P$, given an EC point P, using Horner's scheme. To calculate $[n]P$, we will use another table T^P which contains the *precomputed* values of $[2^a 3^b]P$, where $0 \leq a \leq w_2$ and $0 \leq b \leq w_3$, i.e. $T^P(a, b) = [2^a 3^b]P$. Observe that, since negation of an EC point can be obtained almost free, we omit its cost in our calculation. Using T^P, we can find $[n]P$ as follows.

1. Compute ρ. Then we calculate $M'_j s$, where $n = \sum_{j=1}^{\rho} s_{\rho-j}M_{\rho-j}(2^{w_2}3^{w_3})^{\rho-j}$, where M_j, s_j's are as in Proposition 3.2. (Algorithm 2).
2. Now, we find out $[M_j]P$, (Algorithm 3). To obtain this we first find representation of M_j in DBNS using Algorithm 2, say $\sum_{j=1}^{l} s_j 2^{b_j} 3^{t_j}$. Then looking

at table T^P, we find the values of $s_j[2^{b_j}3^{t_j}]P$ for all $j = 1, \ldots, l$ and adding these points gives the value of $[M_j]P$.

3. After getting the values of all $[M_j]P$ in the representation of n, we evaluate $[n]P$ by applying Horner's scheme (Algorithm 4).

Algorithm 3. calculation of $[m]P$, for m in the window

Input : an integer m such that $0 \leq m \leq 2^{w_2}3^{w_3}$, a point P
 on an elliptic curve E, tables T and T^P.
Output : $[m]P$
1: $A \leftarrow$ **Algorithm 1**(m, w_2, w_3, T)
2: $L \leftarrow length(A)$
3: $P \leftarrow O$ (point at infinity on elliptic curve E)
4: $i \leftarrow 1$
5: **while** $i \leq L$ **do**
6: $(s_i, b_i, t_i) \leftarrow A[i]$
7: $P \leftarrow P + s_i T^P(b_i, t_i)$
8: $i \leftarrow i + 1$
9: **return** P

It is not hard to check that the number of terms in the DBNS representation of m lying in the window is at most $c(w_2 + log3w_3)$ for $c < 1$. Perhaps a much better estimate can be obtained. Thus we have.

Proposition 3.3. *Algorithm 3 correctly computes $[m]P$ for $0 \leq m \leq 2^{w_2}3^{w_3}$ using at most $c(w_2 + log3w_3)$ additions. The table T^P stores $(w_2 + 1)(w_3 + 1)$ EC points.*

Algorithm 4 calculates $[n]P$.

Algorithm 4. Calculation of $[n]P$

Input : an integer n such that $0 \leq n \leq (2^{w_2}3^{w_3})^\rho$, a point P
 on an elliptic curve E, partition length ρ, tables T and T^P.
Output : $[n]P$
1: $A \leftarrow$ **Algorithm 2**(n, w_2, w_3, ρ)
2: $P \leftarrow O$ (point at infinity on elliptic curve E)
3: $i \leftarrow 1$
4: **while** $i \leq \rho$ **do**
5: $(s_{\rho-i}, M_{\rho-i}) \leftarrow A[\rho - i]$
6: $Q \leftarrow$ **Algorithm 3**$(M_{\rho-i}, w_2, w_3, P, T, T^P)$
7: $P \leftarrow P + s_{\rho-i}Q$
8: $P \leftarrow [3^{w_3}]P$
9: $P \leftarrow [2^{w_2}]P$
10: $i \leftarrow i + 1$
11: **return** P

The following is not very hard to check using Horner's scheme, since the probably of each M_j being non-zero is $(1 - \frac{1}{2^{w_2}3^{w_3}})$.

Proposition 3.4. *Algorithm 4 correctly computes $[n]P$ using on an average $(t-1)(c(w_2 + log3w_3) - 1)$ EC additons, $(\rho - 1)w_2$ point doublings and $(\rho - 1)w_3$ point triplings, where $c < 1, \rho = \lceil \frac{log_2 n}{w_2 + w_3 log_2 3} \rceil$, and $t = (1 - \frac{1}{2^{w_2} 3^{w_3}})\rho$. Table T stores $(w_2 + 1)(w_3 + 1)$ integers, while table T^P stores $(w_2 + 1)(w_3 + 1)$ EC points.*

We can reduce the cost of computation if more precomputed points are stored. For that we construct T_{all}^P instead of T^P such that $T_{all}^P(i) = [i]P$ for $1 \le i \le 2^{w_2} 2^{w_3}/2$. Since in the representation of n, maximum value of M_j can be $(2^{w_2} 3^{w_3}/2)$, except $M_{\rho-1}$ which can have maximum value $2^{w_2} 3^{w_3}$, the number of precomputed points is $(2^{w_2} 3^{w_3}/2)$. If $M_{\rho-1} > 2^{w_2} 3^{w_3}/2$ then $[M_{\rho-1}]P$ can be evaluated by calculating first $\lfloor [M_{\rho-1}/2] \rfloor P$ and then doubling and adding P if $M_{\rho-1}$ is odd. Hence, steps for evaluating $[n]P$ using table T_{all}^P will be same except step 3. In modified step 3, we will evaluate $[M_j]P$ by just looking at table T_{all}^P.

Algorithm 4^0. Calculation of $[n]P$
Input : an integer n such that $0 \le n \le (2^{w_2} 3^{w_3})^\rho$, a point P
$\quad$ on an elliptic curve E, partition length ρ, table T_{all}^P.
Output : $[n]P$
1: $A \leftarrow$ **Algorithm 3**(n, w_2, w_3, ρ)
2: $P \leftarrow O$ (point at infinity on elliptic curve E)
3: $i \leftarrow 1$
4: **while** $i \le \rho$ **do**
5: $\quad$ $(s_{\rho-i}, M_{\rho-i}) \leftarrow A[\rho - i]$
6: $\quad$ **if** $M_{\rho-i} = 0$ **then**
7: $\quad\quad$ $Q \leftarrow O$ (point at infinity)
8: $\quad$ **else**
9: $\quad\quad$ **if** $i = 1$ **then**
10: $\quad\quad\quad$ **if** $M_{\rho-i} > 2^{w_2} 3^{w_3}/2$
11: $\quad\quad\quad\quad$ $M_{\rho-i} \leftarrow \lfloor M_{\rho-i}/2 \rfloor$
12: $\quad\quad\quad\quad$ $Q \leftarrow 2[T_{all}^P(M_{\rho-i})]$
13: $\quad\quad\quad\quad$ **if** $M_{\rho-i}$ is odd **then**
14: $\quad\quad\quad\quad\quad$ $Q \leftarrow Q + P$
15: $\quad\quad\quad$ **else**
16: $\quad\quad\quad\quad$ $Q \leftarrow T_{all}^P(M_{\rho-i})$
17: $\quad\quad$ **else**
18: $\quad\quad\quad$ $Q \leftarrow T_{all}^P(M_{\rho-i})$
19: $\quad$ $P \leftarrow P + s_{\rho-i}Q$
20: $\quad$ $P \leftarrow [3^{w_3}]P$
21: $\quad$ $P \leftarrow [2^{w_2}]P$
22: $\quad$ $i \leftarrow i + 1$
23: **return** P

The following is now clear.

Proposition 3.5. *Algorithm 4^0 correctly computes $[n]P$ using $(\rho - 1)w_2$ point doublings, $(\rho - 1)w_3$ point triplings and at most ρ point additions. Table T_{all}^P stores $2^{w_2} 3^{w_3}/2$ EC points.*

4   Computation of T^P and T^P_{all}

The algorithms described so far use one or more look-up tables. If the EC point P is known in advance, then the tables can be precomputed and stored; otherwise they have to be computed online. Formation of tables T, T^P, and T^P_{all} may take much computation but it can be reduced if they are formed recursively.

Note that $T(a,b) = 2^a 3^b$ and $T^P(a,b) = [2^a 3^b]P$, so $T^P(a,b) = [T(a,b)]P$. By considering the lexicographic ordering of the tuples (a,b) we can form T, T^P as follows:

1. $T(0,0) = 1$; $T^P(0,0) = P$
2. $T(0,b) = 3T(0,b-1)$; $T^P(0,b) = [3]T^P(0,b-1), b > 0$.
3. $T(a,b) = 2T(a-1,b)$; $T^P(a,b) = [2]T^P(a-1,b), a > 0$.

Algorithm 5 illustrates the method to form table T^P.

Algorithm 5. Table construction for T^P
Input : window lengths w_2, w_3 and an EC point P
Output : an array $T^P(a,b)$ such that $T^P(a,b) = [2^a 3^b]P$
where $0 \le a \le w_2$ and $0 \le b \le w_3$.
1: $T^P(0,0) \leftarrow P$
2: $a \leftarrow 0$
3: $b \leftarrow 0$
4: **while** $b < w_3$ **do**
5:      $T^P(a,b+1) \leftarrow [3]T^P(a,b)$
6:      $b \leftarrow b+1$
7: $b \leftarrow 0$
8: **while** $b < w_3 + 1$ **do**
9:    **while** $a < w_2$ **do**
10:      $T^P(a+1,b) \leftarrow [2]T^P(a,b)$
11:      $a \leftarrow a+1$
12:    $b \leftarrow b+1$
13:**return** T^P

Proposition 4.1. *Algorithm 5 correctly computes T^P used in Algorithm 4 using w_3 triplings and $w_2(w_3 + 1)$ doublings.*

Remarks: If tripling is less expensive than doubling(as in the case of EC over fields of characteristic 3), we form T^P as follows:

1. $T^P(0,0) = P$
2. $T^P(a,0) = 2[T^P(a-1,0)], a > 0$ $T^P(a,b) = 3[T^P(a,b-1)], b > 0$.

One can then appropriately modify Algorithm 5, using the above recursive relation. This will involve w_2 doubling and $w_3(w_2 + 1)$ triplings

Finally, Table T^P_{all} can be formed as follows, if T^P is given:

1. $T^P_{all}(1) = P$ $T^P_{all}(m+1) = T^P(m+1)$, if $m+1$ is a $\{2,3\}$-integer,
2. $T^P_{all}(m+1) = T^P_{all}(m) + P$, otherwise.

Clearly this requires $2^{w_2} 3^{w_3}/2 - (w_1 w_2 + w_1 + w_2)$ EC point additions.

5   Comparison

The present method for scalar multiplication is comparable or performs better in terms of both storage and computation in many cases.

(a) Storage - Methods for scalar multiplication in [10], [11] and [6] use a table T of large size to find the nearest representation of n, but in our method the table size required to find nearest representation of n is comparatively very small.

(b) Computation - In this method, we use a table T^P or T_{all}^P of precomputed points, which reduces the overall computation in scalar multiplication. We have computed cost of $[n]P$ using existing algorithm for $[2^{w_2}]P$ ([4]) and $[3]P$ ([2]) in affine coordinates for curves over characteristic 2. Since square is almost free in affine coordinates, we have not taken the cost of squaring. On the other hand, cost for computing $[n]P$ has been calculated using algorithm for $[2^{w_2}]P$ ([9]), $[3^{w_3}]P$ ([6]) and mixed addition ([3]) in Jacobian coordinates. Table 1 summarizes the cost of operation required.

We calculated cost of field operations for different window lengths in Table 2. We compared our results with some earlier methods in Table 3.

Table 1. Cost of operation required in different point addition algorithm. Here **[I]**, **[S]** and **[M]** denote cost of field inversion, squaring and multiplication respectively.

Operation	cost	
	Affine	Jacobian
$P + Q$	$1[\mathbf{I}] + 2[\mathbf{M}]$	$4[\mathbf{S}] + 12[\mathbf{M}]$
mixed-$(P + Q)$	-	$3[\mathbf{S}] + 8[\mathbf{M}]$ ($cf[3]$)
$[2^w]P$	$1[\mathbf{I}] + (4w - 2)[\mathbf{M}]$ ($cf[4]$)	$(4w + 2)[\mathbf{S}] + 4w[\mathbf{M}]$ ($cf[9]$)
$[3]P$	$1[\mathbf{I}] + 7[\mathbf{M}]$ ($cf[2]$)	$6[\mathbf{S}] + 10[\mathbf{M}]$ ($cf[6]$)
$[3^w]P$	-	$(4w + 2)[\mathbf{S}] + (11w - 1)[\mathbf{M}]$ ($cf[6]$)

Table 2. Cost of scalar multiplication for 160 bit scalar

w_2	w_3	using T^P			using T_{all}^P		
		# storage	Affine $[\mathbf{I}]/[\mathbf{M}] = 8$	Jacobian $[\mathbf{S}]/[\mathbf{M}] = 0.8$	# storage	Affine $[\mathbf{I}]/[\mathbf{M}] = 8$	Jacobian $[\mathbf{S}]/[\mathbf{M}] = 0.8$
1	1	4	2042.3 [M]	1973.7[M]	3	2066.7[M]	2000.5[M]
1	2	6	2030.0[M]	1966.8[M]	9	1878.9[M]	1809.6[M]
1	3	8	1993.5[M]	1932.9[M]	27	1750.0[M]	1674.6[M]
2	1	6	1716.0[M]	1812.8[M]	6	1679.3[M]	1774.7[M]
2	2	9	1775.0[M]	1823.2[M]	18	1665.4[M]	1708.4[M]
2	3	12	1800.3[M]	1822.7[M]	54	1584.9[M]	1598.6[M]
3	1	8	1578.7[M]	1766.9[M]	12	1490.4[M]	1678.8[M]
3	2	12	1637.8[M]	1760.3[M]	36	1504.4[M]	1623.8[M]
3	3	16	1689.4[M]	1774.6[M]	108	1459.1[M]	1535.0[M]
4	1	10	1485.2[M]	1732.7[M]	24	1310.2[M]	1550.7[M]
4	2	15	1584.4[M]	1764.8[M]	72	1362.5[M]	1534.0[M]
4	3	20	1632.3[M]	1768.1[M]	216	1385.6[M]	1511.6[M]

Table 3. Comparison among different proposed methods

Algorithm	size of Table T	# Precomputed points	Affine $[\mathbf{I}]/[\mathbf{M}] = 8$	Jacobian $[\mathbf{S}]/[\mathbf{M}] = 0.8$
for 160 bit scalar				
Mishra-Dimitrov method [11]	48384	5	1469.0[M]	1502.0[M]
Mishra-Dimitrov method [10] for window length (3,2)	4332	26	-	1692.2[M]
$w-$NAF (for $w = 3$)	0	3	2016[M]	-
$w-$NAF (for $w = 4$)	0	5	1894[M]	-
Our method (using T^P) for window length (4,1)	10	10	1485.2[M]	1732.7[M]
Our method (using T^P_{all}) for window length (4,1)	10	24	1310.2[M]	1550.7[M]
Our method (using T^P) for window length (4,2)	15	15	1584.4[M]	1764.8[M]
Our method (using T^P_{all}) for window length (4,2)	15	72	1362.5[M]	1534.0[M]
for 200 bit scalar				
Doche-Imbert method [7] for window length (1,1)	313	3	-	2019[M]
Our method (using T^P) for window length (4,1)	10	10	-	2312.6[M]
Our method (using T^P) for window length (4,2)	15	15	-	2272.9[M]
Our method (using T^P_{all}) for window length (4,1)	10	24	-	1938.4[M]

References

1. Avanzi, R.M., Cohen, H., Doche, C., Frey, G., Langue, T., Nguyen, K., Vercauteren, F.: Handbook of Elliptic and Hyperelliptic Curve Cryptography. CRC press, Boca Raton, USA (2005)
2. Ciet, M., Lauter, K., Joye, M., Montgomery, P.L.: Trading inversions for multiplications in elliptic curve cryptography. Designs, Codes and Cryptography 39(2), 189–206 (2006)
3. Cohen, H., Miyaji, A., Ono, T.: Efficient Elliptic Curve Exponentiation Using Mixed coordinates. In: Ohta, K., Pei, D. (eds.) ASIACRYPT 1998. LNCS, vol. 1514, pp. 51–65. Springer, Heidelberg (1998)
4. Dahab, R., Lopez, J.: An Improvement of Guajardo-Paar Method for Multiplication on non-supersingular Elliptic Curves. In: SCCC 1998. Proceedings of the XVIII International Conference of the Chilean Computer Science Society, November 12-14, pp. 91–95. IEEE Computer Society Press, Los Alamitos (1998)
5. Dimitrov, V., Gullien, G.A., Miller, W.C.: An algorithm for modular exponentiation. Information Processing Letters 66(3), 155–159 (1998)
6. Dimitrov, V., Imbert, L., Mishra, P.K.: Efficient and Secure Curve Point Multiplication Using Double Base Chain. In: Roy, B. (ed.) ASIACRYPT 2005. LNCS, vol. 3788, pp. 59–79. Springer, Heidelberg (2005)

7. Doche, C., Imbert, L.: Extended Double-Base Number System with Applications to Elliptic Curve Cryptography. In: Barua, R., Lange, T. (eds.) INDOCRYPT 2006. LNCS, vol. 4329, pp. 335–348. Springer, Heidelberg (2006)
8. Hankerson, D., Menezes, A., Vanstone, S.: Guide to Elliptic Curve Cryptography. Springer, New York (2004)
9. Itoh, K., Takenaka, M., Torii, N., Temma, S., Kurihara, Y.: Fast implementation of public-key cryptography on a DSP TMS320C6201. In: Koç, Ç.K., Paar, C. (eds.) CHES 1999. LNCS, vol. 1717, p. 6172. Springer, Heidelberg (1999)
10. Mishra, P.K., Dimitrov, V.: Window-Based Elliptic Curve Scalar Multiplication using Double Base Number Representation, Short Papers, Inscrypt (2007)
11. Mishra, P.K., Dimitrov, V.: Efficient Quintuple Formulas for Elliptic Curves and Efficient Scalar Multiplication using Multibase Number Representation, ePrint archive. In report 2007/040 (2007), http://www.iacr.org

Extended Multi-Property-Preserving and ECM-Construction

Lei Duo and Chao Li

Department of Science, National University of Defense Technology,
Changsha, China
duoduolei@gmail.com

Abstract. For an iterated hash, it is expected that, the hash transform inherits all the cryptographic properties of its compression function. This means that the cryptanalytic validation task can be confined to the compression function. Bellare and Ristenpart [3] introduced a notion Multi-Property preserving (MPP) to characterize the goal. In their paper, the MPP was collision resistance preserving (CR-pr), pseudo random function preserving (PRF-pr) and pseudo random oracle preserving (PRO-pr). The probability distribution of hash transform influences the randomness and adversary's advantage on collision finding, we expect that the hash transform is almost uniformly distributed and this property is inherited from its compression function and call it Almost-Uniform Distribution preserving (AUD-pr). However, AUD-pr is not always true for MD-strengthening Merkle-Damgård [7,12] transform. It is proved that the distribution of Merkle-Damgård transform is not only influenced by output distribution of compression function, but also influenced by the iteration times. Then, we recommend a new construction and give proofs of satisfying MPP that is CR-pr, PRO-pr, PRF-pr and AUD-pr.

Keywords: Hash functions, random oracle, Merkle-Damgård, collision resistance, pseudo random function, almost uniform distribution.

1 Introduction

Most of hash functions are iterated hash function with Merkle-Damgård construction [7,12], it has been proven [7,12] to be collision-resistance preserving (CR-Pr): if its compression function is collision resistant (CR). However, Coron et. al [6] pointed out the MD-strengthening padded Merkle-Damgård construction is not indifferentiable, which was first introduced by Maurer [11], from random oracle even when its compression function is indifferentiable from random oracle. As pointed out by Bellare and Ristenpart [3], current usage makes it obvious that CR no longer suffices as the security goal for hash functions. Then, towards the goal of building strong, multi-purpose hash functions, they introduced the notion of a multi-property preserving (MPP), which is that, if we want a hash function with properties $P_1, \ldots, P_n$, then we should design a compression function with the goal of having properties $P_1, \ldots, P_n$, and apply an iteration transform (domain extension transform) that provably preserves P_i for every

K. Srinathan, C. Pandu Rangan, M. Yung (Eds.): Indocrypt 2007, LNCS 4859, pp. 361–372, 2007.

$i \in [1..n]$. And they call a compression function a multi-property one, and call such a transform a multi-property-preserving transform. In their paper, the MPP was collision resistance preserving (CR-pr), pseudo random function preserving (PRF-pr) and pseudo random oracle oracle preserving (PRO-pr). Then, they recommended a new construction called EMD (Enveloped Merkle-Damgård), which is the first to meet their MPP.

The distribution of a hash construction is an important character to evaluate the randomness. The adversary's advantage on collision finding is also influenced by the distribution bound. We think a good hash construction should have good output distribution that is almost uniform distribution. To make one hash transform fitted with different compression function, we expect that the compression function has almost uniform distribution and the iteration preserves the almost uniform distribution call it almost uniform distribution preserving (AUD-pr). We add this new property to the multi-property preserving and the MPP becomes CR-pr, PRF-pr, PRO-pr and AUD-pr.

The almost uniform distribution is decided by the distribution bounds. Surprisingly, the distribution bound of Merkle-Damgård construction with MD-strengthening may reach 1, even when the compression function has good distribution bound, where the bound of hash is not only related with the distribution bound of compression function, but also related with the iteration times.

Since MD5 and SHA-1 were attacked by Wang et. al [13,4], increased attention has been paid to security of hash, and some improved structures are given. Lucks [10] given two imported structure of Merkle-Damgård construction called wide-pipe hash and double-pipe hash. Gauravaram et. al recommend 3c hash [9]. Biham and Dunkelman recommend a new structure called HAIFA [5]. Bellare and Ristenpart [3] presented EMD construction, which is first structure proven to be MPP (CR-pr, PRF-pr, and PRO-pr). The wide-pipe hash, EMD and HMAC are not AUD-pr and the 3C is AUD-pr, but the 3c is not PRO-pr. Then, we give a new construction, which is called ECM (Enveloped CheckSum Merkle-Damgård) that is MPP(CR-pr, PRF-pr, PRO-pr and AUD-pr). At last a generalized enveloped MPP construction is given.

The section 2 is basic notions. Section 3 is an example of that output of hash compression function has almost uniform distribution, however, iterated by the Merkle-Damgård construction the hash does not have good distribution. Section 4 gives the proof of distribution bounds on Merkle-Damgård construction. New recommendation construction ECM and its proofs of MPP are in section 5. Generalized construction is in section 6.

2   Notation

Capital calligraphic letters (e.g. $\mathcal{X}$) denote sets and the corresponding capital letter (e.g. X) denotes a random variable taking values in the set. Concrete values for X are usually denoted by the corresponding small letter x. The distribution of a random variable X is denoted P_X, we use $P_X(x) \overset{def}{=} P(X = x)$, similarly for conditional probabilities $P_{Z|X=x}(z) \overset{def}{=} P(Z = z|X = x)$.

Let message block $y \in \{0,1\}^{\kappa}$, $\tilde{y} \in \{0,1\}^{\kappa-n}$, padded message $m = y_1 \ldots y_i \in \cup_{l=1}^{*}\{0,1\}^{\kappa \cdot l}$ and $\tilde{m} = y_1\| \ldots \|y_t\|\tilde{y}$, where the padded message m or $\tilde{m}$ means MD-strengthening padded message. Let $\mathbf{0}$ be n bits $'0'$.

Let function $\mathbf{F} : \mathcal{X} \times \mathcal{Y} \to \{0,1\}^n$ be $Z = \mathbf{F}(X,Y)$. For function $Z = \mathbf{F}(X,Y)$, the distribution of Z is decided by distribution of X and Y. We give a definition of derived probability distribution.

Definition 1. *For function* $\mathbf{F} : \mathcal{X} \times \mathcal{Y} \to \mathcal{Z}$, *let* $Z = \mathbf{F}(X,Y)$. X *and* Y *are independent. Let* $\chi_{\mathbf{F}(x,y)}(z) = \begin{cases} 1, z = \mathbf{F}(x,y) \\ 0, z \neq \mathbf{F}(x,y) \end{cases}$. *The derived probability of random variable* Z *is defined as*

$$P_Z(z) = \sum_{x \in \mathcal{X}} \sum_{y \in \mathcal{Y}} P_X(x)P_Y(y)\chi_{\mathbf{F}(x,y)}(z).$$

The derived conditional probability of z *with* X *taking value* x *is defined as* $P_{Z|X=x}(z) = \sum_{y \in \mathcal{Y}} P_X(x)\chi_{\mathbf{F}(x,y)}(z)$. *If* $\mathcal{Z} \subseteq \{0,1\}^n$, $z \notin \mathcal{Z}$ *and* $z \in \{0,1\}^n$, *then we assume* $P_Z(z) := 0$.

Definition 2. *For function* $\mathbf{F} : \mathcal{X} \times \mathcal{Y} \to \{0,1\}^n$ *signified* $Z = \mathbf{F}(X,Y)$. *We saying output of* $\mathbf{F}$ *satisfies an ϵ-almost-uniform distribution, if*
$$\max_{z \in \{0,1\}^n} \left| P_Z(z) - \tfrac{1}{2^n} \right| \leq \epsilon.$$
where X *and* Y *are uniformly distributed in* $\mathcal{X}$ *and* $\mathcal{Y}$, *respectively. And we saying output of* $\mathbf{F}$ *satisfies an ϵ-almost-uniform distribution on condition of* $X = x$, *if*
$$\max_{z \in \{0,1\}^n} \left| P_{Z|X=x}(z) - \tfrac{1}{2^n} \right| \leq \epsilon.$$

Definition 3. *Merkle-Damgård construction* $\mathbf{H}^m : \{0,1\}^{\kappa \cdot *} \to \{0,1\}^n$ *is defined as* $x_0 = iv, x_i = \mathbf{F}(x_{i-1}, y_i)(i = 1, \ldots, t)$, $\mathbf{H}^m(iv, m) = x_t$, *where* $m = y_1\| \ldots \|y_t$. *Function* $\mathbf{F} : \{0,1\}^n \times \{0,1\}^{\kappa} \to \{0,1\}^n$ *is called compression function signified* $Z = \mathbf{F}(X,Y)$. $iv \in \{0,1\}^n$ *is a constant value.*

Definition 4. *The CheckSum Hash* $\mathbf{H}^s : \{0,1\}^{\kappa \cdot *} \to \{0,1\}^n$ *is defined as* $x_0 = iv, x_i = \mathbf{F}(x_{i-1}, y_i)(i = 1, \ldots, t)$, $\mathbf{H}^s(iv, m) = \bigoplus_{i=0}^{t} x_i$.

Definition 5. *The EMD-Hash* $\mathbf{H}^e : \{0,1\}^{\kappa \cdot *+\kappa-n} \to \{0,1\}^n$ *is defined as* $\mathbf{H}^e(iv_1, iv_2, \tilde{m}) = \mathbf{F}(iv_2, \mathbf{H}^m(iv_1, m)\|\tilde{y})$, *where* $\mathbf{H}^m$ *is Merkle-Damgård hash and* $\tilde{m} = m\|\tilde{y}$.

We write $\mathcal{F} = RF_{d,n}$ to signify $\mathcal{F}$ is random oracle from $\{0,1\}^d$ to $\{0,1\}^n$. The compression function $\mathbf{F}$ is also considered as domain extension function $\mathbf{F} : \{0,1\}^{n+\kappa} \to \{0,1\}^n$ and denoted $Z = \mathbf{F}(X\|Y)$.

Let $C^{\mathbf{F}_1,\ldots,\mathbf{F}_l} : \{0,1\}^* \to \{0,1\}^n$ be a function for random oracles $\mathbf{F}_1, \ldots, \mathbf{F}_l = RF_{d,n}$. Then let $S^{\mathcal{F}} = (S_1, \ldots, S_l)$ be a simulator Oracle Turing Machines with access to a random oracle $\mathcal{F} : \{0,1\}^* \to \{0,1\}^n$ and which exposes interfaces for each random oracle utilized by $\mathcal{C}$. The PRO-advantage of an adversary $\mathcal{A}$ against $\mathcal{C}$ is

$$Adv_{\mathcal{C},S}^{pro}(\mathcal{A}) = Pr[\mathcal{A}^{C^{\mathbf{F}_1,\ldots,\mathbf{F}_l},\mathbf{F}_1,\ldots,\mathbf{F}_l} \Rightarrow 1] - Pr[\mathcal{A}^{\mathcal{F},S^{\mathcal{F}}} \Rightarrow 1].$$

The advantage of adversary using resources as that, the total number of left queries q_L (which are either to $\mathcal{C}$ or $\mathcal{F}$), the number of right queries q_i made to each oracle $\mathbf{F}_i$ or simulator interface S_i, the total number of queries q_S to $\mathcal{F}$ and the maximum running time t_S.

Let $\mathbf{F} : \{0,1\}^n \times \{0,1\}^\kappa \to \{0,1\}^n$ be function family and $\mathcal{F} = RF_{n,n}$. For each $k \xleftarrow{\$} \{0,1\}^\kappa$, $\mathbf{F}(\cdot, k)$ is function mapping from $\{0,1\}^n$ to $\{0,1\}^n$. The PRF-Advantage of $\mathbf{F}$ is defined as

$$Adv_{\mathbf{F}}^{prf}(\mathcal{A}) = Pr\left[k \xleftarrow{\$} \mathcal{K}; \mathcal{A}^{\mathbf{F}(\cdot,k)} \Rightarrow 1\right] - Pr\left[\mathcal{A}^{\mathcal{F}} \Rightarrow 1\right].$$

3   An Example of Merkle-Damgård Construction

This section gives an example that, a compression function has good distribution bound, however, iterated by Merkle-Damgård construction, the upper bound of distribution is 1. Let function $\mathbf{F}_1 : \{0,1\}^n \times \{0,1\}^n \to \{0,1\}^n$ be

$$\mathbf{F}_1(x,y) = \begin{cases} (x \ll 1) \boxplus \mathbf{E}_{k_0}(y), & \mathbf{E}_{k_0}(y) = 0 \\ (x \lll 1) \boxplus \mathbf{E}_{k_0}(y), & \mathbf{E}_{k_0}(y) \neq 0 \end{cases},$$

where $x \ll s$: left shift of binary sequence x by s positions; $x \lll s$: circular left shift of binary sequence x by s positions; $\boxplus$: addition of positive integers, reduced modulo 2^n; $\mathbf{E}_{k_0}$ is a block cipher with key k_0 and the k_0 is secret to all (The $\mathbf{E}_{k_0}$ is one way permutation.). Then to get y_0 to satisfy $y_0 = \mathbf{E}_{k_0}^{-1}(0)$ takes 2^{n-1} computation. Let $\mathbf{E}_{k_0}(y_0) = \mathbf{0}$.

The distribution of $\mathbf{F}_1$ is follows. Let $Z = \mathbf{F}_1(X,Y)$, we have

$$P_{Z|Y=y}(z) = \begin{cases} \begin{cases} \frac{2}{2^n} & z \text{ is even} \\ 0 & z \text{ is odd} \end{cases}, & y = y_0 \\ \\ \frac{1}{2^n}, & y \neq y_0 \end{cases},$$

$$P_{Z|X=x}(z) = \begin{cases} \begin{cases} \frac{2}{2^n}, & z = (x \ll 1) \\ 0, & z = (x \ll 1) \oplus 1 \\ \frac{1}{2^n}, & else \end{cases}, & (x \ll 1) \neq (x \lll 1) \\ \\ \frac{1}{2^n}, & (x \ll 1) = (x \lll 1) \end{cases}.$$

The output of function $\mathbf{F}_1$ satisfies $\frac{1}{2^n}$-uniform distribution on condition of $X = x$ or $Y = y$, which means $\mathbf{F}_1$ has distribution bounds $0 \leq P_{Z|X=x}(z) \leq \frac{2}{2^n}$ and $0 \leq P_{Z|Y=y}(z) \leq \frac{2}{2^n}$. Let $M \stackrel{def}{=} Y_1\|\dots\|Y_s\|Y_{s+1}\|\dots\|Y_{s+t}$, $M_A = Y_1\|\dots\|Y_s$, and $M_B = Y_{s+1}\|\dots\|Y_{s+t}$. Iterating the function $\mathbf{F}_1$ by Merkle-Damgård transform denoted $Z = \mathbf{H}_1(iv, M)$, we get distribution bounds

$$\begin{cases} \frac{1}{2^n} \leq P_{Z|M_A=m_A}(z) \leq \frac{2}{2^n} \\ 0 \leq P_{Z|M_B=m_B}(z) \leq 1 \\ P_{Z|M_B=y_0\|\dots\|y_0}(0) \equiv 1 \end{cases}, \quad \text{in which } \underline{y_0\|\dots\|y_0} \stackrel{def}{=} \underbrace{y_0\|\dots\|y_0}_{n}.$$

Then $\forall m, m' \in \{0,1\}^{\kappa \cdot *}$, $\mathbf{H}_1(iv, m\|\underline{y_0\|\dots\|y_0}) \equiv \mathbf{H}_1(iv, m'\|\underline{y_0\|\dots\|y_0}) \equiv \mathbf{0}$.

Although the compression function $\mathbf{F}_1$ has good distribution bounds, which does not guarantee $\mathbf{H}_1$ having a good distribution bounds. The reason is that the distribution bound of Merkle-Damgård construction is influenced by that of compression function and the iterated times, proof of which is given in next section. Finding y_0 to satisfy $\mathbf{E}_{k_0}(y_0) = \mathbf{0}$ takes 2^{n-1} encryption, however, we should remember that, finding y_0 will destroy the hash $\mathbf{H}_1$, it is far more serious than finding collision on it.

4   Distribution of Merkle-Damgård Construction

The distribution bound of Merkle-Damgård Construction is presented in Theorem 1. Since, the output distribution is derived from input distribution, to analysis the output distribution, we have to select the input distribution. Taken the distribution of Merkle-Damgård construction into consideration, the input set can be randomly selected, which is some of input message block can be fixed. To make thing simple, we only consider the distributions of pre- or post- part of message blocks are fixed. The reason is that, the case other message blocks being fixed can be deduced from this two distributions.

We make following assumption. Let $M \overset{def}{=} M_A \| M_B$. Let M_A and M_B be uniformly distributed in $\{0,1\}^{\kappa \cdot s}$ and $\{0,1\}^{\kappa \cdot t}$, respectively, with $s > 0$ and $t > 0$. Then let $M_A \overset{def}{=} Y_1 \| \ldots \| Y_s$, and $M_B \overset{def}{=} Y_{s+1} \| \ldots \| Y_{s+t}$.

Let compression function $\mathbf{F}$ signified as $Z = \mathbf{F}(X, Y)$ have distribution bounds $\max_{x,z \in \{0,1\}^n} P_{Z|X=x}(z) = \frac{\Delta_1}{2^\kappa}$, $\max_{z \in \{0,1\}^n, y \in \{0,1\}^\kappa} P_{Z|Y=y}(z) = \frac{\Delta_2}{2^n}$, where Δ_1 and Δ_2 be two constant values.

Theorem 1. *Merkle-Damgård Hash $Z = \mathbf{H}^m(iv, M_A \| M_B)$ satisfies,*

$$P_{Z|M_A=m_A}(z) \leq \frac{\Delta_1}{2^\kappa}, \qquad P_{Z|M_B=m_B}(z) \leq \min\{1, \frac{\Delta_1(\Delta_2)^t}{2^\kappa}\}.$$

Proof. Let $M_B \overset{def}{=} M'_B \| Y_{s+t}$ and $M_A \overset{def}{=} Y_1 \| M'_A$,

$$P_{Z|M_A=m_A}(z) = \sum_{m_B \in \{0,1\}^{\kappa \cdot t}} P_{M_B}(m_B) \chi_{\mathbf{H}^m(iv, m_A \| m_B)}(z)$$

$$= \sum_{m'_B \in \{0,1\}^{\kappa \cdot (t-1)}} \sum_{y_{s+t} \in \{0,1\}^\kappa} P_{M'_B}(m'_B) P_{Y_{s+t}}(y_{s+t}) \chi_{\mathbf{F}(\mathbf{H}^m(iv, m_A \| m'_B), y_{s+t})}(z)$$

$$\leq \sum_{m'_B \in \{0,1\}^{\kappa \cdot (t-1)}} P_{M'_B}(m'_B) \frac{\Delta_1}{2^\kappa} = \frac{\Delta_1}{2^\kappa}.$$

The proof of $P_{Z|M_B=m_B}(z) \leq \min\{1, \frac{\Delta_1(\Delta_2)^t}{2^\kappa}\}$ is given by deduction of s and t. And it is not influenced by the selection of iv, we assume it be any fixed value $v \in \{0,1\}^n$. When $t = 1, s = 1$, for any $v \in \{0,1\}^n$

$$P_{Z|M_B=m_B}(z) = \sum_{m_A \in \{0,1\}^\kappa} P_{M_A}(m_A) \chi_{\mathbf{F}(\mathbf{H}^m(v, m_A), m_B)}(z)$$

$$= \sum_{u \in \{0,1\}^n} \chi_{\mathbf{F}(u, m_B)}(z) \sum_{m_A \in \{0,1\}^\kappa} P_{M_A}(m_A) \chi_{\mathbf{F}(v, m_A)}(u) \leq \frac{\Delta_1 \Delta_2}{2^\kappa}.$$

It is clear that, when $v = iv$ the inequality is also true. Suppose $t = l_t$, $s = l_s$ the inequality be true for any $v \in \{0,1\}$, when $s = l_s + 1$, $t = l_t$, we have:

$$P_{Z|M_B=m_B}(z) = \sum_{m'_A \in \{0,1\}^{\kappa \cdot l_s}} \sum_{y_1 \in \{0,1\}^{\kappa}} P_{M'_A}(m'_A) P_{Y_1}(y_1) \chi_{\mathbf{H}^m(v,y_1 \| m'_A m_B)}(z)$$

$$= \sum_{u \in \{0,1\}^n} \sum_{y_1 \in \{0,1\}^{\kappa}} P_{Y_1}(y_1) \chi_{\mathbf{F}(v,y_1)}(u) \sum_{m'_A \in \{0,1\}^{\kappa \cdot l_s}} P_{M'_A}(m'_A) \chi_{\mathbf{H}^m(u,m'_A \| m_B)}(z)$$

$$\leq \frac{\Delta_1 (\Delta_2)^{l_t}}{2^{\kappa}} \sum_{y_1 \in \{0,1\}^{\kappa}} P_{Y_1}(y_1) \sum_{u \in \{0,1\}^n} \chi_{\mathbf{F}(v,y_1)}(u) = \frac{\Delta_1 (\Delta_2)^{l_t}}{2^{\kappa}}.$$

When $s = l_s$ and $t = l_t + 1$, let $M_B \overset{def}{=} M'_B \| Y_{l_t+1}$, we have,

$$P_{Z|M_B=m_B}(z) = \sum_{m_A \in \{0,1\}^{\kappa \cdot l_s}} P_{M_A}(m_A) \chi_{\mathbf{H}^m(v,m_A \| m_B)}(z)$$

$$= \sum_{m_A \in \{0,1\}^{\kappa \cdot l_s}} \sum_{u \in \{0,1\}^n} P_{M_A}(m_A) \chi_{\mathbf{H}^m(v,m_A \| m'_B)}(u) \chi_{\mathbf{F}(u,y_{l_t+1})}(z)$$

$$\leq \sum_{u \in \{0,1\}^n} \frac{\Delta_1 (\Delta_2)^{l_t}}{2^{\kappa}} \chi_{\mathbf{F}(u,y_{l_t+1})}(z) \leq \frac{\Delta_1 (\Delta_2)^{l_t+1}}{2^{\kappa}}.$$

From induction principle we get the conclusions. □

Note 1. The distribution of Z on conditioned with $M_B = m_B$ is not only related with distribution of compression function, but also related with the message block length of m_B. Surprisingly, in the worst case the probability $P_{Z|M_B=m_B}(z)$ may increases in exponential way when $|m_B|/\kappa$ increases.

Note 2. Although the complexity of finding m_B, which makes the probability $P_{Z|M_B=m_B}(z)$ getting the maximum value, may pass the complexity of finding collision on $\mathbf{H}^m$. We have to remember that, if we find m_B, which makes $P_{Z|M_B=m_B}(z_0)$ close to 1, then we destroy the whole hash function, it is far more serious than finding one collision on it.

Theorem 2. *For function $\mathbf{F}$, let $\mathbf{F}^{(2)}(x,y \| y') \overset{def}{=} \mathbf{F}(\mathbf{F}(x,y),y')$, if for any $m \in \{0,1\}^{\kappa \cdot l}$, $\mathbf{F}(iv, Y_1)$ is independent from $\bigoplus_{i=2}^{l+1}(\mathbf{F}^{(i)}(iv, Y_1 \| m)$, then CheckSum Hash $\mathbf{H}^s$ satisfies*

$$P_{Z|M_A=m_A}(z) \leq \frac{\Delta_1}{2^{\kappa}}, \qquad P_{Z|M_B=m_B}(z) \leq \frac{\Delta_1}{2^{\kappa}}.$$

Proof. Let $M_B = M'_B \| Y_{s+t}$ and $M_A = Y_1 \| M'_A$.

$$P_{Z|M_A=m_A}(z) = \sum_{m_B \in \{0,1\}^{\kappa \cdot t}} P_{M_B}(m_B) \chi_{\mathbf{H}^s(iv,m_A \| m_B)}(z)$$

$$= \sum_{m_B \in \{0,1\}^{\kappa \cdot t}} P_{M'_B}(m'_B) P_{Y_{s+t}}(y_{s+t}) \chi_{\mathbf{F}(\mathbf{H}^m(iv,m_A \| m'_B),y_{s+t})}(z \oplus \mathbf{H}^s(iv,m_A \| m'_B))$$

$$\leq \sum_{m'_B \in \{0,1\}^{\kappa \cdot (t-1)}} P_{M'_B}(m'_B) \frac{\Delta_1}{2^{\kappa}} = \frac{\Delta_1}{2^{\kappa}}.$$

$$P_{Z|M_B=m_B}(z) = \sum_{m_A \in \{0,1\}^{\kappa \cdot s}} P_{M_A}(m_A) \chi_{\mathbf{H}^s(iv,m_A \| m_B)}(z)$$

$$= \sum_{m_A \in \{0,1\}^{\kappa \cdot s}} P_{M_A}(m_A) \chi_{\mathbf{F}(iv,y_1)}(z \oplus \mathbf{H}^s(iv,m_A \| m_B) \oplus \mathbf{F}(iv,y_1))$$

$$= \sum_{m_A \in \{0,1\}^{\kappa \cdot s}} P_{M_A}(m_A) \chi_{\mathbf{F}(iv,y_1)}(z \oplus \bigoplus_{i=2}^{s+t} \mathbf{F}^{(i)}(iv,y_1 \| \ldots \| y_l) \oplus iv) \leq \frac{\Delta_1}{2^{\kappa}}.$$

5   New Construction and MPP

The CheckSum Hash satisfies AUD-pr. EMD-Hash is indifferentiability from random oracle model with MD-strengthening padding, when compression function is instanced by random oracle model. We recommend a new structure called ECM-Hash (Enveloped Checksum Merkle-Damgård Hash), which is combine of the EMD-Hash and CheckSum Hash, to inherit cryptographic properties of EMD and CheckSum Hash.

Definition 6 (ECM-Hash). *The ECM-Hash* $\mathbf{H}^{ecm} : \{0,1\}^{\kappa \cdot * + \kappa - n} \to \{0,1\}^n$ *is defined as*

$$\mathbf{H}^{ecm}(iv, \widetilde{m}) = \mathbf{F}(\mathbf{H}^s(iv, m), \mathbf{H}^m(iv, m) \| \widetilde{y}),$$

where $m = y_1 \| \ldots \| y_t$, $\widetilde{m} = m \| \widetilde{y}$, $\mathbf{F}$ *is compression function,* $\mathbf{H}^s$ *is CheckSum Hash,* $\mathbf{H}^m$ *is Merkle-Damgård Hash and* $iv \neq \mathbf{0}^1$ *and message padding is MD-strengthening padding.*

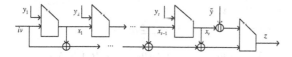

The ECM-Hash is CR-pr, PRO-pr, PRF-pr and AUD-pr, the proofs are as follows, full version of proofs can be found in [8].

AUD PRESERVING. The AUD-pr of ECM relies on AUD-pr of CheckSum Hash. Let $\widetilde{M} = M \| \widetilde{Y}$.

Theorem 3. *The check sum Hash* $\mathbf{H}^s$ *satisfies Theorem 2. Then the ECM-hash satisfies,*

$$P_{Z|M_A = m_A}(z) \leq \frac{\Delta_1 \Delta_2}{2^\kappa}, \qquad P_{Z|M_B \| \widetilde{Y} = m_B \| \widetilde{y}}(z) \leq \frac{\Delta_1 \Delta_2}{2^\kappa}.$$

Proof. We have,

$$P_{Z|M_A = m_A}(z)$$
$$= \sum_{m_B \| \widetilde{y} \in \{0,1\}^{\kappa \cdot t + \kappa - n}} P_{M_B \| \widetilde{Y}}(m_B \| \widetilde{y}) \chi_{\mathbf{F}(\mathbf{H}^s(iv, m_A \| m_B), H^m(iv, m_A \| m_B) \| \widetilde{y})}(z)$$
$$= \sum_{m_B \| \widetilde{y}} P_{M_B \| \widetilde{Y}}(m_B \| \widetilde{y}) \sum_{u \in \{0,1\}^n} \chi_{\mathbf{H}^s(iv, m_A \| m_B)}(u) \chi_{\mathbf{F}(u, \mathbf{H}^m(iv, m_A \| m_B) \| \widetilde{y})}(z)$$
$$\leq \Delta_2 \max_{u \in \{0,1\}^n} \sum_{m_B \in \{0,1\}^{\kappa \cdot t}} P_{M_B}(m_B) \chi_{\mathbf{H}^s(iv, m_A \| m_B)}(u) \leq \frac{\Delta_1 \Delta_2}{2^\kappa}.$$

$$P_{Z|M_B \| \widetilde{Y} = m_B \| \widetilde{y}}(z) = \sum_{m_A \in \{0,1\}^{\kappa \cdot s}} P_{M_A}(m_A) \chi_{\mathbf{F}(\mathbf{H}^s(iv, m), \mathbf{H}^m(iv, m_A \| m_B) \| \widetilde{y})}(z)$$
$$= \sum_{m_A \in \{0,1\}^{\kappa \cdot s}} P_{M_A}(m_A) \sum_{u \in \{0,1\}^n} \chi_{\mathbf{H}^s(iv, m_A \| m_B)}(u) \chi_{\mathbf{F}(u, \mathbf{H}^m(iv, m_A \| m_B) \| \widetilde{y})}(z)$$
$$\leq \Delta_2 \max_{u \in \{0,1\}^n} \sum_{m_A \in \{0,1\}^{\kappa \cdot s}} P_{M_A}(m_A) \chi_{\mathbf{H}^s(iv, m_A \| m_B)}(u) \leq \frac{\Delta_1 \Delta_2}{2^\kappa}. \qquad \square$$

The proof of Theorem 3 indicates that, AUD-pr is inherited from CheckSum Hash.

---

[1] $iv \neq \mathbf{0}$ is required in indifferentiability.

CR PRESERVING. Let compression function $\mathbf{F}$ be a collision resistant compression function. Then any adversary which finds collisions against ECM-Hash (two messages $\tilde{m} \neq \tilde{m}'$ for which $\mathbf{H}^{ecm}(iv, \tilde{m}) = \mathbf{H}^{ecm}(iv, \tilde{m}')$) will necessarily find collisions against $\mathbf{F}$. This can be proven using a slightly modified version of the proof that Merkle-Damgård construction with MD-strengthening is collision-resistant [12,7].

PRO PRESERVING. Now we show that ECM is PRO-pr. Bellare and Ristenpart [3] gave very nice presentation on PRO-pr of EMD. Our resulting scheme is similar to EMD, so we inherits some idea of EMD proofs. However, our transform has a liner checksum sequence, which is fixed value iv_2 in EMD, we can not utilize the proof of EMD in our proof.

Theorem 4 (ECM-Hash is PRO-pr). *Fix n, κ, and let $iv \in \{0,1\}^n$, where $iv \neq \mathbf{0}$. Let $\mathbf{F} = RF_{\kappa+n,n}$ be a random oracle. Let $\mathcal{A}$ be an adversary that asks at most q_L left queries with maximal length $(l-1) \cdot \kappa + (\kappa - n)$ bits for $l \geq 2$, q_R right queries, and runs in time T. Then*

$$Adv^{pro}_{\mathbf{H}^{ecm}, SA}(\mathcal{A}) \leq \frac{(lq_L + q_R)^2}{2^n} + \frac{2lq_L q_R + 2lq_L + 2q_R}{2^n}.$$

where the simulator SA makes $q_{SA} \leq q_R$ queries and runs in time $\mathcal{O}(q_R^2)$.

Proof. We utilize a game-playing argument. The simulator SA is to mimic $\mathbf{F}$ in a way that convinces any adversary that $\mathcal{F}$ is actually $\mathbf{H}^{ecm}$. The simulator SA is defined as follows.

On query $SA(x, y)$:

$z \xleftarrow{\$} \{0,1\}^n$,   Parse y into $u\|w$ s.t. $|u| = n, |w| = n - \kappa$

if $y_1 \ldots y_i \leftarrow \text{GETPATHR}(x)$ **then**

    if $y_1 \ldots y_i \leftarrow \text{GETPATHL}(u)$ **then**   ret $\mathcal{F}(y_1 \ldots y_i y)$

    else   ret z

if $x = iv$ **then** $\text{NEWNODE}(y) \leftarrow (z, z \oplus iv)$

if $y_1 \ldots y_i \leftarrow \text{GETPATHL}(x)$ **then**

    $z' \leftarrow \text{GETNODER}(y_1 \ldots y_i)$, $\text{NEWNODE}(y_1 \ldots y_i y) \leftarrow (z, z' \oplus z)$

ret z

The simulator SA exposes interface that accept $(n+\kappa)$-bit inputs and reply with n bit outputs. The functions GETPATHL, GETPATHR, EWNODE and GETNODER are used to access and modify a tree structure, initially with only a root node labeled with (iv, iv), in which iv is n-bit. The notation $\text{GETPATHL}(z)$, $\text{GETPATHR}(z')$ for $z, z' \in \{0,1\}^n$ returns the sequence of edge labels on a path from the root to a node with left n-bit labeled by z, and right n-bit labeled by z', respectively (if there are duplicate such nodes, return a fitted one, if there are none then return false). The notion $\text{GETNODER}(y_1 \ldots y_i)$ returns the right n-bit of node, which is following the path starting from the root and following the edges labeled by $y_1 \ldots y_i$. The notation $\text{NEWNODE}(y_1 \ldots y_i y) \leftarrow (x, \text{GETNODER}(y_1 \ldots y_i) \oplus x)$ for $y \in \{0,1\}^\kappa, x \in \{0,1\}^n$ and $y_i \in \{\varepsilon\} \cup \{0,1\}^\kappa$ means (1) locate the node found by following the path starting from the root and following the edges labeled by

$y_1, y_2, \ldots, y_i$, and (2) add an edge labeled by y from this found node to a new node left part labeled by x and right n-bit are labeled by $\text{GetNodeR}(y_1 \ldots y_i) \oplus x$.

Let $\mathcal{A}$ be an adversary attempting to differentiate between $\mathbf{H}^{ems}, \mathbf{F}$ and $\mathcal{F}, SA^{\mathcal{F}}$. We replaces the oracles by games that simulate them, in which games show in Fig. 1 to perform the reduction. The Game G0 is simulates exactly the pairs of oracles $\mathbf{H}^{ecm}$ and $\mathbf{F}$.

The G1, which does not include the underline statements, is a correct simulation of $\mathcal{F}, SA^{\mathcal{F}}$. The G0 and G1 are identical until bad returns.

We replace our tracking of $\mathbf{F}$ by two multisets $\mathcal{B}_L$ and $\mathcal{B}_R$ and defer the setting of bad until the finalization step, which is G2. We initially have $\mathcal{B}_L = \{iv\}$ and $\mathcal{B}_R = \emptyset$. It is clear that $Pr[\mathcal{A}^{G1} \text{ sets bad}] \leq Pr[\mathcal{A}^{G2} \text{ sets bad}]$.

The bad being set in query in G2 equals collision occurs in $\mathcal{B}_L$ or $\mathcal{B}_R$, which is collision occurs in left query at line 107, line 108, 114 and 115, and in right query at line 205,208 and 209. In i-th left and right queries, we have

$$Pr[Bad_{107}] \leq \frac{\sum_{t=0}^{l-2}((l+1)(i-1)+t+1+2q_R)}{2^n}, \qquad Pr[Bad_{205}] \leq \frac{(l-1)q_L}{2^n},$$

$$Pr[Bad_{108}] \leq \frac{\sum_{t=0}^{l-2}((l-1)(i-1)+t)}{2^n}, \qquad Pr[Bad_{208}] = 0,$$

$$Pr[Bad_{114 \vee 115}] \leq \frac{(l+1)(i-1)+(l-1)+1+2q_R}{2^n}, \qquad Pr[Bad_{209}] \leq \frac{(l+1)q_L+2i+1}{2^n}.$$

Then, we get the conclusion

$$Adv_{\mathbf{H}_{\mathbf{F}}^{ecm},SA}^{pro}(\mathcal{A}) \leq Pr[\mathcal{A}^{G_0} \Rightarrow 1] - Pr[\mathcal{A}^{\mathcal{F},SA^{\mathcal{F}}} \Rightarrow 1] = Pr[\mathcal{A}^{G_1} \text{ set bad}]$$

$$\leq Pr[\mathcal{A}^{G_2} \text{ set bad}] \leq \sum_{i=1}^{q_L}(Pr[Bad_{107 \cup 108 \cup 114 \cup 115}]) + \sum_{i=1}^{q_R}(Pr[Bad_{205 \cup 208 \cup 209}])$$

$$\leq \frac{(lq_L + q_R)^2}{2^n} + \frac{2q_L q_R + 2q_L + 2q_R}{2^n}. \qquad \square$$

PRF PRESERVING. PRF preserving is that, if an appropriately keyed version of the compression function is a PRF then the appropriately keyed version of the hash function must be a PRF too. Our resulting scheme is very similar to NMAC, which we know to be PRF-Pr [1]. Our transform has a liner checksum sequence, which is key value K_2 in NMAC, we can not directly utilize the proof of NMAC. The majority of the proof of NMAC is captured by following two conclusions [1], the first shows that keyed Merkle-Damgård iteration is computationally almost universal(cAU), the second shows that, composition of a PRF and a cAU function is a PRF. In our schemes we just need to prove that, the composition of a checksum and Merkle-Damgård construction is a PRF. We omit the detail.

Theorem 5. *Fix n, κ and let $\mathbf{F} : \{0,1\}^n \times \{0,1\}^\kappa \rightarrow \{0,1\}^n$ be a function family keyed via the low n bits of its input. Let $\mathcal{A}$ be a prf-adversary against keyed ECM-Hash using q queries of length at most l blocks and running in time T. Then there exists prf-adversaries $\mathcal{A}_1$ and $\mathcal{A}_2$ against $\mathbf{F}$ such that*

$$Adv_{\mathbf{FH}^s\tilde{\mathbf{H}}^m}^{prf}(\mathcal{A}) \leq Adv_{\mathbf{F}}^{prf}(\mathcal{A}_1) + \binom{q}{2}\left[2l Adv_{\mathbf{F}}^{prf}(\mathcal{A}_2) + \frac{1}{2^n}\right].$$

Game G0, Game G1 Response to the t-th query

A left query $L(m^t)$:

000 $y_1^t \ldots y_{k-1}^t \widetilde{y}_k^t \xleftarrow{d} m^t$;

001   ret $LSub(t, y_1^t \ldots y_{k-1}^t \widetilde{y}_k^t)$

SUBROUTINE $LSub(t, y_1^t \ldots y_{k-1}^t \widetilde{y}_k^t)$

100 Let s be min value s.t

$\quad y_1^s \ldots y_{k-1}^s \widetilde{y}_k^s = y_1^s \ldots y_{k-1}^t \widetilde{y}_k^t$

101 if $s < t$ then ret x_k^s

102 $x_0^t \leftarrow iv,\ x_0^{\prime t} \leftarrow iv$

103 for $1 \leq i \leq k-1$

104 $\quad x_i^t \xleftarrow{\$} \{0,1\}^n$

105 $\quad$ if $(x_{i-1}^t, y_i^t) \in Dom(\mathbf{F})$ then

106 $\quad\quad x_i^t \leftarrow \mathbf{F}(x_{i-1}^t, y_i^t)$

107 $\quad \mathbf{F}(x_{i-1}^t, y_i^t) \leftarrow x_i^t$

108 $\quad x_i^{\prime t} \leftarrow x_{i-1}^{\prime t} \oplus \mathbf{F}(x_{i-1}^t, y_i^t)$

109 $x_k^t \xleftarrow{\$} \{0,1\}^n$

110 if $(x_{k-1}^{\prime t}, x_{k-1}^t \| \widetilde{y}_k^t) \in Dom(\mathbf{F})$ then

111 $\quad$ bad$\leftarrow$ true

112 $\quad x_k^t \leftarrow \mathbf{F}(x_i^{\prime t}, x_i^t \| \widetilde{y}_k^t)$

113 $\mathbf{F}(x_{k-1}^{\prime t}, x_{k-1}^t \| \widetilde{y}_k^t) \leftarrow x_k^t$

114 ret x_k^t

A right query $R_{\mathbf{F}}(x^t, y^t)$:

200 $z^t \xleftarrow{\$} \{0,1\}^n$

201 Parse y^t into $u^t \| w^t$ s.t.

$\quad |w^t| = \kappa - n, |u^t| = n$

202 if $y_1 \ldots y_i \leftarrow \mathrm{GetPathR}(x^t)$ then

203 $\quad$ if $\mathrm{GetPathR}(x^t) \leftarrow \mathrm{GetPathL}(u^t)$ then

204 $\quad\quad$ ret $LSub(t, y_1 \ldots y_i y^t)$

205 $\quad$ if $(x^t, y^t) \in Dom(\mathbf{F})$ then

206 $\quad\quad$ bad$\leftarrow$ true

207 $\quad\quad \underline{z^t \leftarrow \mathbf{F}(x^t, y^t)}$

208 if $(x^t, y^t) \in Dom(\mathbf{F})$ then

209 $\quad z^t \leftarrow \mathbf{F}(x^t, y^t)$

210 if $x^t = iv$ then

211 $\quad \mathrm{NewNode}(y) \leftarrow (z^t, z^t \oplus iv)$

212 if $y_1 \ldots y_i^t \leftarrow \mathrm{GetPathL}(x^t)$ then

213 $\quad z^{\prime t} \leftarrow \mathrm{GetNodeR}(\mathrm{GetPathL}(x^t))$

214 $\quad \mathrm{NewNode}(y_1^t \ldots y_i^t y^t) \leftarrow (z^t, z^{\prime t} \oplus z^t)$

215 $\mathbf{F}(x^t, y^t) \leftarrow z^t$

216 ret z^t

Game G2, Response to the t-th query

A left query $L(m^t)$:

000 $y_1^t \ldots y_{k-1}^t \widetilde{y}_k^t \xleftarrow{d} m^t$;

001   ret $LSub(t, y_1^t \ldots y_{k-1}^t \widetilde{y}_k^t)$

SUBROUTINE $LSub(t,, y_1^t \ldots y_{k-1}^t \widetilde{y}_k^t)$

100 Let s be min value s.t

$\quad y_1^s \ldots y_{k-1}^s \widetilde{y}_k^s = y_1^s \ldots y_{k-1}^t \widetilde{y}_k^t$

101 if $s < t$ then ret x_k^s

102 $x_0^t \leftarrow iv,\ x_0^{\prime t} \leftarrow iv$

103 for $1 \leq i \leq k-1$

104 $\quad x_i^t \xleftarrow{\$} \{0,1\}^n$

105 $\quad$ if $(x_{i-1}^t, y_i^t) \in Dom(\mathbf{F})$ then

106 $\quad\quad x_i^t \leftarrow \mathbf{F}(x_{i-1}^t, y_i^t)$

107 $\quad$ else $\mathcal{B}_L \xleftarrow{\cup} x_i^t$

118 $\quad\quad \mathcal{B}_R \xleftarrow{\cup} x_{i-1}^{\prime t} \oplus x_i^t$

109 $\quad \mathbf{F}(x_{i-1}^t, y_i^t) \leftarrow x_i^t$

110 $\quad x_i^{\prime t} \leftarrow x_{i-1}^{\prime t} \oplus \mathbf{F}(x_{i-1}^t, y_i^t)$

111 $x_k^t \xleftarrow{\$} \{0,1\}^n$

112 if $(x_{k-1}^{\prime t}, x_{k-1}^t \| \widetilde{y}_k^t) \in Dom(\mathbf{F})$ then

113 $\quad x_k^t \leftarrow \mathbf{F}(x_i^{\prime t}, x_i^t \| \widetilde{y}_k^t)$

114 $\quad \mathcal{B}_L \xleftarrow{\cup} x_{k-1}^{\prime t}$

115 else $\mathcal{B}_L \xleftarrow{\cup} x_k^t$,

116 $\quad \mathcal{B}_L \xleftarrow{\cup} x_{k-1}^{\prime t}$

117 $\mathbf{F}(x_{k-1}^{\prime t}, x_{k-1}^t \| \widetilde{y}_k^t) \leftarrow x_k^t$

118 ret x_k^t

A right query $R_{\mathbf{F}}(x^t, y^t)$:

200 $z^t \xleftarrow{\$} \{0,1\}^n$

201 Parse y^t into $u^t \| w^t$ s.t.

$\quad |w^t| = \kappa - n, |u^t| = n$

202 if $x^t \in \mathcal{B}_R$ then

203 $\quad$ if $\mathrm{GetPathR}(x^t) \leftarrow \mathrm{GetPathL}(u^t)$ then

204 $\quad\quad$ ret $LSub(t, y_1 \ldots y_i y^t)$

205 $\quad \mathcal{B}_L \xleftarrow{\cup} x^t$

206 if $(x^t, y^t) \in Dom(\mathbf{F})$ then

207 $\quad z^t \leftarrow \mathbf{F}(x^t, y^t)$

208 else $\mathcal{B}_L \xleftarrow{\cup} x^t$

209 $\quad \mathcal{B}_L \xleftarrow{\cup} z^t$

210 if $x = iv$ then

211 $\quad \mathrm{NewNode}(y) \leftarrow (z, z \oplus iv)$

212 if $y_1^t \ldots y_i^t \leftarrow \mathrm{GetPathL}(x^t)$ then

213 $\quad z^{\prime t} \leftarrow \mathrm{GetNodeR}(\mathrm{GetPathL}(x^t))$

214 $\quad \mathrm{NewNode}(y_1^t \ldots y_i^t y^t) \leftarrow (z^t, z^{\prime t} \oplus z^t)$

215 ret z^t

Finalization:

300 $bad \leftarrow \exists x, x' \in \mathcal{B}_L \, s.t. x = x'$

301 $bad \leftarrow \exists x, x' \in \mathcal{B}_R \, s.t. x = x'$

Fig. 1. Games utilized in the proof of Theorem 4. Initially, $\mathcal{B}_L = \{iv\}$, $\mathcal{B}_R = \emptyset$.

where $\mathcal{A}_1$ utilizes q queries and runs in time at most T and $\mathcal{A}_2$ utilizes at most two oracle queries and runs in time $O(lT_{\mathbf{F}})$ where $T_{\mathbf{F}}$ is the time for one computation of $\mathbf{F}$.

6 Generalized Enveloped MPP Transformation

The paper presented ECM transformation to satisfy MPP, in which the liner CheckSum selected as envelped hash is the easiest way of attending AUD-pr.

A generalized Enveloped MPP transformation $\mathbf{H}^g : \{0,1\}^{\kappa \cdot * + \kappa - n} \to \{0,1\}^n$ can be defined as $\mathbf{H}^G(iv, \widetilde{m}) = \mathbf{F}(\mathbf{G}(iv, m), \mathbf{H}^m(iv, m)\|\widetilde{y})$,

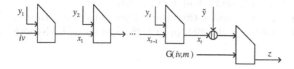

in which $m = y_1\| \ldots \|y_t$, $\widetilde{m} = m\|\widetilde{y}$, $\mathbf{F}$ is compression function, $\mathbf{H}^m$ is Merkle-Damgård Hash and $\mathbf{G} : \{0,1\}^{\kappa \cdot *} \to \{0,1\}^n$ is AUD-pr transformation.

Acknowledgments

The authors are supported in part by NSFC 60573028 and by NUDT FBR JC07-02-02. The first author would like to thank Dr. Wu Wenling for suggestion on formal definition of almost uniform distribution on reviewing the PhD thesis [8].

References

1. Bellare, M.: New proofs for NMAC and HMAC: Security without collision-resistance. In: Dwork, C. (ed.) CRYPTO 2006. LNCS, vol. 4117, pp. 602–619. Springer, Heidelberg (2006)
2. Bellare, M., Canetti, R., Krawczyk, H.: Keyed Hash Functions for Message Authentication. In: Koblitz, N. (ed.) CRYPTO 1996. LNCS, vol. 1109, pp. 1–15. Springer, Heidelberg (1996)
3. Bellare, M., Ristenpart, T.: Multi-Property-Preserving Hash Domain Extension and the EMD Transform. In: Lai, X., Chen, K. (eds.) ASIACRYPT 2006. LNCS, vol. 4284, pp. 299–314. Springer, Heidelberg (2006)
4. Biham, E., Chen, R.: Near-Collisions of SHA-0. In: Franklin, M. (ed.) CRYPTO 2004. LNCS, vol. 3152, pp. 290–305. Springer, Heidelberg (2004)
5. Biham, E., Dunkelman, O.: A Framework for Iterative Hash Functions—HAIFA, http://www.csrc.nist.gov/pki/HashWorkshop/2006/Papers/
6. Coron, J.S., Dodis, Y., Malinaud, C., Puniya, P.: Merkle-damgard revisited: How to construct a Hash Function. In: Shoup, V. (ed.) CRYPTO 2005. LNCS, vol. 3621, pp. 430–448. Springer, Heidelberg (2005)
7. Damgå, I.: A design principle for hash functions. In: Brassard, G. (ed.) CRYPTO 1989. LNCS, vol. 435, pp. 416–427. Springer, Heidelberg (1990)

8. Duo, L.: Analysis of block cipher to design of Hash function, PhD thesis, National University of Defence Technology (2007)
9. Gauravaram, P., Millan, W., Neito, J.G., Dawson, E.: Constructing Secure Hash Functions by Enhancing Merkle-Damgård Construction. In: Batten, L.M., Safavi-Naini, R. (eds.) ACISP 2006. LNCS, vol. 4058, pp. 407–420. Springer, Heidelberg (2006)
10. Lucks, S.: A Failure-Friendly Design Principle for Hash Functions. In: Roy, B. (ed.) ASIACRYPT 2005. LNCS, vol. 3788, pp. 474–494. Springer, Heidelberg (2005)
11. Maurer, U., Renner, R., Holenstein, C.: Indifferentiability, Impossibility Results on Reductions, and Applications to the Random Oracle Methodology. In: Naor, M. (ed.) TCC 2004. LNCS, vol. 2951, pp. 21–39. Springer, Heidelberg (2004)
12. Merkle, R.C.: One Way Hash Functions and DES. In: Brassard, G. (ed.) CRYPTO 1989. LNCS, vol. 435, pp. 428–446. Springer, Heidelberg (1990)
13. Wang, X., Yu, H.: How to Break MD5 and Other Hash Functions. In: Cramer, R.J.F. (ed.) EUROCRYPT 2005. LNCS, vol. 3494, pp. 19–35. Springer, Heidelberg (2005)

Design of a Differential Power Analysis Resistant Masked AES S-Box

Kundan Kumar[1], Debdeep Mukhopadhyay[2], and Dipanwita RoyChowdhury[3]

[1] MTech Student, Department of Computer Science and Engg., Indian Institute of
Technology, Kharagpur, India
kundankr@gmail.com

[2] Assistant Professor, Department of Computer Science and Engg., Indian Institute
of Technology, Madras, India
debdeep@cse.iitm.ernet.in

[3] Professor, Department of Computer Science and Engg., Indian Institute of
Technology, Kharagpur, India
drc@cse.iitkgp.ernet.in

Abstract. Gate level masking is one of the most popular countermeasures against Differential Power Attack (DPA). The present paper proposes a masking technique for AND gates, which are then used to build a balanced and masked multiplier in $GF(2^n)$. The circuits are shown to be computationally secure and have no glitches which are dependent on unmasked data. Finally, the masked multiplier in $GF(2^4)$ is used to implement a masked AES S-Box in $GF(2^4)^2$. Power measurements are taken to support the claim of random power consumption.

1 Introduction

Rijndael-AES (Advanced Encryption Standard) has become the worldwide choice in the field of symmetric key cryptography since October 2001. Since then lots of research work have been carried out on the design and implementations of the AES block cipher. With the imposing threat of side-channel attacks, which exploit weakness in the implementation, design of all cryptographic algorithms required a revisit. Hence for AES, various design architectures have been reported [1,2,3] to make it more and more secured against various side channel attacks. The side channel attacks based on the power consumption of the crypto-device are the hardest to tackle. This attack was first introduced by Paul Kocher et. al. in [4] and subsequently extended by many researchers[5,6,7,8,9].

In [10] the first practical power analysis of AES hardware implementation was proposed. Several research works have been carried out to develop design alternatives to overcome power based side-channel leakages from the AES implementations. One way of tackling the problem is the use of masked gates in AES implementations to prevent side-channel leakage. Several patents exist on the gate level masking strategies [11,12]. Various techniques for random masking in hardware has been presented in [13]. Although as is shown in [14], masked cryptographic circuits may cause leakage against "higher order DPAs", masking

K. Srinathan, C. Pandu Rangan, M. Yung (Eds.): Indocrypt 2007, LNCS 4859, pp. 373–383, 2007.

is still one of the most popular safeguards. Also as proved in [15], using distinct mask values one can protect against higher order DPAs. However, in case of practical CMOS circuits it is natural that the output of internal gates switch more than once before stabilizing depending on the path delay inside the circuit [16]. The results on DPA attacks of masked gates, reported in [17] demonstrate that all proposed masked gates are vulnerable to power based side-channel leakage in the presence of glitches. The work in [18] discusses three different attacks, zero-offset DPA, toggle count DPA and zero input DPA on masked AES hardware implementations. Thus glitches in masked circuits pose a serious threat to masking implementations against power attack. The analysis pinpoints that the switching characteristics of XOR gates in a multiplier are responsible for the power based side channel leakage when there are glitches inside the circuit. Thus, the above observations lead to the following objectives for a secured hardware implementation with respect to power based attacks: *The gates of the circuit to be protected against power based analysis should solely consume power which is independent of the unmasked value* and *the gates should be arranged in a regular and balanced architecture to minimize the glitches which are dependent on the unmasked data.* The authors in [17] show that XOR gates in the masked $GF(2^n)$ multiplier of the S-box are the main sources of side-channel leakage. The multiplier architecture commonly used for masked S-box implementations ([1,2,3]) are analyzed in [18]. In these lines, the present paper proposes a balanced masked AND gate, which avoids any glitches dependent on unmasked data. The paper also shows a balanced and regular architecture for a masked multiplier in $GF(2^n)$. Using the two building blocks the work then develops a masked AES S-Box, when the S-Box is implemented in tower field $GF(2^4)^2$. The design has been prototyped on a Xilinx XCV FPGA platform and the power curves have been analyzed to evaluate that the power consumption of the masked circuits are not correlated with the unmasked inputs.

The outline of the paper is as follows: The design of the proposed masked AND Gate is described in *section 2*. The architecture of a secured masked multiplier in $GF(2^n)$ has been proposed in *section 3*. *Section 4* presents the experimental results of FPGA implementations of a $GF(2^4)$ multiplier and the resultant masked AES S-Box. The work is concluded in *section 5*.

2   Design of Glitch Free Masked AND Gate

In this section, we discuss a masking technique for a basic 2 input AND gate. The two inputs to the AND Gate are x and y. The output is denoted by $z = xy$. A straight-forward implementation of the AND gate is prone to power based side channel attacks, as the power consumption of the gate is proportional to the inputs, x and y. The masking technique uses two random bits or masks, denoted by r_x and r_y. The masked values are derived from the unmasked values, x and y using the relation, $x' = x \oplus r_x$ and $y' = y \oplus r_y$. The assumption is that the random bits are uniformly distributed and mutually statistically

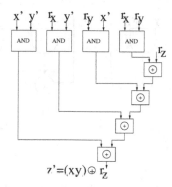

Fig. 1. Unbalanced architecture of a masked multiplier

independent. The masked AND gate is hence obtained by the following equation:
$$z' = r_z \oplus z = r_z \oplus xy = r_z \oplus (x' \oplus r_x)(y' \oplus r_y) = (((r_z \oplus r_x r_y) \oplus r_y x') \oplus r_x y') \oplus x' y'.$$

The solution is depicted in *Fig 1*. The computations are secure on the logic gate level if r_x, r_y and r_z are uniformly distributed and are mutually statistically independent. This may be concluded from the fact that, for each fixed value of (x, y), the two inputs to each AND gate are uniformly distributed and mutually statistically independent. Due to r_z, one of the inputs to the XOR gate is uniformly distributed and statistically independent of the other input. As the other input is also statistically independent of (x, y), the computations remain secure on the logic gate level[13].

However, in [20,17] it was shown that there is another strong assumption behind the above analysis, that the CMOS gates switch once in a clock cycle. That is there are no internal glitches in the circuit, the occurence of which are dependent on the unmasked data, x and y. Because of the unbalanced nature of *Fig 1*, there can be glitches in the circuit, which shall lead to transitions and hence power consumptions which are correlated to x and y. It was pin-pointed in [17], that the XOR gates were responsible of performing different transitions depending on the glitches. Hence, observing the above facts we augment our security conditions against DPA to the following:

C1: If all the inputs to a logic gate are statistically independent of the original data, then the outputs of any set of logic gates in the circuit are also jointly statistically independent of the original data.
C2: The glitches occurring in the circuit should not be dependent on the unmasked values.

The condition for a secured computation of a logic circuit would be hence $C1 \wedge C2$.

Thus, we propose the following circuit which satisfies both the conditions C1 and C2, and thus performs secured computation. The circuit depicted in *Fig 2* is based on the following decomposition:

$$z' = ((x' r_y) \oplus r_s) \oplus ((y' r_x) \oplus r_t)$$
$$r_z = ((x' y') \oplus r_s) \oplus ((r_x r_y) \oplus r_t)$$

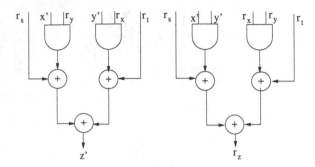

Fig. 2. Data dependent glitch free masked AND gate

In the proposed implementation, r_x, r_y, r_s and r_t are the random bits used. They are assumed to be uniformly distributed and statistically independent. It may be observed, that the decomposition has been made in a fashion such that the signals inside a circuit are mutually statistically independent and also uncorrelated to the unmasked data. Thus in the computation of z', the signal values, $x'r_y$ and $y'r_x$ are mutually statistically independent. This may be observed from the fact that given the value of any one of them does not give any information about the other value, better than a random guess. Also, the values are uncorrelated to the unmasked values. Thus even if some signals reach some nodes before others, no correlation can be made with the unmasked data. Also the balanced nature of the circuit ensures no glitches, if the input signals arrive at the same time. Thus it may be safely concluded that the circuit satisfies both conditions C1 and C2. It satisfies C1 because, the inputs to the AND gates are statistically independent of each other and also uniformly distributed. Thus the computations of the AND gates are secured. The random bits, r_s and r_t ensure that the inputs to the XOR gates are not only statistically independent but also randomly distributed. Also, because of the balanced nature of the circuit there are no glitches in the circuit which are dependent on the unmasked data. Thus, condition C2 is also satisfied. The extra cost for the implementation is an extra random bit. The outputs, z' and r_z appear at the same time. In future computations, z' shall be used, along-with its corresponding mask value r_z.

The proposed architecture may be compared with the previous architecture depicted in *Fig 1*. The proposed architecture has a critical delay proportional to 1 AND gate and 2 XOR gates, compared to the previous circuit which has a delay of 1 AND gate and 4 XOR gates. Using the fact that the delay of 1 XOR gate is equal to 1.5 AND gates, the delay of the proposed architecture is about half than the previous architecture. The number of gates required is 4 AND gates and 6 XOR gates, compared to 4 AND gates and 4 XOR gates required in the previous architecture.

3   Data Dependent Glitch Free Masked $GF(2^n)$ Multiplier

In this section, we present a regular structure of a finite field multiplier operating in $GF(2^n)$. First we present the architecture of the multiplier. Next we present the technique used to mask the multiplier.

3.1   A Regular $GF(2^n)$ Multiplier Architecture

The inputs to the multiplier are denoted by the polynomials, $a(x)$ and $b(x)$, both elements of $GF(2^n)$. The primitive polynomial of the field is denoted by $p(x)$.
   The multiplier has two blocks:

1. *Generate Companion Matrix:* This block generates an n^2 square matrix in $GF(2)$. The matrix is denoted by $T_{a,p}$.
2. *AND-XOR plane:* The AND-XOR plane is a regular switch, which gets configured by the Generate Companion Matrix block.

   We first present the following theorem to explain the working of the algorithm. The proof is omitted for the lack of space.

Theorem 1. *The product of two polynomials $a(x)$ and $b(x)$ (both belongs to $GF(2^n)$) modulo a polynomial $p(x)$ (primitive polynomial in $GF(2^n)$) is represented by the equation $c(x) = [a(x) * b(x)]mod(p(x))$. The equation can be simulated in two stages: First, $a(x), p(x)$ generate the matrix, $T_{a,p}$. The output $c(x)$ is then equal to $c(x) = T_{a,p}b(x)$, where the transition matrix is denoted by*

$$
\mathbf{T_{a,p}} = \begin{bmatrix} | & | & \cdots & | \\ | & | & \cdots & | \\ a(x) & xa(x)modp(x) & \cdots & x^{n-1}a(x)modp(x) \\ | & | & \cdots & | \\ | & | & \cdots & | \end{bmatrix}
$$

   The above theorem leads to the following algorithm, *Generate Companion Matrix*, for the multiplier.

Algorithm 1. *Input: Multiplicand: $a(x)$, Multiplier: $b(x)$, Modulo polynomial: $p(x)$. The above elements belong to $GF(2^n)$.*
   Output: The product $c(x) = a(x) * b(x)modp(x)$
   The Companion Matrix, $T_{a,p}$.
   Step1: Express $a(x)$ in binary notation.
   col[0]=binary representation of $a(x)$
   /* The array col[.] is an eight bit array*/
   Step2: for($i = 0; i < n; i++$)
   {
   {overflow, col[i]}=col[i-1]<<1;
   if(overflow==1)
   col[i]=col[i]⊕prim;
   /*prim is the binary representation of p(x)* /

```
/*⊕ represents bitwise XOR*/
}
```
Step3: The Transition Matrix $T_{a,p}=[col[0], col[1], \ldots, col[n-1]]$
Step 4: $c(x) = T_{a,p}b(x)$

Next we describe the technique used to mask the above computation, so that the computations are secure with respect to power based analysis. The multiplier desires to compute, $c(x) = a(x)b(x)modp(x)$, where $a(x), b(x), c(x) \in GF(2^n)$. The polynomial $p(x)$ is a primitive poynomial of $GF(2^n)$. However, due to masking the multiplier masks the values of $a(x)$ and $b(x)$ with $r_a(x)$ and $r_b(x)$ respectively. Both the masking values belong to $GF(2^n)$ and are statistically independent and uniformly distributed. The masked values are $a'(x) = a(x) \oplus r_a(x)$ and $b'(x) = b(x) \oplus r_b(x)$.

3.2 The Masked Implementation of the *Generate Companion Matrix*

It may be followed from the above description, that the Generate Companion Matrix step is a linear step. Thus, the corresponding masking can be done by processing the masked data and the masking value in separate circuits.

The Generate Companion matrix hence operates as two separate circuits. One of them operates on the mask value, $r_a(x) \in GF(2^n)$ and the other operates on the masked value $a'(x) = a(x) \oplus r_a(x)$. *Fig 3* depicts the block for $GF(2^4)$. We have used the fact, that the value of $p(x) = x^4+x+1$. The block loads the vector col[0] with the input polynomial and generates the vectors, col[1-3] through the circuit. For masking, the block has 2 copies, one operating on the value, $a'(x)$ and the other on $r_a(x)$. It is straight forward to see that both the circuits are thus uncorrelated to the unmasked value, $a(x)$. The blocks generate two square matrices, $T'_{a,p}$ and r_T, both of order n^2. It is evident because of the linearity of the blocks, $T_{a,p} = T'_{a,p} + r_T$, where the operation + is an element by element modulo 2 addition. The number of gates required for the Generate Companion Matrix block for $GF(2^4)$ is 12 number of 2 input XOR gates, as to generate each of the two matrices number of 2 input XOR gates is 6 (refer *Fig 3*). The critical delay is that of 3 XOR gates. After the completion of the operation, a control signal is raised high to commence the operation of the AND-XOR plane.

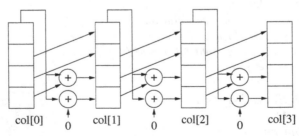

Fig. 3. The Generate Companion Matrix Block for $GF(2^4)$

3.3   The Masked Implementation of the AND-XOR Plane

Masking of the AND-XOR plane is depicted in *Fig 4*. As described previously, the AND-XOR plane is an $n^2 \times n^2$ array, which gets programmed by the value of the Generate Companion Matrix block. The AND-XOR plane operates when the control signal goes high. The functionality of the plane is to compute the value of $c(x) = T_{a,p}b(x)$. To prevent power based side channel attacks, the plane uses the masked values, $b'(x)$ and the elements of the matrix $T'_{a,p}$. The corresponding masking values are $r_b(x)$ and the elements of the matrix r_T.

The plane has n^2 masked AND gates, which are masked as described previously. The masked AND gate at the location i, j, where $0 \le i, j < n$, takes as inputs, $T'_{a,p}[i][j]$ and $r_T[i][j]$. The other inputs are $b'[j]$ and $r_b[j]$. Also as previously discussed, for the masked AND gates two random bits r_t and r_s are used. Each masked AND gate produces two outputs, $z'[i][j]$ and $r_z[i][j]$.

The output of the multiplier has two components: the masked product $c'(x)$ and the corresponding mask $r_c(x)$.

The output equations for the ith $(0 \le i < n)$ bit slice are stated below. $c'[i] = \bigoplus_{j=0}^{n-1}(z'[i][j]); r_c[i] = \bigoplus_{j=0}^{n-1}(r_z[i][j]) \Rightarrow c'[i] \oplus r_c[i] = \bigoplus_{j=0}^{n-1}(z'[i][j] \oplus r_z[i][j]) = \bigoplus_{j=0}^{n-1}((T'_{a,p}[i][j] \oplus r_T[i][j])b[j]) = \bigoplus_{j=0}^{n-1}(T_{a,p}[i][j]b[j]) - c[i]$(using, $T_{a,p} = T'_{a,p} + r_T$).

Thus the masked product $c'(x)$ and $r_c(x)$ may be xored to obtain $c(x)$, which is the product required from Theorem 1. Also, the computations performed by the masked AND gate are secure by our arguments in the previous section. The inputs to all the gates in the masked multiplier are thus mutually statistically independent, uniformly distributed and uncorrelated to the unmasked values.

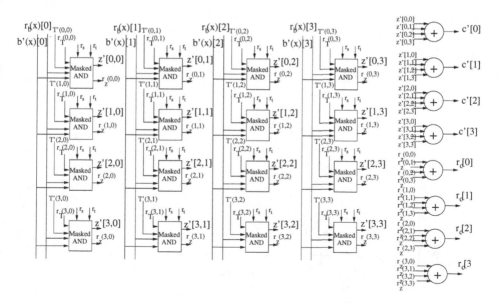

Fig. 4. The AND XOR Plane for $GF(2^4)$

Also, the regular nature of the circuit ensures that there are no unmasked data dependent glitches in the circuit. The linear layer operates separately upon the masked values and the masking values and thus the computations performed are always uncorrelated to the unmasked values. Thus we conclude that the computations in the masked multiplier are not correlated to the unmasked values. Hence, the masked multiplier is secure against power based attacks.

We have implemented the masked multiplier in the field $GF(2^4)$. The total number of gates involved is 116 AND gates and 222 XOR gates. The critical delay is that due to 1 AND and 6 XOR gates. The multiplier is used to implement the AES S-Box in $GF(2^4)^2$. We present the results in the next section to demonstrate that the computations in the proposed masked multiplier and the consequent AES S-Boxes are not correlated with the unmasked data and thus prevent DPA attacks.

4   Experimental Set-Up and Results

The scheme described above was coded in verilog and then downloaded onto a modified Spartan-III FPGA kit. The connection between the V_{dd} of Spartan-III XC3S400 FPGA kit and the V_{dd} pin of the FPGA is cut and a resistor of 0.1 ohm is inserted in between them. The Tektronix TCPA300 Amplifier AC/DC current probe is used to measure the current through the register, which is assumed to be proportional to the power consumption of the circuit. The value of this

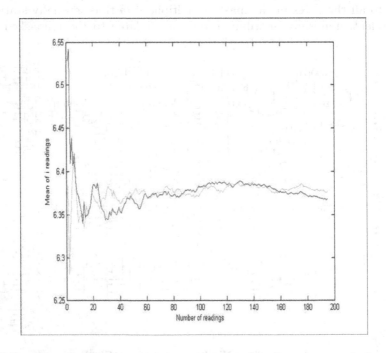

Fig. 5. Mean of peak power consumption vs number of readings for a masked multiplier

Table 1. Comparision of a Masked AND Gate

	Conventional Masked AND gate	Proposed Masked AND gate
#AND Gates	4	4
#XOR Gates	4	6
#Delay	1 AND Gate + 4 XOR Gate	1 AND Gate + 2 XOR Gate
#Random Bits	3	4

current is passed to a Tektronix TDS50328 Digital Phospor storage oscilloscope for computing the peak values of the current. The input to the circuit is changed at the rate of 1 MHz. Thus the average of the peak power consumed by the device were observed and also plotted.

If a masking scheme is secure, the power consumption of the device for all inputs should be randomized. In other words the expected values for peak power consumption should be equal and not correlated to the input transitions. *Fig 4* plot the mean of peak power consumptions for a masked multiplier in $GF(2^4)$. The traces are plotted in the following manner: For a value i in the x-axis, the corresponding value in the y-axis is the mean of first i readings. The darker line in each of the plots represent the transition from 0×0 to $F \times F$, while the other line denotes the power consumption from 0×0 to 0×1. The plots show that the scheme was able to randomize the power consumption. *Fig 4* plots the same

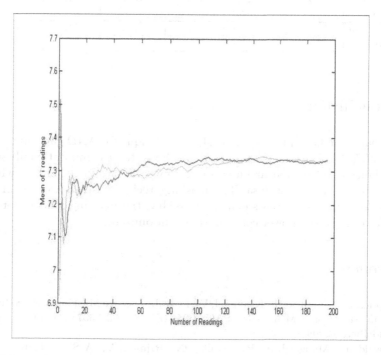

Fig. 6. Mean of peak power consumption vs number of readings for a masked AES S-Box

thing for a masked AES S-Box. The plots show that the masking scheme was indeed able to make the power consumption of AES S-Box randomized.

But the benefit in security comes at the cost of hardware cost and number of random bits required. Table 1 compares the hardware cost, in terms of the AND and XOR gates required to realize a masked AND gate using the proposed technique as compared to a conventional method [3,19]. The table also compares the critical delay of the masked AND gates and the number of random bits required.

Table 2 tabulates the hardware cost, in terms of XOR gates and AND gates for a $GF(2^4)$ multiplier and also the AES S-Box. We also mention the maximum number of random bits, assuming that all the random bits are independent. We compare the results to a conventional masking technique for the components.

Table 2. Comparision of the Masked Circuits

	UnMasked Circuit		Conventional Masking[3,19]		Proposed Masking	
	$GF(2^4)$ Multiplier	AES S-Box	$GF(2^4)$ Multiplier	AES S-Box	$GF(2^4)$ Multiplier	AES S-Box
#AND Gates	16	78	64	312	116	736
#XOR Gates	15	162	124	336	222	1124
#Random bits	–	–	16	78	32	142

5   Conclusion

In this paper we have proposed a masking technique for AND gates and a multiplier in $GF(2^n)$. The schemes proposed are shown to be computationally secured. The balanced and regular architectures remove glitches in the circuit which are dependent on unmasked data. The masking techniques are used to mask the AES S-Box. Practical results demonstrate that the masking schemes presented are able to make the power consumption randomized.

References

1. Blomer, J., Guajardo, J., Krummel, V.: Provably Secure Masking of AES. In: Handschuh, H., Hasan, M.A. (eds.) SAC 2004. LNCS, vol. 3357, pp. 69–83. Springer, Heidelberg (2004)
2. Oswald, E., Mangard, S., Pramstaller, N., Rijmen, V.: A Side-Channel Analysis Resistant Description of the AES S-box. In: Gilbert, H., Handschuh, H. (eds.) FSE 2005. LNCS, vol. 3557, pp. 413–423. Springer, Heidelberg (2005)

3. Trichina, E., Korkishko, T., Lee, K.H.: Small Size, Low Power, Side Channel-Immune AES Coprocessor: Design and Synthesis Results. In: Dobbertin, H., Rijmen, V., Sowa, A. (eds.) Advanced Encryption Standard – AES. LNCS, vol. 3373, pp. 113–127. Springer, Heidelberg (2005)
4. Kocher, P., Jaffe, J., Jun, B.: Introduction to differential power analysis and related attacks (1998), http://www.cryptography.com/
5. Fahn, P.N., Pearson, P.K.: IPA: A New Class of Power Attacks. In: Koç, Ç.K., Paar, C. (eds.) CHES 1999. LNCS, vol. 1717, pp. 173–186. Springer, Heidelberg (1999)
6. Goubin, L., Patarin, J.: DES and Differential Power Analysis - The "Duplication" Method. In: Koç, Ç.K., Paar, C. (eds.) CHES 1999. LNCS, vol. 1717, Springer, Heidelberg (1999)
7. Akkar, M.-L., Bevan, R., Dischamp, P., Moyart, D.: Power Analysis, What is Now Possible. In: Okamoto, T. (ed.) ASIACRYPT 2000. LNCS, vol. 1976, pp. 489–502. Springer, Heidelberg (2000)
8. Schindler, W.: A Combined Timing and Power Attack. In: Naccache, D., Paillier, P. (eds.) PKC 2002. LNCS, vol. 2274, pp. 263–279. Springer, Heidelberg (2002)
9. Yen, S.-M.: Amplified Differential Power Cryptanalysis on Rijndael Implementations with Exponentially Fewer Power Traces. In: Safavi-Naini, R., Seberry, J. (eds.) ACISP 2003. LNCS, vol. 2727, pp. 106–117. Springer, Heidelberg (2003)
10. Ors, S.B., Gurkaynak, F., Oswald, E., Preneel, B.: Power-analysis attack on an ASIC AES implementation. Proceedings ofInformation Technology: Coding and Computing 2, 546–552 (2004)
11. Menicocci, R., Pascal, J.: Elaborazione Crittografica di Dati Digitali Mascherati, Italian Patent IT MI0020031375A (July 2003)
12. Messerges, T.S., Dabbish, E.A., Puhl, L.: Method and Apparatus for Preventing Information Leakage Attacks on a Microelectronic Assembly. US Patent 6,295,606 (September 2001), Available online at http://www.uspto.gov/
13. Golić, J.D.: Random Masking in Hardware. IEEE Transactions on Circuits and Systems-I 54(2) (2007)
14. Waddle, J., Wagner, D.: Towards Efficient Second-Order Power Analysis. In: Joye, M., Quisquater, J.-J. (eds.) CHES 2004. LNCS, vol. 3156, pp. 1–15. Springer, Heidelberg (2004)
15. Chari, S., Jutla, C.S., Rao, J., Rohtagi, P.: Towards Sound Approaches to Counteract Power-Analysis Attacks. In: Wiener, M.J. (ed.) CRYPTO 1999. LNCS, vol. 1666, pp. 398–412. Springer, Heidelberg (1999)
16. Jan, M.: Digital Integrated Circuits. Prentice-Hall, Englewood Cliffs (1996)
17. Mangard, S., Schramm, K.: Pinpointing the Side-Channel Leakage of Masked AES Hardware Implementations. In: Goubin, L., Matsui, M. (eds.) CHES 2006. LNCS, vol. 4249, Springer, Heidelberg (2006)
18. Mangard, S., Pramstaller, N., Oswald, E.: Successfully Attacking Masked AES Hardware Implementations. In: Rao, J.R., Sunar, B. (eds.) CHES 2005. LNCS, vol. 3659, Springer, Heidelberg (2005)
19. Trichina, E., De Seta, D., Germani, L.: Simplified Adaptive Multiplicative Masking for AES. In: D.Walter, C., Koç, Ç.K., Paar, C. (eds.) CHES 2003. LNCS, vol. 2779, pp. 187–197. Springer, Heidelberg (2003)
20. Mangard, S., Popp, T., Gammel, B.M.: Side-Channel Leakage of Masked CMOS Gates. In: Menezes, A.J. (ed.) CT-RSA 2005. LNCS, vol. 3376, pp. 351–365. Springer, Heidelberg (2005)

LFSR Based Stream Ciphers Are Vulnerable to Power Attacks

Sanjay Burman[1], Debdeep Mukhopadhyay[2], and Kamakoti Veezhinathan[3]

[1] PhD Student, Department of Computer Science and Engg., Indian Institute of
Technology, Madras, India
sanjayburman@gmail.com
[2] Assistant Professor, Department of Computer Science and Engg., Indian Institute
of Technology, Madras, India
debdeep@cse.iitm.ernet.in
[3] Associate Professor, Department of Computer Science and Engg., Indian Institute
of Technology, Madras, India
kama@cs.iitm.ernet.in

Abstract. Linear Feedback Shift Registers (LFSRs) are used as building blocks for many stream ciphers, wherein, an n-degree primitive connection polynomial is used as a feedback function to realize an n-bit
LFSR. This paper shows that such LFSRs are susceptible to power analysis based Side Channel Attacks (SCA). The major contribution of this
paper is the observation that the state of an n-bit LFSR can be determined by making $O(n)$ power measurements. Interestingly, *neither the
primitive polynomial nor the value of n be known to the adversary launch-
ing the proposed attack.* The paper also proposes a simple countermeasure
for the SCA that uses n additional flipflops.

Keywords: Linear Feed Back Shift Registers, Side Channel Attacks,
Power Analysis, Hamming Distance, Dynamic Power Dissipation.

1 Introduction

Encryption algorithms are used to protect information from unauthorized access
or disclosure and are constructed using key controlled cryptographic primitives.
The security robustness of cryptographic primitives have traditionally been measured under three mathematical models, namely, (1) when an adversary is assumed to have unlimited computational power (*unconditional security*); (2) when
it can be proved that if an adversary is successful in breaking the cryptographic
primitive under attack then, the adversary can also solve another mathematical
problem that is believed to be hard to solve (*provable security*); and, (3) when
the effort required to break a cryptographic primitive is so large that the cryptographic primitive can be considered to be unbreakable (*computational security*).
However, it has been established in the recent past that even if a cryptographic
primitive is robust against attacks under the three mathematical models mentioned above, there exist a class of attacks *against the real life implementations*

K. Srinathan, C. Pandu Rangan, M. Yung (Eds.): Indocrypt 2007, LNCS 4859, pp. 384–392, 2007.

that must be considered to ensure the security robustness of a system implementing the cryptographic primitives. These are referred to as *Side Channel Attacks* (SCA) [1]. This class of attacks against implementations are rather powerful and lead to system breaks with little effort. The adversary in this case exploits the information leaked unintentionally from the system executing the cryptographic primitive, into the environment, to attack the cryptographic system, often leading to catastrophic failure of security. This is possible even on a system whose *theoretical robustness has been established under the mathematical models mentioned above.*

Stream ciphers are an important class of symmetric ciphers used extensively for encryption by hardware-based cryptographic systems. They are popular because of their simplicity, efficiency and performance. The secure realization of stream ciphers is crucial to guard against the SCAs. Some guidelines in this direction are suggested in [2]. An overview on SCAs on stream ciphers and countermeasures is provided in [3]. LFSRs are used as building blocks for many stream ciphers because of their well defined structure and remarkable properties like long period, ideal autocorrelation and statistical properties. The leakage of information and vulnerability of stream ciphers based on Galois LFSRs is investigated in [4].

Though side channel attacks have reportedly been successfully mounted for many years [5], the publication of [6] by Kocher et.al. is a watershed in this area. This spurred a flurry of research and development in the exploitation of and safeguards against information leaked though side channels with an intention to attack the cryptographic mechanisms built into various security systems. There are several types of side channels through which information leaks inadvertently into the environment. The most prominent of them includes the measurement of the time taken or power consumed to perform a cryptographic function, the argument(s) to the function being the secret cryptographic key/data. A number of successful attacks using the above idea have been reported. These attacks can be mounted by using some very standard test and measuring equipment that are widely available. Typically power attacks can be mounted by measuring the electrical current that flows through a small resistor ($10\ \Omega$ to $50\ \Omega$) placed in series with the pin through which power is fed into a device performing a cryptographic computation. If the current being drawn is a function of the cryptographic key/data then the measurements of current during the cryptographic computation will be correlated with the cryptographic key/data. This correlation can then be analyzed to either directly mount the attack to reveal the key/data or be used in conjunction with a brute force attack to reduce the search space. Similar power attacks can also be mounted by measuring the electromagnetic radiations in the vicinity of the device performing the cryptographic computation.

This paper presents a power analysis based SCA technique to precisely determine the state of an n-bit LFSR by measuring the power consumed by the LFSR in each cycle over consecutive cycles linear in n.

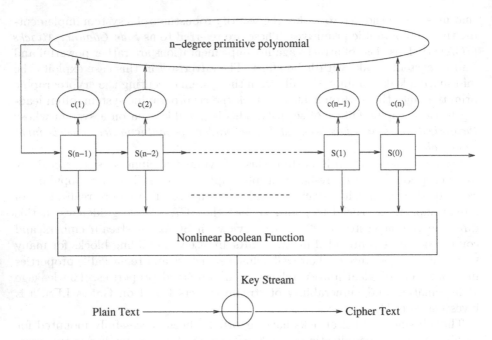

Fig. 1. An n-stage LFSR with a Non-linear filter

2   Preliminaries

2.1   LFSRs

LFSRs are used as primitives in building blocks in many stream ciphers because of their simple structure, guaranteed period and near ideal statistical properties. A general LFSR structure is shown in Figure 1.

The LFSR is a finite state machine that operates over some finite field F_q, where q is a prime or positive power of a prime. For the purposes of this paper we assume that $q = 2^r$, and $r = 1$, i.e. we consider only *binary LFSRs*. An n-stage binary LFSR consists of n consecutive storage elements, called stages. Each stage is a flipflop that stores $S(i)$, such that, $S(i) \in \{0, 1\}, \forall i, 1 \le i \le n$. The content of the n-stages of the LFSR at time t is referred to as the *state of the LFSR* at time t and denoted by ST_t. The state at time $t + 1$ is computed by rightshifting the LFSR by one bit. The value shifted into the first (leftmost) stage, denoted by $S(n)$, is a linear combination of the contents of the n-stages as defined by the feedback polynomial used to realize the LFSR. Therefore, if $ST_t = (S(n-1), \cdots, S(0))$ then, $ST_{t+1} = (S(n), S(n-1), S(n-2), \cdots, S(1))$, where,

$$S(n) = c(1)S(n-1) \oplus c(2)S(n-2) \oplus \cdots \oplus c(n)S(0), \, c(i) \in \{0, 1\}, \forall i, 1 \le i \le n.$$

For more information and background on the LFSRs, the reader is referred to [7]. We shall now present some new and interesting properties of LFSRs.

Theorem 1. *Let HD_t be the* Hamming Distance *between the n-bit vectors, ST_t and ST_{t+1}. Let $PD_t = (HD_t - HD_{t+1})$. Then, $PD_t \in \{-1, 0, 1\}$.*

Proof. Let $ST_t = (S(n-1), \cdots, S(1), S(0))$. Then, $ST_{t+1} = (S(n), S(n-1), \cdots, S(1))$ and $ST_{t+2} = (S(n+1), S(n), S(n-1), \cdots, S(2))$. Let $HW(V)$ denote the Hamming Weight (number of ones) of a bit-vector V. It is straightforward to see the following:

$$HD_t = HW((S(n) \oplus S(n-1)), (S(n-1) \oplus S(n-2)), \cdots, (S(1) \oplus S(0))) \quad (1)$$

$$HD_{t+1} = HW((S(n+1) \oplus S(n)), (S(n) \oplus S(n-1)), \cdots, (S(2) \oplus S(1))) \quad (2)$$

Equations 1 and 2 imply the following:

$$
\begin{align}
PD_t &= HD_t - HD_{t+1} \tag{3} \\
&= HW((S(0) \oplus S(1)) - HW((S(n+1) \oplus S(n))) \tag{4} \\
&= \{0, 1\} - \{0, 1\} \tag{5} \\
&= \{-1, 0, 1\} \tag{6}
\end{align}
$$

Hence, the Theorem.   ◇

Corollary 1. *Let PD'_t be defined as follows: It is equal to 0 when, $HD_t = HD_{t+1}$, else it is 1. Given $S(n+1)$, $S(n)$, $S(1)$ and $S(0)$ as defined in Theorem 1,*

$$PD'_t = S(n+1) \oplus S(n) \oplus S(1) \oplus S(0).$$

Proof. The definition of PD'_t and equation 4 imply the corollary.   ◇

2.2   Dynamic Power Consumption of an LFSR

The dynamic power consumed by a digital circuit is directly proportional to the switching activity (number of components in the circuit that has a state-transition from 0 to 1 or vice-versa) [9]. In the case of LFSRs the dynamic power consumed during the transition in cycle t, that is, from time period t to time period $t + 1$, is proportional to HD_t (refer Theorem 1), as the computed Hamming Distance is a measure of the total number of toggles in the state of the LFSR during the time interval t to $t + 1$. This implies that the *difference in power consumed by the LFSR* between cycle t and cycle $t + 1$ is proportional to PD_t (as defined in Theorem 1). This paper assumes the following:

Assumption 1. *If the number of toggles in the state of an LFSR in cycle t is different than that in cycle $t+1$ (in other words $HD_t \neq HD_{t+1}$), then the power consumed by the LFSR in the two cycles are also different, else they are the same. Therefore, by measuring the power consumption at every cycle, the value of PD'_t as defined in corollary 1, can be computed.*   ◇

3   The Proposed SCA Model

We explain our attack approach for the case where the stream cipher is built using an LFSR with a primitive feedback function and a nonlinear feed forward function as shown in Figure 1. These generators have been widely studied and often used as primitive building blocks in a number of stream ciphers [8]. The n-stage LFSR is initialized with a nonzero state which is the cryptographic key. Every time it is clocked a new key stream bit is generated by filtering the state using a nonlinear Boolean function. The key stream bit is added mod 2 to the plain text to produce the cipher text. The *only* assumption of the proposed SCA model is that the adversary,

– can compute the values of PD'_t by measuring the power consumed by the LFSR, as stated in Assumption 1 above.

It is worthwhile to note that as stated in Assumption 1, the measurement of the power consumed is used *not to compute the number of toggles during any cycle* but *to just indicate whether the number of toggles between any two consecutive cycles is same or different*. Before proceeding further, the following properties of sequences generated by LFSR with primitive connection polynomials are presented. Such sequences are called M-sequences in the literature.

Theorem 2. *[10] The* linear complexity *of an infinite binary sequence s, denoted by $L(s)$, is defined as follows:*

1. *if s is the* zero sequence $s = 0, 0, 0, \ldots$, *then $L(s) = 0$;*
2. *if no LFSR generates s, then $L(s) = \infty$;*
3. *otherwise, $L(s)$ is the length of the shortest LFSR that generates s.*

Let t be a (finite) subsequence of s of length at least $2L(s)$. Then, the Berlekamp-Massey algorithm on input t determines an LFSR of length $L(s)$ which generates s.

Theorem 3. *[11] Given an n-bit LFSR F generating an M-sequence S, a linear combination of the stages of F yields a delayed version (phase) of S. For every delay d, $1 \leq d \leq 2^n - 1$, there exists a linear combination of the stages that yields a version of S that is delayed by d.*   ◇

Let $(S(n-1), S(n-2), \cdots, S(0))$ be the initial unknown state of the given LFSR at time instant 0. Let $S(n + k)$ denote the bit shifted into the LFSR in the kth cycle, $0 \leq k \leq n$. From corollary 1, if the adversary obtains the values of PD'_k, they can be related to the sequence generated by the LFSR as follows:

$$S(n + 1) \oplus S(n) \oplus S(1) \oplus S(0) = PD'_0 \tag{7}$$

$$S(n + 2) \oplus S(n + 1) \oplus S(2) \oplus S(1) = PD'_1 \tag{8}$$

$$\cdots \quad \cdots \quad \cdots \tag{9}$$

$$S(n + k + 1) \oplus S(n + k) \oplus S(k + 1) \oplus S(k) = PD'_k \tag{10}$$

$$\cdots \quad \cdots \quad \cdots \tag{11}$$

The next theorem shows that the PD'_k values computed are a delayed M-sequence generated by the LFSR.

Theorem 4. *The sequence* $PD'_0, PD'_1, \cdot$ *is a delayed sequence of the M-sequence generated by the given LFSR with initial state* $(S(n-1), S(n-2), \cdots, S(0))$.

Proof. Note that by the definition of the LFSR, $S(n)$ is a linear combination of the bits currently stored in the LFSR. Let $S(n) = f_0(S(n-1), S(n-2), \cdots, S(0))$. Now, $S(n+1) = f_0(S(n), S(n-1), \cdots, S(1))$. Since $S(n)$ is a linear combination of $(S(n-1), S(n-2), \cdots, S(0))$, $S(n+1)$ can also be represented as a linear combination of $(S(n-1), S(n-2), \cdots, S(0))$ denoted by $f_1(S(n-1), S(n-2), \cdots, S(0))$.

From equation 7 we see that

$$PD_0 = f_1(S(n-1), \cdots, S(0)) \oplus f_0(S(n-1), \cdots, S(0)) \oplus S(1) \oplus S(0).$$

Hence, PD'_0 is a *linear combination* LC of the bits stored in the LFSR at time instant 0.

From equation 10 we see that

$$PD_k = f_1(S(n+k-1), \cdots, S(k)) \oplus f_0(S(n+k-1), \cdots, S(k)) \oplus S(k+1) \oplus S(k).$$

Note that PD'_k is the *same linear combination*, LC, (as in the case of PD'_0 mentioned above) of the bits stored in the LFSR at time instant k. This and theorem 3 proves this theorem.   ◇

Theorem 5. *Given the length of the LFSR, the primitive connection polynomial and the delayed sequence* $PD'_0, PD'_1, \cdots, PD'_{n-1}$, *the initial state of the LFSR can be determined.*

Proof. As mentioned earlier, let $S(n+k)$ denote the bit shifted into the LFSR in the kth cycle, $0 \leq k \leq n$. Note that by the definition of the LFSR, $S(n)$ is a linear combination of the bits currently stored in the LFSR. Let $S(n) = f_0(S(n-1), S(n-2), \cdots, S(0))$. As the primitive polynomial and the length of the LFSR, n, is known to the adversary imply that the function $f_0()$ is known to the adversary. As mentioned earlier, $S(n+1) = f_0(S(n), S(n-1), \cdots, S(1))$. Since $S(n)$ is a linear combination of $(S(n-1), S(n-2), \cdots, S(0))$, $S(n+1)$ can be represented as the function $f_1(S(n-1), S(n-2), \cdots, S(0))$. In a similar fashion, $S(n+k) = f_k(S(n-1), S(n-2), \cdots, S(0))$. As the primitive polynomial is known to the adversary, all the functions $f_k(S(n-1), S(n-2), \cdots, S(0))$, $0 \leq k \leq n$ are known to the adversary.

Substituting $S(n+k)$ by $f_k(S(n-1), \cdots, S(0))$, $0 \leq k \leq n$, in the equations 7 8 10, we get

$$f_1(S(n-1), \cdots, S(0)) \oplus f_0(S(n-1), \cdots, S(0)) \oplus S(1) \oplus S(0) = PD'_0$$
$$f_2(S(n-1), \cdots, S(0)) \oplus f_1(S(n-1), \cdots, S(0)) \oplus S(2) \oplus S(1) = PD'_1$$
$$\cdots \qquad \cdots \qquad \cdots$$
$$f_n(S(n-1), \cdots, S(0)) \oplus f_{n-1}(S(n-1), \cdots, S(0)) \oplus f_0(S(n-1), \cdots, S(0)) \oplus S(n-1) = PD'_{n-1}$$

Given that PD'_k, $0 \leq k < n$ is known, the above forms a set of n simultaneous equations with n unknowns, namely, $S(n-1), S(n-2), \cdots, S(0)$. Solving the above shall yield the values of $S(n-1), S(n-2), \cdots, S(0)$. Substituting these values in $f_k(S(n-1), S(n-2), \cdots, S(0))$, $0 \leq k \leq n$, gives the values of $S(2n), S(2n-1), \cdots, S(n+1)$.

The fact that the sequence generated by the LFSR is an M-sequence and Theorem 4 imply that the above set of simultaneous equations does have *an unique solution*. This is true from the observation that there should exist states $(S(n-1), S(n-2), \cdots, S(0))$ which is at an unique *delay distance* from the sequence $(PD'_0, PD'_1, \cdots)$; and, there can be only one (state of the LFSR) solution to $(S(n-1), S(n-2), \cdots, S(0))$. If there are more than one solution, it essentially implies that the delayed sequence $(PD'_0, PD'_1, \cdots)$ can be arrived at from two different initial states of the LFSR, with the same amount of delay, which contradicts the fact that the LFSR generates an M-sequence.   ◇

3.1   The Proposed Attack

Let $POW(k)$ denote the dynamic power consumed by the nonlinear filter generator at time instant k.

1. Measure $POW(0)$, for time instant 0;
2. **for** each time instant k, $k \geq 1$ **do**
   (a) Measure the dynamic power, $POW(k)$.
   (b) $PD'_{k-1} = 1$ **if** $POW(k-1) \neq POW(k)$, **else** it is 0.
   (c) Input PD'_{k-1} into the Berlekamp-Massey (BM) Algorithm. **If** BM terminates then **exit this for loop**; **else** repeat Step 2;
3. **Result**
   (a) Berlekamp-Massey algorithm outputs the *length* n of the LFSR F and the connection polynomial realized by F; (as inferred from theorems 2 and 4).
   (b) Now that the length of the LFSR and the connection polynomial realized by the LFSR are known, compute the initial state of the LFSR at the time of launch of the attack using Theorem 5.

4   Countermeasure to the SCA

Figure 2 shows the countermeasure for the SCA. In the circuit, for each flipflop F in the LFSR, there is a corresponding toggle flipflop F' that toggles, if F does not toggle; and, does not toggle, if F toggles. Note that the clock input to F' is the $XNOR$ of the input and output of F AND-ed with the system clock. If the input and output of F is same, that is F does not toggle in the next cycle, the clock is fed into F' essentially toggling it. On the other hand, if the input and output of F are different, that is F toggles in the next cycle, the clock is not fed into F' preventing it from toggling. In this circuit, at each stage there shall be uniformly n toggles, thereby countering the power attack. To avoid clock skews between the LFSR flipflops and their toggle counterparts, the clock path

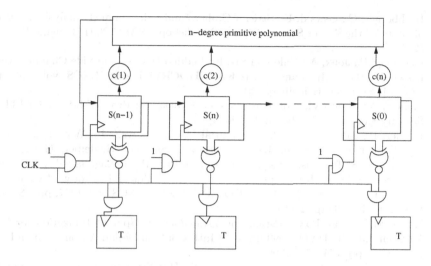

Fig. 2. The Countermeasure to the SCA

to both are balanced by introducing a dummy gate in the clock path driving the
flipflops of the LFSR. The drawbacks of this approach are that it needs double
the number of flipflops and consumes more dynamic power.

5   Conclusions

We have shown an interesting property of LFSRs with respect to the hamming
distance between the state transitions of an LFSR with primitive feedback polynomial. We exploit this property to determine the state of the LFSR by making
$O(n)$ power measurements. In this paper, we have made an ideal assumption that
the power consumed by the LFSR in each of any two consecutive cycles shall remain the same if the number of toggles in the state of the LFSR are also equal in
the two cycles under consideration. A more practical assumption would be that if
the *difference in power consumed across two cycles is less than a threshold* then the
toggles are equal across the cycles, else they are different. Such type of *thresholds*
can be determined by simulating the model of the LFSR, using circuit simulators
like SPICE. The paper also presents a simple countermeasure for the attack.

References

1. Kocher, P., Lee, R., McGraw, G., Raghunathan, A., Ravi, S.: Security as a New
   Dimension in Embedded System Design. In: Proc. of IEEE Design Automation
   Conference - DAC 2004, pp. 753–761. IEEE Computer Society Press, Los Alamitos
   (2004)
2. Kumar, S., Lemke, K., Paar, C.: Some Thoughts about Implementation Properties
   of Stream Ciphers. In: Proc. of State of the Art of Stream Ciphers Workshop -
   SASC 2004, Brugge, Belgium (2004)

3. Rechberger, C., Oswald, E.: Stream Ciphers and Side-Channel Analysis. In: Proc. of State of the Art of Stream Ciphers Workshop - SASC 2004, Brugge, Belgium (2004)
4. Delaunay, P., Joux, A.: Galois LFSR, Embedded Devices and Side Channel Weaknesses. In: Barua, R., Lange, T. (eds.) INDOCRYPT 2006. LNCS, vol. 4329, pp. 436–451. Springer, Heidelberg (2006)
5. Shamir, A.: A Top View of Side Channel Attacks. In: Proc. of L-SEC/CALIT IT Security Congress (October 19-20, 2006)
6. Kocher, P., Jaffe, J., Jun, B.: Differential Power Analysis. In: Wiener, M.J. (ed.) CRYPTO 1999. LNCS, vol. 1666, pp. 388–397. Springer, Heidelberg (1999)
7. Golomb, S.: Shift Register Sequences. Aegean Park Press, Laguna Hills, CA (1981)
8. Bedi, S.S., Pillai, N.R.: Cryptanalysis Of The Nonlinear Feedforward Generator. In: Roy, B., Okamoto, E. (eds.) INDOCRYPT 2000. LNCS, vol. 1977, pp. 188–194. Springer, Heidelberg (2000)
9. Hsiao, M.S.: Peak Power Estimation using Genetic Spot Optimization for large VLSI circuits. In: DATE 1999. Proc. of Intl. Conf. on Design Automation and Test in Europe, pp. 175–179 (1999)
10. Menezes, A., van Oorschot, P., Van stone, S.: Handbook of Applied Cryptography. CRC Press, Boca Raton, USA (1996)
11. Davies, A.C.: Delayed versions of maximal-length linear binary sequences. Electronic Letters 1, 61 (1965)

An Update on the Side Channel Cryptanalysis of MACs Based on Cryptographic Hash Functions*

Praveen Gauravaram[1],** and Katsuyuki Okeya[2]

[1] Department of Mathematics, Technical University of Denmark, Denmark
p.gauravaram@mat.dtu.dk
[2] Hitachi, Ltd., Systems Development Laboratory, Japan
katsuyuki.okeya.ue@hitachi.com

Abstract. Okeya has established that HMAC/NMAC implementations based on only Matyas-Meyer-Oseas (**MMO**) PGV scheme and his two refined PGV schemes are secure against side channel DPA attacks when the block cipher in these constructions is secure against these attacks. The significant result of Okeya's analysis is that the implementations of HMAC/NMAC with the Davies-Meyer (**DM**) compression function based hash functions such as SHA-1 are vulnerable to DPA attacks. In this paper, first we show a partial key recovery attack on NMAC/HMAC based on Okeya's two refined PGV schemes by taking practical constraints into consideration. Next, we propose new hybrid NMAC/HMAC schemes for security against side channel attacks assuming that their underlying block cipher is ideal. We show a hybrid NMAC/HMAC proposal which can be instantiated with **DM** and a slight variant to it allowing NMAC/HMAC to use hash functions such as SHA-1. We then show that M-NMAC, MDx-MAC and a variant of the envelope MAC scheme based on **DM** with an ideal block cipher are secure against DPA attacks.

Keywords: Side channel attacks, DPA, HMAC, M-NMAC, MDx-MAC.

1 Introduction

Okeya [11] has shown that NMAC/HMAC implementations based on eleven out of twelve provably secure Preneel-Govaerts-Vandewalle (PGV) compression functions based on block ciphers [2] are vulnerable to differential power analysis (DPA) and its minor variant reverse DPA (RDPA) attacks even if the underlying block cipher is secure against these attacks. His analysis shows that *only* Matyas-Meyer-Oseas (**MMO**) PGV scheme and his two refined PGV compression function proposals are secure against these attacks and one of the secret keys of NMAC/HMAC schemes can be extracted when they are implemented with

---

* This work is supported by The Danish Research Council for Technology and Production Sciences grant no. 274-05-0151 and partly supported by National Institute of Information and Communications Technology (NICT), Japan.
** Some of this work was performed when the author was at Information Security Institute, Queensland University of Technology, Australia.

K. Srinathan, C. Pandu Rangan, M. Yung (Eds.): Indocrypt 2007, LNCS 4859, pp. 393–403, 2007.
© Springer-Verlag Berlin Heidelberg 2007

the popular Davies-Meyer (**DM**) based hash functions such as MD5, RIPEMD-160, SHA-1, SHA-224/256 and SHA-384/SHA-512. The impact of this attack is that NMAC/HMAC implementations with these hash functions do not always result in a secure MAC as there is no guarantee that these schemes provide secure authentication when the attacker knows any of the secret keys [1,5,4].

In this paper, we show a partial key recovery attack on NMAC/HMAC schemes implemented with two refined PGV compression functions proposed in [11] with different state and block sizes using RDPA attack. We conclude that NMAC/HMAC based on *only* **MMO** PGV scheme with the same/different block and state sizes is secure against side channel attacks. We propose new hybrid NMAC/HMAC functions that use compression functions of different architecture as the inner and outer functions in NMAC/HMAC for security against DPA/RDPA attacks assuming that the block cipher in them is secure. Considering the wide usage of **DM** scheme in many standard hash functions, we propose a DPA/RDPA resistant hybrid NMAC/HMAC scheme based on **DM** and a slight variant of it which follows the design goal of HMAC by calling hash function as a black box. We analysed M-NMAC [5,4], MDx-MAC [14] and a variant of the envelope MAC scheme [14] based on twelve provably secure PGV compression functions against DPA/RDPA attacks assuming that the block cipher of PGV schemes is ideal. We show that these MAC proposals based on **DM** and our two refined PGV compression function proposals are secure against DPA/RDPA attacks. Our work shows that the security of hash based MAC schemes against side channel attacks depends on the architectures of both the MAC function and its underlying compression function.

The DPA/RDPA resistant MAC schemes proposed/analysed in this paper can be used as software programs using a tamper-resistant block cipher coprocessor on the smartcard. These MAC schemes can also be used when a hardware accelerator capable of computing bulk encryption using DPA countermeasure is available but the operations needed for the MAC are computed in a microcontroller without DPA countermeasures.

We organise this paper as follows: In Section 2, we describe iterated hash functions and NMAC/HMAC proposals respectively. In Section 3, we discuss DPA/RDPA attacks on NMAC/HMAC functions. In Section 4, we extend Okeya's attacks on NMAC/HMAC. In Section 5, we propose hybrid MAC proposals following the model of NMAC/HMAC. In Section 6, we review other hash based MAC schemes proposed in the literature. In Section 7, we analyse the security of M-NMAC and its variants based on twelve PGV schemes against DPA/RDPA attacks followed by forgery attacks on these MAC schemes in Section 8. We conclude the paper in Section 9 with some directions to the future work.

2   Cryptographic Hash Functions

Cryptographic hash functions process an arbitrary length message into a fixed length digest. The Merkle-Damgård iterative structure [8,3] has been a popular structure used in the design of many standard hash functions such as MD5,

SHA-1, SHA-224/256 and SHA-384/512. The message x with $|x| \leq 2^l - 1$ bits, to be processed using the hash function H is always padded by appending it with a 1 bit followed by 0 bits until the padded message is l bits short of a block length b of the fixed length input compression function f. The last l bits are filled in with the binary encoded representation of the length of the original unpadded message x to avoid some trivial attacks [7]. This message is an integer multiple of b bits and is represented with b-bit data blocks as $x = x_1, x_2, \ldots x_n$. Each data block x_i is processed using f to compute intermediate states $H_i = f_{H_{i-1}}(x_i)$ where $i = 1$ to n. The final state $H_n = f_{H_{n-1}}(x_n)$ is the message digest of x. In general, the approaches to design compression functions employ block cipher-based constructions. There are twelve out of sixty-four schemes considered by PGV [12] that are provably secure when the underlying block cipher used in them is ideal [2]. This model of PGV uses parameters $y, k, z \in \{H_{i-1}, x_i, H_{i-1} \oplus x_i, 0\}$ and a block cipher G. See Appendix A for these twelve schemes named as f^1 to f^{12}.

2.1 NMAC/HMAC Functions

If k_1 and k_2 are two independent and random secret keys and H is the iterated hash function over the compression function f, then the NMAC function [1] used to process an arbitrary size message x is given by $\text{NMAC}_k(x) = f_{k_1}(H_{k_2}(x))$. The secret key k_2 is the IV of the inner iterated hash function H and the key k_1 is the IV of the outer compression function f. The standard padding procedure is followed for the functions H and f as described in Section 2. HMAC is a "fixed IV" variant of NMAC and uses H as a black box. It is defined by $\text{HMAC}_k(x) = H_{IV}(\overline{k} \oplus \text{opad} \| H_{IV}(\overline{k} \oplus \text{ipad} \| x))$ where opad and ipad are constants as defined in [1], $\overline{k}$ denotes the completion of the key k to a b-bit block by padding k with 0 bits and $\|$ denotes the concatenation operation.

3 Side Channel Attacks on NMAC/HMAC

3.1 DPA and Reverse DPA Attack Models

In the DPA (resp. RDPA) attack, the attacker observes the power consumption of the computing device as side channel information and uses statistical tools such as the average of the power consumption to eliminate the noise in order to extract the secret key [6]. RDPA is a standard DPA where instead of known input, the attacker uses the known output to mount the DPA attack. Following [11], we consider the application of DPA (resp. RDPA) on the target XOR operation in MACs as shown in Figure 1. This analysis is also applicable to the modular addition operation. The XOR operation in Figure 1(a) (resp. Figure 1(b)) has two inputs: y_1, a constant secret and x, a public variable (resp. a secret), and an output y_2, whose value is unknown to the attacker (resp. a public variable).

The attacker uses the DPA (resp. RDPA) attack to recover the secret y_1. The attacker guesses some bit of y_1 and classifies the input x (resp. the output y_2) into two groups depending on the target bit of y_2 (resp. x) which is 1

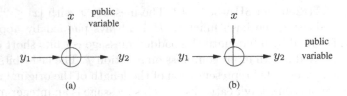

Fig. 1. The DPA and Reverse DPA models

or 0 according to the variable input x (resp. y_2). The attacker then observes the power consumed by the XOR operation due to several inputs of x (resp. outputs of y_2). The attacker then computes the average power consumption for each group. Then the attacker computes the difference between two averages. If positive spikes appear in the difference, the attacker ensures the original guess for the bit of y_1. Otherwise, the attacker performs its bit inversion. Under the Hamming weight model [9], power consumption is correlated with the Hamming weight of the manipulated data. Thus, positive spikes imply that the target bit of y_2 is manipulated as expected, and the averaging eliminates the impact of other bits on the value of y_2 since they behave randomly. By observing sufficient number of power consumptions, re-classifying y_2 and computing average power consumption, the attacker does not have to re-observe them in order to find other target bits. See [11] for the experimental results of these attacks on a XOR operation implemented in an IC chip.

3.2 Okeya's Analysis of NMAC/HMAC

Okeya [11] has analysed HMAC/NMAC based on twelve PGV compression functions f^i where $i \in \{1, 2, \ldots, 12\}$ assuming an ideal block cipher against DPA/RDPA attacks. This analysis assumes that the state size t of H is same as the data block size b of f.

The following setting was considered in the analysis of NMAC [11]: The message block x_1 is public and variable and $H_0(= k_2)$ is secret and fixed. If the compression function contains the XOR operation $x_1 \oplus H_0$, the attacker would directly mount the DPA attack on this operation to extract the secret H_0 following the steps given in Section 3.1. The PGV compression functions f^i where $i \in \{2, 4, 6, 8, 9, 10, 11, 12\}$ have the operation $x_1 \oplus H_0$ and hence the key k_2 can be extracted by mounting the DPA attack. Using k_2, the attacker computes sufficient outputs of the inner hash function $H_{k_2}(x)$ for various messages x. The attacker uses these values as the public and variable input to the XOR operation $H_{k_2}(x) \oplus k_1$ in these compression functions to extract the secret key k_1 by mounting the DPA attack.

Similarly, for NMAC based on **DM**, the secret key k_1 can be extracted using RDPA attack on the target XOR operation between the secret k_1 and a secret variable from the output of G which produces NMAC output. The DPA/RDPA attacks on the compression functions f^3 and f^7 depends on the order of the execution of XOR operation used in these functions. Okeya [11] has refined the compression functions f^3 and f^7 and proposed the new DPA/RDPA resistant compression functions $f^{(3*)}$ and $f^{(7*)}$ as $H_i = (G_{H_{i-1}}(x_i) \oplus H_{i-1}) \oplus x_i$ and $H_i = (G_{x_i}(H_{i-1}) \oplus H_{i-1}) \oplus x_i$ respectively.

4 Extending Okeya's Attack to Different Block/State Sizes

When f^i for $i \in \{2, 4, 6, 8, 9, 10, 11, 12\}$ is used as the inner and outer compression function in NMAC, it is straight forward to mount the DPA attack discussed above on the XOR operation $x_1 \oplus k_2$ to extract the secret key k_2. This attack is independent of the padding and length encoding of the message used in the last block.

Now assume that the output bit length t of the inner function $H_{k_2}(x)$ is less than the block size b of the outer compression function f_{k_1} and $|k_1| = b$. Hence, $H_{k_2}(x)$ is padded to the block size b following the standard procedure discussed in Section 2 and used as input to the outer function f_{k_1}. Let this input be denoted by $H_{k_2}(x) \| \mathsf{pad}$, where pad are the padded and length encoded bits. Even if the attacker computes sufficient count of $H_{k_2}(x) \| \mathsf{pad}$ for various messages x and mounts the DPA attack on the outer function f_{k_1}, the attacker can *only* recover the higher t bits of k_1. The attacker cannot recover the lower $b - t$ bits of k_1 as the padded and length encoded bits appended to the inputs $H_{k_2}(x)$ are unique and their hamming weight cannot be correlated to the power consumption of the device under the Hamming weight model. If $|k_1| < b$ then we pad the key k_1 with $b - |k_1|$ bits that are independent of the secret key k_1. This padding procedure is known to the attacker who mounts the DPA attack to recover the other bits of k_1. The RDPA attack on NMAC based on f^5, f^3 and f^7 [11] is also applicable when the output size of the inner hash function of NMAC is less than the block size of the outer compression function.

4.1 Partial Key Recovery of NMAC/HMAC with $f^{(3*)}/f^{(7*)}$

We analyse NMAC assuming that the output bit length t of its inner function $H_{k_2}(x)$ is less than the block size b of the outer compression function f. Hence, $H_{k_2}(x)$ is padded to the block size b and used as input to the outer function f. Let the inner function of NMAC be any DPA resistant compression function such as f^1, $f^{(3*)}$ and $f^{(7*)}$. Consider the compression function $f^{(3*)}$ for the outer function of NMAC. The output $H_{k_2}(x)$ of the inner function is padded in the standard manner and is denoted by $H_{k_2}(x) \| \mathsf{pad}$ which is used as input to the outer function. The output $\mathrm{NMAC}_k(x)$ is computed as shown in Figure 2.

Here $G(\cdot)$ denotes $G_{k_1}(H_{k_2}(x) \| \mathsf{pad})$. The target XOR operation is $G(\cdot) \oplus k_1$. Since the padding information used in pad is known to the attacker, the attacker can compute the lower $|\mathsf{pad}| (= b - t)$ bits of the result of the first XOR using $\mathrm{NMAC}_k(x)$ and pad. The attacker can now mount the RDPA attack to this case where the lower $|\mathsf{pad}|$ bits of k_1 is a secret constant and the lower $|\mathsf{pad}|$ bits of the result of the first XOR is a public variable. As a result, the attacker can

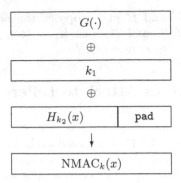

Fig. 2. NMAC with $f^{(3*)}$ as the outer compression function and padded $H_{k_2}(x)$

recover the lower |pad| bits of the key k_1. This attack also applies to NMAC with $f^{(7*)}$ as the outer function.

Once part of k_1 is recovered, its remaining part and the key k_2 can be recovered using brute-force key search. Though this complete key recovery process is impractical for reasonable key sizes, the total work is still less than the work $2^{|k_1|} + 2^{|k_2|}$ of the divide and conquer key recovery attack on NMAC [1]. Note that the knowledge of k_1 does not guarantee NMAC to be a secure MAC [1,5,4]. This attack does not work when the outer function of NMAC is **MMO** scheme as it has no target XOR operation on which the RDPA attack can be mounted.

Illustration: For NMAC with $f^{(3*)}$ or $f^{(7*)}$ as the outer function, if $|H_{k_2}(x)| = 160$ bits, $b = 256$ bits and $|k_1| = 256$ bits, the lower 96 bits of k_1 can be extracted using the above attack.

In order to avoid this attack, we must have $|k_1| \leq t$. If $|k_1| < t$ and $t = b$, we pad the lower $b - |k_1|$ bits of k_1 and ensure that this padding is independent of the secret key. For example, we must avoid padding by repeating the secret key. We can also pad $H_{k_2}(x)$ along with length encoded bits that represent $|H_{k_2}(x)|$ but this requires an extra iteration of f_{k_1} to process the padded block. If $|k_1| < t$ and $t < b$, we pad k_1 with bits independent of the secret key and $H_{k_2}(x)$ as above.

5   Hybrid MAC Proposals

The notion of hybrid MAC functions in the context of NMAC was introduced in [1]. For simplicity of analysis, we assume that block and state sizes of the compression functions are equal. We observe from Section 3 that for NMAC, the resistance against DPA attacks is essential *only* for the inner function and from RDPA attacks is essential *only* for the outer function of these schemes. Due to different security requirements for the inner and outer functions, the hybrid use of PGV schemes provides us more flexible choices to design MACs. To design a secure hybrid NMAC structure, the inner function can be chosen from any of the DPA-resistant compression functions: f^1, $f^{(3*)}$, f^5, $f^{(7*)}$ and the outer function from any RDPA resistant functions: f^1, f^2, $f^{(3*)}$, f^4, f^6, $f^{(7*)}$, f^9, f^{12}.

DPA/RDPA resistant hybrid NMAC/HMAC based on DM

We can use the **DM** scheme f^5 and its slight variant f^6 for the inner and outer functions respectively to design a secure hybrid NMAC/HMAC structure. Note that f^6 and f^5 are related by $f^6_{H_{i-1}}(x) = f^5_{H_{i-1} \oplus x}(x)$. This hybrid use of f^6 and f^5 allows us to design DPA/RDPA resistant NMAC/HMAC using compression functions such as SHA-256. This hybrid NMAC based on **DM** hash function H is defined by $f_{H_{k_2}(x) \oplus k_1}(H_{k_2}(x))$. Similarly, the hybrid HMAC is defined by $H_{IV}(\overline{k} \oplus \text{opad} \oplus H_{IV}(\overline{k} \oplus \text{ipad}||x)||H_{IV}(\overline{k} \oplus \text{ipad}||x))$.

6 Other MAC Proposals Based on Hash Functions

We have found that M-NMAC, MDx-MAC and a variant of the envelope MAC scheme as the DPA/RDPA resistant MAC schemes based on the **DM** scheme after a thorough survey and analysis for DPA/RDPA resistant **DM** based MAC schemes. We treat MDx-MAC and envelope MAC schemes as variants of M-NMAC for an easier presentation of schemes and their analysis.

6.1 M-NMAC Function

M-NMAC [5, 4] is a variant of NMAC which uses the trail key k_1 as an input block to the outer function f in NMAC. M-NMAC is defined by M-NMAC$_k(x) = f_{H_{k_2}(x)}(\overline{k_1})$ where $\overline{k_1}$ ($|k_1| \geq |k_2|$) denotes the key k_1 made to the block size b of the compression function f by concatenating sufficient 0 bits to it if $|k_1| < b$.

6.2 Variants of M-NMAC

The envelope MAC scheme $H(k_2||x||k_1)$ [17], analysed in [13,14,15], has a trail secret key k_1 which either spans across the last two blocks or placed only in the last block. A variant of it proposed by Preneel and van Oorschot [14, p.22] uses the trail secret key k_1 in a separate block. MDx-MAC [13] is another popular MAC scheme which employs MD4 family of hash functions and three keys. The key k_2 replaces the IV of H, the key k_1 exclusive-ored with some constants is appended to the message and the key k_3 influences the internal rounds of the compression function. We note that MDx-MAC can be implemented in such a way that its compression function (which uses k_3 in it) can be called as a black box as the keys k_1 and k_2 are used external to the compression function.

7 Side Channel Attacks on M-NMAC and Its Variants

7.1 Target Compression Functions in M-NMAC

The intermediate state of M-NMAC at any iteration i is given by $H_i = f_{H_{i-1}}(x_i)$ where $i = 1, 2, \ldots, n$ and $H_0 = k_2$. The value $f_{H_n}(\overline{k_1})$ is the authentication tag of M-NMAC. The message x split into blocks $x_1, x_2, \ldots, x_n$ is known to the attacker and is public. At any iteration i in the computation of M-NMAC using the compression function $f_{H_{i-1}}(x_i)$, the value H_{i-1} is secret and fixed and x_i is public and variable whenever the blocks $x_1, \ldots, x_{i-1}$ are fixed. The task of the attacker is to extract H_{i-1}. The attacker uses the DPA attack with the power consumption relating to the output H_i. The other target compression function is the final function $f_{H_n}(\overline{k_1})$ which gives M-NMAC output. Here k_1 and H_n are secrets and the output of M-NMAC is public but is not under the control of the attacker. The attacker's goal is to extract k_1 by mounting RDPA attack using the output of M-NMAC on which the attacker has no control and does not care about the other input H_n.

7.2   DPA/RDPA Attacks on M-NMAC and Its Variants

In this analysis, we set $k_1 = x_i$ and assume that the internal state size of H is the same as the block size of f.

Analysis of M-NMAC against DPA attack

For M-NMAC instantiated with the compression functions f^i where $i \in \{2, 4, 6, 8, 9, 10, 11, 12\}$, the attacker mounts the DPA attack from Section 7.1 to extract the secret key k_2 by choosing a message block x_1 as a public variable value. After recovering k_2, the attacker computes $H_{k_2}(x)$ for any message x. If the outer function f is also one of the eight compression functions f^i where $i \in \{2, 4, 6, 8, 9, 10, 11, 12\}$, it would have the XOR operation $\overline{k_1} \oplus H_{k_2}(x)$ with a public variable $H_{k_2}(x)$ and the fixed secret $\overline{k_1}$. By mounting the DPA attack again, the attacker recovers the secret key k_1.

Analysis of M-NMAC against RDPA attack

The functions f^i where $i \in \{1, 4, 9, 12\}$ directly XORs the key k_1 with the value obtained after processing the chaining value H_n from the last compression function with the block cipher G. In this setting, the key k_1 is secret which the attacker aims to extract, the value obtained as the output of the block cipher is also a secret which the attacker does not care and the XOR of these two values gives the M-NMAC output. For this setting, the attacker mounts the RDPA attack from Section 7.1 to extract k_1.

The applicability of RDPA attack on M-NMAC based on f^3 and f^7 depends on the order of the execution of the XOR operation in them. For f^3 (resp. f^7), we define $G(\cdot)$ by $G_{H_{i-1}}(k_1)$ (resp. $G_{k_1}(H_{i-1})$). These schemes can be implemented in three ways as $f^{3(1)}$ (resp. $f^{7(1)}$): $H_i = (k_1 \oplus H_{i-1}) \oplus G(\cdot)$; $f^{3(2)}$ (resp. $f^{7(2)}$): $H_i = (G(\cdot) \oplus H_{i-1}) \oplus k_1$; $f^{3(3)}$ (resp. $f^{7(3)}$): $H_i = (G(\cdot) \oplus k_1) \oplus H_{i-1}$.

The scheme $f^{3(1)}$ (resp. $f^{7(1)}$) is vulnerable to the DPA attack due to the operation $k_1 \oplus H_{i-1}$. The scheme $f^{3(2)}$ (resp. $f^{7(2)}$) is vulnerable to the RDPA attack due to the operation $y_1 \oplus k_1$ where $y_1 = G(\cdot) \oplus H_{i-1}$. We propose $f^{3(3)}$ and $f^{7(3)}$ as the refined schemes on which DPA/RDPA attacks do not work. This also applies to **DM** as there is no target XOR operation in M-NMAC based on **DM** on which we could mount these attacks. Hence, M-NMAC implemented with $f^{3(3)}$, $f^{7(3)}$ and f^5 is immune against DPA/RDPA attacks. This analysis of M-NMAC also applies to its variants. This shows the significance of the architectural difference of these schemes compared to NMAC/HMAC in using the trail secret key as a block rather than as a state.

8   Forgery Attacks on M-NMAC and Its Variants

For M-NMAC based on f^i for $i \in \{2, 4, 6, 8, 9, 10, 11, 12\}$, our attacks recover both the keys. Hence, the attacker can perform selective forgery on M-NMAC based on these schemes for a message of its choice. The attacker can *only* recover the trail key k_1 of M-NMAC based on f^1, $f^{3(2)}$ and $f^{7(2)}$. In this case, M-NMAC can be forged by querying its oracle for an authentication tag M-NMAC$_k(x) = f_{H_{k_2}(x)}(\overline{k_1})$ of the message x. The attacker then uses the key k_1 to perform existential forgery by computing $f_{\text{M-NMAC}(x)}(\overline{k_1})$ which is the tag for the new forged message $x||\overline{k_1}$.

Similarly, the attacker can selectively forge the variant of the envelope MAC scheme implemented with f^i for $i \in \{2, 4, 6, 8, 9, 10, 11, 12\}$. For this MAC scheme based on f^1, $f^{3(2)}$ and $f^{7(2)}$, the attacker can *only* recover the trail secret key k_1 and can forge it in a similar way as M-NMAC. This analysis also applies to MDx-MAC. Finally, the attacker cannot forge M-NMAC and its variants based on f^5 and our proposed compression functions $f^{3(3)}$ and $f^{7(3)}$ using the techniques from Section 7.2.

Remark 1. M-NMAC based on f^i where $i \in \{2, 3(1), 3(3), 5, 6, 7(1), 7(3), 8, 10, 11\}$ is secure against the RDPA attack. Hence, these schemes can be used as the outer function in M-NMAC. Any of the schemes f^i where $i \in \{1, 3(2), 3(3), 5, 7(2), 7(3)\}$ as the inner function for M-NMAC is secure against the DPA attack. These combinations can be utilized to design hybrid nested MAC schemes in the setting of M-NMAC and its variants.

Remark 2. When the output of any DPA resistant inner compression function of M-NMAC and its variants is less than the key size k_1 then part of k_1 can be recovered in similar to the partial key recovery attack on NMAC and HMAC discussed in Section 4.1.

9   Conclusion

The black box usage of compression function rather than hash function by most of the DPA/RDPA resistant MAC schemes proposed/analysed in this paper agrees with Mironov's concern [10] on proving the security of protocols by assuming Merkle-Damgård hash function as a black box. It is an interesting research problem to formalise models to understand side channel attacks that are directly meaningful to practice for the MAC schemes proposed/analysed in this paper as done for block ciphers [16].

Acknowledgements

Many thanks to the reviewers of CHES 2007 for valuable comments that greatly helped in improving this work and to the reviewers of INDOCRYPT 2007 for their valuable feedback. We also thank Suganya Annadurai, Gary Carter, Ed

Dawson, Choudary Gorantla, Lars Knudsen, William Millan, Juanma González Nieto, Søren Thomsen, Jiri Tuma and Kapali Viswanathan for comments on the previous drafts of this paper.

References

1. Bellare, M., Canetti, R., Krawczyk, H.: Keying hash functions for message authentication. In: Koblitz, N. (ed.) CRYPTO 1996. LNCS, vol. 1109, pp. 1–15. Springer, Heidelberg (1996), available at:
   http://www-cse.ucsd.edu/users/mihir/papers/hmac.html
2. Black, J., Rogaway, P., Shrimpton, T.: Black-Box Analysis of the Block-Cipher-Based Hash-Function Constructions from PGV. In: Yung, M. (ed.) CRYPTO 2002. LNCS, vol. 2442, pp. 320–335. Springer, Heidelberg (2002)
3. Damgård, I.: A Design Principle for Hash Functions. In: Brassard, G. (ed.) CRYPTO 1989. LNCS, vol. 435, pp. 416–427. Springer, Heidelberg (1990)
4. Gauravaram, P.: Cryptographic Hash Functions: Cryptanalysis, Design and Applications. PhD thesis, Information Security Institute, Queensland University of Technogy (June 2007)
5. Gauravaram, P., Hirose, S., Annadurai, S.: An Update on the Analysis and Design of NMAC and HMAC functions. International Journal of Network Security (IJNS) 7(1), 50–61 (July 2008), Online version of the paper is available at http://ijns.nchu.edu.tw/contents/ijns-v7-n1/ijns-v7-n1.html Last access date: 6ᵗʰ of August 2007
6. Kocher, P.C., Jaffe, J., Jun, B.: Differential power analysis. In: Wiener, M.J. (ed.) CRYPTO 1999. LNCS, vol. 1666, pp. 388–397. Springer, Heidelberg (1999)
7. Lai, X., Massey, J.L.: Hash Functions Based on Block Ciphers. In: Rueppel, R.A. (ed.) EUROCRYPT 1992. LNCS, vol. 658, pp. 55–70. Springer, Heidelberg (1993)
8. Merkle, R.: One way Hash Functions and DES. In: Brassard, G. (ed.) CRYPTO 1989. LNCS, vol. 435, pp. 428–446. Springer, Heidelberg (1990)
9. Messerges, T.S.: Using Second-Order Power Analysis to Attack DPA Resistant Software. In: Paar, C., Koç, Ç.K. (eds.) CHES 2000. LNCS, vol. 1965, pp. 238–251. Springer, Heidelberg (2000)
10. Mironov, I.: Hash functions: Theory, attacks, and applications. Technical Report MSR-TR-2005-187, Microsoft Research (November 2005), This technical report is available at the link http://research.microsoft.com/users/mironov/ Last access date: 8ᵗʰ of November 2006
11. Okeya, K.: Side Channel Attacks Against HMACs Based on Block-Cipher Based Hash Functions. In: Batten, L.M., Safavi-Naini, R. (eds.) ACISP 2006. LNCS, vol. 4058, pp. 432–443. Springer, Heidelberg (2006)
12. Preneel, B., Govaerts, R., Vandewalle, J.: Hash Functions Based on Block Ciphers: A Synthetic Approach. In: Stinson, D.R. (ed.) CRYPTO 1993. LNCS, vol. 773, pp. 368–378. Springer, Heidelberg (1994)
13. Preneel, B., van Oorschot, P.C.: MDx-MAC and Building Fast MACs from Hash Functions. In: Coppersmith, D. (ed.) CRYPTO 1995. LNCS, vol. 963, pp. 1–14. Springer, Heidelberg (1995)
14. Preneel, B., van Oorschot, P.C.: On the Security of Two MAC Algorithms. In: Maurer, U.M. (ed.) EUROCRYPT 1996. LNCS, vol. 1070, pp. 19–32. Springer, Heidelberg (1996)

15. Preneel, B., van Oorschot, P.C.: On the Security of Iterated Message Authentication Codes. IEEE Transactions on Information Theory 45(1), 188–199 (1999)
16. Standaert, F.-X., Malkin, T.G., Yung, M.: A formal practice-oriented model for the analysis of side-channel attacks. Cryptology ePrint Archive, Report 2006/139, 2006, this paper is available at http://eprint.iacr.org/2006/139 Last access date: 21st of January 2007
17. Tsudik, G.: Message Authentication with One-Way Hash Functions. In: IEEE Infocom 1992, pp. 2055–2059. IEEE Computer Society Press, Los Alamitos (1992)

A   12 Provably Secure PGV Compression Functions

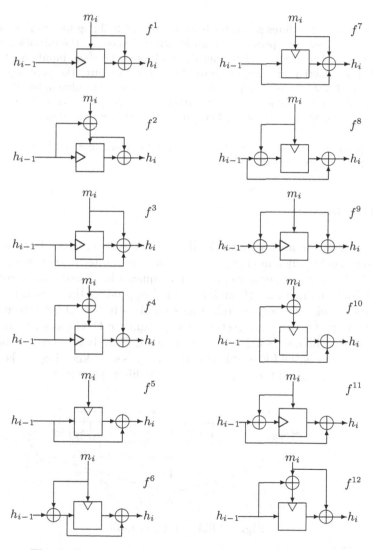

Fig. 3. Compression functions based on PGV construction

Attacking the Filter Generator by Finding Zero Inputs of the Filtering Function

Frédéric Didier[*]

Projet CODES, INRIA Rocquencourt, Domaine de Voluceau,
78153 Le Chesnay cedex
frederic.didier@inria.fr

Abstract. The filter generator is an important building block in many stream ciphers. We present here an attack that recovers the initial state of the hidden LFSR by detecting the positions where the inputs of the filtering function are equal to zero. This attack requires the precomputation of low weight multiples of the LFSR generating polynomial. By a careful analysis, we show that the attack complexity is among the best known and work for almost all cryptographic filtering functions.

Keywords: Stream cipher, filter generator, Boolean functions, low weight multiples, autocorrelation.

1 Introduction

The filter generator uses a linear feedback shift register (LFSR) of length N and characteristic polynomial $g(X)$ that generates a binary sequence $(s_t)_{t\geq 0}$ of period $2^N - 1$. As we can see in Figure 1 this sequence is filtered using a n-variable balanced Boolean function f (from $\mathbf{F}_2^n$ into $\mathbf{F}_2$) to produce the keystream $(z_t)_{t\geq 0}$. The inputs of this function are taken as some bits in the LFSR internal state. We will write $\mathbf{x}_t$ for the n-bit vector corresponding to the inputs of f at time t. Notice that we will always write such elements of $\mathbf{F}_2^n$ in bold. Our goal here is to find the key (that is the LFSR initial state $s_0, \ldots, s_{N-1}$) knowing the keystream sequence $(z_t)_{t\geq 0}$ and all the constituents of the filter generator.

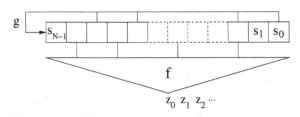

Fig. 1. LFSR filter generator

[*] This work is partially funded by CELAR/DGA.

K. Srinathan, C. Pandu Rangan, M. Yung (Eds.): Indocrypt 2007, LNCS 4859, pp. 404–413, 2007.

The filter generator is one of the simplest stream cipher and it is really interesting to understand what kind of attacks we may perform on it. There is of course a huge literature on the subject and quite a few approaches. Some of the most important ones fall into the category of fast correlation attacks. They were introduced by Meier and Staffelbach [MS88] as an improvement to correlation attack introduced by Siegenthaler in [Sie85]. Since then, many different versions have been proposed (see for instance [CT00], [MFI01],[CF02],[JJ00] and [JJ02]). The other main class is given by the algebraic [CM03] and fast algebraic attacks [Cou03] which can be really efficient if the filtering function is of low algebraic immunity. Recently, Rønjom and Helleseth have proposed a new variant [RH07] which is closely related to the Berlekamp Massey attack. There is also some ideas in [MFI05] and [MFI06] that apply to the two previous categories of attacks. When the inputs positions of f (also known as tapping positions) are not well chosen, one can apply inversion attack or conditional correlation attack (see [Gol96],[GCD00] and [LCPP96]). Finally, there is the very general class of time-memory-data tradeoff attack (see [BS00]) which is often the most efficient if the generator is well designed.

In this paper, we will present a new attack related to vectorial versions of fast correlation attacks (see [LZGB03], [EJ04], [GH05]). This attack has an interesting complexity and appears to be difficult to avoid. The idea is to use the low degree multiples to distinguish the positions corresponding to zero inputs of the filtering function f. Most of the probabilistic analysis is derived from the work of Sabine Leveiller during her PhD [Lev04b] (it is in French but some of it is published in [Lev04a] and [LZGB03]). However, we push it a little further and show that the positions corresponding to zero inputs of f are actually almost always detectable.

The paper is organized as follows. We begin by explaining the attack principle in the first section. Then, in Section 2, we compute the bias at the heart of the attack, this is our main contribution. This also allows us to derive the actual attack complexity in Section 3. We give in Section 4 the time complexity of such an attack on some example filter generators. We finally conclude in the last section.

2   Attack Principle

Our attack uses like many correlation attacks the small weight multiples of the polynomial $g(X)$ generating the LFSR. Each of these multiples induces a linear relation between some points where f is applied to produce the keystream. Namely, for a multiple $p(X) = 1 + \sum_{i=1}^{w} X^{p_i}$ of weight $w + 1$ we have

$$\mathbf{x}_t + \mathbf{x}_{t+p_1} + \cdots + \mathbf{x}_{t+p_w} = \mathbf{0} \quad \forall t \geq 0 \tag{1}$$

In all this paper, we will assume that for a given multiple and a point $\mathbf{x}_t$, the others point $\mathbf{x}_{t+p_1}$ up to $\mathbf{x}_{t+p_w}$ can take with the same probability any value satisfying (1). This is justified by the good properties of an LFSR sequence and appears to be a good working hypothesis since we will see that the experimental results are very close to the predicted ones. With this assumption, we define

$$P_{\mathbf{x}} \overset{\text{def}}{=} \Pr\left(f(\mathbf{x}_1) + \cdots + f(\mathbf{x}_w) = 0 \quad \Big| \quad \sum_{i=1}^{w} \mathbf{x}_i = \mathbf{x} \right). \qquad (2)$$

We did not include the p_i in this expression because they have no real influence in this model. Actually in our model $P_{\mathbf{x}}$ is exactly the probability that for a given multiple and a time t, $z_{t+p_1} + \cdots + z_{t+p_w}$ is equal to zero knowing that $\mathbf{x}_t = \mathbf{x}$. The crux of our attack is based on these probabilities. They can be expressed nicely as we will see in the next section and for an even w they satisfy two interesting properties:

- $P_{\mathbf{0}}$ is always greater than $1/2$ and is greater than or equal to the other $P_{\mathbf{x}}$'s.
- If the function f has a good autocorrelation property then there is always a gap between $P_{\mathbf{0}}$ and the other $P_{\mathbf{x}}$'s.

At this point, one could guess what we are going to do. Using many multiples of g, we will be able to have a good approximation of the probability $P_{\mathbf{x}_t}$ associated to a position t. Now, if the gap between $P_{\mathbf{0}}$ and the others $P_{\mathbf{x}}$ is large enough (depending of the number of multiples we used) we will then be able to detect which time positions are associated with an $\mathbf{x}_t$ equal to $\mathbf{0}$.

Each $\mathbf{x}_t$ equal to $\mathbf{0}$ actually tells us that the n bits of the sequence $(s_t)_{t \geq 0}$ involved in this $\mathbf{x}_t$ are equal to 0. By substituting their linear expression in terms of $s_0, \ldots, s_{N-1}$ we then obtain n linear equations involving the key bits. In the end, merging the equations from all the zero $\mathbf{x}_t$'s, we get a linear system of rank at most $N - 1$ since both the all zero state and the actual LFSR initial state are solutions. We thus hope that given $\lceil N/n \rceil$ such zero $\mathbf{x}_t$, we should get a rank $N - 1$ linear system where the only non trivial solution is the LFSR initial state.

The attack algorithm is summarized here with two parameters D and L that will be discussed later:

1. Compute all the weight $2p + 1$ multiples of $g(X)$ up to degree D. This can be done offline and once for all.
2. Approximate $P_{\mathbf{x}_t}$ for the L first bits of the keystream. In order to do that, for a given position each multiple corresponds to a parity check, and we just have to count how many are satisfied by the keystream bits. Remark that among the L bits, only the ones for which $z_t = f(\mathbf{0})$ have to be considered.
3. Assume that the $\lceil N/n \rceil$ bits with the higher approximated $P_{\mathbf{x}_t}$ correspond to positions where $\mathbf{x}_t = \mathbf{0}$.
4. Solve the linear system induced by the knowledge of $\mathbf{x}_t$ at these positions and retrieve the initial state.

A detailed complexity analysis will be carried in the following sections but we give here a preliminary one. The best complexity for the first step (precomputation) is given by the algorithm of [CJM02] and is in D^p time and $D^{p/2}$ memory. The complexity for Step 2 is in L times the number of multiples and requires the knowledge of $D+L$ keystream bits. The last two steps are negligible in the overall complexity. Remark however that last step may deal with some erroneous answers at Step 3 by trying more than one linear system induced by the positions with a high $P_{\mathbf{x}_t}$.

3   Bias Computation

We will give here a simple expression for the probability

$$P_{\mathbf{x}} \stackrel{\text{def}}{=} \Pr\left(f(\mathbf{x}_1) + \cdots + f(\mathbf{x}_w) = 0 \quad | \quad \sum_{i=1}^{w} \mathbf{x}_i = \mathbf{x} \right) \tag{3}$$

corresponding to an equation of weight $w+1$. We will then use it to compute the gap between P_0 and the other $P_{\mathbf{x}}$ in the case of an even w. In order to do that, let us introduce

$$d_w(\mathbf{x}) \stackrel{\text{def}}{=} \sum_{\mathbf{x}_1,\ldots,\mathbf{x}_{w-1}\in\mathbf{F}_2^n} (-1)^{f(\mathbf{x}_1)+\cdots+f(\mathbf{x}_{w-1})+f(\mathbf{x}+\mathbf{x}_1+\cdots+\mathbf{x}_{w-1})} \tag{4}$$

where $\mathbf{x}+\mathbf{x}_1\cdots+\mathbf{x}_{w-1}$ corresponds to $\mathbf{x}_w$ in (3) since the sum of $\mathbf{x}_1$ up to $\mathbf{x}_w$ must be equal to $\mathbf{x}$. By definition, d_1 is the sign function of f

$$d_1(\mathbf{x}) = (-1)^{f(\mathbf{x})} \tag{5}$$

and the d_w's are directly related to the probability $P_{\mathbf{x}}$ by

$$P_{\mathbf{x}} = \frac{1}{2}\left(1 + \frac{1}{2^{(w-1)n}}d_w(\mathbf{x})\right). \tag{6}$$

Moreover, it is easy to show the following recursive relation

$$d_w(\mathbf{x}) = \sum_{\mathbf{y}\in\mathbf{F}_2^n} (-1)^{f(\mathbf{x}+\mathbf{y})}d_{w-1}(\mathbf{y}) = d_1 * d_{w-1}(\mathbf{x}) \tag{7}$$

where $*$ is the convolution product. Using the properties of the Walsh transform we obtain that $\widehat{d_w}(u) = \widehat{d_1}(u)^w$ where

$$\widehat{d_1}(\mathbf{u}) = \sum_{\mathbf{x}\in\mathbf{F}_2^n} d_1(\mathbf{x})(-1)^{\mathbf{u}.\mathbf{x}} \tag{8}$$

And by using the inverse Walsh transform we finally obtain

$$d_w(\mathbf{x}) = \frac{1}{2^n} \sum_{\mathbf{u}\in\mathbf{F}_2^n} (-1)^{\mathbf{u}.\mathbf{x}}\widehat{d_1}(\mathbf{u})^w. \tag{9}$$

That is

$$P_{\mathbf{x}} = \frac{1}{2}\left[1 + \sum_{\mathbf{u}\in\mathbf{F}_2^n} (-1)^{\mathbf{u}.\mathbf{x}}\left(\frac{\widehat{d_1}(\mathbf{u})}{2^n}\right)^w\right]. \tag{10}$$

This has already be observed by Sabine Leveiller in her PhD thesis [Lev04b], we have just given here another proof of this statement. For reference, one may look at [LZGB03] since her PhD is in French only.

We will show now that the probability P_0 can be distinguished quite well from the others when w is even. For that it is natural to look at the minimum difference between P_0 and P_x, that is to compute $\min_{x \neq 0}(P_0 - P_x)$.

Let us begin by defining Δ to be the minimum of $P_0 - P_x$ when $w = 2$. Using (6) we have

$$\Delta \stackrel{\text{def}}{=} \frac{1}{2}\left[\frac{d_2(\mathbf{0})}{2^n} - \max_{\mathbf{x} \neq \mathbf{0}}\left(\frac{d_2(\mathbf{x})}{2^n}\right)\right] = \frac{1}{2}\left[1 - \max_{\mathbf{x} \neq \mathbf{0}}\left(\frac{d_2(\mathbf{x})}{2^n}\right)\right] \qquad (11)$$

since $d_2(\mathbf{0})$ is equal to 2^n by Parseval's equality. Usually, if f has a good auto-correlation then this Δ is very close to $1/2$. To see that, looking at the formula (4) we have

$$d_2(\mathbf{x}) = \sum_{\mathbf{u}}(-1)^{f(\mathbf{u})+f(\mathbf{u}+\mathbf{x})} \qquad (12)$$

which is nothing more than an autocorrelation coefficient and should be close to 0. Remark that if this is not the case we are confident that we can distinguish quite well other values of P_x from the others.

In the more general case $w = 2p$, we can write the difference between P_0 and P_x as

$$\min_{\mathbf{x} \neq \mathbf{0}}(P_0 - P_x) = \frac{1}{2}\min_{\mathbf{x} \neq \mathbf{0}}\left[\sum_{\mathbf{u} \in \mathbf{F}_2^n}\left(\frac{\widehat{d_1}(\mathbf{u})}{2^n}\right)^{2p} - \sum_{\mathbf{u} \in \mathbf{F}_2^n}(-1)^{\mathbf{u}.\mathbf{x}}\left(\frac{\widehat{d_1}(\mathbf{u})}{2^n}\right)^{2p}\right] \qquad (13)$$

that is,

$$\min_{\mathbf{x} \neq \mathbf{0}}(P_0 - P_x) = \min_{\mathbf{x} \neq \mathbf{0}}\sum_{\mathbf{u}, \mathbf{u}.\mathbf{x}=1}\left(\frac{\widehat{d_1}(\mathbf{u})}{2^n}\right)^{2p}. \qquad (14)$$

Notice that this difference is always greater than or equal to 0 which means that P_0 is always greater than or equal to the other probabilities. In the case $p = 1$ this is nothing more than Δ and using the Hölder inequality (see Appendix A) we can see that the worst case for the others p is when all the $\widehat{d_1}(\mathbf{u})$ are equal. Since there is 2^{n-1} terms in the sum, the worst case is when for $p = 1$ each term in the sum is equal to $\Delta/2^{n-1}$. We thus get this lower bound

$$\min_{\mathbf{x} \neq \mathbf{0}}(P_0 - P_x) \geq 2^{n-1}\left(\frac{\Delta}{2^{n-1}}\right)^p \qquad (15)$$

which corresponds to the bias we will need to detect.

4   Complexity Analysis

Let us look at the complexity of attacking the filter generator when we use multiples of weight $2p + 1$. We will suppose that the function has a good autocorrelation property, meaning that the bias to detect is around

$$\text{bias} \simeq \frac{1}{2^{1+n(p-1)}}. \tag{16}$$

To detect it, we will thus need as many equations as the square of the bias inverse. Looking for multiples of weight $2p + 1$ up to degree D of the LFSR generator polynomial, we know that we will find around

$$\text{degree at most D multiples number} \simeq \binom{D}{2p}\frac{1}{2^N} \simeq \frac{D^{2p}}{(2p)!2^N} \tag{17}$$

of them. This result is well known and is derived as follows. We have $\binom{D}{2p}$ polynomials of weight $2p+1$ and degree at most D. For each of them, we may assume that the rest of the Euclidean division by $g(X)$ is equally distributed among the 2^N possible values. Hence, the formula (17) just express that we get a multiple (a rest equal to 0) one time over 2^N.

Putting the equations (16) and (17) together, to be able to detect our bias we need to choose a degree D such that

$$\frac{D^{2p}}{(2p)!2^N} = 2^{2+2n(p-1)}. \tag{18}$$

That means we will have to compute the weight $2p + 1$ multiples up to a degree D where

$$\log_2 D = \frac{N}{2p} + n\left(1 - \frac{1}{p}\right) + \frac{1}{p}. \tag{19}$$

We neglect the factorial term $(2p)!$ here since in practice p is 2 or 3. Using the algorithm of [CJM02], the complexity to compute them is in D^p time and $D^{p/2}$ memory. Remark that the algorithm is completely parallelizable over many computers. We finally get for the offline part of the attack

$$\log_2(\text{offline time}) = N/2 + (p - 1)n + 1 \tag{20}$$

$$\log_2(\text{offline memory}) = N/4 + (p - 1)n/2 + 1/2. \tag{21}$$

For the online phase, we will need to identify around $\lceil N/n \rceil$ bits corresponding to an $\mathbf{x}$ equal to $\mathbf{0}$. We will thus need to approximate the $P_{\mathbf{x}_t}$ for L bits in average where

$$L = \left\lceil \frac{N}{n} \right\rceil 2^n. \tag{22}$$

This comes from the fact that an $\mathbf{x}$ equal to $\mathbf{0}$ appears in average one time each 2^n keystream bits. We can actually gain a factor 2 because we can skip the positions for which $f(\mathbf{x}_t) \neq f(\mathbf{0})$. For each of these L bits, we will have to compute as many parity checks as the number of multiples. The online phase complexity is then given by

$$\log_2(\text{online time}/L) = 2 + 2n(p - 1) \tag{23}$$

which is in practice really efficient. For the memory we just need to access the stored multiples and a length of keystream equal to $D + L$, that is basically

$$\log_2(\text{keystream length}) = \frac{N}{2p} + n\left(1 - \frac{1}{p}\right) + \frac{1}{p}. \qquad (24)$$

Remark that the overall complexity is quite good. Let us compare it with the time-memory-data tradeoff described in [BS00]. This tradeoff is such that $TM^2K = 2^{2N}$ where T is the online time, M the online memory and K the length of the keystream needed. A good choice is to take $M = K = 2^{N/3}$ which gives an online time of $2^{2N/3}$ and the same precomputation time. For our attack with weight 5 multiples and an n around $N/8$ (which is typical), we have the following: an online time complexity in $2^{N/4}$ and memory in $2^{5N/16}$ for a keystream length of $2^{5N/16}$ bits. This is better than the time-memory-data tradeoff, especially since the precomputation time is a little smaller too ($2^{5N/8}$ compared to $2^{2N/3}$).

5   Experimental Results

We have successfully carried on this attack on some example generators. We give here the timing of our program in C. All computations were performed on a 3.6GHz Pentium4 with 2MB of cache and 2GB of RAM.

We worked on three filter generators of length 53, 59 and 61. In all three cases, the filtering functions used were good cryptographic functions with a maximum Walsh coefficients of respectively 32, 24 and 48. We can see in Table 1 the exact value of the bias for these functions. Notice that it is significantly higher than our lower bound. As a comparison, for a Δ equal to one half and an 8-variable function, our lower bound gives 0.002 for weight 5 and 0.000008 for weight 7.

Table 1. Exact bias to detect the zero inputs for the used functions

N	n	bias for weight 5	bias for weight 7
53	8	0.0039	0.000061
59	8	0.0027	0.000021
61	9	0.0014	0.000006

In Table 2 we can see the timings for some successful attacks. We can see that the online time is really short and that all the computational effort is spent on computing the low weight multiples. We only used weight 5 multiples because we did not have the 2 weeks time needed to precompute enough weight 7 multiples. In all the attacks, the value of L was just chosen to have a very high probability to get enough zero inputs for f in a keystream of length L.

For the first two filter generators, we applied the exact method described in this paper. For the last attack however, we did not want to spend too much time on the multiples computation, so we used a few tricks to improve the practical

Table 2. Successful attack timing on the different generators. Some tricks were used for the last generator as explained below.

N	n	multiples weight	$\log_2 D$	nb of multiples(time)	L	online time
53	8	5	18.6	100000(20min)	3200	10 sec
59	8	5	20.47	330000(1day)	3200	30 sec
61	9	5	21	349034(2days)	4000	1 min

efficiency. Firstly, at the price of doubling the needed keystream, we can get for each multiple $w + 1$ parity check equations (by shifting the multiple by one of its five non null positions). The other improvement is, as explained before, to deal with erroneous zero inputs detection by spending more time on the last phase of the attack. Here for instance, we got 3 erroneous positions among the 10 with the higher $P_{\mathbf{x}}$. To get the correct key, we thus had to try all the $\binom{10}{7}$ linear systems since $\lceil N/n \rceil$ is equal to 7 here. Those tricks helped to perform the attack with less multiples, however we did not really cut down the overall attack complexity.

To conclude this section, notice that the actual value for D is really close to the theoretical one. In order to detect the bias, we theoretically needed respectively around 65746, 137200 and 510000 parity checks for each generator. That gives us, using the approximated formula (17) for the number of multiples, a theoretical $\log_2 D$ of 18.4, 20.16 and 21.14 respectively.

6   Conclusion

As a conclusion, we want to detail some important points about the attack we just presented.

First of all, the probabilistic hypothesis behind the complexity analysis seems quite sound since the simulations are really close to the predicted results. This is actually not always the case with other attacks using a binary symmetric channel model where simulations are usually worse than predicted.

Then, we believe that this attack is difficult to avoid. Using a filtering function with a bad autocorrelation will certainly weaken the cipher. Moreover, in this case other inputs than the all zero one could become detectable. Remark as well that one cannot have a filtering function with too many variables compared to N. This is for performance reasons but also to have tapping positions with good behavior.

Finally, the overall complexity of the attack is quite good as explained at the end of Section 4. In particular, we successfully attacked a length 61 filter generator in a few seconds after a 2 days precomputation on a single computer.

Acknowledgment

The author want to thanks Yann Laigle-Chapuy, Jean-Pierre Tillich and Anne Canteaut for their helpful insight on the subject.

References

[BS00]   Biryukov, A., Shamir, A.: Cryptanalytic time/memory/data tradeoffs for
           stream ciphers. In: Okamoto, T. (ed.) ASIACRYPT 2000. LNCS, vol. 1976,
           Springer, Heidelberg (2000)
[CF02]   Canteaut, A., Filiol, E.: On the influence of the filtering function on the
           performance of fast correlation attacks on filter generators. In: 23rd Sym-
           posium on Information Theory in the Benelux, Louvain-la-Neuve, Belgium
           (May 2002)
[CJM02]   Chose, P., Joux, A., Mitton, M.: Fast correlation attacks: an algorith-
           mic point of view. In: Knudsen, L.R. (ed.) EUROCRYPT 2002. LNCS,
           vol. 2332, pp. 209–221. Springer, Heidelberg (2002)
[CM03]   Courtois, N., Meier, W.: Algebraic attacks on stream ciphers with linear
           feedback. In: Biham, E. (ed.) EUROCRPYT 2003. LNCS, vol. 2656, pp.
           346–359. Springer, Heidelberg (2003)
[Cou03]   Courtois, N.: Fast algebraic attacks on stream ciphers with linear feedback.
           In: Boneh, D. (ed.) CRYPTO 2003. LNCS, vol. 2729, pp. 176–194. Springer,
           Heidelberg (2003)
[CT00]   Canteaut, A., Trabbia, M.: Improved fast correlation attacks using parity-
           check equations of weight 4 and 5. In: Preneel, B. (ed.) EUROCRYPT
           2000. LNCS, vol. 1807, pp. 573–588. Springer, Heidelberg (2000)
[EJ04]   Englund, H., Johansson, T.: A new simple technique to attack filter gen-
           erators and related ciphers. In: Handschuh, H., Hasan, M.A. (eds.) SAC
           2004. LNCS, vol. 3357, pp. 39–53. Springer, Heidelberg (2004)
[GCD00]   Golic, J.D., Clark, A., Dawson, Ed.: Generalized inversion attack on non-
           linear filter generators. IEEE Trans. Comput. 49(10), 1100–1109 (2000)
[GH05]   Golic, J.D., Hawkes, P.: Vectorial approach to fast correlation attacks. Des.
           Codes Cryptography 35(1), 5–19 (2005)
[Gol96]   Golic, J.D.: On the security of nonlinear filter generators. In: Proceedings
           of the Third International Workshop on Fast Software Encryption, pp.
           173–188. Springer, London (1996)
[JJ00]   Johansson, T., Jöhansson, F.: Fast correlation attacks through reconstruc-
           tion of linear polynomials. In: Bellare, M. (ed.) CRYPTO 2000. LNCS,
           vol. 1880, pp. 300–315. Springer, Heidelberg (2000)
[JJ02]   Jönsson, F., Johansson, T.: A fast correlation attack on LILI-128. Informa-
           tion Processing Letters 81(3), 127–132 (2002)
[LCPP96]   Lee, S., Chee, S., Park, S.-J., Park, S.-M.: Conditional correlation attack
           on nonlinear filter generators. In: Kim, K.-c., Matsumoto, T. (eds.) ASI-
           ACRYPT 1996. LNCS, vol. 1163, pp. 360–367. Springer, Heidelberg (1996)
[Lev04a]   Leveiller, S.: A new algorithm for cryptanalysis of filtered lfsrs: the
           "probability-matching" algorithm. ISIT 1978, 234 (2004)
[Lev04b]   Leveiller, S.: Quelques algorithmes de cryptanalyse du registre filtré. PhD
           thesis, Télécom Paris, ENST (November 2004)
[LZGB03]   Leveiller, S., Zémor, G., Guillot, P., Boutros, J.: A new cryptanalytic attack
           for PN-generators filtered by a boolean function. In: Nyberg, K., Heys,
           H.M. (eds.) SAC 2002. LNCS, vol. 2595, pp. 232–249. Springer, Heidelberg
           (2003)
[MFI01]   Mihaljevic, M.J., Fossorier, M.P.C., Imai, H.: A low-complexity and high-
           performance algorithm for the fast correlation attack. In: Schneier, B. (ed.)
           FSE 2000. LNCS, vol. 1978, pp. 45–60. Springer, Heidelberg (2001)

[MFI05]   Mihaljevic, M., Fossorier, M.P., Imai, H.: Cryptanalysis of keystream gen-
           erator by decimated sample based algebraic and fast correlation attacks. In:
           Maitra, S., Madhavan, C.E.V., Venkatesan, R. (eds.) INDOCRYPT 2005.
           LNCS, vol. 3797, pp. 155–168. Springer, Heidelberg (2005)
[MFI06]   Mihaljevic, M., Fossorier, M.P.C., Imai, H.: A general formulation of alge-
           braic and fast correlation attacks based on dedicated sample decimation.
           In: Fossorier, M.P.C., Imai, H., Lin, S., Poli, A. (eds.) AAECC 2006. LNCS,
           vol. 3857, pp. 203–214. Springer, Heidelberg (2006)
[MS88]   Meier, W., Staffelbach, O.: Fast correlation attacks on stream ciphers. In:
           Günther, C.G. (ed.) EUROCRYPT 1988. LNCS, vol. 330, pp. 301–314.
           Springer, Heidelberg (1988)
[RH07]   Rønjom, S., Helleseth, T.: A new attack on the filter generator. In IEEE
           IT (to appear, 2007)
[Sie85]   Siegenthaler, T.: Decrypting a class of stream ciphers using ciphertext only.
           IEEE Trans. Computers 34(1), 81–85 (1985)

A   Lemma Used in the Bias Computation

The proof at the end of Section 3 is based on the following lemma applied with
$m = 2^{n-1}$, $s = \Delta$ and $a_i = \left(\frac{\widehat{d_1}(\mathbf{u})}{2^n}\right)^2$ for the $\mathbf{u}$ in $\mathbf{F}_2^n$ such that $\mathbf{x}.\mathbf{u} = 1$.

Lemma 1. *Given m positive real numbers $(a_i)_{i=1...m}$ and an integer $p > 1$ we
have the lower bound*

$$\sum_{i=1}^{m} a_i^p \geq m \left(\frac{s}{m}\right)^p \tag{25}$$

where $s \stackrel{\text{def}}{=} \sum_{i=1}^{m} a_i$.

Proof. The result comes almost directly from Hölder's inequality

$$\sum_{i=1}^{m} |x_i y_i| \leq \left(\sum_{i=1}^{m} |x_i|^p\right)^{1/p} \left(\sum_{i=1}^{m} |y_i|^q\right)^{1/q} \tag{26}$$

where $(x_i)_{i=1...m}$, $(y_i)_{i=1...m}$, p, q are in $\mathbb{R}$ and such that $\frac{1}{p} + \frac{1}{q} = 1$. If we apply
it with all the y_i equal to 1, the x_i equal to the positive a_i, the p from the lemma
and the corresponding q we obtain

$$\sum_{i=1}^{m} a_i \leq \left(\sum_{i=1}^{m} a_i^p\right)^{1/p} m^{1/q} \quad \text{that is} \quad \left(\sum_{i=1}^{m} a_i^p\right) \geq (s m^q)^p. \tag{27}$$

And since $\frac{1}{q} = \frac{1-p}{p}$ we finally obtain

$$\left(\sum_{i=1}^{m} a_i^p\right) \geq s^p m^{1-p} \geq m \left(\frac{s}{m}\right)^p. \tag{28}$$

Remark that there is an equality when all the a_i are equal to s/m.

Efficient Implementations of Some Tweakable Enciphering Schemes in Reconfigurable Hardware

Cuauhtemoc Mancillas-López, Debrup Chakraborty,
and Francisco Rodríguez-Henríquez

Computer Science Departament,
Centro de Investigación y Estudios Avanzados del IPN,
Av. Instituto Politécnico Nacional No. 2508, México D.F.

Abstract. We present optimized FPGA implementations of three tweakable enciphering schemes, namely, HCH, HCTR and EME using AES-128 as the underlying block cipher. We report performance timings and hardware resources occupied by these three modes when using a fully pipelined AES core and a sequential AES design. Our experimental results suggest that in terms of area HCTR, HCH and HCHfp (a variant of HCH) require more area than EME. However, HCTR performs the best in terms of speed followed by HCHfp, EME and HCH.

1 Introduction

A tweakable enciphering scheme (TES) is a specific kind of block-cipher mode of operation which provides a strong pseudorandom permutation (SPRP). A fully defined TES for arbitrary length messages using a block cipher was first presented in [9]. In [9] it was also stated that a possible application area for such encryption schemes could be low level disc encryption, where the encryption/decryption algorithm resides on the disc controller which has access to the disc sectors but has no knowledge of the disk's high level partitions such as directories, files, etc. Furthermore, it was suggested in [9] that sector addresses could be used as tweaks. Because of the specific nature of this application, a length preserving enciphering scheme is required and under this scenario, a SPRP can provide the highest possible security.

In the last few years there have been numerous proposals for TES. These proposals fall in three basic categories: Encrypt-Mask-Encrypt type, Hash-ECB-Hash type and Hash-Counter-Hash type. CMC [9], EME [10], EME* [7] falls under the Encrypt-Mask-Encrypt group. PEP [3], TET [8], HEH [17] falls under the Hash-ECB-Hash type and XCB [15], HCTR [19], HCH [4] falls under the Hash-Counter-Hash type. Although about nine different constructions of different TES have been proposed, we are not aware of any work reporting experimental performance data of any of these schemes. A comparative performance comparison of these modes is very necessary given the current efforts of IEEE security in storage working group [12] towards standardization of TES.

K. Srinathan, C. Pandu Rangan, M. Yung (Eds.): Indocrypt 2007, LNCS 4859, pp. 414–424, 2007.

A speculative performance comparison of the EME*, XCB, HCH and TET modes of operation in hardware is provided in [8]. This comparison assumes the same hardware implementation setting reported in [1], where a fully-parallel $GF(2^n)$ field multiplier capable of performing one multiplication in one clock cycle was implemented at a hardware cost in area of about three times the cost associated with one AES round function. The AES core was implemented through the computation of ten such modules. However, this analysis might not be quite accurate because, as we will see in the rest of this paper, one can implement a $GF(2^n)$ field multiplier with an efficiency comparable to the one of an AES round function in terms of both, the critical path and the cost in area.

In this paper we present performance data for hardware implementation of three TES. Our implementations are optimized for the application of low level disc encryption. The modes we select for our comparative study are EME, HCH and HCTR. Also we provide performance data for a variant of HCH which is called HCHfp, which is particularly useful for disk encryption. We use AES-128 as the underlying block-cipher, and use a fully parallel Karatsuba-Ofman multiplier to compute the hash functions. We carefully analyze and present our design decisions and finally report hardware performance data of the three modes. Due to lack of space in this paper we discuss in detail the construction and implementation of HCH only, but present performance data of all the modes we implemented. The full implementation details of the three modes will appear in the full version of the paper.

Notations. In the rest of the paper by $E_K(\ )$ we shall mean a n bit block cipher call with key K. By $X||Y$ we shall mean the concatenation of two binary strings X and Y and $\text{bin}_n(|X|)$ will denote the n-bit binary representation of $|X|$, which denotes the length of X. By $\text{pad}_r(X)$ we shall mean concatenation r zeros to the end of X and $\text{drop}_r(X)$ will denote the $r \leq |X|$ most significant bits of X. We will treat n bit strings as polynomials of degree less than n of the field $GF(2^n)$. If X and Y are n bit strings then by $X \oplus Y$ and XY we shall mean addition and multiplication in the field respectively. By xX we would represent the multiplication of X by the polynomial x.

2   The Schemes

As mentioned earlier HCH falls under the category of Hash-Counter-Hash constructions. HCH uses an universal hash function of the form:

$$H_{R,Q}(A_1, \ldots, A_m) = Q \oplus A_1 \oplus A_2 R^{m-1} \oplus \cdots \oplus A_{m-1} R^2 \oplus A_m R \qquad (1)$$

Where $A_1, A_2, \ldots, A_m, R, Q$ are n bit strings. In addition to the hash function HCH requires a counter mode of operation. Given an n-bit string S, the counter mode is defined as

$$\text{Ctr}_{K,S}(A_1, \ldots, A_m) = (A_1 \oplus E_K(S_1), \ldots, A_m \oplus E_K(S_m)). \qquad (2)$$

Algorithm $E_K^T(P_1, \ldots, P_m)$	Algorithm $D_K^T(C_1, \ldots, C_m)$
1. $R \leftarrow E_K(T); Q \leftarrow E_K(R \oplus \mathrm{bin}_n(l));$	1. $R \leftarrow E_K(T); Q \leftarrow E_K(R \oplus \mathrm{bin}_n(l));$
2. $M_m \leftarrow \mathrm{pad}_{n-r}(P_m);$	2. $U_m \leftarrow \mathrm{pad}_{n-r}(C_m);$
3. $M_1 \leftarrow H_{R,Q}(P_1, \ldots, P_{m-1}, M_m);$	3. $U_1 \leftarrow H_{R,xQ}(C_1, \ldots, C_{m-1}, U_m);$
4. $U_1 \leftarrow E_K(M_1); I \leftarrow M_1 \oplus U_1; S \leftarrow E_K(I);$	4. $M_1 \leftarrow E_K^{-1}(U_1); I \leftarrow M_1 \oplus U_1; S \leftarrow E_K(I);$
5. $(C_2, \ldots, C_{m-1}, D_m)$	5. $(P_2, \ldots, P_{m-1}, V_m)$
$\leftarrow \mathrm{Ctr}_{K,S}(P_2, \ldots, P_{m-1}, M_m);$	$\leftarrow \mathrm{Ctr}_{K,S}(C_2, \ldots, C_{m-1}, U_m);$
6. $C_m \leftarrow \mathrm{drop}_{n-r}(D_m); U_m \leftarrow \mathrm{pad}_{n-r}(C_m);$	6. $P_m \leftarrow \mathrm{drop}_{n-r}(V_m); M_m \leftarrow \mathrm{pad}_{n-r}(P_m);$
7. $C_1 \leftarrow H_{R,xQ}(U_1, C_2, \ldots, C_{m-1}, U_m);$	7. $P_1 \leftarrow H_{R,Q}(M_1, P_2, \ldots, P_{m-1}, M_m);$
8. return $(C_1, \ldots, C_m).$	8. return $(P_1, \ldots, P_m).$

Fig. 1. Encryption and decryption using HCH. The tweak is T and the key is K. For $1 \le i \le m-1$, $|P_i| = n$ and $|P_m| = r$ where $r \le n$, and l is the length of the message.

Where $S_i = S \oplus \mathrm{bin}_n(i)$. The complete encryption and decryption algorithm of HCH is given in Fig. 1.

HCH can encrypt arbitrary long messages greater than n bits. It uses a single key which is same as the block-cipher key. It requires $m+3$ block cipher calls and $2m-2$ finite field multiplications to encrypt a m block message. The key for the universal hash is R, which is derived by encrypting the tweak. Thus R changes across encryption calls and this does not allow the use of pre-computations for computing the hash. HCH requires two passes over the data. In [5] a modification of HCH is also proposed which is called HCHfp. HCHfp can only be used in those applications where the message length is fixed. This construction simplifies the general HCH construction and requires one less block-cipher call, but it requires two separate keys for the hash and the block-cipher. As in HCHfp the hash key is not dependent on the tweak so pre-computation for calculating the hash is also possible. HCHfp is particularly of interest for disk encryption applications as here the message length is fixed and same as the sector length. The encryption decryption algorithm using HCHfp can be found in [5].

All variants of HCH are provably secure and the authors guarantee that the advantage of any computationally bounded chosen plaintext chosen ciphertext adversary in distinguishing HCH from a random permutation can be at most $O(\sigma_n^2)/2^n + \delta$ where σ_n denotes the number of n bit plaintexts and/or ciphertext blocks the adversary has access to, and δ denotes the advantage of an adversary to distinguish the underlying block-cipher from a random permutation.

The structure of HCTR is similar to that of HCH with some important differences. HCTR can also encrypt arbitrary long messages. It requires m block cipher calls and $2m + 2$ field multiplications to encrypt an m block message. It utilizes two different keys and it is proved to be secure with a security bound of $O(\sigma_n^3)/2^n + \delta$. Thus it provides lesser security than HCH and it requires three less block cipher calls than HCH and 2 less block cipher calls than HCHfp but it needs four more multiplications than both HCH and HCHfp. A full description of HCTR can be found in [19].

EME stands for ECB-Mask-ECB (EME)[10]. As the name suggests, the mode consists of two electronic code-book layers with a masking layer in between. The structure of EME is quite different from HCH and HCTR. EME falls under the category of Encrypt-mask-Encrypt constructions. It does not use any hash function, but instead uses two layers of encryption. EME requires $2m + 2$ block cipher calls for encrypting a m block message. It requires no multiplication. EME uses a single key same as the block-cipher key. EME has some message length restrictions. If the block length of the underlying block cipher is n then the message length should always be a multiple of n. Moreover, EME cannot encrypt more than n blocks of messages. This means that if an AES-128 is used as the underlying block-cipher then EME cannot encrypt more than 2048 bytes (2 KB) of data. This message length restriction was removed in a construction called EME* which requires more block-cipher calls than EME. But for the purpose of disc encryption EME appears to be sufficient, as generally disk sectors lengths are less than 2KB and their lengths are multiples of 128 bits. EME has a security bound of $O(\sigma^2)/2^n + \delta$. A full description of EME can be found in [10].

3   Design Decisions

For implementing all three schemes we chose the underlying block cipher as AES-128. As mentioned earlier the designs that we present here are directed towards the application of disk sector encryption. In particular, our designs are optimized for applications where the sector length is fixed to 512 bytes. As the sector address is considered to be the tweak, thus the tweak length itself is considered to be fixed and equal to the block length of the block cipher.

The speed of a low level disk encryption algorithm must meet the current possible data rates of disc controllers. With emerging technologies like serial ATA and Native Command Queuing (NCQ) the modern day discs can provide data rates around 3Giga-bits per second[18]. Thus, the design objective should be to achieve an encryption/decryption speed which matches this data rate.

The modes HCH and HCTR use two basic building blocks, namely, a polynomial universal hash and the block-cipher. EME requires only a block-cipher. Since AES-128 was our selection for the underlying block-cipher, proper design decisions for the AES design must meet the desired speed. Out of many possible designs reported in the literature [13,6,2,11] we decided to design the AES core so that a 10-stage pipeline architecture could be used to implement two different functionalities: the counter mode, and the encryption of one single block that we will call in the rest of this paper as single mode. This decision was taken based on the fact that the structure of the AES algorithm admits to a natural ten-stage pipeline design, where after 11 clock cycles one can get one encrypted block in each subsequent clock-cycle. It is worth mentioning that in the literature, several ultra fast designs with up to 70 pipeline stages have been reported [13], but such designs would increase the latency, i.e., the total delay before a single block of cipher-text can be produced. As the message lengths in the target application are specifically small, such pipeline designs are not suitable for our target application.

The main building block needed for implementing the polynomial hash of the HCH and HCTR modes is an efficient multiplier in $GF(2^{128})$. Out of many possible choices we selected a fully parallel Karatsuba-Ofman multiplier which can multiply two 128-bit strings in a single clock-cycle at a sub-quadratic computational cost [16]. This time efficient multiplier occupies about 2 times the hardware resources required by one single AES round. Because of this, the total hardware area required by HCTR and HCH are significantly more than EME (which does not require multipliers). A more compact multiplier selection would yield significantly lower speeds which violates the design objective of optimizing for speed.

The specifications of both the HCTR and HCHfp algorithms imply that one multiplicand is always fixed, thus allowing the usage of pre-computed look up tables that can significantly speed up the multiplication operation. Techniques to speed up multiplication by look-up tables in software are discussed in [14,1]. These techniques can be be extended to hardware implementations also. However, there is a tradeoff in the amount of speed that can be obtained by means of pre-computation and the amount of data that needs to be stored in tables. Significantly higher speeds can be obtained if one stores large tables. This speedup thus comes with an additional cost of area and also the potentially devastating penalty of secure storage. Moreover, if pre-computation is used in a hardware design, then the key needs to be hardwired in the circuit which can lead to numerous difficulties in key setup phases and result in lack of flexibility for changing keys. Because of the above considerations, we chose not to store key related tables for our implementations. Thus the use of an efficient but large multiplier is justified in the scenario under analysis.

We implemented the schemes on a FPGA device which operates at lower frequencies than true VLSI circuits. Thus the throughput that we obtained probably can be much improved if we use the same design strategies on a different technology. Our target device was a XILINX Virtex 4, xc4v1x100-12FF1148.

4   The Design Overviews

In this Section we give a carefully analysis of the data dependencies of HCH and explain how we exploit the parallelism present in the algorithm. Similar analysis for HCTR and EME can be found in the extended version of this paper.

In the analysis which follows we assume the message to be of 512 bytes (32 AES blocks). Also, we assume a single AES core designed with a 10 stage pipeline and a fully parallel single clock cycle multiplier. We also calculate the key schedules for AES on the fly, this computation can be parallelized with the AES rounds. The polynomial universal hash functions are computed using the Horner's rule.

Referring to the Algorithm of Fig. 1 the algorithm starts with the computation of the parameter R in Step 1. For computing R the AES pipeline cannot be utilized and must be accomplished in simple mode, implying that 11 clock cycles will be required for computing R. At the same time, the AES round keys can be computed by executing concurrently the AES key schedule algorithm. The hash function of Step 3 can be written as

$$H_{R,Q}(P_1, P_2, \ldots P_{32}) = P_1 \oplus Q \oplus Z$$

where $Z = R^{31}P_2 \oplus \ldots \oplus RP_{32}$. So, Z and Q can be computed in parallel. For computing Z, 31 multiplications are required and computation of Q takes 11 clock cycles. So the computation of the hash in step 2 takes 31 clock cycles. Then, the computation of Step 4 requires two simple mode encryption which implies 22 more clock cycles. So we need to wait 64 clock cycles before the counter mode starts. The counter mode in step 5 requires 31 block cipher calls which can be pipelined. So computation of step 5 requires a total of $30 + 11 = 41$ clock cycles. The first cipher block C_2 is produced 11 clock cycles after the counter starts. The second hash function computation of Step 7 can start as soon as C_2 is available in the clock cycle 75. Hence the computation of the hash function can be completed at the same time that the last cipher block (C_m) of Step 5 is produced. Figure 2 depicts above analysis. It can be seen that a valid output will be ready after the cycle 75 and a whole disk sector will be ready in the cycle 106. In case of HCHfp the computation of Q is not required, and it uses a hash key which is different from R. Thus R and the hash function can be computed in parallel, which gives rise to a savings of 11 clock cycles. So HCHfp will produce a valid output in 64 clocks and it will take 95 clock-cycles to encrypt the 32 block message.

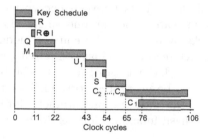

Fig. 2. HCH Time Diagram

A similar analysis can be done in case of HCTR and EME. Exploiting the parallelism present in these algorithms to the full extent we obtain that for HCTR a valid output will be ready after the cycle 55 and a whole disk sector will be ready in the cycle 88. For EME the first block of valid output would be produced after 75 clock cycles and the whole sector would be ready after 106 clock-cycles.

5   Implementation

Due to lack of space, in this Section we only discuss the design details of the basic control unit of HCH. The other implementation details along with the details for HCTR and EME implementation will appear in the full version of this paper, but we shall provide performance data for all the modes in Section 6.

Fig. 3 shows the general architecture of the HCH mode of operation. It can be seen that AES must be implemented both, in counter and in simple mode.

Additionally, a hash function is also required as one of the main building blocks. The architecture operation is synchronized through a control unit that performs the adequate sequence of operations in order to obtain a valid output.

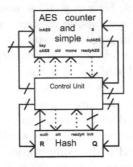

Fig. 3. HCH General Architecture

The HCH control unit architecture is shown in Fig. 4. It controls the AES block by means of four 1-bit signals, namely: **cAES** that initializes the round counter, the **c/d** signal that selects between encryption or decryption mode, the **msms** signal that indicates whether one single block must be processed or rather, multiple blocks by means of the counter mode. Finally, **readyAES** indicates whenever the architecture has just computed a valid output. The AES dataflow is carried out through the usage of three 128-bit busses, namely, **inAES** that receives the blocks to be encrypted, **outAES** that sends the encrypted blocks and **S** that receives the initialization parameter for the counter mode. The communication with the hash function block is done using two signals: **cH** for

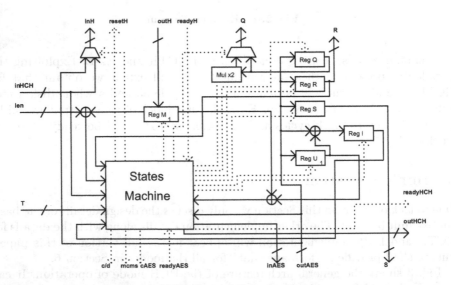

Fig. 4. HCH Control Unit Architecture

initializing the accumulator register and the counter of blocks already processed and **readyH** that indicates that the hash function computation is ready. The data input/output is carried by the **inH** and **outH** busses, respectively. The parameters **R** and **Q** are calculated in the control unit and send through the busses to the hash function.

The HCH control unit implements a finite state automaton that executes the HCH sequence of operations. It uses eight states: *RESET, AES1, AES2, HASH1, AES3, AES4, ECOUNTER* and *HASH2*. In each state, an appropriate control word is generated in order to perform the required operations. The correct algorithm execution requires storing the **R, Q, S, I, U$_1$** and **M$_1$** values. Thus, six registers are needed. In particular the hash function input **inH** can come from the system input or from the output of the AES counter mode. Therefore, a multiplexer is needed for addressing the correct input, where the multiplexer signals are handled by the state machine's control word.

6   Results

In this section we provide the performance results obtained from our implementations. We will measure the performances based on the following criteria: time taken for encrypting 32 blocks of data, the latency, i.e., the time required to produce the first block of output, the size of the circuit in slices, the number of B-RAMs used and the throughput. The performance/area tradeoff is evaluated using the Throughput per Area (TPA) metric, which is computed as, $TPA = [(slices + 128 \cdot BRAMS) \cdot time]^{-1}$. For a given design, a high TPA indicates high efficiency, i.e., a good performance/area tradeoff.

In Table 1 we show the performance of the basic building blocks of the architectures, i.e, the performance of one AES round and one multiplier. Table 1 shows that considering the B-RAMs and slices the size of our multiplier circuit is about two times the size of a AES round. The critical path delay of the AES round is more than the multiplier, so the AES round determines the critical path in all the implementations.

Table 1. Performance of AES round and multiplier

Design	Slices	B-RAM	Critical Path(nS)
AES round	1215	8	10.998
multiplier	3223	-	9.85

Table 2 gives the performance of the full AES (both a sequential and pipelined architecture), it also shows the performance of the two different hash functions for HCH and HCTR. Column 5 of Table 2 shows the throughput. Throughput does not carry the usual meaning in case of the hash functions, as they produce a 128 bit output irrespective of the input size. By throughput of the hash function we mean the number of bits they can process per unit time. The sequential

Table 2. Performance of the AES and Hash implementations

Method	Slices	B-RAM	Frequency (MHz)	Throughput (GBits/s)
AES-Sequential	1301	18	81.967	1.049
AES-Pipeline	6368	85	83.88	10.736
Hash-HCTR	3986	-	101.08	12.15
Hash-HCH	4014	-	101.45	12.98

AES gives significantly poor throughput and both hash functions have better throughput than the AES-pipeline.

In Table 3 we show the experimental results for the four modes of operations implemented using a pipelined AES core. As we implemented both encryption and decryption functionalities in the same circuit and due to the symmetry of the algorithms the timings for encryption and decryption operations are the same. Note that the number of clock-cycles reported in Table 3 are one more than those estimated in Section 4, this is because in the true implementations one clock cycle is lost due to the initial reset operation. From Table 3 it is evident that EME is the most economical mode in terms of area resources, mainly due to the fact that this mode does not utilizes a hash function. The most costly mode in terms of area is HCH since it requires 6 registers in contrast to the three registers required by the HCTR. Additionally, HCH has more possible inputs for its AES building block. In terms of speed, the fastest mode is HCTR since it only utilizes one AES block cipher call in sequential mode, whereas HCH requires a total of four such calls (although only three have consequences in terms of clock cycles since the other one is masked with the computation of the hash function). HCHfp is better than both EME and HCH in terms of speed.

Table 3. Hardware costs of the HCTR, HCH and EME modes with an underlying AES full pipeline core: The time and clock-cycles are the time required to encrypt 32 blocks

Mode	Slices	B-RAM	Frequency (MHz)	Clock Cycles Cycles	Time (μS)	Latency (μS)	Throughput GBits/Sec	TPA
HCH	13755	85	65.939	107	1.622	1.167	2.46	24.42
HCHfp	12970	85	66.500	96	1.443	0.992	2.83	29.05
HCTR	12068	85	79.65	89	1.117	0.703	3.66	39.81
EME	10120	87	67.835	107	1.576	1.120	2.64	29.85

In Table 4 we show the four modes of operation when using a sequential implementation of the AES core. In a sequential architecture, EME is the most inefficient mode in terms of latency due to the two costly block cipher passes that require eleven clock cycles per block. Hence, a significant increment in the required number of clock cycles is observed for the EME mode. This situation does not occur in HCTR or in HCH since they only need one encryption pass. The hash function computation is not affected in this scenario due to the fact that we use a multiplier which is essentially a combinatorial circuit able to produce a result in one clock cycle.

Table 4. Hardware costs of the HCTR, HCH and EME modes with an underlying sequential AES core: The times reported are for 32 blocks

Mode	Slices	B-RAM	Frequency (MHz)	Clock Cycles	Time (μS)	Throughput (Gbits/sec)	TPA
HCH	8688	18	64.026	416	6.497	0.631	14.00
HCHfp	7903	18	64.587	405	6.270	0.653	15.73
HCTR	7006	18	77.737	388	4.991	0.820	21.53
EME	5053	20	65.922	716	10.861	0.377	12.09

Discussion: As we stated in Section 3 the design objective would be to match the data rates of modern day disk controllers which are of the order of 3Gbits/sec. Table 4 shows that using a sequential design it is not possible to achieve such data rates though this strategy provides more compact designs. If we are interested in encrypting hard disks of desktop or laptop computers the area constraint is not that high, but speed would be the main concern. So, a pipelined AES will probably be the best choice for designing disk encryption schemes.

From Table 3 we see that the most efficient mode in terms of speed is HCTR followed by HCHfp, EME and HCH. The full functionality of HCH is not needed for disk encryption schemes as for this application messages would be of fixed length. Thus we can conclude that HCTR and HCHfp are the best modes to use for this application. But, the security guarantees that HCTR provides is quite weak as it have a cubic security bound. Thus, among the different modes that we implemented, and in view of all these constraints, HCHfp should probably be the most preferred mode.

7   Conclusion

We presented optimized implementation of three TES. To our knowledge this is the first work to report real performance data of any TES on hardware. There are many other TES schemes which needs to be implemented and then a true performance comparison would be possible. This performance comparison will of course help in selection of the best mode. This work can be seen as a first step towards this objective.

References

1. Bo Yang, R.K., Mishra, S.: A high speed architecture for galois/counter mode of operation (gcm). Cryptology ePrint Archive, Report 2005/146 (2005), http://eprint.iacr.org/
2. Canright, D.: A Very Compact S-Box for AES. In: Rao, J.R., Sunar, B. (eds.) CHES 2005. LNCS, vol. 3659, pp. 441–455. Springer, Heidelberg (2005)
3. Chakraborty, D., Sarkar, P.: A New Mode of Encryption Providing a Tweakable Strong Pseudo-random Permutation. In: Robshaw, M. (ed.) FSE 2006. LNCS, vol. 4047, pp. 293–309. Springer, Heidelberg (2006)

4. Chakraborty, D., Sarkar, P.: HCH: A New Tweakable Enciphering Scheme Using the Hash-Encrypt-Hash Approach. In: Barua, R., Lange, T. (eds.) INDOCRYPT 2006. LNCS, vol. 4329, pp. 287–302. Springer, Heidelberg (2006)
5. Chakraborty, D., Sarkar, P.: HCH: A new tweakable enciphering scheme using the hash-counter-hash approach. Cryptology ePrint Archive, Report 2007/028 (2007), http://eprint.iacr.org/
6. Good, T., Benaissa, M.: AES on FPGA from the Fastest to the Smallest. In: Rao, J.R., Sunar, B. (eds.) CHES 2005. LNCS, vol. 3659, pp. 427–440. Springer, Heidelberg (2005)
7. Halevi, S.: EME*: Extending EME to handle arbitrary-length messages with associated data. In: Canteaut, A., Viswanathan, K. (eds.) INDOCRYPT 2004. LNCS, vol. 3348, pp. 315–327. Springer, Heidelberg (2004)
8. Halevi, S.: TET: A wide-block tweakable mode based on Naor-Reingold. Cryptology ePrint Archive, Report 2007/014 (2007), http://eprint.iacr.org/
9. Halevi, S., Rogaway, P.: A tweakable enciphering mode. In: Boneh, D. (ed.) CRYPTO 2003. LNCS, vol. 2729, pp. 482–499. Springer, Heidelberg (2003)
10. Halevi, S., Rogaway, P.: A parallelizable enciphering mode. In: Okamoto, T. (ed.) CT-RSA 2004. LNCS, vol. 2964, pp. 292–304. Springer, Heidelberg (2004)
11. Hsiao, S.F., Chen, M.C.: Efficient Substructure Sharing Methods for Optimising the Inner-Product Operations in Rijndael Advanced Encryption Standard. IEE Proceedings on Computer and Digital Technology 152(5), 653–665 (2005)
12. IEEE Security in Storage Working Group (SISWG). PRP modes comparison IEEE p1619.2. IEEE Computer Society (March 2007), Available at http://siswg.org/
13. Jarvinen, K., Tommiska, M., Skytta, J.: Comparative survey of high-performance cryptographic algorithm implementations on FPGAs. Information Security, IEE Proceedings 152(1), 3–12 (2005)
14. McGrew, D., Viega, J.: The galois/counter mode of operation (GCM), submission to nist modes of operation process (January 2004), Available at http://csrc.nist.gov/CryptoToolkit/modes/proposedmodes/gcm/gcm-revised-spec.pdf
15. McGrew, D.A., Fluhrer, S.R.: The extended codebook (XCB) mode of operation. Cryptology ePrint Archive, Report 2004/278 (2004), http://eprint.iacr.org/
16. Rodríguez-Henríquez, F., Koç, Ç.: On fully parallel karatsuba multipliers for GF(2^m). In: International Conference on Computer Science and Technology CST 2003, pp. 405–410. Acta Press (May 2003)
17. Sarkar, P.: Improving upon the TET mode of operation. Cryptology ePrint Archive, Report 2007/317 (2007), http://eprint.iacr.org/
18. Seagate Technology. Internal 3.5-inch (sata) data sheet, Available at: http://www.seagate.com/docs/pdf/datasheet/disc/ds_internal_sata.pdf
19. Wang, P., Feng, D., Wu, W.: HCTR: A variable-input-length enciphering mode. In: Feng, D., Lin, D., Yung, M. (eds.) CISC 2005. LNCS, vol. 3822, pp. 175–188. Springer, Heidelberg (2005)

Author Index

Lecture Notes in Computer Science

Sublibrary 4: Security and Cryptology

Vol. 4266: H. Yoshiura, K. Sakurai, K. Rannenberg, Y. Murayama, S.-i. Kawamura (Eds.), Advances in Information and Computer Security. XIII, 438 pages. 2006.

Vol. 4258: G. Danezis, P. Golle (Eds.), Privacy Enhancing Technologies. VIII, 431 pages. 2006.

Vol. 4249: L. Goubin, M. Matsui (Eds.), Cryptographic Hardware and Embedded Systems - CHES 2006. XII, 462 pages. 2006.

Vol. 4237: H. Leitold, E.P. Markatos (Eds.), Communications and Multimedia Security. XII, 253 pages. 2006.

Vol. 4236: L. Breveglieri, I. Koren, D. Naccache, J.-P. Seifert (Eds.), Fault Diagnosis and Tolerance in Cryptography. XIII, 253 pages. 2006.

Vol. 4219: D. Zamboni, C. Krügel (Eds.), Recent Advances in Intrusion Detection. XII, 331 pages. 2006.

Vol. 4189: D. Gollmann, J. Meier, A. Sabelfeld (Eds.), Computer Security – ESORICS 2006. XI, 548 pages. 2006.

Vol. 4176: S.K. Katsikas, J. López, M. Backes, S. Gritzalis, B. Preneel (Eds.), Information Security. XIV, 548 pages. 2006.

Vol. 4117: C. Dwork (Ed.), Advances in Cryptology - CRYPTO 2006. XIII, 621 pages. 2006.

Vol. 4116: R. De Prisco, M. Yung (Eds.), Security and Cryptography for Networks. XI, 366 pages. 2006.

Vol. 4107: G. Di Crescenzo, A. Rubin (Eds.), Financial Cryptography and Data Security. XI, 327 pages. 2006.

Vol. 4083: S. Fischer-Hübner, S. Furnell, C. Lambrinoudakis (Eds.), Trust and Privacy in Digital Business. XIII, 243 pages. 2006.

Vol. 4064: R. Büschkes, P. Laskov (Eds.), Detection of Intrusions and Malware & Vulnerability Assessment. X, 195 pages. 2006.

Vol. 4058: L.M. Batten, R. Safavi-Naini (Eds.), Information Security and Privacy. XII, 446 pages. 2006.

Vol. 4047: M.J.B. Robshaw (Ed.), Fast Software Encryption. XI, 434 pages. 2006.

Vol. 4043: A.S. Atzeni, A. Lioy (Eds.), Public Key Infrastructure. XI, 261 pages. 2006.

Vol. 4004: S. Vaudenay (Ed.), Advances in Cryptology - EUROCRYPT 2006. XIV, 613 pages. 2006.

Vol. 3995: G. Müller (Ed.), Emerging Trends in Information and Communication Security. XX, 524 pages. 2006.

Vol. 3989: J. Zhou, M. Yung, F. Bao (Eds.), Applied Cryptography and Network Security. XIV, 488 pages. 2006.

Vol. 3969: Ø. Ytrehus (Ed.), Coding and Cryptography. XI, 443 pages. 2006.

Vol. 3958: M. Yung, Y. Dodis, A. Kiayias, T.G. Malkin (Eds.), Public Key Cryptography - PKC 2006. XIV, 543 pages. 2006.

Vol. 3957: B. Christianson, B. Crispo, J.A. Malcolm, M. Roe (Eds.), Security Protocols. IX, 325 pages. 2006.

Vol. 3956: G. Barthe, B. Grégoire, M. Huisman, J.-L. Lanet (Eds.), Construction and Analysis of Safe, Secure, and Interoperable Smart Devices. IX, 175 pages. 2006.

Vol. 3935: D.H. Won, S. Kim (Eds.), Information Security and Cryptology - ICISC 2005. XIV, 458 pages. 2006.

Vol. 3934: J.A. Clark, R.F. Paige, F.A.C. Polack, P.J. Brooke (Eds.), Security in Pervasive Computing. X, 243 pages. 2006.

Vol. 3928: J. Domingo-Ferrer, J. Posegga, D. Schreckling (Eds.), Smart Card Research and Advanced Applications. XI, 359 pages. 2006.

Vol. 3919: R. Safavi-Naini, M. Yung (Eds.), Digital Rights Management. XI, 357 pages. 2006.

Vol. 3903: K. Chen, R. Deng, X. Lai, J. Zhou (Eds.), Information Security Practice and Experience. XIV, 392 pages. 2006.

Vol. 3897: B. Preneel, S. Tavares (Eds.), Selected Areas in Cryptography. XI, 371 pages. 2006.

Vol. 3876: S. Halevi, T. Rabin (Eds.), Theory of Cryptography. XI, 617 pages. 2006.

Vol. 3866: T. Dimitrakos, F. Martinelli, P.Y.A. Ryan, S. Schneider (Eds.), Formal Aspects in Security and Trust. X, 259 pages. 2006.

Vol. 3860: D. Pointcheval (Ed.), Topics in Cryptology – CT-RSA 2006. XI, 365 pages. 2006.

Vol. 3858: A. Valdes, D. Zamboni (Eds.), Recent Advances in Intrusion Detection. X, 351 pages. 2006.

Vol. 3856: G. Danezis, D. Martin (Eds.), Privacy Enhancing Technologies. VIII, 273 pages. 2006.

Vol. 3786: J.-S. Song, T. Kwon, M. Yung (Eds.), Information Security Applications. XI, 378 pages. 2006.

Vol. 3108: H. Wang, J. Pieprzyk, V. Varadharajan (Eds.), Information Security and Privacy. XII, 494 pages. 2004.

Vol. 2951: M. Naor (Ed.), Theory of Cryptography. XI, 523 pages. 2004.

Vol. 2742: R.N. Wright (Ed.), Financial Cryptography. VIII, 321 pages. 2003.